ENVIRONMENT AND ECOLOGY I
THE LONG NINETEENTH-CENTU

ENVIRONMENT AND ECOLOGY IN THE LONG NINETEENTH-CENTURY

Edited by Mark Frost

Volume I

Scientific and Professional Perspectives on Environment, 1789–1858

Routledge
Taylor & Francis Group
LONDON AND NEW YORK

First published 2022
by Routledge
4 Park Square, Milton Park, Abingdon, Oxon OX14 4RN

and by Routledge
52 Vanderbilt Avenue, New York, NY 10017

Routledge is an imprint of the Taylor & Francis Group, an informa business

British Library Cataloguing-in-Publication Data
A catalogue record for this book is available from the British Library

Library of Congress Cataloging-in-Publication Data
Names: Frost, Mark, editor.
Title: Environment and ecology in the long nineteenth-century / edited by Mark Frost.
Description: Milton Park, Abingdon, Oxon ; New York, NY : Routledge, 2022. | Includes bibliographical references and index.
Identifiers: LCCN 2021056587 (print) | LCCN 2021056588 (ebook) | ISBN 9780367377007 (volume 1 ; hardback) | ISBN 9780429355653 (ebook)
Subjects: LCSH: Ecology—History—19th century. | Environmental sciences—History—19th century.
Classification: LCC QH540.8 .E58 2022 (print) | LCC QH540.8 (ebook) | DDC 577.09—dc23/eng/20211217
LC record available at https://lccn.loc.gov/2021056587
LC ebook record available at https://lccn.loc.gov/2021056588

ISBN: 978-0-367-37700-7
eISBN: 978-0-429-35565-3

DOI: 10.4324/9780429355653

Typeset in Times New Roman
by Apex CoVantage, LLC

TO CHRISTINE, ONE OF THE GREATEST WONDERS OF THE NATURAL WORLD, WHO MAKES MY LIFE COMPLETE.

CONTENTS

GENERAL INTRODUCTION

The present edition sets out on a daunting endeavour, but one that has especial relevance in an era of severe ecocrisis. The core task is to convey, through a wide range of primary materials, the evolving attitudes to environment, the ongoing treatment of environments, and the state of thought and opinion (scientific and cultural) on environment during the long nineteenth century, defined here as the years 1780–1920. The original brief of this edition was to provide a survey of ecology in this period, but it was immediately clear that such a work was not sustainable over four volumes and that there was an opportunity to pursue a more ambitious goal. Certainly, proto-ecological thought significantly pre-dates the establishment of ecological science, generally associated with the works of Ernst Haeckel, George Perkins Marsh, and others in the 1860s, but even if one were to augment this focus by turning to the much broader history of environmentalism (however loosely conceived), two problems remain. The first is finding sufficient materials for a work of this length. The second, and more important, lies in what is neglected by such an approach. The yet-broader subject of environments and ecology in the long nineteenth century offers greater opportunities to provide an overview of the complexities, range, and forms of responses to environment in the period. As we shall see, many of the extracts included in these volumes evince indifference or hostility to environment, which is often figured purely as exploitable resource. In the context of our Anthropocene era, and of the multiple environmental crises that we face, exploring their deeper historical roots is a vital task, as is understanding the wider background in which countervailing opinions from ecological, conservationist, and environmental thought struggled for articulation and coherence. If the long nineteenth century marks an intensification of the logic of environmental violence that has always underpinned human civilisation and that has led us to our current ecocrises, the more comprehensive gaze of this anthology is surely valuable.

It is evident that even within the generous allowance of pages provided by a four-volume anthology, much will be left out, and readers may have different complaints about particular exclusions. Any selection will be particular to, and will reflect both the strengths and flaws of, the editor. The aim is to be both as comprehensive and as representative as possible – in terms of coverage of key topics and preoccupations but also in terms of gender, class, and ethnicity (a tricky task given the social conditions of the times, as I will suggest in a moment) and in international reach. The initial publishing brief – to create an anthology focused on British writers and sources – has, as a result, been followed with deliberate indiscipline, but it is beyond the possibilities of this work to be as truly representative as I would like. One element of creating the anthology has involved focusing on scientific perspectives in some detail while placing them in wider contexts. Important as scientific perspectives are, however, such an approach is insufficient because of the obvious fact that engagements with nature were not confined to scientific works, which in any case do not have a monopoly on wisdom, knowledge, or subjective response. These volumes therefore also trace a broad range of social, cultural, and political responses to environments. This reflects the fact that nature was of wider social and public interest but also that different perspectives yield a richer, more nuanced picture.

The structure of the volumes replicates this reasoning while also attempting to reflect the considerable changes taking place over the 140-year period. Volumes I and III trace scientific perspectives, while Volumes II and IV trace broader engagements – but it must be noted that the division between scientific and non-scientific writing is far from clear and that this has meant some difficult (and perhaps contentious) decisions about what to include in different volumes. This is particularly true of the period we are covering, in which leading scientific works were comprehensible to educated general readers to a much greater extent than today. As a result, some works in Volume I might have been included instead in Volume II, and vice versa, reflective of the fact that scientific and popular writings existed on a spectrum rather than in always clearly distinct realms. The volumes are also chronologically divided, the first two covering the period 1780–1858 and the latter two covering 1859–1920, the dividing point being Charles Darwin's *Origin of Species* (1859), the most significant work of natural history in the long nineteenth century and more or less marking the mid-point of the period. Each volume is divided into subject sub-sections to permit various foci on important topics. There will always be a mixture of sound reasoning and idiosyncrasy in the formation of sections, and they are not meant to reflect any definitive truths but rather to provide readers with an overview of key ideas, interventions, and preoccupations but also with the means to delve into particular areas of interest and to make comparisons within and between sub-sections and volumes. The extracts in each section are organised chronologically to enable understanding of developments in particular areas over the period covered.

It is perhaps worth pausing to consider the period under review and the challenges and opportunities this generates. The period is book-ended by revolutions – in America and France at the opening of the period and in Russia, Germany, and Mexico at its close. It is also the period of the establishment of modern nation states and the era at which Great Britain's dominance was challenged by Germany, the United States, and others. Independence movements were afoot in South and Central America. At the start of this period, industrial revolutions were already underway in a number of countries, and a global network of capitalist exchange was in place, dominated by the British and Dutch East India Companies. By the end of this period, the 'scramble for Africa' was complete and anti-colonial movements were already gathering force. The 1780s were the heyday of cavalry war, but by the 1860s, modern technological warfare was established, played out in the American Civil War and the Franco-Prussian War and with a grim culmination in World War I. Britain in the 1780s was a realm of rotten boroughs and limited representation, but after a slow history of parliamentary reform, the aim of universal male and female suffrage was almost achieved by 1918. In the arts, Europe and America witnessed Romanticism and a great era of poetry, the establishment of the novel as the dominant literary form, and the rise of Naturalism and Modernism. Industrial and social changes saw widening readerships and an insatiable appetite for art and entertainment, popular and elite. The late eighteenth century was the age of Goethe, Wordsworth, Scott, Goya, Mozart, and Kant, while, by stark contrast, the first decades of the twentieth century were the age of Joyce, Woolf, T.S. Eliot, Picasso, Stravinsky, and Wittgenstein. At the start of the period, industries and households relied on wood, peat, and coal, the dominance of the latter a defining feature of the entire period in its economic, social, and environmental implications. By its close, the dawn of the petro-chemical age promised yet greater environmental challenges ahead. Populations expanded rapidly during the course of our 140-year period, driving social and environmental change and a concomitant process of urbanisation with complex effects, including pollution, education, deprivation, medical and sanitary crises, political expression, and social mobility. The British countryside also changed at a rapid pace, characterised by a general process of intensification which was multi-faceted and followed to different degrees in other parts of Europe. Globally, environmental exploitation was already an established feature of European contact with the 'new world' – the despoliation of Mauritius, St Helena, the West and East Indies, and parts of India being only the most egregious examples of the consequences of European establishment of cash crops, slave economies, and imported species in fragile environments. By 1920, the environmental results of the resource exploitation of an intensive technological capitalist system were strongly evident in the skies, waters, and soils of 'developed' and 'developing' nations alike, but the decades that followed our period have only seen further intensification of the disastrous paths pursued between 1780 and 1920. Even a piecemeal and inadequate survey of the period such as this should offer some indication of the rapidity of change; the pressures on environments, societies, and peoples; and the opportunities to delve into a range of important issues.

It is not the aim of this work, nor is it within the capacity of any single editor of a wide-ranging work covering a period of 140 years, to provide specialist analysis of every topic included within the volumes. The Headnotes included at the start of each subject subsection within each volume are instead designed to give readers an introductory overview of key issues and contexts. The aim is not to engage in detailed analysis of secondary literature but to provide guidance on further reading for those wishing to explore topics in more detail. They also include specific analyses of each extract included in the subsections, placing them in relevant contexts and drawing attention to points of particular interest, as well as providing explanatory notes as required. These Headnotes are more detailed than envisaged in the initial brief, but it is hoped that they make a valuable guide, particularly to non-specialists seeking to understand an array of different topics.

The Headnotes, and the volumes more generally, trace a series of complex issues and developments during a period of immense environmental, social, demographic, technological, cultural, and political change and upheaval but repeatedly circle around some key factors, including the increasing professionalisation and specialisation of science, the state of particular environments and how these change, and the diversity and changefulness of attitudes to environment. On one level, the contents of the anthology reflect a dual impulse of the period: on the one hand, a set of claims about human sovereignty over environment – the right of our species to do as it wishes to everything labelled 'non-human', to commodify, commercialise, and exploit that which is constructed as such; and, on the other, the slow, often tentative or limited development of resistance to this dominant thread in the long history of *Homo sapiens*. In both cases, these impulses were strongly articulated in British, North American, and European contexts: precisely those places in which the mechanisms, processes, infrastructure, and social and conceptual systems that enabled exploitation of environments for social and economic gain were centred. Likewise, the development of countervailing attitudes – animal rights movements; vegetarianism and veganism; Romantic re-conceptualisations of environment; conservationist and environmentalist thought; and the sciences of biogeography, evolutionary theory, ecology, and so on – also saw their most powerful manifestations in these locations, precisely because these parts of the world were the imperial centres of environmental violence.

Although the original brief encouraged a British focus, it was immediately apparent that it was impossible to do justice to the subject without turning to European and American contributions. The creation of pan-European scientific networks is a marked feature of the age and a key enabler of scientific discoveries and new perspectives. It has also become apparent, however, that an expanded western purview is also insufficient, for a number of reasons. As Richard H. Grove has demonstrated in *Green Imperialism* (1995), the emergence of early modern scientific ideas owed much to the influence of Indian, Arabic, and other centres of learning, and such global influences on the supposedly purely European project of the Enlightenment are still routinely occluded in studies. It is also the case that the globalisation of European power in the long nineteenth century, and its attempts to gain mastery over global environments and peoples, requires us to engage with the intersections of imperialism and colonialism with the underlying nexus of environmental violence. This has proved a particularly challenging task. While it has been straightforward enough to trace some of the comings-together of science and imperialism via western works of expeditionary travel, global studies of flora and fauna, biogeographical works, and so forth, it is far more difficult to find counter-voices on environments from the colonised, enslaved, or displaced, largely because of the relative paucity of their textual traces but also because of the limited availability of accessible, copyright-free English editions of those works that do exist. It is hoped that some progress has been made in this direction, particularly in the final two volumes, and this anthology certainly endorses the decolonising imperatives of initiatives like the V21 Collective. It is also hoped that the work done here is a spur to others to go further in the future.

This anthology proceeds, therefore, in the spirit that studying the broader environmental-social impacts of western nations in the period is a crucial endeavour. Similar issues attend the question of gender. In Volume I (covering science and environment in the first half of our period), only four women – Mary Anning, Etheldred Benett, Katherine Kane, and Anna Russell – are represented amongst a collection of 106 extracts. This is largely a reflection of what is in itself an important topic, the deliberate exclusion of women from the male bastions of science, and it should be noted that Benett, Kane, and Russell still struggle for representation in modern coverage of their respective fields. It should not be forgotten that for much of this early period, it was virtually impossible for women to gain entry to elite scientific societies or institutions of learning or that progress after mid-century remained slow. At the same time, and in ways that reflect the gender constructions of the period, women contributed in countless ways to science, primarily as popularisers, collectors, and illustrators, and to a range of other engagements with environment. Much greater coverage of women has therefore been possible from Volume II onwards, and it has been pleasing to include a large number of valuable but routinely overlooked contributions by women in the anthology.

Because covering the range of representations of environments calls for coverage of scientific and wider cultural manifestations, and because the task of understanding how environments of various kinds were understood, engaged with, used and abused, thought about and represented, one must turn to the varied discourses through which all of this was articulated. In a practical sense, this results in a wide range of sources, including scientific books and papers, popularisations of scientific knowledge, newspaper and periodical journalism, novels, short stories, poetry, travel literature, diaries, letters, et al. No attempt has been made to correct or regularise the various spellings used in the extracts, and the aim throughout is to preserve the specific flavour of the language. The use of long dashes and of quotation mark formatting has, however, been regularised, as has the formatting of sub-titles to extracts.

It is worth adding here some explanation of the additional scholarly apparatus included at the end of each volume of the anthology. These include a People Index, comprising short notes on figures mentioned within the volume, hopefully a considerable aid given frequent references to now-obscure individuals. Because of the inclusion of this resource, sub-section Headnotes include little or no biographical information on authors. The apparatus for each volume also includes full bibliographies of specific editions used for the extracts and Subject Indexes.

In closing, I would like to acknowledge the invaluable assistance (and patience) of the excellent staff at Routledge, including Rachel Douglas as commissioning editor, and Simon Alexander (senior development editor) and his team. I would also like to thank those who have kindly helped out in various ways, including Christine Berberich (University of Portsmouth), Arjun Jain and Bikram Grewel (New Delhi), Professor Liz Miller (UCD), and Catherine Layton (Wollongong).

INTRODUCTION TO VOLUME I

This volume uses key works to trace the emergence or development of a range of major scientific fields in the seven decades prior to Darwin's *Origin of Species*. It is arranged in sections that gather some of the most important scientific interventions of the period, as well as other minor but valuable sources. Many of these works were seminal in their day but are difficult to access now, while others were relatively obscure at the time but significant in their overall contribution to their particular field. The aim of the various sections is to demonstrate the ways in which:

i) scientific engagements became more professional and more materialistic, in ways that reflect a growing crisis in Natural Theological attitudes;
ii) environment became a major focus in a staggering number of ways, spawning discrete and important new fields of enquiry, as well as interdisciplinary methods;
iii) conservationist, environmentalist, and proto-ecological ideas gather energy during this period.

Volume II, covering the same period, will examine popularisations of science as well as cultural, social, and political engagements with environment. Volumes I and II exist in a state of dialogue, and readers are advised to cross-reference between them. In some cases, topics covered in Volume I from a scientific perspective are then treated from a popular perspective in the second and subsequent volumes, while some sections are unique to particular volumes, reflecting predominant preoccupations of particular periods. The Headnotes included at the start of each section of this volume are designed to provide readers with a detailed guide to key issues, as well as of environmental contexts of the period. This Introduction aims simply to provide a brief overview of contents and of the logic behind the volume.

Part 1 (Precursors) offers some coverage of important interventions in seventeenth- and eighteenth-century natural science. While it is impossible to provide adequate coverage of this enormously influential period, its significance to what follows in the long nineteenth century is worthy of exploration and contextualisation. Accordingly, I have included key works on those areas of investigation that would in time

DOI: 10.4324/9780429355653-1

cohere as distinct disciplines (geology, botany, zoology) as well as examples of early conservationist thought, travel writing, and imperialist perspectives on environment. The authors included are amongst the key contributors to their respective fields, and the section illustrates that the long nineteenth century reflected continuities and discontinuities with what came before.

Part 1 anticipates the concerns of Part 2 (Natural Theology and the Great Chain of Being), chosen to open the coverage of the long nineteenth century because of the status of Natural Theology as the orthodoxy of natural historians prior to the 1850s. The key claim of Natural Theology (that evidence of God's wisdom and design are traceable in the natural world) is explored and articulated from various angles, via key works (including William Paley's *Natural Theology* and two extracts from *The Bridgewater Treatises*) and a range of other texts, some of which modify or query the Natural Theological project or demonstrate the increasing strains placed upon it by ongoing scientific discoveries. In 1834, in an extract included in this section, Adam Sedgwick could write with confidence that:

> All nature is but the manifestation of a supreme intelligence, and to no being but him, to whom is given the faculty of reason, can this truth be known. By this faculty he comes the lord of created beings, and finds all matter, organic and inorganic, subservient to his happiness, and working together for his good.
>
> (*On the Studies of the University*)

One should note the permission given here by the assumptions of Natural Theology: that the non-human is 'subservient' to *Homo sapiens* and designed for our benefit, but it is more broadly indicative of the vision of divine environmental harmony which Natural Theology conjured. The two extracts (Philip Gosse, Hugh Miller) indicate that by the 1850s, Sedgwick's confidence was not so widely shared: the project of Natural Theology had become beleaguered by the onward march of materialist science in ways that anticipated the grand division between religious and secular perspectives that greeted the emergence of Charles Darwin's *Origin of Species* at the end of that decade.

Part 3 (Geology) follows because it was established in this period as a major science and because its discoveries and debates were key to the vicissitudes of Natural Theology. The development of the modern earth sciences are traced, first via their roots in James Hutton, Baron Cuvier, and William Smith, and then through their manifestations in the heyday of Victorian geology via works by major figures (William Buckland, Roderick Murchison, and Charles Lyell). The emergence of distinct fields of stratigraphy and palaeontology are important aspects, and the contributions (and difficulties) of female geologists, including Mary Anning and Etheldred Benett, are considered alongside the work of pioneers such as Gideon Mantell. The section closes with two extracts from the 1850s that illustrate the closing phase of a key issue running through many of the works included here: the increasingly fraught relationship between science and scripture as it played out in debates about the earth's prehistory and the concept of 'deep time' and began to query the centrality of humankind in the cosmos.

New ways of thinking, centred around the idea that it was necessary to study animals as functioning wholes existing within, dependent upon, and interacting with their immediate environments, emerged strongly within the discipline that forms the focus of Part 4 (Comparative Anatomy). While anatomy may not seem an obvious candidate as an environmental science, it was precisely these new concepts, arising from the work of Baron Cuvier but developed by Richard Owen and Louis Agassiz, that made it a crucial area for the emergence of new concepts of environment. That these increasingly materialistic accounts of animal life and environmental relationships would also provide support for evolutionary thinking in subsequent decades is a particular irony given the status of Cuvier, Owen, and Agassiz as Natural Theologians. This section also intersects with the previous one because of the focus of some extracts on the application of anatomical analysis to palaeontology.

Part 5 (Botany) turns to a significant, long-established field of natural science, doing so in ways that reflect the wide-ranging aspects of plant study in the period and tracing a gradual move from more informal (and often cultural) engagements with flora to the establishment of a highly professional discipline centred on morphology and taxonomy. Debates about classification systems and about the role of plant science (in the wider economy, agriculture, culture, and society) are addressed in various ways, while the extracts also inadvertently provide alarming evidence of declines in flora since the nineteenth century. As well as featuring major botanists (James Sowerby, Gilbert White, Sir William Jackson Hooker, John Lindley, Darwin, and Joseph Dalton Hooker), the section also turns to significant but often-overlooked contributions (Johann Wolfgang von Goethe, George Luxford, Anna Russell) and to those who approached botany from an aesthetic angle (John Ruskin) or from a purely economic and exploitative angle (Thomas Ewbank). The significance of botany as a key nineteenth-century environmental science is underlined in its national, European, and global aspects.

Part 6 (Zoology) turns to another well-established natural science of the period but is somewhat shorter because of a strong focus on this topic in Volume II. Covering a range of sub-fields (entomology, ornithology, microscopic life), the focus is again national and global, turning to localised British studies as well as accounts of international exploration. The status of animals and their treatment provides one issue, but this section also explores the impacts of comparative anatomy in shaping the ways in which animals began to be conceptualised in ways that co-existed uncomfortably with Natural Theological modes and began to demand new ways of conceptualising environments as complex, dynamic, functioning wholes.

Part 7 ('New World' Environments and Scientific Exploration) overlaps substantially with the previous two sections, given the focus in places on flora and fauna, but the key element here is the manner in which European science intensified its project to globalise knowledge by exploring 'new' environments.

This section, comprising a range of major and less familiar accounts of global scientific travel, underlines the ways in which science interacted with the imperialist project, sometimes reflecting and endorsing it and sometimes resisting or challenging its impulses. The section focuses on the formation of new knowledge but also on thc ways in which the various travellers engaged with the peoples and environments that they encountered.

There is inevitably a strong overlap between Part 7 and Part 8 (Demographics, Geography, and Biogeography), particularly as some of the works in both sections are travel literature. The establishment of geography and biogeography as powerful fields in this period was, on one level, concomitant with the broader European imperial project – a process of gaining command over lands, people, and knowledge – but, on another, a spur to new ways of thinking that often ran in different directions, sometimes spurring critiques of the disproportionate and damaging impacts of humanity on environments. Given the rootedness of ecology in geography and biogeography, this section reflects the complexities of these sciences. Alexander von Humboldt's enormous contributions in this area are reflected in Parts 7 and 8, as are those of Darwin, W.J. Hooker and J.D. Hooker, and John Gould, but both sections also feature less well-known but extremely fascinating contributions from less prestigious or now more obscure writers (Carl Thunberg, Charles Waterton, George Gardner, Thomas Thomson). The related issue of demographics is largely represented by Thomas Malthus's enormously influential *Essay on Population* (1798).

Part 9 (Evolutionary Thought Before *Origin of Species*) is in many ways a culmination of many threads traced in previous sections – the re-conceptualisation of the relationship between organisms and environment in particular – but also the significance of the geological concept of 'deep time' and a growing understanding of the dynamic 'economy of nature' (the harmonious and balanced state of environmental systems unafflicted by human activity) – as well as being the most significant contribution to the environmental sciences in the period. While the post-*Origin* history of evolutionary thought is traced in Volume III, this section traces its roots, going back to early work by Erasmus Darwin and Jean-Baptiste Lamarck, moving forward through some significant contributions to proto-evolutionary thought, and culminating with the announcement of the theory of evolution by natural selection by Darwin and Alfred Russel Wallace in 1858.

The final section (Agricultural Science and Land Management) is somewhat distinct from the others, but it is also amongst the most important, given the enormous impacts of agriculture on biodiversity, climate change, habitat loss, and other environmental crises. Although it is hardly the case that most of the work included in this section can be described as science, the various extracts reflect the gradual emergence of a recognisably more scientific approach to the land. Many of the extracts focus on the practical work of farmers, suggestions for 'improvement' of agriculture, and accounts of the state of agricultural land in the period. These are contextualised in terms of what they reveal about human attitudes to the land and of the broader logic of human sovereignty over environment that they largely articulate.

The Headnotes that follow offer some contextual overviews of the condition of environments in the period and the pressures placed upon them from various directions, so it is unnecessary to replicate that coverage here. More broadly, though, this volume is about offering some sense of the enormous productivity of the natural sciences in the period, the growth or establishment of disciplines, and the ways in which science both followed and transformed attitudes to the natural world. The Volume I extracts also demonstrate the complexity of scientific engagements in the period, sometimes articulating, supporting, and enabling instrumental and exploitative approaches to the non-human but also, in their quest for knowledge, opening up new avenues and possibilities. That one of the important destinations of science in the nineteenth century was the establishment of ecology, a science that rapidly generated a wider field of social, cultural, and political engagement, is a subject taken up in more detail from Volume II onwards.

Part 1

PRECURSORS

Precursors

The selections offered here of seventeenth- and eighteenth-century work provide some indication, however inadequate, of environmental engagements and representations before the long nineteenth century. To some extent, the period 1780–1920 represents a distinctive era in the development of new attitudes to nature, science, and culture, but at the same time, it is impossible to trace a definitive break from its precursors. While there are many ways in which the extracts in this section differ from those that follow, there are also significant continuities: Michel Foucault's reading of scientific history as a series of discrete 'epistemes' in his otherwise astonishing *The Order of Things* is not borne out by close examination of sources.

While it was possible to devote this section to works of the earlier period that exemplify starkly 'unscientific' approaches to nature, our task – to contextualise the works covered in the anthology's first two volumes – is most effectively achieved by tracing some of their earlier foundations. These extracts have been chosen, then, to demonstrate a) a growing zeal to engage with the natural world in ways that involve empirical information gathering, organisation, and analysis; b) evidence of conservationist or environmentalist thought; and c) the establishment of Natural Theology, a philosophical method of enquiry that confidently attempted to ally a rational methodology to Christian faith and to demonstrate that the existence, benign purpose, and active creativity of God were achievable by investigating the *natura codex* (or book of nature).

Tracing Natural Theology's early roots is helpful because the relationship between science and religion slowly became a key point of tension in the European natural sciences. John Ray is included here as the most significant early contributor, but most of the writers in this section broadly share his approach, though it is worth noting that the extracts by Robert Hooke and Georges Buffon already begin to exemplify the lurking danger that attempting to prove God's design by direct investigation of 'His works' might lead to findings disruptive of a religious cosmogony. Because Natural Theology continued to play a significant role in early nineteenth-century science, Part 2 of this volume is devoted to its closer examination.

Part 1 also demonstrates that environmentalist or conservationist consciousness significantly pre-dates the nineteenth century and nods to its longer history. Readers seeking detailed exploration of the earlier, global roots of environmentalism can turn to Richard H. Grove's *Green Imperialism* (1995) and Donald Worster's *Nature's Economy* (1994). It should also be noted that while the extracts in this section are focused on Europe, one must acknowledge the enormous prior contributions of Classical, Mesopotamian, Indian, and Islamic scholarship. The importance of these predecessors is often obscured within histories of Enlightenment science, and while it lies beyond the remit of the current volumes (and the expertise of its editor) to do full justice to these cultural contexts, the anthology as a whole will seek a decidedly more global coverage of the long nineteenth century than is often the case, and to explore the varied, complex intersections of European imperialism with global environments. While there is, in the broadest sense, an imperialist element to most of the extracts included in this section – insofar as they rest on assumptions about the right to achieve mastery over environments, peoples, and knowledge – Emmerich de Vattel's *Law of Nations* (1760) has been included as one of the most direct statements of imperial intent in these contexts. European imperialism also meant the formation of a body of scientific knowledge and a proliferation of specimen collections resulting from global voyages and expeditionary travel. The expansion of 'New World' travel in the long nineteenth century, including many scientific expeditions, forms the subject of Part 7 of this volume.

It would be tempting to begin a selection of precursor texts with Francis Bacon's *Novum Organum* (1620), sometimes seen as the founding text of the 'scientific revolution', or with the earlier Italian physician-naturalist Andrea Cesalpino, whose *De Plantis Libri XVI* (1583) anticipated Linnaeus in classifying plants according to fruits and seeds rather than medicinal properties, astrology, or arcane systems and whose *De Metallicis Libri Tres* (1596) offered ground-breaking insights into mineralogy and fossils. Only a lack of English translations prevents his inclusion.

We begin instead with two works by John Evelyn. While Bacon's 'inductive method' of systematic investigation was influential, Evelyn's work is of greater environmental significance. *Fumifugium: or, The Inconvenience of the Aer and Smoke of London Dissipated* (1661) and *Sylva; or a Discourse of Forest-Trees and the Propagation of Timber in His Majesty's Dominions* (1664) deserve consideration as pioneering examples of conservationist thinking, although the reality is more complex and more interesting. Evelyn was certainly novel in analysing the impacts of coal smoke on London, in critiquing deforestation, and in his visionary solutions to these problems, but he also advocated massive colonial exploitation of overseas timber, and, in seeking to influence government, his forestry recommendations prioritise the Royal Navy's timber requirements.

Fumifugium opens with Evelyn's appeal to Charles II to attend to proposals to alleviate the frightful effects of London's smoke, a reminder that such problems were evident in urban areas long before the nineteenth century and that while the Industrial Revolution is generally dated to the eighteenth century, coal had already become a significant element of the British economy in the preceding century. Evelyn relates an incident of smoke causing inconvenience to the Court itself before summarising the solutions that are subsequently outlined in detail in Parts 2 and 3. Writing long before mechanisms of air pollution were fully understood, Evelyn and his fellow Londoners could nonetheless see, feel, and taste the by-products of industrial production at first hand (on this, see Toby Travis, Further Reading). Neither Evelyn's proposal to remove industries to London's outskirts, or to create plantations of sweet-smelling verdure on its perimeters, were ever pursued in Britain, but they demonstrate that the environmental impacts of human activity were already causing alarm. Given recent debates about dating the commencement of the Anthropocene (the geological period in which human activities have left tangible marks in the stratigraphical record), Evelyn's work is a timely reminder of the difficulties of conceptualising a 'beginning' to processes of environmental violence that recede into antiquity. It is also true, however, that the long nineteenth century witnessed a marked acceleration of anthropogenic impacts on the planet. Regarded in this light, works like Evelyn's can be likened, appropriately, to the proverbial canary in the (global) coal mine, with the caveat that Evelyn's warnings, like so many of those that followed, were ignored.

One of the earliest guides to silviculture, *Sylva* is refreshingly wide ranging, covering forestry, orchards, standing groves, gardening, and geology. Our extract focuses on Evelyn's proposals for the preservation and increase of timber forests. His 'Introduction' evokes the deeply English image of the nation's 'wooden walls' (symbolising both the Royal Navy and the forest timber on which it relied), thus eliding imperial and environmental causes. The maintenance of the navy during a period of intense European rivalries and of expanded global ambitions reliant on naval power should be set against a backdrop of declining forest cover in the British Isles. As Oliver Rackham (1987, 16) points out, of the lands surveyed by the Domesday Book in the 1080s, only 15 per cent were forests, and further deforestation took place up to Evelyn's time and beyond, so that by the commencement of World War I, British woods were at 'a nadir of around 4 per cent' (Simmons, 2006, 97). As Evelyn goes on to point out, forest legislation was one response to concerns about deforestation from ancient to early modern times. In Britain, parliamentary acts in 1258–9, 1290, and 1543 made inadequate and widely ignored provision to protect woodlands from agricultural expansion, which throughout history has proved a far greater danger than naval extraction or iron industries. Evelyn, well aware of 'the disproportionate spreading of *Tillage*', offers a series of novel suggestions for the protection of woodlands in Chapters 6 and 7. Forests, 'the greatest *Magazines* of the *Wealth* and *Glory* of this nation', and particularly oaks – which had already gained an especial symbolic role in the formation of English national identity – should be more widely planted and protected from browsing cattle and the health of mature plantations ensured by planting cereal crops between stands. Signalling endemic early modern tensions between those seeking legislative protection for woodlands and those, including farmers, iron-smelters, '*Foresters*, and *Borderers*' who relied on woodlands for their living, Evelyn proposes a 'high resolution to reduce these Men to their due *Obedience*'. As Groves (1995, 56–61) shows, however, legislative protection was far more successful in France than in Britain over the next century. Eighteenth-century British forests changed less by deforestation and more by the slow loss of traditional mixed coppice-and-canopy broadleaved woodland to conifer plantations, with a resulting decline in biodiversity. All the same, Evelyn's anxieties were still evident in *The Eleventh Report of the Commissioners Appointed to Inquire into The State and Condition of the Woods, Forests, and Land Revenues of The Crown* (1792), which complained that woodlands, 'so valuable a Part of the Landed Property of the Country should not be suffered longer to continue in its present unproductive State' but should be protected by legislation 'which may render those Forests useful Nurseries of

Timber for the Navy'. The limits and contexts of Evelyn's environmentalism are clear both in his references to production and in his proposal that 'the Exorbitance and Increase of devouring *Iron-mills*' be alleviated at home by transferring all iron production (and hence all of the timber extraction on which iron industries relied for charcoal) to 'the *Holy-land* of *New-England*'. Avoiding 'the *exhaust*[*ion*] [of] our *Woods* at *Home*', for Evelyn, could be purchased by exploiting the seemingly bottomless resources of 'New World' environments.

Evelyn was amongst the founders, in 1660, of The Royal Society of London for Improving Natural Knowledge, which quickly became the premier venue for the exchange of 'scientific' ideas in the period and a centre for the 'scientific revolution' inspired by Bacon. Other founders included Robert Boyle and Christopher Wren, and many of the extract authors throughout this volume were Fellows. While there were other global centres of scholarship, Europe in this period achieved an academic pre-eminence that manifested its confidence and ambition on the global stage and that went hand in hand with its growing economic and military might. From the seventeenth century, cross-fertilisation of ideas from major European centres of thought intensified, with London, Edinburgh, Paris, Leiden, Berlin, and St Petersburg particularly prominent. Manifestations of pan-European (and American) exchange of ideas and the creation of natural history networks will be evident throughout this volume, while American contexts will become more important from Volume II onwards.

The third extract is by another early Royal Society Fellow, John Ray. His *Wisdom of God Manifested in the Creation* (1691) is a founding text of Natural Theology and worth reading alongside William Derham's contemporaneous *Physico-Theology* (see Further Reading). The culmination of Ray's decades of wide-ranging studies across natural history, *Wisdom of God* yielded significant studies of fish, birds, and plants. A pioneering taxonomist, Ray introduced the term 'species', and many of his botanical names are still in use today. He was also the first, in *Historia Plantarum Species* (1685–1703) to divide herbaceous plants into Cryptograms (non-flowering) and Perfect (flowering plants), further dividing the latter into Monocotyledons and Dicotyledons.

Ray was thus a significant figure in a shift towards rational investigation of the natural world. At the same time, this was allied, as the Preface proclaims, to a project to prove the wisdom of God 'by Arguments drawn from the Light of Nature, and Works of the Creation'. Defining the subsequent direction of Natural Theology, Ray combines scripture and material investigation, ranging from celestial bodies to insect life in order to offer multiple 'proofs' of Divine design. Ray is the first of many figures in Volume I who combined a clerical career with scientific pursuits, a subject taken up in detail by Patrick Armstrong (see Further Reading).

The second part of the extract, Ray's discussion of the care of young amongst avian species, exemplifies his methods and his sympathetic observation of nature. As part of his self-education in natural history, Ray travelled widely in Europe, but in his references to India, the ornithological information resulted from his connections to the East India Company surgeon and specimen collector Edward Bulkley via London apothecary and Royal Society fellow Jacob Petiver. For a 1713 page on the birds of Fort St. George or Madras (Chennai) which involves a collaboration between Ray, Bulkley, Petiver, and an Indian artist, see 'Madras Birds' in Further Reading at the end of this section. The most likely candidate for the bird to which Ray refers is the bayā weaver (*Ploceus philippinus*), an indigenous Indian species whose nesting habits match Ray's descriptions. I wish to thank Mr Arjun Jain and Mr Bikram Grewal of New Delhi for assistance in pursuing this subject. Already in the early days of the 'scientific revolution', writers like Ray were drawing on imperial encounters – via the East India Company's early penetration of India (to which we will return in a later context in Part 10 of this volume) or by travellers' tales from South America – to expand knowledge and to thus claim power over the globe.

The fourth extract is from the voluminous *Posthumous Works of Robert Hooke*, collated and edited by Richard Waller for the Royal Society in 1705. Another early fellow of the Royal Society, Hooke contributed enormously to seventeenth-century 'natural philosophy', and the extracts from his work give some sense of the breadth of his interests. Hooke is regularly cited by Evelyn in *Sylva.*

In 'First General' and 'Second General', Hooke outlines principles through which natural history should be pursued. Like Bacon, Hooke favours 'objective', unbiased enquiries based on observation and experimentation, but his methodology is less hidebound and arcane. His various headings represent a range of topics impossible for even a driven individual like Hooke to pursue in a lifetime, but they represent a prospectus of the physical and natural sciences that existed or were subsequently developed. Another noteworthy feature of Hooke's principles is their lack of Natural Theological intent: while there is no reason to suspect atheism, the tone of the work is in striking contrast with Ray.

The remainder of the extract involves Hooke's significant engagements with palaeontology and his argument – in alignment with Cesalpino but against the current of early modern opinion – that fossils were animal and vegetable remains. Hooke rejects theories that stones were formed into figured shapes by 'some Plastick Vertue' inherent in the soil or by mysterious 'Celestial Aspect of Influence of the Planets'. His conclusions – that fossils are 'Animal or Vegetable Substances . . . converted into Stone, by having their Pores fill'd up with some petrifying liquid Substance', or impressions of organic bodies left in earth and subsequently hardened – have subsequently been vindicated. Hooke's further conclusion – 'that many very Inland Parts of this Island, if not all, may have been heretofore all cover'd with the Sea, and have had Fishes swimming over it' in the distant past – anticipates James Hutton (see Part 3 of this volume) in positing the role of vulcanicity in raising sedimentary deposits formed in marine environments. Like Hutton, Hooke draws attention to the curious presence of shell fossils at upland elevations. Most incendiary of all, from a religious standpoint, Hooke claims that 'there have been many other Species of Creatures in former Ages, of which we can find none at present', thus foreshadowing debates about strata and scripture that form an abiding theme of Part 3 of this volume. The correspondence also included in the extract was inserted by Waller as relevant to Hooke's geology and therefore placed after Hooke's further analysis of fossil finds (not included in the extract) in which he describes and illustrates 'snails' (ammonites), nautilids, 'Helmet-stones' (bivalves), 'Button-stones' (sea urchins), belemnites, and shark's teeth.

Hooke's archaic language in the extract requires some explanation for non-specialists. A placit is a decree or dictum; Supellex refers in this context to apparatus; voragoes are chasms; morses are walruses; Opacousness is opacity; Spars are crystals (feldspars, barytes, and calcites) with discernible faces, while Tooth-sparr refers to dogtooth spar (speleothem), a calcite formation of limestone caves. 'Coperas Stones' is another term for iron pyrites or fool's gold (bisulphide of iron, FeS^2). Throughout the seventeenth and eighteenth centuries, the copperas industry, one of the earliest examples of chemical manufacturing, converted iron pyrites into green vitriol (ferrous sulphate), essential for making inks, dyes, and sulphuric acid.

The fifth extract is from *Lachesis Lapponica* (first published as *Flora Lapponica* in 1737), a journal written by the greatest naturalist-taxonomist of his age, Carl von Linne (Linnaeus), following his five-month Lapland expedition. Chosen in preference to his many taxonomic works because of its greater insight into his techniques, attitudes, and achievements, it is also as an absorbing work of travel literature. Opening sections from the early stages of Linnaeus's journey are typical of its blend of botanical and zoological observations with medical, topographical, meteorological, geological, agricultural, ethnographic, and cultural commentary. In this, it somewhat anticipates both the biogeography of nineteenth-century traveller-naturalists like Alexander von Humboldt, Charles Darwin, and Joseph Dalton Hooker (see subsequent sections of this volume), as well as the establishment of Human Geography in the nineteenth century (see Part 9 and subsequent volumes). Likewise, the later sections from Linnaeus's time within Lapland closely examine the native inhabitants, their reindeer, and their relationship to the surrounding environment and to the Swedish settlers. The Lapland section also underlines the hazards of Linnaeus's extensive journey (May to September 1732) and the remoteness of the places he visited.

Linnaeus's early reference to 'having been appointed by the Royal Academy of Sciences to travel through Lapland' is perhaps a mistranslation: the Swedish Royal Academy of Sciences was formed in 1739, after Linnaeus's journey and publication. Instead, he received a grant from the Uppsala Royal Society of Sciences, the oldest Swedish scientific society (founded 1710). Consequently, the journey began in Uppsala, Sweden's ecclesiastical capital. Linnaeus's dates are 'old style': he employs the Swedish calendar, in use from March 1700 until February 1712. It was initially one day ahead of the Julian calendar and ten days behind the Gregorian, which Sweden was particularly slow to adopt: first used elsewhere in 1582, Sweden introduced it by excluding leap days in the decades prior to 1740.

Some explanation of the places visited in the extracts is required: 'Helsingland' or Hälsingland (sometimes Helsingia) is a historical province or 'landskap' in mid-Sweden, roughly 200 miles north of Stockholm and 150 miles north of Uppsala. The oldest surviving Sami settlement in Swedish Lapland, Lycksele, saw the establishment of the first Swedish Sami school in 1634. By the time he reached it, Linnaeus had travelled over 400 miles (650 km). 'Lapmark' or Lappmarken, was the traditional term for the northerly Sami regions of Sweden, divided into territories or 'Lappmark'. 'Jockmock' (or Jockmokk) in Lulean Lappland is 590 miles (950 km) from Uppsala.

During the journey, Linnaeus observed and collected plants, animals, and minerals, employing his own newly-coined binomial system of classification for the first time. He also demonstrates interests in ferns, mosses, lichen, and bacteria. In his disputations with Messrs Malming and Högling, for example, Linnaeus speaks of 'Nostoc', a genus of cyanobacteria: on the ground, a Nostoc colony is not normally evident, but after rains, it swells into a jelly-like substance: because of this association, it was often thought to have fallen from the sky.

The next pair of extracts is from a towering figure of eighteenth-century French natural history, Georges Louis Leclerc de Buffon. From his epic encyclopaedia of science, *Histoire Naturelle* (1749–1804, 36 vols), they reflect his importance to geology and biology and demonstrate his characteristic synthesis of scientific methods and Natural Theology. Just as the Royal Society was the primary centre for natural history in England, so the French academy in Paris came to assume a leading position in the eighteenth century: many of its leading figures appear in various subsequent extracts in this volume.

The first extract is from Buffon's 'Theory of the Earth', where his interest in the 'history of nature on the terrestrial globe' sees him reaching for existing disciplinary categories (geography, physics, anatomy) while half-envisaging others yet to fully emerge (geology, biogeography, ecology). 'Theory' encapsulates Buffon's role in catastrophism, a powerful geological theory during the next half-century that claimed the earth's strata resulted from enormous, relatively brief catastrophic events like the Biblical deluge. While this reconciled science and scripture, the extract demonstrates that Buffon is uncertain that all available geological evidence could be explained by a single deluge, leading him and others to postulate a series of large-scale catastrophes. Moreover, Buffon's estimate of the earth's age (elsewhere in his work) at 75,000 years was already irreconcilable with Biblical literality: Bishop Ussher in *Annals of the World* (1654) used scriptural exegesis to determine what became the orthodox belief that the earth was created in 4004 BCE. In these respects, Buffon's catastrophism is less dogmatic than that of many other proponents, especially the nineteenth-century English 'scriptural geologists' (see Part 3 of this volume).

The second half of the extract, a study of the hare, exemplifies Buffon's methods of close observation and accumulation of facts. Basing much on Jacques du Fouilloux's hunting text, *La Vénerie* (1560), Buffon goes much further by reflecting upon both the hare's habits and animal demographics, which he explicitly links to scarcity in ways that anticipate John Bruckner (see later extract in this section), William Smellie (see Part 2), Thomas Malthus, and evolutionary theory (see Parts 8 and 9). Speaking of the 'two unchangeable wheels; the one, unbounded fecundity of every species; the other, the innumerable causes of destruction which are perpetually reducing the produce of that fecundity to a determinate measure', the intertwining of demographics and environment tantalisingly invite, but do not yield, an evolutionary reading. Rather, like his fellow Natural Theologians, Buffon's account of species distribution seeks to prove the harmonious regulation of a designed Creation that maintains the 'Great Chain of Being' (see Part 2 of this volume).

The next extract, from Emmerich de Vattel's *The Law of Nations, or Principles of the Law of Nature, Applied to the Conduct and Affairs of Nations and Sovereigns* (1758) is included, despite not being a work of natural history, because it illuminates the intersections of imperialism and environment in the period. The 1835 American fourth edition used here is a sign of Vattel's long-term influence and his significance to the founding fathers: *The Law of Nations* was amongst overdue library books found in George Washington's study after his death. Its significance is usually related to its contribution to international law and its aspiration to ground law in ethics. In the context of its time, Vattel's work is politically progressive – an attempt, as he says in the Preface (not extracted here) to 'conceive the idea of a system of natural law of nations, which should claim the obedience of states and sovereigns' and achieve the happiness of citizens. Vattel's argument that 'there certainly exists a natural law of nations, since the obligations of the law of nature are no less binding on states . . . than on individuals' can, however, also be viewed through environmental and social lenses. While Vattel's treatment of the broader legal subject is beyond our purview, this coming together of nature and law is significant, not least because it initially seems to resist the age-old separation of human culture from nature and of human and natural histories, in which the former pair is always privileged over (as well as separated from) the subordinate latter (in ways that have proved significant to Anthropocene scholars such as Dipesh Chakrabaty – see Further Reading). The reality of Vattel's concept of human-natural relations is less idealistic than this, however, and, crucially, is rooted in an exploitative model of environment.

As the extract from Chapter 17 demonstrates, Vattel rests his view of human activities on a model of human sovereignty over nature: 'the earth belongs to mankind in general; destined by the Creator to be their common habitation, and to supply them with food'. Therefore, 'they all possess a natural right to inhabit it, and to derive from it whatever is necessary for their subsistence'. Vattel exemplifies what Timothy Morton describes as 'agrilogistics': 'the highly addictive logistics of agriculture that began in the Fertile Crescent . . . whose logical structure is never examined, though it underpins the machinations that resulted in the Agricultural and then Industrial Revolutions' (2015, 1). Human 'civilisation', as Morton suggests and Vattel contends, rests on unsupported assertions that *Homo sapiens* alone has an unchallengeable right to exploit environments. Moreover, cultivation of the soil and pastoralisation of animals, rather than dwelling on the land, are 'the origin of the rights of *property* and *dominion*' (ideas also explored a few decades later by Thomas Paine: see Volume II of this anthology). For Vattel, territorial sovereignty is established wherever people occupy and cultivate land, even if there are already 'wandering tribes of men continuing to possess it in common'. Vattel argues that a nation simply taking possession of territory is guilty of 'an absolute infringement of the natural rights of men', but in turning to the 'celebrated question' of whether 'a nation may lawfully take possession of some part of a vast country, in which there are none but erratic nations whose scanty population is incapable of occupying the whole', his answer is different. Because indigenous peoples do not (he claims) cultivate land, 'their unsettled habitation in those immense regions cannot be accounted a true and legal possession'. Thus enabling an ongoing project of imperial conquest, Vattel's international law is founded on an entitled Eurocentric vision that is uninterested in the very different perspectives of indigenous peoples:

> The people of Europe, too closely pent up at home, finding lands of which the savages stood in no particular need, and of which they made no actual and constant use, were lawfully entitled to take possession of it.

Vattel's praise of the 'laudable example' of the settlement of Pennsylvania by the purchase of land from Indians (rather than direct conquest) is merely a gloss on the underlying imperial imperative. Vattel's grounding of law in nature, then, produces an influential reading of environment as exploitable resource that disqualifies native populations from equal human status by 'naturalising' them as animal-like non-cultivators. Vattel thus fulfils the subtitle's attempt to '*Display the True Interest of Powers*'. Having deemed it vital to include this extract as a statement of Eurocentric sovereignty over environment and colonised peoples, various extracts in subsequent sections (particularly 7 and 8) and volumes will delve further into these crucial global engagements.

The issues of animal demographics in Buffon's study of the hare are yet more central in our final extract, from John Bruckner's *A Philosophical Survey of the Animal Creation* (1768). Its central argument is that God's creation of carnivores adds to the happiness and fecundity of Creation by enabling an extra level of life. His varied sources include earlier and contemporary scientific works, journals, travel literature, and the Bible. While this is a classic work of Natural Theology, its central purpose being praise of wise design, it also anticipates Malthusian and biogeographical thought.

Specific explanations are helpful in reading Bruckner. He refers to infestations of 'May-Bugs' (cockchafer beetle) in England and unnamed insects in Angoumois (Angoulême), citing evidence from *Journal des Savants* (1742), the earliest European academic journal, running from 1665 until 1792 and 1816 to the present. Bruckner's work provides strong evidence of the cross-fertilisation of European scientific knowledge. José Mouthaan (Further Reading) provides a detailed account of Bruckner's references to the mysterious appearance of 'sea worms' (*Teredo navalis*, naval shipworms) that ravaged Dutch sea defences in the 1730s. Despite such examples, Bruckner's emphasis is, in classic Natural Theological fashion, on proving the overall equilibrium existing in the natural world.

Although no set of brief extracts can do justice to the task of examining the predecessor-thought prior to the long nineteenth century, this section at least points to some foundations, leading ideas and interventions, and directions of travel. The following further reading provides many avenues to explore these various issues, and the sections that follow provide the means to explore the various manifestations of science in the long nineteenth century.

Further reading

Armstrong, Patrick (2000), *The English Parson-Naturalist: A Companionship Between Science and Religion* (Leominster: Gracewing Publishing).

Chakrabaty, Dipesh. 'The Climate of History: Four Theses', *Critical Inquiry* 35:2 (2009), 197–222.

Derham, William, *Physico-Theology: Or, A Demonstration of the Being and Attributes of God, From His Works of Creation* (1713).

Foucault, Michel, *The Order of Things: An Archaeology of the Human Sciences* (London: Tavistock, 1989).

Grove, Richard H., *Green Imperialism: Colonial Expansion, Tropical Islands, Edens and the Origins of Environmentalism, 1600–1860* (Cambridge: Cambridge University Press, 1995).

Holmes, Richard, *The Age of Wonder: How the Romantic Generation Discovered the Beauty and Terror of Science* (London: Harper, 2008).

Huxley, Robert, *The Great Naturalists* (London: Thames & Hudson/Natural History Museum, 2007).

Jardine, Lisa, *The Curious Life of Robert Hooke* (New York: HarperCollins, 2003).

Koerner, Lisbet, *Linnaeus: Nature and Nation* (Cambridge: Harvard University Press, 2009).

'Madras Birds': https://commons.wikimedia.org/wiki/File:Madras_Birds.jpg

McKusick, James C, 'John Evelyn: The Forestry of Imagination', *Wordsworth Circle* 44:2/3 (2013), 110–4.

Morton, Timothy, 'The Biosphere Which Is Not One: Towards Weird Essentialism', *The Journal of the British Society for Phenomenology* 46 (2015), 141–55.

Mouthaan, José, 'The Appearance of a Strange Kind of "Sea-Worm" at the Dutch Coast, 1731–1735', *Dutch Crossing* 27:1 (2003), 3–22.

Rackham, Oliver, *The History of the British Countryside* (London: Dent, 1987).

Simmons, I.G., *An Environmental History of Great Britain From 10,000 Years Ago to the Present* (Edinburgh: Edinburgh University Press, 2006).

The Eleventh Report of the Commissioners Appointed to Inquire Into the State and Condition of the Woods, Forests, and Land Revenues of the Crown and to Sell or Alienate Fee Farm and Other Unimprovable Rents (London: J. Debrett, 1792).

Travis, Toby, '"Belching It Forth Their Sooty Jaws": John Evelyn's Vision of a "Volcanic" City', *London Journal* 39:1 (2014), 1–20.

White, Lynn White, Jr., 'The Historical Roots of Our Ecologic Crisis', in Cheryll Glotfelty, and Harold Fromm (eds.), *The Ecocriticism Reader: Landmarks in Literary Ecology*, Athens: University of Georgia Press, 1996), 3–14.

Worster, Donald, *Nature's Economy: a History of Ecological Ideas* (Cambridge: Cambridge University Press, 1994).

1

JOHN EVELYN, *FUMIFUGIUM: OR, THE INCONVENIENCE OF THE AER AND SMOKE OF LONDON DISSIPATED. TOGETHER WITH SOME REMEDIES HUMBLY PROPOSED BY J.E. ESQ.; TO HIS SACRED MAJESTIE, AND TO THE PARLIAMENT NOW ASSEMBLED*

(London: Gabriel Bedel, and Thomas Collins, 1772 [1661])

TO THE KINGS MOST SACRED MAJESTY.

SIR,

IT was one day, as I was Walking in Your MAJESTIES Palace at *WHITE-HALL,* (where I have sometimes the honour to refresh myself with the Sight of Your Illustrious Presence, which is the Joy of Your Peoples hearts) that a presumptuous Smoake issuing from one or two Tunnels neer *Northumberland-house,* and not far from *Scotland-yard,* did so invade the Court; that all the Rooms, Galleries, and Places about it were filled and infested with it; and that to such a degree, as Men could hardly discern one another for the Clowd, and none could support, without manifest Inconveniency. It was not this which did first suggest to me what I had long since conceived against this pernicious Accident, upon frequent observation; But it was this alone, and the trouble that it must needs procure to Your Sacred Majesty, as well as hazzard to Your Health, which kindled this Indignation of mine, against it, and was the occasion of what it has produced in these Papers.

[. . .]

Sir, I prepare in this short Discourse, an expedient how this pernicious *Nuisance* may be reformed; and offer at another also, by which the *Aer* may not only be freed from the present Inconveniency; but (that removed) to render not only Your Majesties Palace, but the whole City likewise, one of the sweetest, and most delicious Habitations in the World; and this, with little or no expence; but by

DOI: 10.4324/9780429355653-3

improving those Plantations which Your Majesty so laudably affects, in the moyst, depressed and Marshy Grounds about the Town, to the Culture and Production of such things, as upon every gentle emission through the *Aer,* should so perfume the adjacent places with their breath; as if, by a certain charm, or innocent *Magick,* they were transferred to that part of *Arabia,* which is therefore styled the *Happy,* because it is amongst the Gums and precious Spices. Those who take notice of the Scent of the *Orange*-flowers from the Rivage of *Genoa,* and St. *Pietro dell' Arena*; the Blossomes of the *Rosemary* from the Coasts of *Spain* many Leagues off at Sea; or the manifest and odoriferous wafts which flow from *Fontenay* and *Vaugirard,* even to *Paris,* in the season of *Roses,* with the contrary Effects of those less pleasing Smells from other accidents, will easily consent to what I suggest: And, I am able to enumerate a Catalogue of native *Plants,* and such as are familiar to our Country and *Clime,* whose redolent and agreeable Emissions would even ravish our senses, as well as perfectly improve and meliorate the *Aer* about *London*; and that, without the least prejudice to the Owners and Proprietors of the Land to be employed about it.

Part II

WE know (as the *Proverb* commonly speaks) that, as *there is no Smoake without Fire*; so neither is there hardly any *Fire* without *Smoake,* and that the ἀκαπνα ξύλα, materials which burn clear are very few, and but comparatively so tearmed: That to talk of serving this vast City (though *Paris* as great, be so supplied) with *Wood,* were madnesse; and yet doubtlesse it were possible, that much larger proportions of Wood might be brought to *London,* and sold at easier rates, if that were diligently observed, which both our *Laws* enjoyn, as faisible and practised in other places more remote, by Planting and preserving of *Woods* and *Copses,* and by what might by Sea, be brought out of the *Northern Countries,* where it so greatly abounds, and seems inexhaustible. But the *Remedy* which I would propose, has nothing in it of this difficulty, requiring only the Removal of such *Trades,* as are manifest *Nuisances* to the City, which, I would have placed at farther distances; especially, such as in their Works and Fournaces use great quantities of *Sea-Coale,* the sole and only cause of those prodigious Clouds of *Smoake,* which so universally and so fatally infest the *Aer,* and would in no City of *Europe* be permitted, where Men had either respect to Health or Ornament. Such we named to be *Brewers, Diers, Sope* and *Salt-boylers, Lime-burners,* and the like: These I affirm, together with some few others of the same *Classe* removed at competent distance, would produce so considerable (though but partial) a Cure, as Men would even be found to breath a new life as it were, as well as *London* appear a new City, delivered from that, which alone renders it one of the most pernicious and insupportable abodes in the World, as subjecting her Inhabitants to so infamous an *Aer,* otherwise sweet and very healthful: For, (as we said) the *Culinary* fires (and which *charking* would greatly reform) contribute little, or nothing in comparison to these foul mouth'd Issues, and Curles of *Smoake,* which (as the Poet[1] has it) do *Cœlum subtexere fumo,* and draw a sable Curtain over Heaven. Let any man observe it, upon a *Sunday,* or such time as these Spiracles cease, that the Fires are generally extinguished, and he shall sensibly conclude, by the clearnesse of the Skie, and universal serenity of the *Aer* about it, that all the Chimnies in *London,* do not darken and poyson it so much, as *one* or two of those Tunnels of *Smoake*; and, that, because the most imperceptible transpirations, which *they* send forth, are ventilated, and dispersed with the least breath which is stirring: Whereas the *Columns* and Clowds of *Smoake,* which are belched forth from the sooty Throates of those Works, are so thick and plentiful, that rushing out with great impetuosity, they are capable even to resist the fiercest winds, and being extremely surcharg'd with a fuliginous Body, fall down upon the City, before they can be dissipated, as the more thin and weak is; so as two or three of these *fumid vortices,*[2] are able to whirle it about the whole City, rendering it in a few Moments like the Picture of *Troy* sacked by the *Greeks,* or the approches of *Mount-Hecla.*

I propose therefore, that by an *Act* of this present *Parliament,* this infernal *Nuisance* be reformed; enjoyning, that all those *Works* be removed five or six miles distant from *London* below the River of *Thames*; I say, five or six Miles, or at the least so far as to stand behind that *Promontary* jetting out, and securing *Greenwich* from the pestilent *Aer* of *Plumstead*-Marshes: because, being placed at any lesser Interval beneath the *City,* it would not only prodigiously infect that his *Majesties* Royal Seat (and as *Barclay* calls it) *pervetusta Regum Britannicorum domus*; but during our nine Months *Etesians* (for so we may justly name our tedious Western-winds) utterly darken and confound one of the most princely, and magnificent Prospects that the World has to shew: Whereas, being seated behind that Mountain, which seems to have been thus industriously elevated, no winds, or other accident whatever can force it through that solid obstacle

Part III

THERE is yet another expedient, which I have here to offer (were *This* of the poisonous and filthy *smoake* remov'd) by which the *City* and environs about it, might be rendred one of the most pleasant and agreeable places in the world. In order to this I propose,

That all low-grounds circumjacent to the City, especially *East* and *South-west,* be cast and contriv'd into square plots, or Fields of twenty, thirty, and forty *Akers,* or more, separated from each others by Fences of double *Palisads,* or *Contr'spaliers,* which should enclose a Plantation of an hundred and fifty, or more, feet deep, about each Field; not much unlike to what His *Majesty* has already begun by the wall from Old *Spring-garden* to *St. James's* in that *Park*; and is somewhat resembled in the new *Spring-garden* at *Lambeth.* That these *Palisads* be elegantly planted, diligently kept and supply'd, with such *Shrubs,* as yield the most fragrant and odoriferous *Flowers,* and are aptest to tinge the *Aer* upon every gentle emission at a great distance: Such as are (for instance amongst many others) the *Sweet-brier,* all the *Periclymenas* and *Woodbinds*; the Common *white* and *yellow Jessamine,* both the *Syringas* or *Pipe trees*; the *Guelder-rose,* the *Musk,* and all other *Roses; Genista Hispanica:* To these may be added the *Rubus odoratus, Bayes, Juniper, Lignum-vitæ, Lavender:* but above all, *Rosemary,* the *Flowers* whereof are credibly reported to give their scent above thirty Leagues off at Sea, upon the coasts of *Spain*: and at some distance towards the Meadow side, *Vines,* yea, *Hops*.

Notes

1 Virgil.
2 Pliny.

2

JOHN EVELYN, *SYLVA; OR A DISCOURSE OF FOREST-TREES AND THE PROPAGATION OF TIMBER IN HIS MAJESTY'S DOMINIONS*

(London: J. Walthoe, 1729 [1664])

Introduction

1. SINCE there is nothing which seems more fatally to threaten a *Weakning,* if not a *Dissolution* of the Strength of this famous and flourishing *Nation,* than the sensible and notorious Decay of her *Wooden Walls,* when either through *Time, Negligence,* or other *Accident,* the present *Navy* shall be worn out and impair'd; it has been a very *worthy* and seasonable *Advertisement* in the Honourable the principal *Officers* and *Commissioners,* what they have lately suggested to this *Illustrious Society* for the timely Prevention and Redress of this intolerable Defect. For it has not been the late Increase of *Shipping* alone, the Multiplication of *Glass-Works, Iron-Furnaces,* and the like, from whence this impolitick Diminution of our *Timber* has proceeded; but from the disproportionate spreading of *Tillage,* caused through that prodigious Havock made by such as lately professing themselves against *Root* and *Branch* (either to be re-imburs'd their *Holy* Purchases, or for some other sordid Respect) were tempted, not only to *fell* and *cut* down, but utterly to *extirpate, demolish,* and *raze,* as it were, all those many goodly *Woods* and *Forests,* which our more prudent *Ancestors* left standing, for the Ornament and Service of their *Country*. And this *Devastation* is now become so *epidemical,* that unless some favourable *Expedient* offer it self, and a Way be seriously and speedily resolv'd upon, for a future Store, one of the most glorious and considerable *Bulwarks* of this *Nation,* will, within a short Time, be totally wanting to it.

2. To attend now a *spontaneous* Supply of these decay'd *Materials* (which is the vulgar and natural Way) would cost (besides the *Inclosure*) some entire *Ages* repose of the *Plow,* though *Bread* indeed require *our* first Care: Therefore, the most expeditious and obvious *Method* would doubtless be *one* of these *two* Ways, *Sowing* or *Planting*. But, first, it will be requisite to agree upon the *Species*; as what *Trees* are likely to be of greatest *Vse,* and the fittest to be cultivated; and

DOI: 10.4324/9780429355653-4

then, to consider of the *Manner* how it may be best effected. Truly, the *Waste* and *Destruction* of our *Woods* has been so universal, that I conceive nothing less than an *universal Plantation of all the sorts of Trees* will supply, and well encounter the Defect; and therefore I shall here adventure to speak something in general of them all; though I chiefly insist upon the Propagation of *such* only as seem to be the most wanting and serviceable to the End proposed.

Chap. VI

Of the *Laws* and *Statutes* for the *Preservation* and *Improvement* of *Woods* and *Forests*

'TIS not to be passed by, that the very first *Law* we find which was ever *promulged,* was concerning *Trees*; and that *Laws* themselves were first *written* upon them, or *Tables* composed of them; and after that Establishment in *Paradise,* the *next* we meet withal are as antient as *Moses:* You may find the *Statute* at large in *Deut*. xx. 19, 20. Which though they chiefly tended to *Fruit-trees,* even in an *Enemy*'s Country, yet you will find a Case of *Necessity* only alledged for the *Permission* to destroy any other.

[. . .]

7. But to the *Laws:* It were to be wished that our tender and improvable *Woods,* should not admit of *Cattle* by any Means, till they were quite grown out of reach; the *Statutes* which connive at it, in favour of *Custom,* and for the satisfying of a few clamorous and rude *Commoners,* being too indulgent; since it is very evident, that less than a fourteen or fifteen Years *Enclosure,* is in most Places too soon; and our most *material* Trees would be of *infinite* more Worth and Improvement, were the *Standards* suffered to grow to *Timber,* and not so frequently cut, at the next *felling* of the *Wood,* as the general Custom is. In 22 *Edw*. IV. the Liberty arrived but to *seven* Years after a felling of a *Forest* or *Purlieu*; and but *three* Years before, without special *License:* This was very narrow; but let us then look on *England* as an over-grown Country.

8. *Wood* in *Parks* was afterwards to be *four* Years *fenced,* upon *felling*; and yearling *Colts* and *Calves* might be put into inclosed *Woods* after *two:* By the 13 *Eliz*. five Years, and no other *Cattle* till six, if the Growth was under *fourteen* Years; or until *eight,* if exceeding that Age till the last *feeling*. All which *Statutes* being by the *Act* of *Hen*. VIII. but *temporal,* this *Parliament* of *Elizabeth* thought fit to make *perpetual*.

9. Then, to prevent the destructive razing and converting of *Woods* to *Pasture:* No *Wood* of two *Acres,* and above two *Furlongs,* from the *Mansion-house,* should be indulged: And the *Prohibitions* are good against *Assarts* made in *Forests, &c.* without *License:* The *Penalties* are indeed great; but how seldom inflicted? And what is now more easy, than compounding for such a *License?*

In some Parts of *Germany,* where a single *Tree* is observed to be extraordinary fertile, a constant and plentiful *Mast-bearer,* there are *Laws* to prohibit their *felling* without special Leave: And it was well *enacted* amongst us, that even the

Owners of *Woods* within *Chases,* should not cut down the *Timber* without View of *Officers*; this *Act* being in *Affirmance* of the *Common-law,* and not to be violated without *Prescription:* See the *Case* cited by my Lord *Coke* in his *Comment* on *Littleton. Tenure Burgage,* Lib. ii. Sect. 170. Or if not within *Chases,* yet where a common Person had Liberty of Chase, *&c.* and this would be of much Benefit, had the Regarders performed their Duty, as 'tis at large described in the *Writ* of the 12 *Articles*; and that the *Surcharge* of the *Forests* had been honestly inspected with the due *Perambulations,* and antient *Metes*. Thus should the *Justices* of *Eyre* dispose of no *Woods* without express *Commission,* and in convenient Places: *Minuti blaterones quercuum, culi, & curbi,* as our Law terms *Wind-falls, Dotterels, Scrags, &c.* and no others.

10. Care is likewise, by our Laws, to be taken that no unnecessary *Imbezilment* be made by Pretences of Repair of *Paling, Lodges, Browse* for *Deer,* &c. *Wind-falls, Root-falls,* dead and *Sear-trees,* all which is subject to the Inspection of the Warders, Justices, Itinerants, *&c.* and even *Trespasses* done *de Viridi* on Boughs of *Trees, Thickets,* and the like; which (as has been shewed) are very great Impediments to their Growth and Prosperity, and should be duly looked after, and punished; and the great Neglect of *Swainmote-Courts* reformed, *&c.* See *Consuet. & Assis. Fores. Pannagium,* or *Pastura pecorum & de Glandibus, Fleta,* &c. *Manwood*'s *Forest-laws: Coke pla. fol.* 366. *Lib.* viii. *fol.* 138.

11. Finally, that the Exorbitance and Increase of devouring *Ironmills* were looked into, as to their *Distance* and *Number* near the Seas, or navigable *Rivers*; and what if some of them were even *removed* into another World? the *Holy-land* of *New-England* (there to build Ships, erect Saw-Mills, near their noble *Rivers*) for they will else ruin *Old-England.* 'Twere better to purchase all our *Iron* out of *America,* than thus to *exhaust* our *Woods* at *Home,* although (I doubt not) they might be so ordered, as to be rather a Means of conserving them. There was a Statute made by Queen *Eliz.* to prohibit the converting of *Timber-trees* to *Coal,* or other *Fuel,* for the Use of *Iron-mills,* if the *Tree* were of one Foot square, and growing within fourteen Miles of the Sea, or the greater Rivers, *&c.* 'Tis pity some of those Places in *Kent, Sussex* and *Surrey* were excepted in the *Proviso,* for the Reason expressed in a Statute made 23 *Eliz.* by which even the employing of any *Vnder-wood* as well as great Trees, was prohibited within twenty-two Miles of *London.*

[. . .]

Chap. VII

The Parænefis *and* Conclusion, *containing some* Encouragements *and* Proposals *for the* Planting *and* Improvement *of His* Majesty'*s* Forests, *and other* Amœnities *for* Shade *and* Ornament.

1. SINCE our *Forests* are undoubtedly the greatest *Magazines* of the *Wealth* and *Glory* of this *Nation,* and our *Oaks* the truest *Oracles* of its *Perpetuity* and *Happiness,* as being the only Support of that *Navigation* which makes us fear'd Abroad, and flourish at Home; it has been strangely wondered at by some good *Patriots,*

how it comes to pass, that many *Gentlemen* have frequently repaired, or gained a sudden *Fortune,* with *plowing* part of their *Parks,* and setting out their fat Grounds to *Gardeners, &c.* and very wild *Wood-land* Parcels (as may be instanced in several Places) to Dressers of *Hopyards,* &c. whiles the *Royal Portion* lies folded up in a *Napkin,* uncultivated and neglected, especially those great and ample *Forests*; where, though *plowing* and *sowing* have been forbidden, a *Royal Command* and *Design* may well dispense with it, and the breaking up of those *Intervals,* advance the Growth of the Trees to an incredible *Improvement.*

2. It is therefore insisted on, that there is not a cheaper, easier, or more prompt *Expedient* to advance *Ship-timber,* than to sollicit, that in all his *Majesty's Forests, Woods* and *Parks,* the spreading *Oak, &c.* (which we have formerly described) be cherished, by plowing and sowing *Barley, Rye, &c.* (with due Supply of *Culture* and *Soil,* between them) as far as may (without danger of the *Plow-share*) be broken up. But this is only where these Trees are arrived to some Magnitude, and stand at competent Distances; a hundred, or fifty Yards (for their *Roots* derive Relief far beyond the Reach of any *Boughs*) as do the *Walnut-trees* in *Burgundy,* which stand in their best *Plow'd-lands.*

3. But, that we may particularise in his *Majesty's Forests* of *Deane, Sherewood, Enfield-Chase, &c.* and in some sort gratify the *Quœries* of the Honourable the Principal *Officers* and *Commissioners* of the *Navy*; I am advised by such as are every way *judicious,* and of long Experience in those Parts, that to *enclose* would be an excellent Way: But it is to be considered, that the People, *viz. Foresters,* and *Borderers,* are not generally so civil and reasonable as might be wished; and therefore to design a solid *Improvement* in such Places, his Majesty must assert his *Power,* with a firm and high Resolution to reduce these Men to their due *Obedience,* and to a Necessity of submitting to their own and the publick Utility, though they preserved their *Industry* this Way, at a very tolerable Rate, upon that Condition; while some Person of Trust and Integrity did regulate and supervise the *Mounds* and *Fences,* and destine some Portions frequently set apart for the raising and propagating of *Wood,* till the whole *Nation* were furnished for *Posterity.*

3

JOHN RAY, *THE WISDOM OF GOD MANIFESTED IN THE WORK OF CREATION*

The Seventh Edition, Corrected (London: William Innys, 1717 [1691])

Preface

First, the Belief of a Deity being the Foundation of all Religion, (Religion being nothing but a devout Worshipping of God, or an Inclination of Mind to serve and worship Him) for he that cometh to God, must believe that he is, it is a Matter of the highest concernment, to be firmly settled and established in a full Persuasion of this main Point: Now this must be demonstrated by Arguments drawn from the Light of Nature, and Works of the Creation: For as all other Sciences, so Divinity proves not, but supposes its Subjects, taking it for granted that by Natural Light, Men are sufficiently convinced of the Being of a Deity. There are indeed supernatural Demonstrations of this fundamental Truth, but not common to all Persons or Times, and so liable to Cavil and Exception by Atheistical Persons, as inward Illuminations of Mind, a Spirit of Prophecy and Foretelling future Contingents, illustrious Miracles, and the like. But these Proofs taken from Effects, and Operations, exposed to every Man's view, not to be denied or questioned by any, are most effectual to convince all that deny or doubt of it. Neither are they only convictive of the greatest and subtlest Adversaries, but intelligible also to the meanest Capacities: For you may hear illiterate Persons of the lowest Rank of the Commonalty affirming, That they need no Proof of the Being of a God, for that every Pile of Grass, or Ear of Corn, sufficiently proves that: For, say they, all the Men of the World cannot make such a thing as one of these; and if they cannot do it, who can, or did make it but God? To tell them, that it made itself, or sprung up by Chance, would be as ridiculous as to tell the greatest Philosopher so.

Secondly, the Particulars of this Discourse serve not only to demonstrate the Being of a Deity, but also to illustrate some of his principal attributes; namely, his infinite Power and Wisdom. The vast Multitude of Creatures, and those not only small, but immensely great, the Sun and Moon, and all the Heavenly Host, are Effects and Proofs of His Almighty Power. The Heavens declare the Glory of God, and the Firmament sheweth His Handy-Work, *Psal.* xix. 1. The admirable

DOI: 10.4324/9780429355653-5

Contrivance of all and each of them, the Adapting all the Parts of Animals to their several Uses, the Provision that is made for their Sustenance, which is often taken Notice of in Scripture, *Psal.* cxlv. 15, 16. The Eyes of all wait upon Thee: Thou givest them their Meat in due Season. Thou openest Thy Hand, and satisfiest the Desire of every living Thing. *Matth.* vi. 26. Behold the Fowls of the Air, for they sow not, neither do they reap nor gather into Barns yet your Heavenly Father feedeth them. *Psal.* cxlvli. 9. He giveth to the Beast his Food, and to the young ravens when they cry. And, Lastly, Their mutual Subserviency to each other, and unanimous conspiring to promote and carry on the Publick Good, are evident Demonstrations of His Sovereign Wisdom.

[. . .]

Of Bodies Endued with a Sensitive Soul, or Animals'

I proceed now to the Consideration of Animate Bodies endue'd with a sensitive Soul, call'd *Animals*. Of these I shall only make some general Observations, not curiously consider the Parts of each particular *Species*, save only as they serve for Instances or Examples.

First of all, because it is the great Design of Providence to maintain and continue every species, I shall take Notice of the great Care and abundant Provision that is made for the securing this End [. . .] Why can we imagine all Creatures should be made Male and Female but to this Purpose? Why should there be implanted in each Sex such a vehement and inexpungable Appetite of Copulation? Why in viviparous Animals, in the Time of Gestation, should the Nourishment be carried to the *Embryo* in the Womb, which at other Times goeth not that way? When the Young is brought forth, how comes all the Nourishment be transfer'd from the Womb to the Breasts or Paps, leaving its former Channel, the Dam at such Time being for the most Part lean and ill-favor'd? To all this I might add, as a great Proof and Instance of the Care that is taken, and Provision made for the Preservation and Continuance of the Species, the lasting Fœcundity of the Animal, Seed or Egg in the Females of Man, Beasts and Birds.

[. . .]

That flying Creatures of the greater Sort, that is *Birds*, should all lay Eggs, and none bring forth live Young, is a manifest Argument of Divine Providence, designing thereby their Preservation and Security, that there might be the more Plenty of them; and that neither the Birds of Prey, the Serpent nor the Fowler, should straiten their Generations too much. For if they had been viviparous, the Burden of their Womb, if they had brought forth any competent Number at a Time, had been so great and heavy, that their Wings would have fail'd them, and they become an easie Prey to their Enemies. Or, if they had brought but one or two at a Time, they would have been troubled all the Year long with feeding their Young, or bearing them in their Womb [. . .]

The marvellous speedy Growth of Birds that are hatch'd in Nests, and fed by the Old ones there, 'till they be fledg'd, and come almost to their full Bigness, at which Perfection they arrive within the short Term of about one Fortnight, seems to me an Argument of Providence, designing thereby their Preservation, that they might not lie long in a Condition expos'd to the Ravine of any Vermine that may find them, being utterly unable to escape or shift for themselves.

Another and no less effectual Argument may be taken from the Care and Providence us'd for the Hatching and Rearing their Young: And first, they search out a secret and quiet Place where they may be secure and undisturb'd in their incubation; then they make themselves Nests every one after his kind, that so their Eggs and Young may lie soft and warm, and their exclusion and growth be promoted. These Nests some of them so elegant and artificial, that it is hard for Man to imitate them and make the like. I have seen Nests of an Indian Bird so artificially compos'd of the Fibres, I think, of some Roots, so curiously interwoven and platted together, as is admirable to behold: Which Nests they hang on the end of the Twigs of Trees over the Water, to secure their Eggs and Young from the ravage of Apes and Monkeys, and other Beasts that might else prey upon them. After they have laid their Eggs, how silently and patiently do they sit upon them 'till they be hatch'd, scarce affording themselves time to go off to get them Meat? Nay, with such an ardent and impetuous desire of sitting are they inspir'd, that if you take away all their eggs, they will sit upon an empty Nest: And yet one would think that sitting were none of the most pleasant Works. After their Young are hatch'd, for sometimes they do almost constantly brood them under their Wings, lest the Cold and sometimes perhaps the Heat should harm them. All this while also they labour hard to get them Food, sparing it out of their own Bellies, and pining themselves almost to death rather than they should want. Moreover it is admirable to observe, with what courage they are at that time inspir'd, that they will even venture their own Lives in defence of them. The most timorious, as Hens and Geese become then so couragious, as to dare to fly in the Face of a Man that shall molest or disquiet their Young, which would never do so much in their own defence. These things being contrary to any motions of Sense, or instinct of self-preservation, and so eminent pieces of self-denial, must needs be the Works of Providence, for the continuation of the Species and upholding of the World. Especially if we consider that all these pains is bestow'd upon a thing which takes no notice of it, will render them no thanks for it, nor make them any requital or amends; as also, that after the *Young* is come to some growth, and able to shift for itself, the old one [. . .] takes no further care of it, but will fall upon it, and beat it indifferently with others.

[. . .]

I shall take notice of the various strange Instincts of Animals; which will necessarily demonstrate, that they are directed to Ends unknown to them, by a wise Superintendant. As, 1. That all Creatures should know how to defend themselves, and offend their Enemies; where their natural Weapons are situate, and how to make use of them. A *Calf* will so manage his Head as tho' he would push with his Horns even before they shoot. A *Boar* knows the use of his Tushes; a *Dog* of his Teeth; a *Horse* of his Hoofs; a *Cock* of his Spurs; a *Bee* of her Sting; a *Ram* will butt with his Head, yea tho' he be brought up tame, and never saw that Manner of Fighting. Now, why another Animal, which hath no Horns should not make a Shew of pushing, or no Spurs, of striking with his Legs, and the like, I know not, but that every Kind is providentially directed to the Use of its proper and natural Weapons. 2. That those Animals that are weak, and have neither Weapons nor Courage to fight, are for the most Part created swift of Foot or Wing, and so being naturally timorious, are both willing and able to save themselves by Flight. 3. That Poultry, Partridge, and other Birds, should at the first Sight know Birds of Prey, and make Sign of it by a peculiar Note of their Voice to their Young, who presently thereupon hide themselves: That the Lamb should acknowledge the Wolf its Enemy, tho' it had never seen one before, as is taken for granted by most Naturalists, and may, for ought I know, be true, argues the Providence of Nature, or more truly the God of Nature, who, for their Preservation hath put such an Instinct into them. 4. That young Animals, so soon as they are brought forth, should know their Food: As for Example; such as are nourish'd with Milk presently find their Way to the Paps, and suck at them, whereas none of those that are not design'd for that Nourishment ever offer to suck, or seek out any such Food. Again, 5. That such Creatures as are whole-footed, or Fin-toed, *viz* some Birds, and Quadrupeds, are naturally directed to go into the Water, and swim there, as we see *Ducklings*, tho' hatch'd and led by a Hen, if she brings them to the Brink of a Rover or Pond of Water, they presently leave her, and in they go, tho' they never saw any such Thing done before; and tho' the Hen clucks and calls, and doth what she can to keep them out [. . .] So that we see every Part in Animals is fitted to its Use, and the Knowledge of this Use put into them.

4

ROBERT HOOKE, *A GENERAL SCHEME, OR IDEA OF THE PRESENT STATE OF NATURAL PHILOSOPHY, AND HOW ITS DEFECTS MAY BE REMEDIED BY A METHODICAL PROCEEDING IN THE MAKING EXPERIMENTS AND COLLECTING OBSERVATIONS WHEREBY TO COMPILE A NATURAL HISTORY, AS THE SOLID BASIS FOR THE SUPERSTRUCTURE OF TRUE PHILOSOPHY*

(London: Royal Society, 1705)

First General

Present State of *Natural Philosophy* and Wherein it is Deficient

THE Business of Philosophy is to find out a perfect Knowledge of the Nature and Proprieties of Bodies, and of the Causes of Natural Productions, and this Knowledge is not barely acquir'd for it self, but in order to the inabling a Man to understand how by the joyning of fit Agents to Patients according to the Orders, Laws, Times, and Methods of Nature, he may be able to produce and bring to pass such Effects, as may very much conduce to this well being in this World, both for *Satisfying his Desires*, and the relieving of his *Necessities*: And for advancing his State above the common Condition of Men, and make him able to excel them as much, almost, as they do Brutes or Ideots.

DOI: 10.4324/9780429355653-6

Now though there have been many Men, in divers Ages of the World, which seem to have had some confus'd and imperfect Conception of this Idea of the Business of Philosophy, and accordingly seem to have had some Aims and Designs towards the attaining of their propos'd end; yet having not a right Understanding of the chief end, and failing much more in the Knowledge of the Means, or the manner of making use of them, they have generally left Philosophical Knowledge almost in the Condition they found it: Without making any considerable Increase or Addition to it. Whence this kind of Knowledge has been very little promoted ever since the very first times we have had any History of it.

[. . .]

Nor is the State of Philosophy as yet very much improved by our Modern Writers, who have endeavour'd to illustrate or piece up the old, by adding some Placits of their own: There are yet many Impediments to be removed and many Helps to be supplied before any very great Increase in Knowledge is to be expected. It may be questioned whether piecing or mending will serve the turn, or whether there must not be a new Foundation laid on the Information of our Senses, and more strictly examined and surveyed by accurate and judicious Experiments and Observations. That which hath had the Cultivation of many Hundreds of Years, and by divers very acute Men in all Ages, and yet as to the Inquiry after the Causes of Natural Efficients, has made so little, if any Progress at all, cannot with any Probability be imagined to afford a Method sufficient for this Inquiry. I do not here altogether reject *Logick*, or the way of Ratiocination already known; as a thing of no use. It has its peculiar Excellencies and Uses, in ordinary Discourse and Conversation: And affords some Helps to some kinds of Invention, especially of Arguments, as well as to the Memory, by its Method: It affords copious Matter for Disputes as well for, as against the Truth, and teaches how to solve as well as how to make a fallacious Assertion [. . .] But as to the Inquiry into Natural Operations, what are the Kinds of secret and subtile Actors, and what the abstruse and hidden Instruments and Engines there made use of may be; It seems not, to me, as yet at all adapted and wholly deficient. For 'tis not to be expected from the Accomplishments the Creator has endowed Man withal, that he should be able to leap, from a few particular Informations of his Senses, and those very superficial at best, and for the most part fallacious, to the general Knowledge of Universals or abstracted Natures, and thence be able, as out of an inexhaustable Fountain, to draw out a perfect Knowledge of all Particulars, to deduce the Causes of all Effects and Actions from this or that Axiom or Sentence, and as it were intuitively, to know what Nature does or is capable of effecting [. . .]

Some other Course therefore must be taken to promote the Search of Knowledge. Some other kind of Art for Inquiry than what hath been hitherto made use of, must be discovered; the Intellect is not to be suffer'd to act without its Helps, but is continually to be assisted by some Method or Engine, which shall be as a Guide to regulate its Actions [. . .]

Second General

Of the True Method of Building a Solid Philosophy, or of a Philosophical Algebra

THIS Method of Philosophical Algebra, I shall divide into two main Branches.

The first shall contain the manner of Preparing the Mind, and Furnishing it with fit Materials to work on.

The second shall contain the Rules and Methods of proceeding or operating with this so collected and qualify'd Supellex.

The former therefore has these Three Parts considerable in it, which shall be treated of in three distinct Sections.

1st. An Examination of the Constitution and Powers of the Soul, or an Attempt of Disclosing the Soul to its self, being an Endeavour of Discovering the Perfections and Imperfections of Humane Nature, and finding out ways and means for the attaining of the one, and of helping the other.

2dly, A Method of making use of, or employing these Means and Assistances of Humane Nature for collecting the Phenomena of Nature and for compiling a Philosophical History. Consisting of an exact Description of all sorts of Natural and Artificial Operations, or a Method of making Experiments and Observations for the Prosecution and Examination of any Philosophical Enquiry.

3dly, A Method of describing, registring and ranging these Particulars so collected, as that they may become the most adapted Materials for the raising of Axioms and the Perfecting of Natural Philosophy.

[. . .]

We will divide the whole Business of Philosophical History into these particular Heads of Inquiry, in which we have no so much proceeded according to the Nature of the things themselves, as according to their Appearance or Respect to us: For though the Earth, in Comparison of the Heavens, be as it were a Point, yet in Relation to its Nearness and Sensibleness to us, it becomes much more considerable, and the Consideration of it and its Parts will take up the greatest Part of this History. We will divide the Subject of Philosophical History into two parts; to wit, into things Natural and things Artificial.

[. . .]

He ought to proceed with the greatest degree of Candour and Freedom from Prejudice, not to be byassed by this or that Opinion in making of Deductions, nor by the Pleasantness or Gainfulness of the Experiment, or any other by Consideration that does not immediately look at the present Discovery he is searching after [. . .] lest like sweet singing Syrens they seduce their Followers out of their right way to their utter Destruction. He ought also to proceed with the greatest Circumspection and Diligence to find out such things, as are Indications of what he seeks, and from those to take Incouragement to prosecute his Intentions [. . .]

The next thing to be considered is, what the Subject of his Enquiry is; which I shall endeavour to explain by setting down the General Scheme of the whole Matter, about which a Philosophical History is to treat.

[. . .]

We may make these particular Heads of Inquiry, which for the Journal of first Book of Entries, will be particular and distinct enough.

1 The History of Comets and Blazing Stars.
2 The History of the Sun, Moon, Stars and Planets.
3 The History of the Aether.
4 The History of the Height, Extent, Figure, *&c.* of the Atmosphere or Air.
5 The History of the Variety of its Parts, or several Climates, and in several Regions or Heights.
6 The History of the various kinds of Mixtures it suffers from Meteors.
7 The History of its various Motions, Breizes, Winds, Storms, Hurricanes.
8 The History of Insects.
9 The History of Birds
10 The History of Beasts.
11 The History of Man [. . .]
12 The History of the Figure, Extent, Bulk, *&c.* of the Water.
13 The History of the Seas, Lakes, Ponds, Rivers, Fountains, Subterraneous Rivers *&c.*
14 The History of the various sourts of Bodies that are found incorporate, or that may be dissolv'd by it, as Salts, Slimes, Gums *&c.*
15 The History of Currents, Ebbings and Flowings, Increase and Decrease, Overflowings, Inundations, and Deserting of several Parts, of Voragoes, Submarine Fountains *&c.*
16 The History of Sea Insects, compar'd with Aerial and Terrestrial.
17 The History of Fish, both of fresh Water and Salt, describing their Internal Structure, and Shapes as well as Outwards.
18 The History of Sea Beasts, *&c.* Morses, Seales, Tortoises *&c.* Anatomiz'd and compar'd with other Creatures.
19 The History of the Extent, Figure, Magnitude *&c.* of the Earth, both in respect to other great Bodies in the World, as the Sun, the Moon, the Sea, *&c.* and in respect also of the Body of Man, or our common Measures.
20 The History of its various Parts, External Mountains, Vales, Plains, Clifts, Places of Reception for the Sea, *&c.*
21 The History of its Mixtures, Metals, Minerals, Stones, Clays, Earths, Sands, Oyls, Salts, *&c.* and the various Constitution of its Parts; the several Regions of it, of what kind of Shells, or Layers of Sand, Stone, Earth, Clay, *&c* it consists at several Depths.
22 The History of its Motions, Diurnal, Annual, Lunar, or Tide-Making.
23 The History of its Internal Motions, Earthquakes, Eruptions, *&c* Transpositions and Transformations.

24 The History of the Magnetism of it.
25 The History of its Gravitating Power.
26 The History of the Subterraneous Fires, Rivers, Caverns, Damps, *&c.*
27 The History of Mushrooms, Mosses and Plants, Roots, *&c.*
28 The History of Shrubs and Trees.
29 The History of Ground Animals and Worms.

Besides these particular Histories of the several parts of the World, there ought to be several Histories compos'd of the prime sensible Qualities, such as may serve afterwards for the finding out of those Properties first which are more simple, such as these.

1 The History of Light and Darkness.
2 The History of Transparency and Opacousness.
3 The History of Colours, commonly distinguish into real and appearing.
4 The History of Sounds, Musical and Harmonious.
5 The History of Tastes.
6 The History of Smells.
7 The History of Heat and Cold.
8 The History of Gravity and Levity.
9 The History of Density and Expansion.
10 The History of Flexibility and Stiffness.
11 The History of Malleability and Brittleness.

[. . .]

Lectures and Discourses of Earthquakes and Subterraneous Eruptions, Explicating The Causes of the Rugged and Uneven Face of the Earth and What Reasons may be given for the frequent finding of Shells and other Sea and Land Petrified Substances, Scattered over the whole Terrestrial Superficies

[. . .]

The obviousness and easiness of knowing many Things in Nature, has been the Cause of their being neglected, even by the more diligent and curious; which nevertheless, if well examined, do very often contain Information of the greatest value. It has been generally noted by common, as well as inquisitive Persons, that divers Stones have been found, formed into the Shape of Fishes, Shells, Fruits, Leaves, Wood, Barks, and other Vegetable and Animal Substances: We commonly know some of them exactly resembling the Shape of Things we commonly find (as the Chymists speak) in the Vegetable or Animal Kingdom; others of them indeed bearing some kind of Similitude, and agreeing in many Circumstances, but yet not exactly figured like any other thing in Nature; and yet of so curious a Shape, that they easily raise both the Attention and Wonder, even of those that are less inquisitive. Of these beautifully shaped Bodies I have observed two sorts: First, some more properly natural, such as have their Figures peculiar to their Substances: Others more improperly so, that is, such as seem to receive their Shape from an external and accidental Mould.

Of the first sort, are all those curiously figured Bodies of Salts, Talks, Spars, Crystals, Diamonds, Rubies, Amethysts, Ores, and divers other Mineral Substances, wherewith the World is adorned and enriched; which I at present omit to describe, as reserving them for a Second Part, they seeming to be, as it were, the Elemental Figures, or the *A B C* of Nature's working, the Reason of whose curious Geometrical Forms (as I may so call them) is very easily explicable Mechanically: And shall proceed to the second sort of Bodies.

Of these are two kinds; either first the very Substances themselves converted into Stone, such are Bones, Teeth, Shells, Fruit, Wood, Moss, Mushrooms, and divers Vegetable and Animal Substances: or secondly such other Mineral and Earthy Substances, as Clays, Sands, Earths, Flinty, Juices, *&c.* which have filled up, and been moulded in divers other Bodies, as Shells, Bones, Fruits, *&c.* but especially Shells. These, according to Representations they bear of other Bodies, have received divers Names [. . .] Of these I shall describe some few, because every one has not the Opportunity of seeing and examining them.

I have designed 15 several sorts of Snail rather than Snake-stones, call'd by some Authors *Cornua Ammonis*, or *Sceleta Serpentum*, all of them, both of different Substances and various Shapes; but yet all of them agreeing in these Properties, that they were made of a Tapering or Pyramidal Body kept exactly in the same Plane. 3. That all of them were ridged or furrow'd with Rings, Furrows, or Protruberances and Depressions, which respected the Center of the Spiral, for the most part, but were moulded and rang'd each of them different ways, all of them very regular, and exceedingly ornamental. 4. That in the coiling the lesser and inner Parts sunk, as it were always into the inside of the greater encompassing Part. 5. That all of them had Diaphragms, or separating Valves, whereby the Parts might oft-times be easily separated. 6. That the *Fimbria*, or Edges of these Diaphragms, were in most of these Stones very visible; in others of them, where they were somewhat more obscure, they might be made apparent, by scraping or rubbing away the outside of them. 7. That these *Fimbria*, or Edges appear'd on the Surface, like the Out-lines of some curious Foliage [. . .] This, upon Examination of them, I found to proceed from the Fulness of the Edges of the Diaphragms whereby the Edges were waved or plaited, somewhat in the manner of a Ruff. 8. That most of them were covered with a very curiously polish'd, as well as curiously carv'd Surface, some of them shining like burnish'd Brass [. . .]; others of them like transparent Horn [. . .] others like Coperas-stones; others like a coarser sort of white Marble; others like black Marble. 9. That from these polish Surfaces one might oftimes easily pick off a Substance exactly resembling the plaited shining Substances of a Shell; and this did very visibly in many of them cover the internal stony Body, with a Coat two or three times as thick as a Snail's Shell. 10. That the biggest end of these Spiral Bodies was always imperfect without any determinate Figure. 11. That many of these Spiral Bodies seem'd, as if they had been broken and shatter'd, and had grown together again in an irregular Posture. 12. That many of them were compounded of several Substances, the Spaces between the Diaphragms being sometimes filled with one kind of Substance, sometimes with another, and sometimes they were found empty, only all the sides of the Diaphragm were covered with a kind of Tooth-Sparr.

[. . .]

Bristol, *Aug*. 17, 1687

Sir,

In answer to some of your Enquiries, as to the *Cornua Ammonis*, and other Shell-like Stones found about *Keinsham*, and other Places, I shall give you this short Information of my Discoveries, and present you with the Draughts of some I happen'd to meet with there, and in other Places not very far distant, that is, in Part of *Gloucester* and *Somersetshire*.

The *Cornua Ammonis*, near *Keinsham*, lie most of them upon a little Hill, or rising Ground, above *Keinsham*-Bridge; the Place, as I take it, is about 18 Foot above the River; The River there runs half round the Foot of the Hill, where they lye very thick almost to touch each other, and are all of the large sort bedded in an hard Rock or Stone; some also I found near a Mile from thence in the Stone-walls of their Fields, and on the way in the Lanes; and at *Stowey*, four of five Miles from *Keinsham*, I saw some Snake-stones, Oyster, and Cockle-shells petrified and bedded in hard Stone, where is also a petrifying Spring incrustating the Moss and Grass, and all the wooden Troughs, by which it is conveyed with a stony Substance [. . .]

These, and the like Shapes, because many of them are curious, have so far wrought on some Men, that they have endeavoured to give us an Explication of the manner of their Formation; in doing of which they have so far rambled from the true and genuine Cause of them, that they have left the Matter much more difficult than they found it. Amongst the rest, *Gaffarel*, a *French* writer, seems not the least mistaken, who has transferr'd them over to the Confirmation, as he thinks, of his Astrological and Magical Fancy; and thinks that as they were produced from some extraordinary Celestial Influence, and that the Aspects and Positions of the fix'd Stars and Planets conduc'd to their Generation, so that they also have in them, a secret Vertue whereby they do at a distance work Miracles on things of the like Shape. But these, as fantastical and groundless, I shall not spend time on at present to refute, nor on the Conjectures and Hypotheses of divers others; which though perhaps somewhat more tolerable than that I last recited, yet most of them have recourse to some vegetative or plastick Vertue inherent in the Parts of the Earth where they were made, or in the very parcels of which they consist, which, to me, seems not at all consonant to the other workings of Nature; for those more curiously carved and beautiful Forms are usually bestow'd on some vegetable or animal Body. But my Business at present shall not be so much to confute other

Conjectures, as to make probably some of my own; which tho' at first hearing they may seem somewhat paradoxical, yet if the Reasons that have induced me thereunto be well consider'd and weigh'd, I hope at least they may seem possible, if not more than a little probable.

The particular Productions of this kind that I have taken notice of my self in my own Enquiries, and which I find dispersed up and down in the Writings of others, may be reduced under some one or other of these General Heads or Propositions.

First, That there are found in most Countries of the Earth, and even in such where it is somewhat difficult to imagine (by reason of their vast distance from the Sea or Waters, how they should come there) great quantities of Bodies resembling both in Substance and Shape the Shells of divers sorts of Shell-fishes; and many of them so exactly, that any one that knew not whence they came, would without the least scruple firmly believe them to be the Shells of such Fishes; But being found in Places so unlikely to have produced them, and not conceiving how else they should come there, they are generally believed to be real Stones form'd into these Shapes, either by some plastick Vertue inherent in those Parts of the Earth, which is extravagant enough, or else by some Celestial Influence or Aspect of the Planets operating at a distance upon the yielding Matter of the Parts of the Earth, which is much more extravagant [. . .]

Secondly, That there often have been, and are still daily found in other Parts of the Earth buried below the present Surface thereof divers sorts of Bodies, besides such as I newly mention'd, resembling both in Shape, Substance, and other Properties, the Parts of Vegetables, having the perfect Rind or Bark, Pith, Pores, Roots, Branches, Gums, and other constituent Parts of Wood, though in another posture, lying for the most part Horizontal, and sometimes inverted, and much differing from that of the like Vegetables when growing, and wanting also, for the most part, the Leaves, smaller Roots and Branches, the Flower and Fruit and the like smaller Parts, which are common to Trees of that kind; of which sort is the *Lignum fossile*, which is found in divers Parts of *England*, *Scotland*, *Ireland*, and divers Parts of *Italy*, *Germany*, the *Low Countries*, and indeed almost in every country of the world.

Thirdly, That there are often found in divers other Parts of the Earth, Bodies resembling the whole Bodies of Fishes, and other Animals and Vegetables, or the Parts of them, which are of a much less permanent Nature than the Shells abovemention'd, such as Fruits, Leaves, Barks, Woods Roots, Mushrooms, Bones, Hoofs, Claws, Horns, Teeth, *&c.* [. . .]

Fourthly, That the Parts of the Earth in which these kinds have been found, are some of them some hundreds of Miles distant from any Sea, as in several of the Hills of *Hungary*, the Mountain *Taurus*, the *Alpes*, *&c.*

Fifthly, That divers of those Parts are many Scores, nay, some many Hundreds of Fathoms above the Level of the Surface of the next adjoining Sea, there having been found of them on some of the most Inland, and on some of the highest Mountains in the World.

Sixthly, That divers other Parts where these Substances have been found, are many Fathoms below the Level both of the Surface of the next adjoining Sea, and of the Surface of the Earth itself, they having been found buried in the bottoms of some of the deepest Mines and Wells, and inclosed in some of the hardest Rocks and toughest Metals [. . .]

Seventhly, That there are often found in the midst of the Bodies of very hard and close stone, such as Marbles, Flints, *Portland*, and *Purbeck*-stone, *&c* which lye upon, or very near to the Surface of the Earth great quantities of these kind of figured Bodies or Shells, and that there are many of such Stones which seem to be made of nothing else.

These Phænomena, as they have hitherto much puzled all Natural Historians and Philosophers to give an Account of them, so in truth are they in themselves so really wonderful, that 'tis not easie without making multitudes of Observations, and comparing them very diligently with the Histories and Experiments that have been already made, to fix upon a plausible Solution of them. For as on the one side, it seems very difficult to imagine that Nature formed all these curious Bodies for no other End, than only to play the Mimick in the Mineral Kingdom, and only to imitate what she had done for some more noble End, and in a greater Perfection in the Vegetable and Animal Kingdoms; and the strictest Survey that I have made both of the Bodies themselves, and of the Circumstances obvious enough about them, do not in the least hint any thing else; they being promiscuously found of any kind of Substance, and having not the least appearance of any internal or substantial Form, but only of an external or figured Superficies. As I say, 'tis something harsh, to imagine that these thus qualified Bodies should, by an immediate plastick Vertue, be thus shaped by Nature contrary to her general Method of acting in all other Bodies; so on the other side, it may seem at first hearing somewhat difficult to conceive how all those Bodies, if they either by the real Shells or Bodies of Fish, or other Animals or Vegetables, which they represent, or an Impression left on those Substances from such Bodies, should be, in such great quantities, transported into Places so unlikely to have received them from any help of Man, or from any other obvious Means.

[. . .]

My first Proposition then is, That all, or the greatest part of these curiously figured Bodies found up and down in divers Parts of the World, are either those Animal or Vegetable Substances they represent converted into Stone, by having their Pores fill'd up with some petrifying liquid Substance, whereby their Parts are, as it were, lock'd up and cemented together in their Natural Position and Contexture; or else they are the lasting Impressions made on them at first, whilst a yielding Substance by the immediate Application of such Animal or Vegetable Body as was so shaped, and that there was nothing else concurring to their Production, save only the yielding of the Matter to receive the Impression, such as heated Wax affords to the Seal [. . .]

Secondly, Next that there seems to have been some extraordinary Cause, which did concur to the promoting of this Coagulation or Petrifaction [. . .]

Thirdly, That the concurrent Causes assisting towards the turning of these Substances into Stone, seem to have been one of these, either some kind of fiery

Exhalation arising from subterraneous Eruptions or Earthquakes; or secondly, a Saline Substance, whither working by Dissolution and Congelation, or Cystallization, or else by Precipitation and Coagulation; or thirdly, some glutinous or bituminous Matter, which upon growing dry or setling grows hard, and unites sandy Bodies together into a pretty hard Stone; or fourthly, a very long continuation of these Bodies under a great degree of Cold and Compression.

[. . .]

Sixthly, That a great part of the Surface of the Earth hath been since the Creation transformed and made of another Nature; namely, many Parts which have been Sea are now Land, and divers other Parts are now Sea which were once a firm Land; Mountains have been turned into Plains, and Plains into Mountains, and the like.

Seventhly, That divers of these kind of Transformations have been effected in these Islands of *Great Britain*; and that 'tis not improbable, but that many very Inland Parts of this Island, if not all, may have been heretofore all cover'd with the Sea, and have had Fishes swimming over it.

Eighthly, That most of those Inland Places, where these kinds of Stones are, or have been found, have been heretofore under the Water; and that either by the departing of the Waters to another part or side of the Earth by the alteration of the Center of Gravity of the whole Bulk, which is not impossible; or rather by the Eruption of some kind of subterraneous Fires or Earthquakes, whereby great quantities of Earth have then been rais'd above the former Level of those Parts, the Waters have been forc'd away from the Parts they formerly cover'd, and many of those Surfaces are now raised above the Level of the Water's Surface many scores of Fathoms.

[. . .]

Eleventhly, That there have been many other Species of Creatures in former Ages, of which we can find none at present; and that 'tis not unlikely also but that there may be divers new kinds now, which have not been from the beginning.

5

LINNAEUS (CARL VON LINNE), *LACHESIS LAPPONICA, OR A TOUR IN LAPLAND*, ED. JAMES EDWARD SMITH, TRANS. CHARLES TROILIUS

(London: Richard Taylor and Co, 1811). (First published as *Flora Lapponica*, Amsterdam, 1737)

Journey to Lapland

HAVING been appointed by the Royal Academy of Sciences to travel through Lapland, for the purpose of investigating the three kingdoms of Nature in that country, I prepared my wearing apparel and other necessaries for the journey as follows.

My clothes consisted of a light coat of Westgothland linsey-woolsey cloth without folds, lined with red shalloon, having small cuffs and collar of shag; leather breeches; a round wig; a green leather cap, and a pair of half boots. I carried a small leather bag, half an ell in length, but somewhat less in breadth, furnished on one side with hooks and eyes, so that it could be opened and shut at pleasure. This bag contained one shirt; two pair of false sleeves; two half shirts; an inkstand, pencase, microscope, and spying-glass; a gauze cap to protect me occasionally from the gnats, a comb; my journal, and a parcel of paper stitched together for drying plants, both in folio; my manuscript Ornithology, *Flora Uplandica*, and *Characteres generici*. I wore a hanger at my side, and carried a small fowling-piece, as well as an octangular stick, graduated for the purpose of measuring. My pocket-book contained a passport from the Governor of Upsal, and a recommendation from the Academy.

May 12, 1732, old style, Upsal

I set out alone from the city of Upsal on Friday May 12, 1732, at eleven o'clock, being at that time within half a day of twenty-five years of age.

At this season Nature wore her most cheerful and delightful aspect, and Flora celebrated her nuptials with Phœbus.

Omnia vere vigent et veris tempore florent,
Et totus fervet Veneris dulcedine mundus.

DOI: 10.4324/9780429355653-7

Spring clothes the fields and decks the flowery grove,
And all creation glows with life and love.

Now the winter corn was half a foot in height, and the barley had just shot out its blade. The birch, the elm, and the aspen-tree began to put forth their leaves.

Upsal is the ancient seat of government. Its palace was destroyed by fire in 1702. With respect to situation, and variety of prospects, scarcely any city can be compared with this. For the distance of a quarter of a Swedish mile it is surrounded with fertile corn-fields, which are bounded by hills, and the view is terminated by spacious forests.

May 15, Helsingland

I had scarcely travelled a quarter of a mile beyond the river when I observed a red earth close to the road, which promises to be very useful in painting, if it should prove sufficiently plentiful, and capable of being cleansed from its impurities. The people at the next post-house informed me that the same earth, but of a much better quality, was found in the parish of Norrbo.

The Common and Spruce Firs (*Pinus Sylvestris* and *P. Abies*) grow here to a very large size. The inhabitants had stripped almost every tree of its bark.

A number of small white bodies were hanging on the plants of Ling (*Erica*), of a globular form, but cut off, as it were, though not open, on the lower side, each about the size of a Bilberry (*Vaccinium Myrtillus*), and consisting of a thin white silky membrane. A small white insect was lodged within.

There were also affixed to some plants ovate white bodies of a silky texture, apparently formed of innumerable silky threads. These contained each a small insect.

A little further on I observed close to the road a rather lofty stone containing in its substance large fragments of mica.

At last to my great satisfaction I found myself at the great river Liusnan. From this part of the forest to the sea the distance is three miles. Here and there in the woods lay blood-red stones, or rather stones which appeared to have been partially stained with blood. On rubbing them I found the red colour merely external, and perfectly distinct from the stone itself. It was in fact a red Byssus (*B. Jolithus*).

Many sepulchral mounds are in this neighbourhood.

Not far from Norrala, situated about a mile from the last post-house, the water in the ditches deposits a thick sediment of ochre.

Several pair of semicircular baskets made of wicker work were placed in the water, intended principally to catch Bream (*Cyprinus Brama*). Here I observed the Lumme, or Black-throated Diver (*Colymbus Arcticus*), which uttered a melancholy note, especially in diving.

From Norrala I proceeded to Enänger, through a heavy fog, as it had rained violently while I rested at the former place. Towards evening it thundered and lightened. In the course of this whole day's journey I observed a great variety in the face of the country as well as in the soil. Here are mountains, hills, marshes, lakes, forests, clay, sand, and pebbles.

Cultivated fields indeed are rare. The greater part of the country consists of uninhabitable mountainous tracts. In the valleys only are to be seen small dwelling-houses, to each of which adjoins a little field. Even in these there is no great proportion of fertile land, the principal part being marshy.

The people seemed somewhat larger in stature than in other places, especially the men. I inquired whether the children are kept longer at the breast than is usual with us, and was answered in the affirmative. They are allowed that nourishment more than twice as long as in other places. I have a notion that Adam and Eve were giants, and that mankind from one generation to another, owing to poverty and other causes, have diminished in size. Hence perhaps the diminutive stature of the Laplanders [. . .]

Brandy is not always to be had here. The people are humane and civilized. Their houses are handsome externally, as well as neat and comfortable within; in which respects they have the advantage of most other places.

The old tradition, that the inhabitants of Helsingland never have the ague, is without foundation. In every parish where I made the inquiry I found many persons who had had that disorder, which appears to be not unfrequent among them.

Here were plenty of Mountain Finches (*Fringilla Montifringilla*); but, what is remarkable, they were all males, known by the orange-coloured spot on the breast.

[. . .]

June 2, Lycksele Lapland

The forest here was full of the noblest pine trees, growing to no purpose with respect to the inhabitants, as the wood is not used even for building huts, nor the bark for food, as it is in some other parts. I wonder they have not contrived to turn these trees to some account, by burning them for tar or pitch.

The colonists who reside among the Laplanders are beloved by them, and treated with great kindness. These good people willingly point out to the strangers where they may fix their abode so as to have access to moist meadows affording good hay, which they themselves do not want, their herds of reindeer preferring the driest pastures. They expect in return that the colonists should supply them with milk and flour.

Ovid's description of the silver age is still applicable to the native inhabitants of Lapland. Their soil is not wounded by the plough, nor is the iron din of arms to be heard; neither have mankind found their way to the bowels of the earth, nor do they engage in wars to define its boundaries. They perpetually change their abode, live in tents, and follow a pastoral life, just like the patriarchs of old.

Among these people the men are employed in the business of cookery, so that the master of a family has no occasion to speak a good word to his wife, when he wishes to give a hospitable entertainment to his guests.

The dress of these Laplanders is as follows.

On the head they wear a small cap, like those used at my native place of Stenbrohult, made with eight seams covered with strips of brown cloth, the cap itself being of a greyish colour. This reaches no lower than the tips of the ears.

Their outer garment, or jacket, is open in front half way down the bosom, below which part it is fastened with hooks, as far as the pit of the stomach. Consequently the neck is bare, and from the effects of the sun abroad and the smoke at home, approaches the complexion of a toad. The jacket when loose reaches below the knees; but it is usually tied up with a girdle, so as scarcely to reach so far, and is sloped off at the bottom. The collar is of four fingers' breadth, thick, and stitched with thread.

All the needle-work is performed by the women. They make their thread of the sinews in the legs of the reindeer, separating them, while fresh, with their teeth, into slender strings, which they twist together. A kind of cord is also made of the roots of spruce fir.

The country bordering on the sea coast hereabouts, in some places consists of grassy pastures, in others of pebbly or sandy tracts. Large stones are rare.

The river of Umoea now began to swell, the weather having been for some days very warm, so as to melt the ice and snow in the frozen regions above. The stream was now so deep and strong that it was not to be navigated without difficulty. In general the strongest flood does not set-in till Midsummer.

This river, as I was informed, has its source in the alps about a mile from the sea of Norway, and empties itself into the gulf of Bothnia at Umoea.

No colonists are to be met with north of this river.

After proceeding for a while up the stream, we went on shore to repose a little at a cottage. The wind blew very cold from the north.

[. . .]

A tree close to one of the tents was adorned with more than a dozen pair of horns of the male reindeer, or Brunren. When castrated, the same animal is called Ren oxe. The female is denominated Kiælfja.

[. . .]

I made some inquiries here concerning the diseases of the people.

They are subject to the ullem, or colic, of which I have already spoken, for which they use soot, snuff, salt, and other remedies. The pain sometimes seizes them so violently that they crawl on the ground while it lasts, not being able to stand or lie still. They are also afflicted with the asthma, the epilepsy, and a swelling of the uvula. The husband of a woman who had the last-mentioned disorder, cut away a part of the swelling, but it grew as large again in the course of a twelvemonth. The prolapsus uteri also sometimes occurs.

Many persons have the pleurisy, and others rheumatic complaints in the back, which descend down the hips and legs, leaving the part first attacked. These complaints happen in summer as well as in winter.

We continued our course up the river of Umoea. At length, quitting the main stream, we proceeded along a branch to the right, which bears the name of Juita, and left Lycksele church at about four miles distance, as near as I could guess, for the Laplanders know nothing about the matter.

The inhabitants of this country no longer use bows and arrows, but rifle-guns loaded with bullets, not with small shot.

They wear no stockings. Their breeches, made of the coarse and slight woollen cloth of the country called *walmal*, reach down to their feet, tapering gradually to the bottom, and are tied with a bandage over their half boots.

I observed the Red Whortle-berries (*Vaccinium Vitis Idœa*) were here of a larger size than in the country lower down; but Juniper on the contrary was very diminutive, and grew mostly in fens or watery places. The Crake berries (*Empetrum nigrum*) were as large as the Black Bilberry. Close to a waterfall in Juita Rotogviek or Rootforsen, in a marsh on the right hand, I found Herb Paris (*Paris quadrifolia*), *Aconitum lycoctonum* and *Thalictrum* (*flavum*). But what most surprised and pleased me was the little round-leaved Yellow Violet, with a branched stem, and narrow, smooth, not bearded, petals, described by Morison, which had not before been observed in Sweden (*Viola biflora*).

Several kinds of Willows grew every where near the water, but had not yet displayed their leaves.

I came to a hut, consisting of eighteen posts, covered with walmal, or coarse cloth, ten feet long and eight broad. Also some winter huts, the poles of which the Laplanders remove with them from place to place. Each hut is formed with three poles, forked at the top. Under the shelter of these huts or tents were suspended dried fish, cheese, clothes, pots and various utensils. There were neither walls nor doors, consequently no locks to protect them.

At length meeting with a very long shelvy contraction in the river, we were obliged to quit our boat, and go by land in search of a Laplander to serve as my guide further on, whom we expected to find at a place a mile distant. But it appeared to me full a mile and half, over hills and valleys, rivulets and stones. The hills were clad with Ling and with *Empetrum*, which entangled our feet at every step; not to mention the trees lying in all directions in our way, and over which we were obliged to climb. The marshy spots were not less difficult to pass over. The Bog-moss (*Sphagnum*) afforded but a treacherous support for our feet, and the Dwarf Birch (*Betula nana*) entangled our legs.

I could not help remarking that all the fibres of the full-grown pine trees seemed to be obliquely twisted, and in a contrary direction to the diurnal motion of the sun. I leave this to the consideration of the curious physiologist; whether it may arise from any thing in the soil or air, or from any polar attraction. [. . .].

At length we came to a sort of bay or creek of the river, which we were under the necessity of wading through. The water reached above our waists, and was very cold. In the midst of this creek was so deep a hole that the longest pole could scarcely fathom it. We had no resource but to lay a pole across it, on which we passed over at the hazard of our lives; and indeed when I reached the other side, I congratulated myself on having had a very narrow escape. A neighbouring mountain affords grey slate, but of a loose and brittle kind.

We had next to pass a marshy tract, almost entirely under water, for the course of a mile, nor is it easy to conceive the difficulties of the undertaking. At every step we were knee-deep in water; and if we thought to find a sure footing on some grassy tuft, it proved treacherous, and only sunk us lower. Sometimes we came where no bottom was to be felt, and were obliged to measure back our weary steps. Our half boots were filled with the coldest water, as the frost, in some places, still remained in the ground. Had our sufferings been inflicted as a capital punishment, they would, even in that case, have been cruel, what then had we to complain of? I wished I had never undertaken my journey, for all the elements seemed adverse. It rained and blowed hard upon us. I wondered that I escaped with life, though certainly not without excessive fatigue and loss of strength.

After having thus for a long time gone in pursuit of my new Lapland guide, we reposed ourselves about six o'clock in the morning, wrung the water out of our clothes, and dried our weary limbs, while the cold north wind parched us as much on one side as the fire scorched us on the other, and the gnats kept inflicting their stings. I had now my fill of travelling.

The whole landed property of the Laplander who owns this tract consists chiefly of marshes, here called stygx. A divine could never describe a place of future punishment more horrible than this country, nor could the Styx of the poets exceed it. I may therefore boast of having visited the Stygian territories.

We now directed our steps towards the desert of Lapmark, not knowing where we went.

June 30, Lulean Lapland

THE clergyman of Jockmock, Mr Malming, who is the schoolmaster, and Mr Högling the curate, tormented me with their consummate and most pertinacious ignorance. I could not but wonder how so much pride and ambition, such scandalous want of information, with such incorrigible stupidity, could exist in persons of their profession, who are commonly expected to be men of knowledge; yet any school-boy twelve years of age might be better informed. No man will deny the propriety of such people as these, at least, being placed as far as possible from civilized society.

The learned curate began his conversation with remarks on the clouds in this country, setting forth how they strike the mountains as they pass, carrying away stones, trees and cattle. I ventured to suggest that such accidents were rather to be attributed to the force of the wind, for that the clouds could not of themselves lift, or carry away, any thing. He laughed at me, saying surely I had never seen any clouds. For my part, it seemed to me that he could have never been any where but in the clouds. I replied, that whenever the weather is foggy I walk in clouds, and when the fog is condensed, and no longer supported in the air, it immediately rains beneath my feet. At all such reasoning, being above his comprehension, he only laughed with a sardonic smile. Still less was he satisfied with my explanation how watery bubbles may be lifted up into the air, as he told me the clouds were solid bodies. On my denying this, he reinforced his assertion with a text of scripture, silencing me by authority, and then laughing at my ignorance. He next condescended to inform me that after rain a phlegm is always to be found on the mountains, where the clouds have touched them. Upon my replying that this phlegm is a vegetable called *Nostoc*, I was, like St Paul, judged to be mad, and that too much learning had turned my brain. This philosopher, who was as fully persuaded of his own complete knowledge of nature, as Sturmius was of being able to fly by means of hollow globes, was pleased to be very facetious at my expense. At length he graciously advised me to pay some regard to the opinions of people skilled in these abstruse matters, and not, at my return home, to expose myself by publishing such absurd and preposterous opinions as I had now advanced.

The other, the pedagogue, lamented that people should bestow so much attention upon temporal vanities, and consequently, alas! neglect their spiritual good; and he remarked that many a man had been ruined by too great application to study.

Both these wise men concurred in one thing. They could not conceal their wonder that the Royal Academy should expressly have appointed a mere student for the purposes for which I was sent, without considering that there were already as competent men resident in the country, who would have undertaken the business. They declared they would either of them have been ready to accept of the charge. In my opinion, however, they would but have exhibited a fresh illustration of the proverb of the ass and the lyre.

The number of pupils under the care of the gentleman above mentioned at this time amounted to four only. The church is but a small one.

It is a practice here with some persons who have the headache, from excessive drinking or any other cause, to hold their foreheads before the fire till they smart violently. Others apply to the temples young shoots of spruce fir bruised.

Half a mile from the church I gathered the *Cirsium minus* (*Serratula alpina*), the *Cacalia* (*Tussilago frigida*), the latter not in flower, and one kind of Botsko of the Laplanders, called Biœrnstut in Westbothnia (*Angelica sylvestris*), which is the narrow-leaved species of *Angelica*, and resembles the larger kind. Its general umbel is destitute of an involucrum. My Lapland companion seized it immediately, and peeling the stalk, which had not yet flowered, ate it like a turnip, as a great delicacy. Indeed it tasted not unpleasantly, especially the upper part, which is the most tender. This dainty is in great request amongst the Laplanders.

We arrived at length at Purkijau, a small island, the northern side of which is planted with forests of spruce fir, and the others with woods of birch, by way of protection to the corn. The colonist who resides here informed me that the corn never suffered from cold, as, besides the shelter afforded by these plantations, the circumjacent water moderated the degree of frost. The situation of this island is pleasant. I found in some bushy parts of it the *Sceptrum Carolinum*, and another species of *Pedicularis*, with narrow leaves and a tuft of purple flowers (this seems to have been *P. sylvatica* only).

The river Karax, where is a pearl fishery, runs not far from hence. On its banks I remarked the *Sceptrum carolinum*, which became very common as I advanced further on my journey.

Another mile brought us to the lake of Randiau; on approaching which we saw nothing before us but lofty mountains of an oblong obtuse form, lifting their summits one above another, and on the most distant of these snow was to be seen, though half melted away like snow in the spring.

6

GEORGES LOUIS DE BUFFON, *BUFFON'S NATURAL HISTORY: A THEORY OF THE EARTH, A GENERAL HISTORY OF MAN, OF THE BRUTE CREATION, AND OF VEGETABLES, MINERALS, & C. & C., FROM THE FRENCH, WITH NOTES BY THE TRANSLATOR, IN TEN VOLUMES*, ED. AND TRANS. WILLIAM SMITH BARR, VOL VI

(London: S. D. Symonds, 1797 [1756])

Volume 1 (1749)

'The Theory of the Earth'

NEITHER the figure of the earth, its motion, nor its external connections with the rest of the universe, pertain to our present investigation. It is the internal structure of the globe, its composition, form, and manner of existence which we purpose to examine. The general history of the earth should doubtless precede that of its productions, as a necessary study for those who wish to be acquainted with Nature in her variety of shapes, and the detail of facts relative to the life and manners of animals, or to the culture and vegetation of plants, belong not, perhaps, so much to Natural History, as to the general deductions drawn from the observations that have been made upon the different materials which compose the terrestrial globe: as the heights, depths, and inequalities of its form; the motion of the sea, the direction of mountains, the situation of rocks and quarries, the rapidity and effects of currents in the ocean, &c. This is the history of nature in its most ample extent, and these are the operations by which every other effect is influenced and produced. The theory of these effects constitutes what may be termed a primary science,

DOI: 10.4324/9780429355653-8

upon which the exact knowledge of particular appearances as well as terrestrial substances entirely depends. This description of science may fairly be considered as appertaining to physics; but does not all physical knowledge, in which no system is admitted, form part of the History of Nature?

In a subject of great magnitude, whose relative connections are difficult to trace, and where some facts are but partially known, and others uncertain and obscure, it is more easy to form a visionary system, than to establish a rational theory; thus it is that the Theory of the Earth has only hitherto been treated in a vague and hypothetical manner

[. . .]

In the history of the Earth, we shall therefore begin with those facts that have been obtained from the experience of time, together with what we have collected by our own observations.

This immense globe exhibits upon its surface heights, depths, plains, seas, lakes, marshes, rivers, caverns, gulphs, and volcanos; and upon the first view of these objects we cannot discover in their dispositions either order or regularity. If we penetrate into its internal part, we shall there find metals, minerals, stones, bitumens, sands, earths, waters, and matters of every kind, placed as it were by chance, and without the smallest apparent design. Examining with a more strict attention, we discover sunk mountains, caverns filled, rocks split and broken, countries swallowed up, and new islands rising from the ocean; we shall also perceive heavy substances placed above light ones, hard bodies surrounded with soft; in short, we shall there find matter in every form, wet and dry, hot and cold, solid and brittle, mixed in such a sort of confusion as to leave room to compare them only to a mass of rubbish and the ruins of a wrecked world.

We inhabit these ruins however with a perfect security. The various generations of men, animals, and plants, succeed each other without interruption; the earth produces fully sufficient for their subsistence; the sea has its limits; its motions and the currents of air are regulated by fixed laws: the returns of the seasons are certain and regular; the severity of the winter being constantly succeeded by the beauties of the spring: every thing appears in order, and the earth, formerly a CHAOS, is now a tranquil and delightful abode, where all is animated, and regulated by such an amazing display of power and intelligence as fills us with admiration, and elevates our minds with the most sublime ideas of an all-potent and wonderful Creator.

Let us not then draw any hasty conclusions upon the irregularities of the surface of the earth, nor the apparent disorders in the interior parts, for we shall soon discover the utility, and even the necessity of them; and, by considering them with a little attention, we shall, perhaps, find an order of which we had no conception, and a general connection that we could neither perceive nor comprehend, by a slight examination: but in fact, our knowledge on this subject must always be confined. There are many parts of the surface of the globe with which we are entirely unacquainted, and have but partial ideas of the bottom of the sea, which in many places we have not been able to fathom. We can only penetrate into the coat of the earth; the greatest caverns and the deepest mines do not descend above the eight thousandth part of its diameter, we can therefore judge only of the external and mere superficial part; we know, indeed, that bulk for bulk the earth weighs four times heavier than the sun, and we also know the proportion its weight bears with other planets; but this is merely a relative estimation; we have no certain standard nor proportion; we are so entirely ignorant of the real weight of the materials, that the internal part of the globe may be a void space, or composed of matter a thousand times heavier than gold; nor is there any method to make further discoveries on this subject; and it is with the greatest difficulty any rational conjectures can be formed thereon.

We must therefore confine ourselves to a correct examination and description of the surface of the earth.

[. . .]

The changes and alterations which have happened to the earth, in the space of the last two or three thousand years, are very inconsiderable indeed, when compared with those important revolutions which must have taken place in those ages which immediately followed the creation; for as all terrestrial substances could only acquire solidity by the continued action of gravity, it would be easy to demonstrate that the surface of the earth was much softer at first than it is at present, and consequently the same causes which now produce but slight and almost imperceptible changes during many ages, would then effect great revolutions in a very short space. It appears to be a certain fact, that the earth which we now inhabit, and even the tops of the highest mountains, were formerly covered with the sea, for shells and other marine productions are frequently found in almost every part; it appears also that the water remained a considerable time on the surface of the earth, since in many places there have been discovered such prodigious banks of shells, that it is impossible so great a multitude of animals could exist at the same time: this fact seems likewise to prove, that although the materials which composed the surface of the earth were then in a state of softness, that rendered them easy to be disunited, moved and transported by the waters, yet that these removals were not made at once; they must indeed have been successive, gradual, and by degrees, because these kind of sea productions are frequently met with more than a thousand feet below the surface, and such a considerable thickness of earth and stone could not have accumulated but by the length of time. If we were to suppose that at the Deluge all the shell-fish were raised from the bottom of the sea, and transported over all the earth; besides the difficulty of establishing this supposition, it is evident, that as we find shells incorporated in marble and in the rocks of the highest mountains, we must likewise suppose that all these marbles and rocks were formed at the same time, and that too at the very instant of the Deluge; and besides, that previous to this great revolution there were neither mountains, marble, nor rocks, nor clays, nor matters of any kind similar to those we are at present acquainted with, as they almost all contain shells and other productions of the sea. Besides, at the time of the Deluge, the earth must have acquired a considerable degree of solidity, from the action of gravity for more than sixteen centuries, and consequently it does not appear possible that the waters, during the short time the Deluge lasted, should have overturned and dissolved its surface to the greatest depths we have since been enabled to penetrate.

But without dwelling longer on this point, which shall hereafter be more amply discussed, I shall confine myself to well-known observations and established facts. There is no doubt but that the waters of the sea at some period covered and remained for ages upon that part of the globe which is now known to be dry land; and consequently the whole continents of Asia, Europe, Africa, and America, were then the bottom of an ocean abounding with similar productions to those which the sea at present contains: it is equally certain that the different strata which compose the earth are parallel and horizontal, and it is evident their being in this situation is the operation of the waters which have collected and accumulated by degrees the different materials, and given them the same position as the water itself always assumes. We observe that the position of strata is almost universally horizontal: in plains it is exactly so, and it is only in the mountains that they are inclined to the horizon, from their having been originally formed by a sediment deposited upon an inclined base. Now I insist that these strata must have been formed by degrees, and not all at once, by any revolution whatever, because strata, composed of heavy materials, are very frequently found placed above light ones, which could not be, if, as some authors assert, the whole had been mixed with the waters at the time of the Deluge, and afterwards precipitated; in that case every thing must have had a very different appearance to that which now exists. The heaviest bodies would have descended first, and each particular stratum would have been arranged according to its weight and specific gravity, and we should not see solid rocks or metals placed above light sand any more than clay under coal.

Volume 6 (1756)

'The Hare'

THE species of animals which are most numerous are not the most useful. Nothing can be more noxious than the multitudes of rats, mice, locusts, caterpillars, and many other insects, of which it would seem that Nature rather admitted than ordained the extraordinary increase. But those of the hare and rabbit are advantageous to us both from the number and utility. Hares are abundantly spread over the face of the earth; and rabbits, though originally natives of particular climates, multiply so prodigiously in almost every place to which they are transported, that instead of being extirpated, no small art is required in order to diminish their too-often inconvenient number. When we reflect on the astonishing fecundity of each particular species, on the quick and prodigious multiplication of certain animals which come into existence, as it were, to desolate the fields and ravage the earth, we are astonished they do not oppress Nature with their numbers, and after having devoured her productions become themselves victims to the destruction they have made. We cannot view without terror those thick clouds, those winged phalanxes of famished insects which seem to menace the whole globe, and whether lighting on the fruitful plains of Egypt, or of India, in an instant destroy the labours and hopes of a whole people; and sparing neither grain, fruit, herbs, nor leaves, strip the earth of its verdure, and change the richest countries into barren desarts. We behold rats descending from the northern mountains, in innumerable multitudes, rushing like a deluge of living matter, overflow the plains, spread themselves over the southern provinces, and after having destroyed in their passage every thing that lives, or vegetates, finish their career with infecting the earth and air with their putrid carcasses. We behold in the southern regions myriads of ants issuing from the desarts, which, like an exhaustless torrent, arrive in thick and successive columns, take possession of every spot, drive away men and animals from their habitations, and never retire till they have caused a general devastation. And in those times when man himself was but half civilized, and subject to all the laws and even excesses of Nature, were there not similar inundations of the human species? Have there not been Normans, Huns, and Goths, whole nations, or rather tribes of animals bearing the human form without dwellings, and without distinction, who have suddenly rushed from their caves, and marched in tumultuous herds, and without any force but what consists in numbers, overthrown empires, destroyed nations, and having ransacked the earth, concluded by repeopling it with a race not less barbarous than themselves?

These æras, these great events, though so strongly marked in the History of Mankind, are yet only slight vicissitudes in the ordinary course of animated nature, which is in general always uniform and the same; its movements are regulated by two unchangeable wheels; the one, unbounded fecundity of every species; the other, the innumerable causes of destruction which are perpetually reducing the produce of that fecundity to a determinate measure, so as to preserve nearly the same number of individuals in each species. And as these multitudinous animals, which appear suddenly, disappear in the same manner, without augmenting their race, so does the human species always remain the same; the variations only are more slow, because the life of man being longer than that of small animals, the alternate changes of increase and diminution must necessarily require a greater portion of time. But time itself is only an instant in the succession of ages, and only strikes us the more forcibly, from having been accompanied with horror and destruction; for, taking all the inhabitants of the globe together, the number of the human race, like that of other animals, will, at all times, appear to be nearly the same; as this depends entirely upon an equilibrium of physical causes, – an equilibrium to which every thing has long been reduced, and which neither the efforts of man, nor any moral circumstances whatever, can dissolve; those circumstances themselves being dependant on physical causes. Whatever care man may bestow on his own species, he will never be able to render it more numerous in one place without destroying or diminishing it in another. As soon as any one country is overstocked with inhabitants they diffuse themselves over other countries, or destroy each other, and not unfrequently establish laws and customs calculated to prevent an excess of multiplication. In climates of exuberant fertility, as China,

Egypt, and Guinea, they banish, mutilate, drown, or sell their infants; in Catholic countries they condemn them to perpetual celibacy. Those who actually exist find no difficulty in arrogating to themselves the disposal of the rights of those who have no existence. Considering themselves as necessary, they annihilate contingent beings, and scruple not to suppress future generations for their own ease and convenience. Mankind, without perceiving it, treat their own species exactly in the same manner as they do other animals; they cherish and multiply, or neglect and destroy them, according as it suits their purpose; and as all moral effects depend upon physical causes, which ever since the earth assumed its form, are fixed and permanent, it follows that in the human, as well as in the other animal species, the number must likewise be uniform and unalterable. It is to be observed that this fixed state, this permanent number, are not to be considered in an absolute sense; all physical and moral causes, and all the effects which flow from them, are comprised and balanced within certain limits, more or less extended, but never so large as to destroy the equilibrium. As the whole universe is in a state of perpetual motion, and as all the forces of matter act against and counter-balance each other, so every thing is brought about in a kind of oscillation, to the middle points of which we refer the ordinary course of Nature, and whose extremes are the furthest removed from that course. In effect, therefore, we find that an excess of fecundity, either in animals or vegetables, is the usual fore-runner of sterility. Plenty and scarcity present themselves so alternately, and often follow so close upon each other, that a tolerable judgment may be formed of the product of one year by that of the preceding. The apple, plum, oak, beech, and indeed most fruit and forest trees, do not bear plentifully two years together. So likewise it is with caterpillars, May-bugs, flies, field mice, and many other animals, who if they multiply to excess one year, they will produce but a very small number the next. What, indeed, would become of all the fruits of the earth, of the most useful animals, or even of man himself, if these insects were to be proportionally increased after a fertile season? But no; the causes of destruction and sterility immediately follow those of an excessive multiplication. Independent of contagion, a necessary consequence of too great a mass of living matter assembled in one place, there are in every species, certain causes of death, as we shall hereafter have occasion to mention, and which are sufficient to counter-balance any preceding excess of fecundity. I must again observe that this is not to be taken in an absolute or strict sense, especially with respect to those species which do not remain entirely in a state of nature. Those which man takes care to rear are more abundant than they otherwise would be; but as his attention has its limits, so the increase which flows from it has long since been confined by unalterable bounds; and though in civilized countries, the human species and domestic animals, are more numerous than in other climates, they are never so to excess; because the very power which calls them into existence, destroys them when they become troublesome.

In those districts which are reserved for the chace, four or five hundred hares are sometimes killed in the course of one day's sport. These animals multiply amazingly; they engender at all seasons, and are in a condition to propagate before the first year of their life is expired. The females do not go with young above thirty or thirty-one days; they produce three or four, and are immediately after ready to receive the male; they likewise receive him during the time of gestation, and by a particular formation of their organs are often found to have a super-foetation; for the vagina and the matrix are continuous, and the latter has neither neck or orifice in the womb, as in other animals; yet each horn has an orifice which opens into the vagina and dilates during the time of bringing forth; and which forming two distinct uteri, act independently of each other; so that the females of this species are capable of conceiving and bringing forth by each matrix at different times; and consequently super-foetation must be as common among these animals, as it is rare among those which have not this double organ. It is plain, therefore, that the females may be impregnated at all times. By another singularity in their conformation they are found to be as lascivious as they are fruitful; the gland of the clitoris is prominent and almost as large as the sexual distinction of the male; and as the vulva is hardly visible, and the males when young have no exterior marks, it is often difficult to distinguish the sexes. It is these circumstances which have given rise to the opinions that there are many hermaphrodites among these animals, that the males sometimes bring forth, and that some are alternately males and females, and perform the office of either sex; because the females being more lascivious than the males will get upon them, and because they so much resemble each other externally, that unless very closely examined one sex may be mistaken for the other.

The young ones have their eyes open when brought forth; the mother suckles them about twenty days, after which they separate and provide for themselves; they do not wander far from each other, nor from the place of their birth; yet they live in solitude, each composing itself a form at the distance of sixty or eighty paces; thus when we find a leveret in any place, we are almost certain of finding one or two more in the neighbourhood. They feed more by night than day; and chiefly upon herbs, leaves, fruits, and grain, but above all they prefer those plants which yield a milky juice; they even eat the bark of trees in winter, except that of the alder and lime, neither of which they ever touch. When reared at home they are fed with lettuces and other herbs; but the flesh of these domestic fed hares has always a bad taste. They sleep and repose themselves in their forms during the day, and only live, as it were, in the night, when they range about, feed, and copulate; they may be seen by moonlight playing, leaping, and pursuing each other, but the smallest noise, even the rustling of a falling leaf is sufficient to alarm them; they fly, and in their flight take different ways.

[. . .]

Hares sleep much, but always with their eyes open. They have neither eye-lids, nor cilia, and seem to have bad eyes; but as if for a recompence of that defect, their hearing is exceedingly acute, and their ears are very large in proportion to the size of their bodies. They move these long ears with great facility, and use them as an helm to direct their course, which is so rapid that they easily outstrip all other animals. Their fore legs being much shorter than their hind ones they can more easily mount than descend, for which reason when pursued they always make towards the rising grounds. Their running is a kind of leaping gallop, and they proceed without making the smallest noise, as their feet, even underneath, are covered with hair, and perhaps they are the only animals which have hair growing inside of their mouths. The hare does not live above seven or eight years; he completes his growth in one, and the duration of its life is proportioned to this period, for he lives to about seven times that space. Some indeed assert that the males live longer than the females, but that I much doubt. They pass their lives in solitude and silence, and never exert their voices but when seized or wounded; their cry is sharp and strong, and not unlike the human voice. They are not so savage as by their habits and manners might be supposed; they are gentle, and susceptible of a species of improvement. They are easily tamed, but never acquire that degree of attachment which is requisite to render them domestic, for those which are taken very young, and brought up in a house, will take the first opportunity to escape and fly into the country. As they have a good ear, as they sit of their own accord upon their hind legs, and use the fore legs like arms, some have been so tutored as to beat a drum, to perform gestures in cadence, &c.

In general the hare possesses sufficient instinct for its preservation, and sagacity to escape its enemies. It prepares itself a form, or nest; in winter he chuses a spot exposed to the south, and in summer one to the north. To conceal himself from view he hides among hillocks of the same colour with his own hair. 'I have seen', says du Fouilloux, 'a hare so cunning, that upon hearing the huntsman's horn he started from his form, and though at the distance of a quarter of a league, hasted to a pond, and there hid himself among the rushes in the middle of it, and thus escaped the pursuit of the dogs. I have seen a hare, which after running more than two hours before the dogs, has dislodged another, and took possession of his form. I have seen others, swim over two or three ponds, of which the smallest was not less than eighty paces broad. I have seen others, after a chace of two hours, enter a sheep cot, and remain among the cattle. I have seen others, when closely pursued, take refuge among a flock of sheep, from which they would not be separated. I have seen others, upon hearing the noise of the hounds, conceal themselves in the earth. I have seen others, which have gone along one side of the hedge, and returned by the other, so that there was only the thickness of the hedge between them and the dogs; and I have seen others, after a chace of half an hour, mount an old wall six feet high, and take refuge in a hole covered with ivy'. But these facts are doubtless the greatest efforts of their instinct, for their common resources are less refined and intricate. They, in general, when pursued, content themselves with running rapidly, and afterwards tracing and retracing their own steps. They never direct their course against the wind, but always run with it. The females do not run so far out as the males, but they double more frequently. Hares, in general, if hunted upon their native spot, do not remove a great way from it, but return to their form, and if chaced for two successive days, they make exactly the same doublings on the second as they did on the first. If a hare runs straight forward, and to a great distance, it is a proof of his being a stranger to that spot, and that he was only there by accident. This generally happens during their most particular times of rutting, which are in January, February, and March, when the male hares finding but few females in their own districts, will roam for several leagues in search of them; but immediately upon being roused by the dogs, they make towards their native abodes, and never return again. The females do not thus go abroad; they are larger than the males, but have less strength and agility, and are more timid, for they never allow the dogs to come so near their forms as the males, and make use of more doublings and artifice. They are also more delicate, and more susceptible of the impressions of the air; they dread the water, and even avoid the dews; whereas among the males there is a kind which are fond of water, and are chaced in marshy and watery grounds, but the flesh of this sort has a very bad taste; and, in general, the flesh of all those which inhabit low valleys is whitish and insipid, while those in elevated countries, where the wild thyme, and other fine herbs abound, are delicious to the palate. It has also been remarked, that those which live in the centre of the woods, even in the same countries, are not so good as those that inhabit the borders, or live among the cultivated fields and vineyards; and that the flesh of the female is always more delicate than that of the male.

The nature of the soil has a great influence on hares, as well as on all other animals. The hares of the mountains are larger and fatter than those of the plains, and are also of a different colour, the former being browner, and having more white under the neck than the latter which are inclined to red. On high mountains, and in northern countries, they become white in winter, and recover their ordinary colour in the summer; there are but a few, and those perhaps very old ones, that continue always white, for all of them change more or less white as they advance in years.

7

EMMERICH DE VATTEL, *THE LAW OF NATIONS, OR PRINCIPLES OF THE LAW OF NATURE, APPLIED TO THE CONDUCT AND AFFAIRS OF NATIONS AND SOVEREIGNS*

(Philadelphia: P.H. Nicklin & T. Johnson, 1835)
(first published in French, 1758,
in English 1760)

Of the Establishment of a Nation in a Country

HITHERTO we have considered the nation merely with respect to itself, without any regard to the country it possesses. Let us now see it established in a country which becomes its own property and habitation. The earth belongs to mankind in general; destined by the Creator to be their common habitation, and to supply them with food, they all possess a natural right to inhabit it, and to derive from it whatever is necessary for their subsistence, and suitable to their wants. But when the human race became extremely multiplied, the earth was no longer capable of furnishing spontaneously, and without culture, sufficient support for its inhabitants; neither could it have received proper cultivation from wandering tribes of men continuing to possess it in common. It therefore became necessary that those tribes should fix themselves somewhere and appropriate to themselves portions of land, in order that they might, without being disturbed in their labour, or disappointed of the fruits of their industry, apply themselves to render those lands fertile, and thence derive their subsistence. Such must have been the origin of the rights of *property* and *dominion:* and it was a sufficient ground to justify their establishment. Since their introduction, the right which was common to all mankind is individually restricted to what each lawfully possesses. The country which a nation inhabits, whether that nation has emigrated thither in a body, or

DOI: 10.4324/9780429355653-9

the different families of which it consists were previously scattered over the country, and, there uniting, formed themselves into a political society, – that country, I say, is the settlement of the nation, and it has a peculiar and exclusive right to it.

This right comprehends two things: 1. The *domain,* by virtue of which the nation alone may use the country for the supply of its necessities, may dispose of it as it thinks proper, and derive from it every advantage it is capable of yielding. – 2. The *empire,* or the right of sovereign command, by which the nation directs and regulates at its pleasure every thing that passes in the country.

When a nation takes possession of a country to which no prior owner can lay claim, it is considered as acquiring the *empire* or sovereignty of it, at the same time with the *domain.* For, since the nation is free and independent, it can have no intention, in settling in a country, to leave to others the right of command, or any of those rights that constitute sovereignty. The whole space over which a nation extends its government, becomes the seat of its jurisdiction, and is called its *territory.*

If a number of free families, scattered over an independent country, come to unite for the purpose of forming a nation or state, they altogether acquire the sovereignty over the whole country they inhabit: for, they were previously in possession of the domain – a proportional share of it belonging to each individual family: and since they are willing to form together a political society, and establish a public authority, which every member of the society shall be bound to obey, it is evidently their intention to attribute to that public authority the right of command over the whole country.

All mankind have an equal right to things that have not yet fallen into the possession of any one; and those things belong to the person who first takes possession of them. When, therefore, a nation finds a country uninhabited, and without an owner, it may lawfully take possession of it; and after it has sufficiently, made known its will in this respect, it cannot be deprived of it by another nation. Thus, navigators going on voyages of discovery, furnished with a commission from their sovereign, and meeting with islands or other lands in a desert state, have taken possession of them in the name of their nation: and this title has been usually respected, provided it was soon after followed by a real possession.

But it is questioned whether a nation can, by the bare act of taking possession, appropriate to itself countries which it does not really occupy, and thus engross a much greater extent of territory than it is able to people or cultivate. It is not difficult to determine that such a pretension would be an absolute infringement of the natural rights of men, and repugnant to the views of nature, which, having destined the whole earth to supply the wants of mankind in general, gives no nation a right to appropriate to itself a country, except for the purpose of making use of it, and not of hindering others from deriving advantage from it. The law of nations, will, therefore, not acknowledge the property and sovereignty of a nation over any uninhabited countries, except those of which it has really taken actual possession, in which it has formed settlements, or of which it makes actual use.

[. . .]

There is another celebrated question, to which the discovery of the new world has principally given rise. It is asked whether a nation may lawfully take possession of some part of a vast country, in which there are none but erratic nations whose scanty population is incapable of occupying the whole? We have already observed (§ 81), in establishing the obligation to cultivate the earth, that those nations cannot exclusively appropriate to themselves more land than they have occasion for, or more than they are able to settle and cultivate. Their unsettled habitation in those immense regions cannot be accounted a true and legal possession; and the people of Europe, too closely pent up at home, finding land of which the savages stood in no particular need, and of which they made no actual and constant use, were lawfully entitled to take possession of it, and settle it with colonies. The earth, as we have already observed, belongs to mankind in general, and was designed to furnish them with subsistence: if each nation had, from the beginning, resolved to appropriate to itself a vast country, that the people might live only by hunting, fishing, and wild fruits, our globe would not be sufficient to maintain a tenth part of its present inhabitants. We do not, therefore, deviate from the views of nature in confining the Indians within narrower limits. However, we cannot help praising the moderation of the English puritans who first settled in New England; who, notwithstanding their being furnished with a charter from their sovereign, purchased of the Indians the land of which they intended to take possession. This laudable example was followed by William Penn, and the colony of quakers that he conducted to Pennsylvania.

8

JOHN BRUCKNER, *A PHILOSOPHICAL SURVEY OF THE ANIMAL CREATION, AN ESSAY*, TRANS. FROM THE FRENCH

(London: S. Highley, 1791)

Preface

Nothing contributes so much to the welfare of men, as the progress they make in the study of nature. Other researches may serve to amuse, but those which have for their object the actual state of things, their properties, tendency, order, and connection, besides the amusement they afford, lay a foundation for very useful discoveries relative to the necessities and conveniences of life. As all the powers of nature are subject to fixt, immutable laws, and necessarily produce their effects, they must of consequence contribute to the happiness of men, wherever men have skill enough to direct them to such an end, this they are taught to a certain degree by the study of nature [. . .] By discovering the connection that necessarily subsists betwixt the different classes of beings, and the general laws by which they all conspire to one principal end, it conveys those enlarged conceptions to the understanding which alone can enable us to form just ideas of the Deity [. . .]

Part the Second

Section I, Of the Perpetual Opposition that Subsists Between Animals of a Different Race

It is very evident that Providence not only permits, but has designed, that animals should devour each other. If any one doubts this, let him consult the common impulses of his nature. From whence proceed those involuntary horrors at the approach of some animals, and that mechanical desire, as it were, of falling upon them? Persons of delicacy, and polished manners, will assert, perhaps, that they are strangers to these emotions, and they do not find in other examples enough to constitute a general rule. But if such propensities are unknown to them, it is

DOI: 10.4324/9780429355653-10

because they have been subdued by those gentle manners, that mildness of disposition, which they have acquired in civil life, yet they still lie concealed in their nature, and there are seasons when they will break forth in spite of all these barriers [. . .] From whence arises that desire apparent in most men to feed upon flesh? Or if you suspect human nature to be depraved in this respect, examine the other species. See how some animals thirst after the blood of others, how nature has armed them with claws and teeth to put their bloody purpose in execution, while she has only endowed the victims of their fury with vigilance and activity, and to others she has left no other means of defence than cries and groans. Consider the voraciousness of the eagle, the surprizing strength of its bill, and its piercing eye, that darts swift as lightning upon the most distant objects [. . .] It is evident, I say, that animals are in a state of perpetual war, and that it is the will of their Creator that one should live upon another. And what is the consequence? That the works of the Omnipotent are defective? Or that the world, which was created perfect, has since fallen into general depravity? These by no means follow. Proofs of the depravity of nature may be sought after elsewhere. It is no less certain, that the law which enjoins the destruction of one animal for the advantage of another, contributes to the increase and happiness of life.

[. . .]

Part the Second

Section VI, Of the Consequences which Necessarily Result from the Law of Multiplication

What is the consequence of this astonishing increase, common to every species, when they are protected from injuries, and able to supply their own wants? It is that all, without exception, require some coercive force to suppress their progress, and to prevent their exceeding a due proportion relative to the other species. For it is apparent, that without this, the law of multiplication, so necessary for their preservation, would threaten their immediate destruction. For whenever one species is predominant, it must necessarily deprive the others of their nutriture. It would also infect the air with noxious exhalations; especially when such innumerable multitudes come to be mown down by death and from all these causes united, a universal languor and debility of the animal system, must unavoidably take place. The reader needs not to be reminded of the plagues with which God visited Egypt, particularly the frogs and insects, or of the mortality which followed. Or was he to recollect what we have just related concerning the grasshopper in Barbary, he would easily conceive the miseries of a country, where the trees themselves are devoured to the quick, where the earth, spoiled of its treasures, presents to the view nothing but the footsteps of mournful desolation. Of all the scourges which God afflicts his creatures, none perhaps are more terrible than this. There is in Augenmois in France, a small insect that attaches itself to the corn, both in granaries, and in the fields; and which has increased to such a degree, as to consume the greatest part of it, so that the common people were destitute of bread. May-bugs, some years ago, laid waste several districts in England. And a sea-worm increased beyond its due proportion, has, more than once, brought Holland to the very brink of destruction.

[. . .]

Part the Second

Section IX, Of the Equilibrium Established by Nature Betwixt the Different Species in the Animal System

Of so great utility is the Law of Nature, which enjoins, that one part of the living substance should feed upon the other, that it is the basis of the order and well-being of the whole animal system. We accordingly find that Providence has extended its empire over every species. They are all of them from the smallest to the greatest, exposed to enemies ready to check their progress, and furnished with the arms and faculties necessary for this effect. To those swarms of insects and reptiles that cover the surface of the earth, are opposed an army of birds, an active, vigilant, voracious race, which seem created for their destruction. Hares, rabbits, rats, mice, field-mice, and all those animals, which multiply with so much facility, are exposed both to the depradations of birds and quadrupeds, active as themselves in all their motions, endowed with superior strength, and with a more penetrating eye. The amazing bulk of the horned cattle, the lightness of the deer, the strength and swiftness of the horse, all these qualities do not exempt those that enjoy them from the order of Nature. She has formed numberless other animals which possess the same faculties to a greater degree, and which employ them in the destruction of the former. The carnivorous animals find in their turn, that the human race are furnished with numberless resources to check their increase, and prevent their multiplying beyond certain limits. Man is appointed to watch over all the other species, and maintain the balance betwixt them. In this consists his empire over the other animals. This is one reason why Providence has so closely united mankind together, and given them that understanding which she has refused to the rest of the creation.

Part 2

NATURAL THEOLOGY AND THE GREAT CHAIN OF BEING

Natural Theology and the Great Chain of Being

FOLLOWING on from the focus on John Ray and the emergence of Natural Theology in the seventeenth century, the extracts provided in this section survey some key works from this school (from 1791–1855), as well as examples of one of its key concepts, 'the great chain of being'. This is a story of tension, decline, and crisis in a major natural history tradition that went hand-in-hand with a gradual decline of the clergyman-naturalist tradition with which it is often associated. This section explores its final phase of significant influence: while anti-Darwinist Creationism exists today, it represents a beleaguered outlier in contemporary thought. Natural Theology, by contrast, was the mainstream of scientific opinion up until the mid-nineteenth century.

At its heart, Natural Theology rests on the propositions, outlined with particular directness by the Reverend William Buckland and included in an extract in this section, that 'no reasonable man can doubt that all the phenomena of the natural world derive their origin from God' and that 'no one who believes the Bible to be the word of God, has cause to fear any discrepancy between his word, and the results of any discoveries respecting the nature of his works'. If Natural Theology aims to provide further proof of God's wisdom and goodness via 'His' works, and to trace evidence of design, these tasks became increasingly problematic as nineteenth-century science progressed. Indeed, so mired did attempts to reconcile science and scripture become that Buckland's confidence proved entirely misplaced. Moreover, the arguments for design forwarded by Natural Theologians like William Paley, Buckland, and Thomas Malthus were often precisely those used by evolutionary theorists to provide a very different account of the earth's history and inhabitants. In terms of historical contextualisation, it is worth noting that UK Natural Theology was an overwhelmingly Anglican tradition, its chief proponents often Church of England clerics, and that their confidence in the project was by no means shared by Roman Catholics and nonconformists, who were reluctant to 'prove' God's existence through science and preferred to claim that faith required no rational explanation.

The first extract, from William Smellie's *The Philosophy of Natural History* (1791), shares much with the final extract (by Bruckner) in Part 1 in arguing that God ensures overall harmony within creation through conflictual relationships between herbivores and carnivores. Smellie's work is also often seen as an inadvertent precursor of evolutionary theory because of its account of what would come to be known as 'the struggle for existence', but he is also included in Part 6 of this volume to give a flavour of his zoological studies.

An extract from Chapter XV, 'Of the Progressive Scale or Chain of Beings in the Universe', is included for Smellie's statement of this key Natural Theological idea, a hierarchical and 'graduated scale or chain of existence, not a link of which, however seemingly insignificant, could be broken without affecting the whole'. In one sense a highly traditional argument for the pre-eminence of *Homo sapiens* within creation, there are also proto-ecological insights in Smellie's insistence that even seemingly inconsequential species play a role within nature: 'all of them possess degrees of perfection or of excellence proportionate to their station in the universe'. Although Smellie defends a hierarchical and anthropocentric view of natural history, he also insists on the importance of species that are 'inconvenient, and even noxious to man'. While 'man is unquestionably the chief or capital link' in the Great Chain, Smellie's alarming vision of the 'system of devastation' that would proceed from species losses anticipates the Darwinian-ecological concept of the 'economy of nature', as well as the rhetorical strategies by which environmentalist warnings are constructed. What is most obviously (and understandably) lacking in ecological terms, however, is a sense of the natural world as dynamic and changeable. Smellie's is a fixed and divine cosmos, but while his purposes are matched to those of Natural Theology, he – as so many other Natural Theologians do – introduces ideas that ultimately destabilise its project.

In his discussion of 'animalcules', Smellie reflects developments in microscopic technology during the previous century. First coined by the seventeenth-century Dutch pioneer of microscopy Antonie van Leeuwenhoek to refer to the seemingly miraculous creatures he observed in rainwater, 'animalcule' ('little animal') is a now-defunct term for microscopic organisms, including bacteria and protozoans.

Smellie's vision of a fixed natural order is matched by claims about the fixity and harmony of social difference and rank in human society, in which the class system is naturalised as innate and divinely appointed. His work involves what would become comparative anthropology or ethnology, although his attitudes to race (in his references to 'Hottentots' and Gentoos) do not coincide with modern thought: now regarded as a racist term, Hottentot was used in Smellie's day by Dutch and British South African colonists to refer to the Khoi, Khoikhoi, or Khoisan, non-Bantu indigenous nomadic pastoralists. Later used to indiscriminately refer to black Africans, it gathered colonial connotations of savagery. 'Gentoo', a historic and now defunct British term for indigenous inhabitants of India, has been superseded by terms that more specifically indicate particular groups (Hindus, Moslems, Tamils, etc). Again, the term was used in an increasingly pejorative and racist manner. While Smellie refers to Hottentots as 'brutal' and 'savage', he speaks approvingly of the caste system of the 'gentoos' as a social order that accords with his own views of animal hierarchy.

The second extract is from Chapter 18 of Thomas Malthus's *Essay on the Principle of Population* (1798). Malthus's famous theory, outlined in an extract included in Part 8 of this volume, argues that population will eventually outstrip the available subsistence and that this may lead to famine and population decline. In this extract, and in ways that resemble Bruckner's work, Malthus argues that his law of population and subsistence is ultimately positive. Just as Bruckner argues that God's creation of carnivorous species provides benefits by creating new layers of organic being, so Malthus argues that the pressures of the law of population have a salutary effect on humankind by fostering invention and creativity. The extract is Natural Theology because he sees demographic laws as divinely created and understanding them as providing a means to 'vindicate the ways of God to man' (Alexander Pope, *Essay on Man* [1733–4], 1.26). Quoting from *Isaiah* 55. 8–9 and *Job* 11.7, Malthus urges readers to redouble their 'feeble efforts' to understand God's purposes but also, in a distinctive Natural Theological manoeuvre, to turn to nature in order to understand its creator and thus become more spiritual, knowledgeable, and industrious.

The third extract, from William Paley's *Natural Theology* (1802), is probably the most quintessential and influential example of the school. As J.W. Barrow (1985, 23) points out, it was still taught as part of the tripos at Cambridge until the early 1900s. Paley was certainly required reading for naturalists in the first half of the nineteenth century. The extract is the celebrated part of Paley's work, a meticulously built analogy between a watch found on a heath and the process of understanding the intention behind divine Creation. In this 'argument from design', demonstrative of the continuing influence of mechanistic, Newtonian readings of environment, God (the ultimate watchmaker) is a profound and flawless artificer. Paley remains confident in the project of reconciling science and scripture and in the inherently religious status of natural philosophy, and his watch analogy relies on the accumulation of seemingly irrefutable evidence and the processes of rational inference that characterise the scientific method. Within sixty years, however, and not least because of the efforts of one of Paley's keenest readers, Charles Darwin, this confidence had been shattered, leaving science and religion in uneasy, antagonistic, or irreconcilable relations. In this, Paley's work represents a key feature of nineteenth-century Natural Theology: the manner in which its evidences proved crucial in overturning its claims.

The fourth extract, from Peter Mark Roget's 'Animal and Vegetable Physiology Considered with Reference to Natural Theology' (*The Bridgewater Treatises*, Treatise 5, 1834), is another example of confident pre-Victorian natural theology. Sponsored by the Earl of Bridgewater, the eight *Treatises* were designed to illustrate 'the Power, Wisdom, and Goodness of God, as manifested in the Creation' (see Barrow, 1985, 21). Three extracts are included in this volume.

Roget writes persuasively in a pleasing, rhetorical style in attempting to investigate 'the Book of Nature' in order to gain 'some insight, however limited, into the order and arrangements of creation'. Like Malthus, Roget argues that it is the duty of humankind, the highest species in creation, to strive for knowledge and betterment. There are clear parallels between Roget, Bruckner, and Malthus, especially evident in Roget's claims that

> as animals are ultimately dependent on the vegetable kingdom for the materials of their subsistence, and as the quantity of these materials is, in a state of nature, necessarily limited by the extent of surface over which vegetation is spread, a time must arrive when the number of animals thus continually increasing is exactly such as the amount of food produced by the earth will maintain.

Broadly Malthusian, Roget's remarks on 'The Functions of Life' reflect Bruckner's arguments about the providential existence of carnivora, while the argument is also rooted in a notion of harmonious balance between species that recapitulates the concept of the Great Chain of Being.

While confident in his Natural Theological project, Roget acknowledges the greater complexity of the study of organic nature in comparison to physics and astronomy, and his references to the 'boundless variety, inscrutable complexity [and] perpetual mutation' of organisms inadvertently anticipate the evolutionary and ecological direction of nineteenth-century science. Particularly interesting is the manner in which, in Chapter 2 of the Introduction, Roget narrows the gap between humans and other animals by emphasising their shared attributes and feelings.

The fifth extract, from Adam Sedgwick's *On the Studies of the University* (1834), offers a stout, often intemperate defence of the conformity of science and religion and follows the preceding extracts in believing that creation is underpinned by divine 'laws of nature' that create 'harmony and order' and a 'condition of equilibrium'. There are also clear markers of anxiety, however, particularly in Sedgwick's reference to having heard the promulgation of atheistical and materialist views within the sacred walls of Cambridge University. In response, Sedgwick, a major geologist of the period, outlines his argument for the primary role of a designer God in creating the laws of the universe as well as the creatures that dwell therein. The scientist, he says, may legitimately pursue the physical and natural sciences, but when it comes to origins – the 'vital power' with which God set the universe in motion and breathed organic life into existence – we enter the ineffable realms beyond science. In the decades prior to Darwin, the forerunners of the 'development hypothesis' were already evident (see Part 9 of this volume). In rejecting the theory of 'transmutation of species', which he describes as 'no better than a phrensied dream', Sedgwick refers to Jean-Baptiste Lamarck, and Sedgwick went on to oppose Darwinism after 1859.

The second part of the extract, from the Appendix added to the second edition, is a lengthy footnote to a remark in the lectures that 'the religion of Christ does not oppose, but lends support, to all those high faculties that give its only true elevation to the character of man'. Here, Sedgwick attacks the inadequacy of 'scriptural geology'. Its adherents, a 'class of men who pursue Geology by a nearer road', were primarily concerned with defending scripture rather than undertaking science and believed that geologists such as Sedgwick, Buckland, and Lyell were creating a dangerously impious, secular, and materialist science. Amongst the scriptural geologists to whom Sedgwick refers or alludes are George Bugg, Walter Forman, Frederick Nolan, Granville Penn, and George Young.

By contrast, Sedgwick speaks warmly of Buckland and Dr Thomas Chalmers, quoting his speech at a meeting of the British Association of the Advancement of Science (University of Cambridge, June 1833). Chalmers forwarded 'gap creationism', the theory that there were a series of creations prior to the one described in *Genesis*. In closing, Sedgwick also quotes from his own 'Anniversary Address to the Geological Society' (*Annals of Philosophy*, April 1830). In positioning himself alongside Chalmers and Buckland, Sedgwick's aim, like many of his professional geological contemporaries, was to steer a course between an atheistical, secular science and an unscientific account of the earth based purely on the Bible. Thus, he acknowledges (in passing) that the short duration of the earth derived from scripture is false, but he does not believe this removes God from the equation. Like Buckland and Charles Lyell (see Part 3), Sedgwick accepts a much longer account of the earth's history than a literal reading of the Mosaic narrative would permit but argues that God oversaw a series of creations in the ages prior to the creation of man, which event took place broadly in line, in temporal terms, with the account in *Genesis*.

Sedgwick's willingness to combat scriptural geology and promote an increasingly professional geological science tells us much about its status in the 1830s: during this decade, with Sedgwick as chair, and Buckland, Murchison, and Lyell all influential members, the Geological Society had moved from its earlier status as a dining club for bookish dilettantes to a powerful group of practical scientists. Sedgwick's splenetic attack on those followers of his own religion who insisted on a literal reading of scripture is a remarkable indicator, however, of the degree to which his attempt to steer a middle course was in difficulties: the evidence of science had begun, in the minds of practitioners like Sedgwick, to outweigh scripture: the latter now had to conform to the former and not, as the scriptural geologists would have it, the other way round. The task, Sedgwick implied, was to explain what scripture 'really meant' when its account of Creation did not match the evidence of nature.

The scriptural geologists were less convinced than Sedgwick that 'Geology gives its aid to natural religion'. The sixth extract in this section, from the Reverend Henry Cole's *Popular Geology Subversive of Divine Revelation!* (1834) attacks Sedgwick and the 'new science' of geology as impious meddlers who questioned the primacy of 'divine revelation' (the Bible) as the source of truth about the earth's past and organic life. Cole's response to Sedgwick indicates the developing polarisation between science and religion. Cole represents the more conservative tendency amongst churchmen to treat the Bible as a literal document and to altogether reject geology, in particular the claim, soon to be voiced by Roderick Murchison in *The Silurian System* (see Part 3 of this volume), that 'as the globe passed from one condition to another, whole races of animals perished, and were succeeded by others with organizations adapted to the altered state of our planet'. As scientific geology gradually cohered in the 1830s and 1840s around a consensus indicating a much longer time-span for the earth's history than scripture would permit, figures like Cole felt no choice but to declare their implacable opposition to the new science. Cole is particularly enraged that Sedgwick's address took place in the Chapel of Trinity College and before Cambridge's Regius Professor of Divinity, Reverend Thomas Turton (1780–1864). Placed together, the extracts from Sedgwick and Cole vividly enact the tense debates within the powerful Anglican establishment of the time.

The next extract, from Chapter 2 of Buckland's contribution to the *Bridgewater Treatises*, is invested in the debates so far described. Buckland argues that rather than being 'hostile to revealed religion', geology in fact 'abound[s] with proofs of some of the highest attributes of the Deity'. In a pattern many would follow, he attempts to square the evidence of scripture with that of the rocks – to 'read' nature as a book of God and to find evidence of 'His' design. Ultimately, however, the conservative Anglican objection to meddling in geological science was vindicated by the direction in which it was being taken by Charles Lyell, whose incendiary *Principles of Geology* was in its early editions at the time Buckland was writing. Lyell's uniformitarianism opposed the flawed catastrophist theories which reconciled scripture and science by positing a series of huge cataclysmic events (such as the Mosaic Deluge) that supposedly created the great upheavals evident in the earth's geological record. Like Sedgwick, Buckland has geology's religious detractors more in mind in his writing than uniformitarians like Lyell. His attempts to use the avowedly Natural Theological *Bridgewater Treatises* to persuade fellow clerics of a non-literal reading of the Bible leads him to simultaneously argue that 'the lapse of very long periods of time [has] been an essential condition' of the physical and chemical condition of the earth's strata and surface *and* that there is 'no inconsistency between our interpretation of the phenomena of nature and of the Mosaic narrative'. Deeming catastrophism's attempts 'to ascribe the formation of all the stratified rocks to the effects of the Mosaic Deluge [. . .] irreconcilable with the enormous thickness and almost infinite subdivisions of these strata', Buckland is closer here to Lyell. Like Sedgwick, Buckland endorses the 'gap creationism' of Chalmers, Benjamin Silliman, and others, breaking the Gordian knot of science and scripture by positing that the 'days' of *Genesis* were indeterminate but very long periods of geological time prior to the Biblical creation and arguing that the opening line of *Genesis* – 'In the beginning' – represents a vast period of time before the direct subject of its anthropocentric history, which is therefore 'passed over in silence by the sacred historian'. In support, he quotes from his own inaugural Oxford lecture as Reader in Geology (15 May, 1819), published as *Vindiciae Geologicae; or the Connexion of Geology with Religion Explained* (1820). The extract illustrates the earnest circumlocutions with which Natural Theologians were increasingly forced to involve themselves, but the many problems of such attempted reconciliations were pointed out with increasing frequency in the decades that followed, notably in C.W. Goodwin's chapter on the 'Mosaic Cosmogony' in the incendiary 1860 volume of Biblical criticism, *Essays and Reviews* (see Volume III).

Tantalisingly, Buckland also anticipates (but does not participate in) evolutionary theories in his observations that catastrophism is incompatible with evidence that 'the numerous and regular successions which [the strata] contain of the remains of animals and vegetables, differing more and more widely from existing species, as the strata in which we find them are placed at greater depths' and in the focus on the existence of innumerable animal and vegetable fossils of extinct species that 'could therefore have formed no part of the creation with which we are immediately concerned'. This subject is taken up again in Parts 3 and 9 of this volume.

The next extract, from the correspondence of a youthful member of the early Geological Society and Oxford student who assisted at Buckland's Oxford lectures, is a yet more compelling example of the difficulties of the project of reconciling science and scripture. John Ruskin's *Letters Addressed to a College Friend* were written to the Reverend Edward Clayton, who, as Ruskin's editors observe, was 'one of Ruskin's friends at Christ Church'. Ruskin, they note, 'adopts the familiar style common among College friends, and condescends to occasional use of slang'. The letters were not published until Ruskin approved an edition in 1894.

Although Ruskin would make his name as a critic of art, architecture, and politics, his letter and article from 1843 demonstrate the depth of his engagement with contemporary science and scientific methods. Ruskin went on to write several works of natural history (see Part 5 of this volume, and subsequent volumes). The Clayton letters are particularly interesting because their forensic argument for death in Eden, a claim that contradicts religious orthodoxy at the time, rests on a thorough grounding in a range of sciences. Ruskin is a curious product, created from three key influences that existed in a rarely stable synthesis: he was at once deeply devout in his adherence to his parents' evangelicalism; strongly influenced by Romantic poetry and its intense engagements with nature; *and* versed in modern, materialist science. At the Geological Society, he rubbed shoulders with leading figures of the day, and in 1843 he was reading (in the original French), Louis Agassiz's *Poissons Fossiles* (1833–43). Both of these scientific influences are evident in the extract, but Ruskin was also clearly familiar (probably via Bruckner or Roget) with arguments about the providential benefits of carnivora, as well as modern works of anatomy, chemistry, and agricultural science. All contribute to a remarkable and unremitting assault on literal readings of *Genesis*. In discussing Eden, Ruskin departs from literalist evangelical exegesis, deploying materialist methodologies to argue that there was death there prior to the Fall. His early adherence to such methods is in stark contrast to his later anti-Darwinian assault on such positions (see Volume IV of this anthology). The letters to Clayton show that while Ruskin publicly supported Natural Theology in his early published works, here he anticipates the anti-literalist 'higher criticism' of the 1860s, a period in which Ruskin lent support to one of its principal figures, Bishop Colenso (see Volume III).

Ruskin's critique of Clayton's apparently orthodox position centres upon an investigation of Eden's flora and fauna which, according to literal readings, were immortal and at peace prior to the Fall. There is something remorselessly materialist in his descriptions of organic life – of leaves as 'an instrument for depriving carbonic acid of its oxygen'; in his claim that 'the very meaning of the word flower is – something to supply *death*' (because it exists to create seeds to produce plants to replace the parent); and especially in his conclusion that if there were trees and flowers in Eden, death existed there – in contradiction to the Bible message. Likewise, Ruskin uses Baron Cuvier's anatomical principles (see Part 4 of this volume) to prove that the supposedly peaceful lions of Eden must have been carnivorous because their anatomy was set up to catch and digest meat.

In geological terms, Ruskin's letter and essay go far beyond gap creationism, demonstrating support for Lyell's uniformitarianism. Referring to arguments in Volume 1, Chapter 14, of *Principles of Geology* (1830), Ruskin challenged Clayton with 'geological evidence of death extending for an infinite series of ages before man' and spoke of Lyell's discovery of 'the bones of the mastodon, the most recent of all fossils, in a bed cut through by the ancient course of the Niagara, three hundred feet above its present bed, and three miles and a half below the falls', a finding which meant that 'the river by the very lowest calculation must have been occupied 15,000 years'. Ruskin's uniformitarian acceptance of this dating was irreconcilable to literalist exegesis, and yet his conclusions – like those of Buckland, Sedgwick, and others – seek to maintain faith by reinterpreting the Bible, as the closing sections of the extract show. Ruskin's letter is a fascinatingly atypical example of Natural Theology and a clear indicator of the broadness and internal tensions of this school of thought.

The penultimate extract is from Edward Hitchcock's *The Religion of Geology and Its Connected Sciences* (1851) (for further extracts, see Volume II of this anthology). It is a significant late contribution to Natural Theology from a scholar rooted in its heyday and coming towards the end of an illustrious career in American education at a time when science was becoming increasingly materialist and when Darwin's *Origin of Species* was less than a decade away. It can be read as a swansong to the golden age of Natural Theology.

The Dedication to Hitchcock's work is included to provide a window into an often-overlooked aspect of nineteenth century science – the careers and contributions of women. Hitchcock dedicates his work to his wife, Orra White Hitchcock, who made substantial contributions to science in her own right within the field of botany and illustration, and he speaks warmly and movingly of their relationship.

Parts of Hitchcock's Preface have been included for their insights into the contemporary contexts within education and nonconformism, as well as the debates already described between secular and non-secular science. Hitchcock speaks of the fact that the UK seminaries of the Congregationalists and Free Church of Scotland included chairs in Natural Science but bemoans the absence of professors of Natural Theology and of adequate scientific education within religious institutions. His stated aim is to harmoniously combine the insights of hermeneutics, revealed religion, geology, and other sciences. Like Sedgwick, Hitchcock is critical of scriptural geology, and like Sedgwick, he criticises the 'development hypothesis' in its Lamarckian manifestation as well as that of *Vestiges of the Natural History of Creation* (1844) (see Part 9 of this volume).

The main part of the extract is from Hitchcock's first lecture, where he runs through familiar arguments by which apparent conflicts between science and scripture are explained. These include the idea that the Bible's seeming inaccuracies can be explained by the argument that as

> the Scriptures were generally addressed to men in the earliest and most simple states of society, with very limited views of the extent of creation, we ought to suppose that, in all cases where no new fact is revealed, the language was adapted to the narrow ideas which then prevailed.

In all, he offers a rosy picture of the possibilities of religiously inspired science that were by no means borne out in the decades to come.

In some senses, the final extract, from Thomas Ewbank's *The World a Workshop* (1855), shares the aim of Sedgwick and Hitchcock of writing to appeal to Christians to see the value of science and to argue for its accord with scripture, but in other ways, Ewbank's approach is markedly and fascinatingly different. Also written just a few years before *Origin of Species*, it upholds belief in God's design and the central place of humankind within a divinely wrought creation but also depicts creation as deeply mechanistic. While its figure of the world as a workshop of organic and inorganic manufactories to be exploited by a skilful engineer is somewhat in the image of the original artificer from Paley's watchmaker, it is also stark and unappealing, a clear exemplar of the manner in which Natural Theology participates in the age-old vision of unquestioned human sovereignty over environment. Ewbank's work is unusual within Natural Theology not merely for his speculations about life on other planets but also for the firmness of its commitment to materialist science: it is in this a curious late manifestation of the school.

At times, Ewbank also reaches towards somewhat environmental positions – for example, in his muted concern for resource management and depletion (something perhaps more evident in further extracts from this work in Parts 5 and 6 of this volume) but also in his slightly ecological reading of the complex intersections of the Great Chain of Being. The volatile coming-together of all of these disparate forces makes it tempting to read Ewbank's work as a culminating moment in the collision between scripture and science, in which the pressure of the latter has not yet quite overwhelmed the perspectives of the former. Natural Theology is intact here, but its days were numbered and the future foreshadowed.

Although this selection offers key moments, interventions, and complications in Natural Theology, its ideas emerge again in subsequent sections. Many of the leading scientists of the period – up to and beyond *Origin of Species* – continued to believe in the possibility of reconciling science and scripture. Many figures (including Richard Owen and Louis Agassiz; Part 4) began their investigations when Natural Theology was a near-universal orthodoxy, but as their long careers continued, they encountered increasing support for various formulations of 'the development hypothesis' as well as the accumulation of evidence – from geology, botany, zoology, and anatomy – which raised troubling questions about relations between species and the development of species over time. That many leading Natural Theologians (especially geologists and anatomists) provided much of the evidence, and pioneered new branches of investigation, that only strengthened evolutionary arguments is an especial irony. As this section has hopefully demonstrated, the assumption that science and religion were in harmony was increasingly challenged before 1858. It would not be widely held thereafter.

Further reading

Armstrong, Patrick, *The English Parson-Naturalist: A Companionship Between Science and Religion* (Leominster: Gracewing, 2000).

Barber, Lynn, *The Heyday of Natural History 1820–1870* (London: Cape, 1980).

Barrow, J.W., 'Editor's Introduction', Charles Darwin, *The Origin of Species* (Harmondsworth: Penguin, 1985), 11–48.

Brooke, John H., *Science and Religion: Some Historical Perspectives* (Cambridge: Cambridge University Press, 1991).

Cosslett, Tess (ed.), *Science and Religion in the Nineteenth Century* (Cambridge: Cambridge University Press, 1984).

Dixon, Thomas, *Science and Religion: A Very Short Introduction* (Oxford: Oxford University Press, 2008).

Emblen, Donald Lewis, *Peter Mark Roget: The Word and the Man* (London: Longman, 1970).

Frost, Mark, 'A Vital Truth: Ruskin, Science and Dynamic Materiality', *Journal of Victorian Literature and Culture*, 39:2 (2011), 367–83.

———. 'Of Trees and Men: The Law of Help and the Formation of Societies in *Modern Painters V*', *Nineteenth Century Prose*, 38:2 (2011), 85–108.

Jenkins, Bill (2019), 'Natural History in Edinburgh, 1779–1832', *Evolution Before Darwin: Theories of the Transmutation of Species in Edinburgh, 1804–1834* (Edinburgh: Edinburgh University Press), 37–74.

Lawrence, Philip J., 'Edward Hitchcock: The Christian Geologist', *Proceedings of the American Philosophical Society* 116:1 (1972), 21–34.

Mortenson, T.J., 'British Scriptural Geologists in the First Half of the Nineteenth Century', Unpublished PhD Thesis (Coventry: Coventry University in collaboration with Wycliffe Hall, Oxford, 1996).

Rupke, Nicolaas, *The Great Chain of History: William Buckland and the English School of Geology 1814–1850* (Oxford: Oxford University Press, 1983).

Worster, Donald, *Nature's Economy: A History of Ecological Ideas* (Cambridge: Cambridge University Press, 1994).

excellence proportionate to their station in the universe. Even among mankind, which is a particular species, the scale of intellect is very extensive. What a difference between an enlightened philosopher and a brutal Hottentot! Still, however, nature observes for the wisest purposes, her uniform plan of gradation. In the human species the degrees of intelligence are extremely varied. Were all men philosophers, the business of life could not be executed, and neither society nor even the species could long exist. Industry, various degrees of knowledge different dispositions, and different talents, are great bonds of society. The Gentoos, from certain political and religious institutions, have formed their people into different castes or ranks, out of which their posterity can never emerge. To us such institutions seem to be tyrannical and restraints on the natural liberty of man. In some respects they are so; but they seem to have been originally results of wisdom and observation; for, independently of all political institutions, nature herself has formed the human species into castes or ranks. To some she gives superior genius and mental abilities; and even of these, the views, the pursuits, and the tastes are most wonderfully diversified.

[. . .]

Let man therefore be contented with the powers and sphere of action assigned him. There is an exact adaptation of his powers, capacities, and desires, both bodily and intellectual, to the scene in which he is destined to move. His station in the scale of nature is fixed by wisdom. Let him study the works of nature, and find in the contemplation of all that is beautiful, curious, and wonderful in them, proofs of the existence and attributes of his Creator. Let him see in his own structure, and that of all other animals, and in the whole economy of the universe, animate and inanimate, the evidences of the wisdom, the skill, the benevolence, and the justice of that great and overruling Intelligence, who has made all things, and who upholds all things. Let him find in the contemplation of the final destiny which is promised him, a source of consolation for the imperfections, pains, and trials of the present state of being. Let him fill up his rank here with dignity, and consider every partial evil as a cause, or an effect, of general ultimate good; and let him adore and worship that great and good Being, who has, even in this state of discipline and probation, dispensed so many blessings to alleviate its necessary and unavoidable evils.

9

WILLIAM SMELLIE, *THE PHILOSOPHY OF NATURAL HISTORY*

(Edinburgh, 1791)

Chapter XV, Of the Progressive Scale or Chain of Bei in the Universe

To men of observation and reflection it is apparent that all the beings or whether animals or vegetables, have a mutual connection and a mut dence on each other. There is a graduated scale or chain of existence, of which, however seemingly insignificant, could be broken without a whole. Superficial men, or, which is the same thing, men who avoid of serious thinking, wonder at the design of producing certain insects a But they do not consider that the annihilation of any one of these spec some of them are inconvenient, and even noxious to man, would mak nature, and prove destructive to other species, which feed upon them their turn, would be the cause of destroying other species, and system tion would gradually proceed, till man himself would be extirpated, an earth destitute of all animation.

In the chain of animals, man is unquestionably the chief or capita highly-rational animal, improved with science and arts, he is, in son related to beings of a superior order, wherever they exist. By conten works of nature, he even rises to some faint ideas of her great Author. been asked, are not men endowed with the capacity and powers of ang of whom we have not even a conception. With the same propriety it m 'Why have not beasts the mental powers of men?' Questions of this results of ignorance, which is always petulant and presumptuous. Ev is perfect, according to its destination. Raise or depress any order of whole system, of course, will be deranged, and a new world would be support them. Particular orders of beings should not be considered se by the rank they hold in the general system. From man to the minutes which can be discovered by the microscope, the chasm seems to be that chasm is actually filled up with sentient beings, of which the lines nation are almost imperceptible. All of them possess degrees of per

DOI: 10.4324/9780429355653-12

10

THOMAS MALTHUS, *AN ESSAY ON THE PRINCIPLE OF POPULATION, AS IT AFFECTS THE FUTURE IMPROVEMENT OF SOCIETY WITH REMARKS ON THE SPECULATIONS OF MR GODWIN, M. CONDORCET, AND OTHER WRITERS*

(London: J. Johnson, 1798)

Chapter 18

THE view of human life which results from the contemplation of the constant pressure of distress on man from the difficulty of subsistence, by shewing the little expectation that he can reasonably entertain of perfectibility on earth, seems strongly to point his hopes to the future. And the temptations to which he must necessarily be exposed, from the operation of those laws of nature which we have been examining, would seem to represent the world in the light in which it has been frequently considered, as a state of trial and school of virtue preparatory to a superior state of happiness. But I hope I shall be pardoned if I attempt to give a view in some degree different of the situation of man on earth, which appears to me to be more consistent with the various phenomena of nature which we observe around us and more consonant to our ideas of the power, goodness, and foreknowledge of the Deity.

It cannot be considered as an unimproving exercise of the human mind to endeavour to 'vindicate the ways of God to man' if we proceed with a proper distrust of our own understandings and a just sense of our insufficiency to comprehend the reason of all we see, if we hail every ray of light with gratitude, and, when no light appears, think that the darkness is from within and not from without, and bow with humble deference to the supreme wisdom of him whose 'thoughts are above our thoughts' 'as the heavens are high above the earth'.

DOI: 10.4324/9780429355653-13

In all our feeble attempts, however, to 'find out the Almighty to perfection', it seems absolutely necessary that we should reason from nature up to nature's God and not presume to reason from God to nature. The moment we allow ourselves to ask why some things are not otherwise, instead of endeavouring to account for them as they are, we shall never know where to stop, we shall be led into the grossest and most childish absurdities, all progress in the knowledge of the ways of Providence must necessarily be at an end, and the study will even cease to be an improving exercise of the human mind. Infinite power is so vast and incomprehensible an idea that the mind of man must necessarily be bewildered in the contemplation of it. With the crude and puerile conceptions which we sometimes form of this attribute of the Deity, we might imagine that God could call into being myriads and myriads of existences, all free from pain and imperfection, all eminent in goodness and wisdom, all capable of the highest enjoyments, and unnumbered as the points throughout infinite space. But when from these vain and extravagant dreams of fancy, we turn our eyes to the book of nature, where alone we can read God as he is, we see a constant succession of sentient beings, rising apparently from so many specks of matter, going through a long and sometimes painful process in this world, but many of them attaining, ere the termination of it, such high qualities and powers as seem to indicate their fitness for some superior state. Ought we not then to correct our crude and puerile ideas of infinite Power from the contemplation of what we actually see existing? Can we judge of the Creator but from his creation? And, unless we wish to exalt the power of God at the expense of his goodness, ought we not to conclude that even to the great Creator, almighty as he is, a certain process may be necessary, a certain time (or at least what appears to us as time) may be requisite, in order to form beings with those exalted qualities of mind which will fit them for his high purposes?

A state of trial seems to imply a previously formed existence that does not agree with the appearance of man in infancy and indicates something like suspicion and want of foreknowledge, inconsistent with those ideas which we wish to cherish of the Supreme Being. I should be inclined, therefore, as I have hinted before, to consider the world and this life as the mighty process of God, not for the trial, but for the creation and formation of mind, a process necessary to awaken inert, chaotic matter into spirit, to sublimate the dust of the earth into soul, to elicit an ethereal spark from the clod of clay. And in this view of the subject, the various impressions and excitements which man receives through life may be considered as the forming hand of his Creator, acting by general laws, and awakening his sluggish existence, by the animating touches of the Divinity, into a capacity of superior enjoyment. The original sin of man is the torpor and corruption of the chaotic matter in which he may be said to be born.

It could answer no good purpose to enter into the question whether mind be a distinct substance from matter, or only a finer form of it. The question is, perhaps, after all, a question merely of words. Mind is as essentially mind, whether formed from matter or any other substance. We know from experience that soul and body are most intimately united, and every appearance seems to indicate that they grow from infancy together. It would be a supposition attended with very little probability to believe that a complete and full formed spirit existed in every infant, but that it was clogged and impeded in its operations during the first twenty years of life by the weakness, or hebetude, of the organs in which it was enclosed. As we shall all be disposed to agree that God is the creator of mind as well as of body, and as they both seem to be forming and unfolding themselves at the same time, it cannot appear inconsistent either with reason or revelation, if it appear to be consistent with phenomena of nature, to suppose that God is constantly occupied in forming mind out of matter and that the various impressions that man receives through life is the process for that purpose. The employment is surely worthy of the highest attributes of the Deity.

[. . .]

The first great awakeners of the mind seem to be the wants of the body. [. . .] They are the first stimulants that rouse the brain of infant man into sentient activity, and such seems to be the sluggishness of original matter that unless by a peculiar course of excitements other wants, equally powerful, are generated, these stimulants seem, even afterwards, to be necessary to continue that activity which they first awakened. The savage would slumber for ever under his tree unless he were roused from his torpor by the cravings of hunger or the pinchings of cold, and the exertions that he makes to avoid these evils, by procuring food, and building himself a covering, are the exercises which form and keep in motion his faculties, which otherwise would sink into listless inactivity. From all that experience has taught us concerning the structure of the human mind, if those stimulants to exertion which arise from the wants of the body were removed from the mass of mankind, we have much more reason to think that they would be sunk to the level of brutes, from a deficiency of excitements, than that they would be raised to the rank of philosophers by the possession of leisure. In those countries where nature is the most redundant in spontaneous produce the inhabitants will not be found the most remarkable for acuteness of intellect. Necessity has been with great truth called the mother of invention. Some of the noblest exertions of the human mind have been set in motion by the necessity of satisfying the wants of the body. Want has not unfrequently given wings to the imagination of the poet, pointed the flowing periods of the historian, and added acuteness to the researches of the philosopher, and though there are undoubtedly many minds at present so far improved by the various excitements of knowledge, or of social sympathy, that they would not relapse into listlessness if their bodily stimulants were removed, yet it can scarcely be doubted that these stimulants could not be withdrawn from the mass of mankind without producing a general and fatal torpor, destructive of all the germs of future improvement.

[. . .]

The necessity of food for the support of life gives rise, probably, to a greater quantity of exertion than any other want, bodily or mental. The Supreme Being has ordained that the earth shall not produce food in great quantities till much preparatory labour and ingenuity has been exercised upon its surface. There is no conceivable connection to our comprehensions, between the seed and the plant or tree that rises from it. The Supreme Creator might, undoubtedly, raise up plants of all kinds, for the use of his creatures, without the assistance of those little bits of matter, which we call seed, or even without the assisting labour and attention of man. The processes of ploughing and clearing the ground, of collecting and sowing seeds, are not surely for the assistance of God in his creation, but are made previously necessary to the enjoyment of the blessings of life, in order to rouse man into action, and form his mind to reason.

To furnish the most unremitted excitements of this kind, and to urge man to further the gracious designs of Providence by the full cultivation of the earth, it has been ordained that population should increase much faster than food. This general law (as it has appeared in the former parts of this Essay) undoubtedly produces much partial evil, but a little reflection may, perhaps, satisfy us, that it produces a great overbalance of good. Strong excitements seem necessary to create exertion, and to direct this exertion, and form the reasoning faculty, it seems absolutely necessary, that the Supreme Being should act always according to general laws. The constancy of the laws of nature, or the certainty with which we may expect the same effects from the same causes, is the foundation of the faculty of reason. If in the ordinary course of things, the finger of God were frequently visible, or to speak more correctly, if God were frequently to change his purpose (for the finger of God is, indeed, visible in every blade of grass that we see), a general and fatal torpor of the human faculties would probably ensue; even the bodily wants of mankind would cease to stimulate them to exertion, could they not reasonably expect that if their efforts were well directed they would be crowned with success. The constancy of the laws of nature is the foundation of the industry and foresight of the husbandman, the indefatigable ingenuity of the artificer, the skilful researches of the physician and anatomist, and the watchful observation and patient investigation of the natural philosopher. To this constancy we owe all the greatest and noblest efforts of intellect. To this constancy we owe the immortal mind of a Newton.

As the reasons, therefore, for the constancy of the laws of nature seem, even to our understandings, obvious and striking; if we return to the principle of population and consider man as he really is, inert, sluggish, and averse from labour, unless compelled by necessity (and it is surely the height of folly to talk of man, according to our crude fancies of what he might be), we may pronounce with certainty that the world would not have been peopled, but for the superiority of the power of population to the means of subsistence. Strong and constantly operative as this stimulus is on man to urge him to the cultivation of the earth, if we still see that cultivation proceeds very slowly, we may fairly conclude that a less stimulus would have been insufficient. Even under the operation of this constant excitement, savages will inhabit countries of the greatest natural fertility for a long period before they betake themselves to pasturage or agriculture. Had population and food increased in the same ratio, it is probable that man might never have emerged from the savage state. But supposing the earth once well peopled, an

Alexander, a Julius Caesar, a Tamberlane, or a bloody revolution might irrecoverably thin the human race, and defeat the great designs of the Creator. [. . .] The principle, according to which population increases, prevents the vices of mankind, or the accidents of nature, the partial evils arising from general laws, from obstructing the high purpose of the creation. It keeps the inhabitants of the earth always fully up to the level of the means of subsistence; and is constantly acting upon man as a powerful stimulus, urging him to the further cultivation of the earth, and to enable it, consequently, to support a more extended population. But it is impossible that this law can operate, and produce the effects apparently intended by the Supreme Being, without occasioning partial evil. Unless the principle of population were to be altered according to the circumstances of each separate country (which would not only be contrary to our universal experience, with regard to the laws of nature, but would contradict even our own reason, which sees the absolute necessity of general laws for the formation of intellect), it is evident that the same principle which, seconded by industry, will people a fertile region in a few years must produce distress in countries that have been long inhabited.

It seems, however, every way probable that even the acknowledged difficulties occasioned by the law of population tend rather to promote than impede the general purpose of Providence. They excite universal exertion and contribute to that infinite variety of situations, and consequently of impressions, which seems upon the whole favourable to the growth of mind. It is probable, that too great or too little excitement, extreme poverty, or too great riches may be alike unfavourable in this respect. The middle regions of society seem to be best suited to intellectual improvement, but it is contrary to the analogy of all nature to expect that the whole of society can be a middle region. The temperate zones of the earth seem to be the most favourable to the mental and corporal energies of man, but all cannot be temperate zones. A world, warmed and enlightened but by one sun, must from the laws of matter have some parts chilled by perpetual frosts and others scorched by perpetual heats [. . .]

In the same manner, though we cannot possibly expect to exclude riches and poverty from society, yet if we could find out a mode of government by which the numbers in the extreme regions would be lessened and the numbers in the middle regions increased, it would be undoubtedly our duty to adopt it. [. . .] If no man could hope to rise or fear to fall, in society, if industry did not bring with it its reward and idleness its punishment, the middle parts would not certainly be what they now are. In reasoning upon this subject, it is evident that we ought to consider chiefly the mass of mankind and not individual instances. There are undoubtedly many minds, and there ought to be many, according to the chances out of so great a mass, that, having been vivified early by a peculiar course of excitements, would not need the constant action of narrow motives to continue them in activity. But if we were to review the various useful discoveries, the valuable writings, and other laudable exertions of mankind, I believe we should find that more were to be attributed to the narrow motives that operate upon the many than to the apparently more enlarged motives that operate upon the few.

11

WILLIAM PALEY, *NATURAL THEOLOGY; OR, EVIDENCES OF THE EXISTENCE AND ATTRIBUTES OF THE DEITY, COLLECTED FROM THE APPEARANCES OF NATURE*, 2ND ED.

(London: R. Paulder, 1802)

Chapter I, State of the Argument

In crossing a heath, suppose I pitched my foot against a stone, and were asked how the stone came to be there, I might possibly answer, that for any thing I knew to the contrary it had lain there for ever; nor would it, perhaps, be very easy to show the absurdity of this answer. But suppose I had found a watch upon the ground, and it should be inquired how the watch happened to be in that place, I should hardly think of the answer which I had before given, that for any thing I knew the watch might have always been there. Yet why should not this answer serve for the watch as well as for the stone; why is it not as admissible in the second case as in the first? For this reason, and for no other, namely, that when we come to inspect the watch, we perceive – what we could not discover in the stone – that its several parts are framed and put together for a purpose, e.g. that they are so formed and adjusted as to produce motion, and that motion so regulated as to point out the hour of the day; that if the different parts had been differently shaped from what they are, or placed after any other manner or in any other order than that in which they are placed, either no motion at all would have been carried on in the machine, or none which would have answered the use that is now served by it. To reckon up a few of the plainest of these parts and of their offices, all tending to one result: We see a cylindrical box containing a coiled elastic spring, which, by its endeavour to relax itself, turns round the box. We next observe a flexible chain – artificially wrought for the sake of flexure – communicating the action of the spring from the box to the fusee. We then find a series of wheels, the teeth of which catch in and apply

 DOI: 10.4324/9780429355653-14

to each other, conducting the motion from the fusee to the balance and from the balance to the pointer, and at the same time, by the size and shape of those wheels, so regulating that motion as to terminate in causing an index, by an equable and measured progression, to pass over a given space in a given time. We take notice that the wheels are made of brass, in order to keep them from rust; the springs of steel, no other metal being so elastic; that over the face of the watch there is placed a glass, a material employed in no other part of the work, but in the room of which, if there had been any other than a transparent substance, the hour could not be seen without opening the case. This mechanism being observed – it requires indeed an examination of the instrument, and perhaps some previous knowledge of the subject, to perceive and understand it; but being once, as we have said, observed and understood, the inference we think is inevitable, that the watch must have had a maker – that there must have existed, at some time and at some place or other, an artificer or artificers who formed it for the purpose which, we find it actually to answer, who comprehended its construction and designed its use.

I. Nor would it, I apprehend, weaken the conclusion, that we had never seen a watch made – that we had never known an artist capable of making one – that we were altogether incapable of executing such a piece of workmanship ourselves, or of understanding in what manner it was performed; all this being no more than what is true of some exquisite remains of ancient art, of some lost arts, and, to the generality of mankind, of the more curious productions of modern manufacture. Does one man in a million know how oval frames are turned? Ignorance of this kind exalts our opinion of the unseen and unknown artist's skill, if he be unseen and unknown, but raises no doubt in our minds of the existence and agency of such an artist, at some former time and in some place or other. Nor can I perceive that it varies at all the inference, whether the question arise concerning a human agent or concerning an agent of a different species, or an agent possessing in some respects a different nature.

II. Neither, secondly, would it invalidate our conclusion, that the watch sometimes went wrong, or that it seldom went exactly right. The purpose of the machinery, the design, and the designer might be evident, and in the case supposed, would be evident, in whatever way we accounted for the irregularity of the movement, or whether we could account for it or not. It is not necessary that a machine be perfect, in order to show with what design it was made: still less necessary, where the only question is whether it were made with any design at all.

III. Nor, thirdly, would it bring any uncertainty into the argument, if there were a few parts of the watch, concerning which we could not discover or had not yet discovered in what manner they conduced to the general effect; or even some parts, concerning which we could not ascertain whether they conduced to that effect in any manner whatever. For, as to the first branch of the case, if by the loss, or disorder, or decay of the parts in question, the movement of the watch were found in fact to be stopped, or disturbed, or retarded, no doubt would remain in our minds as to the utility or intention of these parts, although we should be unable to investigate the manner according to which, or the connection by which, the ultimate effect depended upon their action or assistance; and the more complex the machine, the more likely is this obscurity to arise. Then, as to the second thing supposed, namely, that there were parts which might be spared without prejudice to the movement of the watch, and that we had proved this by experiment, these superfluous parts, even if we were completely assured that they were such, would not vacate the reasoning which we had instituted concerning other parts. The indication of contrivance remained, with respect to them, nearly as it was before.

IV. Nor, fourthly, would any man in his senses think the existence of the watch with its various machinery accounted for, by being told that it was one out of possible combinations of material forms; that whatever he had found in the place where he found the watch, must have contained some internal configuration or other; and that this configuration might be the structure now exhibited, namely, of the works of a watch, as well as a different structure.

V. Nor, fifthly, would it yield his inquiry more satisfaction, to be answered that there existed in things a principle of order, which had disposed the parts of the watch into their present form and situation. He never knew a watch made by the principle of order; nor can he even form to himself an idea of what is meant by a principle of order, distinct from the intelligence of the watchmaker.

VI. Sixthly, he would be surprised to hear that the mechanism of the watch was no proof of contrivance, only a motive to induce the mind to think so:

VII. And not less surprised to be informed, that the watch in his hand was nothing more than the result of the laws of metallic nature. It is a perversion of language to assign any law as the efficient, operative cause of any thing. A law presupposes an agent; for it is only the mode according to which an agent proceeds: it implies a power; for it is the order according to which that power acts. Without this agent, without this power, which are both distinct from itself, the law does nothing, is nothing. The expression, 'the law of metallic nature', may sound strange and harsh to a philosophic ear; but it seems quite as justifiable as some others which are more familiar to him, such as 'the law of vegetable nature', 'the law of animal nature', or, indeed, as 'the law of nature' in general, when assigned as the cause of phenomena, in exclusion of agency and power, or when it is substituted into the place of these.

VIII. Neither, lastly, would our observer be driven out of his conclusion or from his confidence in its truth, by being told that he knew nothing at all about the matter. He knows enough for his argument; he knows the utility of the end; he knows the subserviency and adaptation of the means to the end. These points being known, his ignorance of other points, his doubts concerning other points, affect not the certainty of his reasoning. The consciousness of knowing little need not beget a distrust of that which he does know.

Chapter II, State of the Argument Continued

SUPPOSE, in the next place, that the person who found the watch should after some time discover, that in addition to all the properties which he had hitherto observed in it, it possessed the unexpected property of producing in the course of its movement another watch like itself – the thing is conceivable; that it contained within it a mechanism, a system of parts – a mould, for instance, or a complex adjustment of lathes, files, and other tools – evidently and separately calculated for this purpose; let us inquire what effect ought such a discovery to have upon his former conclusion.

I. The first effect would be to increase his admiration of the contrivance, and his conviction of the consummate skill of the contriver. Whether he regarded the object of the contrivance, the distinct apparatus, the intricate, yet in many parts intelligible mechanism by which it was carried on, he would perceive in this new observation nothing but an additional reason for doing what he had already done – for referring the construction of the watch to design and to supreme art. If that construction without this property, or which is the same thing, before this property had been noticed, proved intention and art to have been employed about it, still more strong would the proof appear when he came to the knowledge of this further property, the crown and perfection of all the rest.

II. He would reflect, that though the watch before him were in some sense the maker of the watch which was fabricated in the course of its movements, yet it was in a very different sense from that in which a carpenter, for instance, is the maker of a chair – the author of its contrivance, the cause of the relation of its parts to their use. With respect to these, the first watch was no cause at all to the second; in no such sense as this was it the author of the constitution and order, either of the parts which the new watch contained, or of the parts by the aid and instrumentality of which it was produced. We might possibly say, but with great latitude of expression, that a stream of water ground corn; but no latitude of expression would allow us to say, no stretch of conjecture could lead us to think, that the stream of water built the mill, though it were too ancient for us to know who the builder was. What the stream of water does in the affair is neither more nor less than this: by the application of an unintelligent impulse to a mechanism previously arranged, arranged independently of it and arranged by intelligence, an effect is produced, namely, the corn is ground. But the effect results from the arrangement. The force of the stream cannot be said to be the cause or the author of the effect, still less of the arrangement. Understanding and plan in the formation of the mill were not the less necessary for any share which the water has in grinding the corn; yet is this share the same as that which the watch would have contributed to the production of the new watch, upon the supposition assumed in the last section. Therefore,

III. Though it be now no longer probable that the individual watch which our observer had found was made immediately by the hand of an artificer, yet doth not this alteration in anywise affect the inference, that an artificer had been originally employed and concerned in the production. The argument from design remains as it was. Marks of design and contrivance are no more accounted for now than they were before. In the same thing, we may ask for the cause of different properties. We may ask for the cause of the colour of a body, of its hardness, of its heat; and these causes may be all different. We are now asking for the cause of that subserviency to a use, that relation to an end, which we have remarked in the watch before us. No answer is given to this question, by telling us that a preceding watch produced it. There cannot be design without a designer; contrivance, without a contriver; order, without choice; arrangement, without any thing capable of arranging; subserviency and relation to a purpose, without that which could intend a purpose; means suitable to an end, and executing their office in accomplishing that end, without the end ever having been contemplated, or the means accommodated to it. Arrangement, disposition of parts, subserviency of means to an end, relation of instruments to a use, imply the presence of intelligence and mind. No one, therefore, can rationally believe that the insensible, inanimate watch, from which the watch before us issued, was the proper cause of the mechanism we so much admire in it – could be truly said to have constructed the instrument, disposed its parts, assigned their office, determined their order, action, and mutual dependency, combined their several motions into one result, and that also a result connected with the utilities of other beings. All these properties, therefore, are as much unaccounted for as they were before.

IV. Nor is any thing gained by running the difficulty farther back, that is, by supposing the watch before us to have been produced from another watch, that from a former, and so on indefinitely. Our going back ever so far brings us no nearer to the least degree of satisfaction upon the subject. Contrivance is still unaccounted for. We still want a contriver. A designing mind is neither supplied by this supposition nor dispensed with. If the difficulty were diminished the farther we went back, by going back indefinitely we might exhaust it. And this is the only case to which this sort of reasoning applies. Where there is a tendency, or, as we increase the number of terms, – a continual approach towards a limit, there, by supposing the number of terms to be what is called infinite, we may conceive the limit to be attained; but where there is no such tendency or approach, nothing is effected by lengthening the series. There is no difference as to the point in question, whatever there may be as to many points, between one series and another – between a series which is finite, and a series which is infinite. A chain composed of an infinite number of links can no more support itself than a chain composed of a finite number of links. And of this we are assured, though we never can have tried the experiment; because, by increasing the number of links, from ten, for instance, to a hundred, from a hundred to a thousand, etc., we make not the smallest approach, we observe not the smallest tendency towards self support. There is no difference in this respect – yet there may be a great difference in several respects – between a chain of a greater or less length, between one chain and another, between one that is finite and one that is infinite. This very much resembles the case before us. The machine which we are inspecting demonstrates, by its construction, contrivance and design. Contrivance must have had a contriver, design a designer, whether the machine immediately proceeded from another machine or not. That circumstance alters not the case. That other machine may, in like manner, have proceeded from a former machine: nor does that alter the case; the contrivance must have had a contriver. That former one from one preceding it: no alteration still; a contriver is still necessary. No tendency is perceived, no approach towards a diminution of this necessity. It is the same with any and every succession of these machines – a succession of ten, of a hundred, of a thousand; with one series, as with another – a series which is finite, as with a series which is infinite. In whatever other respects they may differ, in this they do not. In all equally, contrivance and design are unaccounted for.

The question is not simply. How came the first watch into existence? which question, it may be pretended, is done away by supposing the series of watches thus produced from one another to have been infinite, and consequently to have had no such first, for which it was necessary to provide a cause. This, perhaps, would have been nearly the state of the question, if nothing had been before us but an unorganized, unmechanized substance, without mark or indication of contrivance. It might be difficult to show that such substance could not have existed from eternity, either in succession – if it were possible, which I think it is not, for unorganized bodies to spring from one another – or by individual perpetuity. But that is not the question now. To suppose it to be so, is to suppose that it made no difference whether he had found a watch or a stone. As it is, the metaphysics of that question have no place; for, in the watch which we are examining, are seen contrivance, design, an end, a purpose, means for the end, adaptation to the purpose. And the question which irresistibly presses upon our thoughts is, Whence this contrivance and design? The thing required is the intending mind, the adapted hand, the intelligence by which that hand was directed. This question, this demand, is not shaken off by increasing a number or succession of substances destitute of these properties; nor the more, by increasing that number to infinity. If it be said, that upon the supposition of one watch being produced from another in the course of that other's movements, and by means of the mechanism within it, we have a cause for the watch in my hand, namely, the watch from which it proceeded – I deny, that for the design, the contrivance, the suitableness of means to an end, the adaptation of instruments to a use, all of which we discover in the watch, we have any cause whatever. It is in vain, therefore, to assign a series of such causes, or to allege that a series may be carried back to infinity; for I do not admit that we have yet any cause at all for the phenomena, still less any series of causes either finite or infinite. Here is contrivance, but no contriver; proofs of design, but no designer.

V. Our observer would further also reflect, that the maker of the watch before him was, in truth and reality, the maker of every watch produced from it: there being no difference, except that the latter manifests a more exquisite skill, between the making of another watch with his own hands, by the mediation of files, lathes, chisels, etc., and the disposing, fixing, and inserting of these instruments, or of others equivalent to them, in the body of the watch already made, in such a manner as to form a new watch in the course of the movements which he had given to the old one. It is only working by one set of tools instead of another.

The conclusion which the first examination of the watch, of its works, construction, and movement, suggested, was, that it must have had, for cause and author of that construction, an artificer who understood its mechanism and designed its use. This conclusion is invincible. A second examination presents us with a new discovery. The watch is found, in the course of its movement, to produce another watch similar to itself; and not only so, but we perceive in it a system or organization separately calculated for that purpose. What effect would this discovery have, or ought it to have, upon our former inference? What, as hath already been said, but to increase beyond measure our admiration of the skill which had been employed in the formation of such a machine? Or shall it, instead of this, all at once turn us round to an opposite conclusion, namely, that no art or skill whatever has been concerned in the business, although all other evidences of art and skill remain as they were, and this last and supreme piece of art be now added to the rest? Can this be maintained without absurdity?

Yet this is atheism.

12

PETER MARK ROGET, *ANIMAL AND VEGETABLE PHYSIOLOGY CONSIDERED WITH REFERENCE TO NATURAL THEOLOGY*, TREATISE 5 (2 VOLS.), *THE BRIDGEWATER TREATISES*, 3RD ED.

(London: William Pickering, 1834)

Introduction: Chapter 1, Final Causes

To investigate the relations which connect Man with his Creator is the noblest exercise of human reason. The Being who bestowed on him this faculty cannot but have intended that he should so exercise it, and that he should acquire, through its means, some insight, however limited, into the order and arrangements of creation; some knowledge, however imperfect, of the divine attributes; and a distinct, though faint, perception of the transcendent glory with which those attributes are encompassed. To Man have been revealed the POWER, the WISDOM, and the GOODNESS of GOD, through the medium of the Book of Nature, in the varied pages of which they are inscribed in indelible characters. On Man has been conferred the high privilege of interpreting these characters, and of deriving from their contemplation those ideas of grandeur and sublimity, and those emotions of admiration and of gratitude, which elevate and refine the soul, and transport it into regions of a purer and more exalted being.

A study which embraces so extensive a range of objects, and which involves questions of such momentous interest to mankind, must necessarily be arduous, and requires for its successful prosecution, the strenuous exertions of the human intellect, and the combined labours of different classes of philosophers, during many ages. The magnitude of the task is increased by the very success of those previous efforts: for the difficulties augment as the object multiply, and the eminence, on which the accumulated knowledge of centuries has placed us, only discloses a wider horizon, and the prospect of more fertile regions of inquiry; till at length

DOI: 10.4324/9780429355653-15

the mind, conscious of the inadequacy of its own powers to the comprehension of even a small part of the system of the universe, is appalled by the overwhelming consideration of the infinity that surrounds us. The reflection continually presents itself that the portion of creation we are here permitted to behold is as nothing when compared with the immensity of space, which, on every side, spreads far beyond the sphere of our vision, and which the power of human imagination is inadequate to comprehend [. . .]

Measured on the vast scale of the universe, the globe we inhabit appears but as an atom; and yet, within the compass of this atom, what an inexhaustible variety of objects is contained; what an endless diversity of phenomena is presented; what wonderful changes are occurring in rapid and perpetual succession! Throughout the whole series of terrestrial beings, what studied arrangements, what preconcerted adaptations, what multiplied evidences of intention, what signal proofs of beneficent design exist to attract our notice, to excite our curiosity, and to animate our enquiries. Splendid as are the monuments of divine power and wisdom displayed throughout the firmament, in objects fitted by their stupendous magnitude to impress the imagination and overpower us by the awful grandeur, not less impressive, nor less replete with wonder, are the manifestations of those attributes in the minuter portions of nature, which are more on a level with our senses, and more within the reach of our comprehension.

[. . .]

In the investigation of the powers which are concerned in the phenomena of living beings we meet with difficulties incomparably greater than those that attend the discovery of the physical forces by which the parts of inanimate matter are actuated. The elements of the inorganic world are few and simple; the combinations they present are, in most cases, easily unravelled; and the powers which actuate their motions, or effect their union and their changes, are reducible to a small number of general laws [. . .] To whatever department of physical science of researches have extended, we everywhere meet with the same regularity in the phenomena, the same simplicity in the laws, and the same uniformity in the results. All is strictly defined, and subjected to rigid rule: all is subordinate to one pervading principle of order. The Greater Creator of the universe has exercised in the construction the severest and most refined geometry, has traced with unerring precision the boundaries of all its parts, and has prescribed to each element and each power its respective sphere and limit.

Far different is the aspect of living Nature. The spectacle here offered to our view is everywhere characterised by boundless variety, by inscrutable complexity, by perpetual mutation. Our attention is solicited to a vast multiplicity of objects, curious and intricate in their mechanism, exhibiting peculiar movements, actuated by new and unknown powers, and gifted with high and refined endowments. In place of the simple combination of elements, and the simple properties of mineral bodies, all organic structures, even the most minute, present exceedingly complicated arrangements, and a prolonged succession of phenomena, so varied and so anomalous, as to be utterly irreducible to the known laws which govern inanimate matter. Les us hasten, with fresh ardour, to explore this new world that here opens to our view.

Turning, then, from the examination of the passive objects of the material world, we now direct our attention to the busy theatre of animated existence, where scenes of wonder and enchantment are displayed in endless variety around us; where life in its ever-changing forms meets the eye in every region to which our researches can extend; and where every element and every clime is peopled by multitudinous races of sensitive beings, who have received from the bounteous hand of their Creator the gift of existence and the means of enjoyment. Our curiosity is powerfully excited by phenomena in which our own welfare is so intimately concerned, as are all those that relate to animal life; and we cannot but take a lively and sympathetic interest in the history of beings in many respects so analogous to ourselves; like us, possessing powers of spontaneous action, impelled by passions and desires, and endowed with capacities of enjoyment and of suffering. Can there be a more gratifying spectacle than to see an animal in the full vigour of health and the free exercise of its powers, disporting in its native element, revelling in the bliss of existence, and testifying by its incessant gambols the exuberance of its joy?

Introduction: Chapter 2, the Functions of Life

THE intentions of the Deity in the creation of the animal kingdom, as far as we are competent to discern or comprehend them, are referable to the following classes of objects. The first relates to the individual welfare of the animal, embracing the whole sphere of its sensitive existence, and the means of maintaining the vitality upon which that existence is dependent. The second comprises the provisions which have been made for repairing the chasms resulting, in the present circumstances of the globe, from the continual destruction of life, by ensuring the multiplication of the species, and the continuity of the race to which each animal belongs. The third includes all those arrangements which have been resorted to in order to accommodate the system to the consequences that follow from an indefinite increase in the numbers of each species. The fourth class relates to that systematic economy in the plans or organization by which all the former objects are most effectually secured.

[. . .]

As animals are ultimately dependent on the vegetable kingdom for the materials of their subsistence, and as the quantity of these materials is, in a state of nature, necessarily limited by the extent of surface over which vegetation is spread, a time must arrive when the number of animals thus continually increasing is exactly such as the amount of food produced by the earth will maintain. When this limit has been attained no further increase can take place in their number, except by resorting to the expedient which we find actually adopted, namely, that of employing the substance of one animal for the nourishment of others [. . .] Hence has the ordinance been issued to a large portion of the animal world that they are to maintain themselves by preying upon other animals; either consuming their substance when already dead, or depriving them of life in order to prolong their own. Such is the command given to the countless hosts of living beings which people the vast expanse of ocean; to the unnumbered tribes of insects which every spot of earth discloses; to the greater number of the feathered race; and also to a more restricted order of terrestrial animals. To many has the commission been given to ravage and to slaughter by open violence; others are taught more insidious, though no less certain arts of destruction; and some appear to be created chiefly for the purpose of clearing the earth of all decomposing animal or vegetable materials [. . .] That a large portion of evil is the direct consequence of this system of extensive warfare, it is in vain to deny. But although our sensibility may revolt at the wide scene of carnage which is so generally presented to our view, our more sober judgment should place in the other scale the great preponderating amount of gratification which is also its result [. . .] It affords a ready and effectual means of preserving the proper balance between different races. Each separate species of animals, far from being isolated and independent, performs the part assigned to it in the system of nature, and, however apparently insignificant, may have a sensible influence on the rest of the animal creation [. . .]

Part I, The Mechanical Functions

Chapter I, Organic Mechanism

LIFE, which consists of a continued series of actions directed to particular purposes, cannot be carried on but by the instrumentality of those peculiar and elaborate structures and combinations of material particles which constitute *organization.* All these arrangements, both as respects the mechanical configuration and the chemical constitution of the element of which the organized body is composed, even when apparently most simple, are, in reality, complex and artificial in the highest possible degree. Let us take as a specimen the crystalline lens, or hard central part, of the eye of a cod fish, which is a perfectly transparent, and to all appearance homogenous spherule. No one, unaccustomed to explore the wonders of nature, would suspect that so simple a body, which he might suppose to be formed of a uniform material cast in a mould, would disclose, when examined under a powerful microscope, and with the skill of a [Sir David] Brewster, the most refined and exquisite conformation. Yet, as I shall have occasion to specify more in detail in its proper place, this little spherical body, scarcely larger than a pea, is composed of upwards of five millions of fibres, which lock into one another by means of more than sixty-two thousand five hundred millions of teeth. If such be the complication of a portion of the eye of that animal, how intricate must be the structure of the other parts of the same organ, having equally important offices! What exquisite elaborations must those textures have received, whose functions are still more refined! What marvellous workmanship must have been exercised in the organization of the nerves and of the brain, those subtle instruments of the higher animal faculties, and of which even the modes of action to us not merely inscrutable, but surpassing all our powers of conception!

13

ADAM SEDGWICK, *ON THE STUDIES OF THE UNIVERSITY*, 2ND ED.

(Cambridge: Pitt Press, 1834)

A Discourse

[. . .]

In speaking of the laws of nature and of harmonious changes resulting from their action, in spite of ourselves we fall into language in which we describe the operations of intelligence: and language, let me observe, was never formed by a convention of learned men, or constructed on the scheme of an hypothesis. It is the true offspring of our intellectual nature, and bears the image of such ideas as rise up of necessity in the mind, from our relation to the things around us. If we forget him in our thoughts, with our lips at least we must do homage to the God of nature. What are the laws of nature but the manifestations of his wisdom? What are material actions but manifestations of his power? Indications of his wisdom and his power co-exist with every portion of the universe. They are seen in great luminaries of heaven – they are seen in the dead matter whereon we trample – they are found in all parts of space, remote as well as near, which we in our ignorance sometimes regard as mere vacuities – they are unceasing – they are unchangeable.

Contemplations such as these lift the soul to a perception of some of the attributes of God; imperfect as it may be, but still suited to the condition of our being. But are thoughts like these to pass through the mind and produce only a cold acquiescence? Are we to dwell in the perpetual presence of God and yet dishonour him by the worship of ourselves, and refuse to him the homage of our humble praise? [. . .] We are told by St Paul that even the Gentiles are without excuse, *for the invisible things of God from the creation of the world are clearly seen, being understood by the things which are made, even his eternal power and Godhead.* Yet I have myself heard it asserted within these very walls, that there is no religion of nature and that we have no knowledge of the attributes of God or even of his existence, independently of revelation. The assertion is, I think, mischievous, because I believe it untrue: and by truth only can a God of truth be honoured, and the cause of true religion be served.

DOI: 10.4324/9780429355653-16

But there is another class of objectors, who not only adopt this cold and unnatural conclusion, but rejecting revelation along with it, banish utterly all thought of God from the world. It is indeed true, as these objectors state, that all material changes are governed by fixed laws, and that the present condition of all material things is but a natural consequence of these laws operating on that condition of matter which preceded the phenomena we contemplate. They rest their strength, as far as I understand their meaning, in this immutability of the laws of nature: and having, with much labour, decyphered a portion of these laws, and having traced the ordained movements of the material world without ever thinking of the Being by whom these movements were directed, they come at length to deify the elements themselves, and to thrust the God of nature from his throne. But where is the reasonableness of this conclusion? The unchangeableness of the laws of nature is not only essential to the well being, but to the very existence of creatures like ourselves. The works of our hands are liable to perpetual change, from caprice, from violence, or from natural decay: but in the material laws ordained by God, there are no such indications, because they partake of the perfections of his attributes, and are therefore unchangeable.

[. . .]

But after all, we do contemplate something more than a succession of material changes. We find that these changes are limited by an adjusting power, and tend to a condition of equilibrium, and that the ultimate results of the laws of nature are harmony and order. We find them operating in different portions of space with endless complexity, and yet producing effects perfectly adapted to each other. We see innumerable portions of matter not only obeying laws common to all matter, but acted on by new laws subservient to vital powers; and by these laws gradually moulded into a beautiful form and mechanism – suited at once to all the complicated functions of a sentient being, and to all the material elements which surround it. Are we to believe that there can be such beautiful and harmonious movements in the vast mechanism of nature, and yet think that the Spirit of God hath not brooded over them, and that his hand hath not guided them? Can we see in every portion of the visible world the impress of wisdom and power, and yet believe that these things were not foreseen in the Divine mind, and these ends not contemplated before he called the universe into being?

[. . .]

Something like this we can trace in the development of organic beings. They are formed of matter, which was created, and governed by its own laws, anterior to their existence: they are matured by a regulated succession of material actions: when perfected, they exhibit an exquisite combination of mechanical contrivances, and organs fitted to carry them into effect. To such a structure are superadded vital functions and appetencies, which (like the moving force of a complicated engine) put all its parts into motion, and compel them to obey the laws of their destination. The external forms of organic bodies we can study, their functions we can observe, their internal mechanism we can partly trace: but when we consider the vital powers connected with their origin and development, we find ourselves among phenomena out of the ken of our senses, and removed beyond our intellectual grasp; and are compelled to acknowledge our utter ignorance. But, on this account to exclude an intelligent contriver, would not be more wise, than for a man to assert the fortuitous concourse of the wheels of a machine, because he knew not the power by which it was set in motion.

[. . .]

By the discoveries of a new science (the very name of which has been but a few years engrafted on our language), we learn that the manifestations of God's power on the earth have not been limited to the few thousand years of man's existence. The Geologist tells us, by the clearest interpretation of the phenomena which his labours have brought to light, that our globe has been subject to vast physical revolutions. He counts his time not by celestial cycles, but by an index he has found in the solid framework of the globe itself. He sees a long succession of monuments each of which may have required a thousand ages for its elaboration. He arranges them in chronological order; observes on them the marks of skill and wisdom, and find within them the tombs of the ancient inhabitants of the earth. He finds strange and unlooked-for changes in the forms and fashions of organic life during each of the long periods he thus contemplates. He traces these changes backwards through each successive era, till he reaches a time when the monuments lose all symmetry, and the types of organic life are no longer seen. He has then entered on the dark age of nature's history; and he closes the old chapter of her records [. . .]

Geology, like every other science when well interpreted, lends its aid to natural religion. It tells us, out of its own records, that man has been but a few years a dweller on the earth; for the traces of himself and his works are confined to the last monuments of its history. Independently of every written testimony, we therefore believe that man, with all his powers and appetencies, his marvellous structure and his fitness for the world around him, was called into being within a few thousand years of the days in which we live – not by a transmutation of species, (a theory no better than a phrensied dream), but by a provident contriving power [. . .]

But this is not the only way in which Geology gives its aid to natural religion. It proves that a pervading intelligent principle has manifested its power during times long anterior to the records of our existence. It adds to the great cumulative argument derived from the forms of animated nature, by shewing us new and unlooked-for instances of organic structure adjusted to an end, and that end accomplished. It tells us that God has not created the world and left it to itself, remaining ever after a quiescent spectator of his own work: for it puts before our eyes the certain proofs, that during successive periods there have been, not only great changes in the external conditions of the earth, but corresponding changes in organic life; and that in every such instance of change, the new organs, as far as we can comprehend their use, were exactly suited to the functions of the beings they were given to. It shews intelligent power not only contriving means adapted to an end: but at many successive times contriving a change of mechanism adapted to a change of external conditions; and thus affords a proof, peculiarly its own, that the first great cause continues a provident and active intelligence.

[. . .]

At succeeding epochs, new tribes of beings were called into existence, not merely as the progeny of those that had appeared before them, but as new and living proofs of creative interference: and though formed on the same plan, and bearing the same marks of wise contrivance, oftentimes as unlike those creatures which preceded them, as if they had been matured in a different portion of the universe and cast upon the earth by the collision of another planet. At length, within a few thousand years of the days in which we live (a period short indeed if measured by the physical monuments of time past), man and his fellow beings are placed upon the earth [. . .] All nature is but the manifestation of a supreme intelligence, and to no being but him, to whom is given the faculty of reason, can this truth be known. By this faculty he comes the lord of created beings, and finds all matter, organic and inorganic, subservient to his happiness, and working together for his good.

Appendix

[. . .]

There is another class of men who pursue Geology by a nearer road, and are guided by a different light. Well-intentioned they may be, but they have betrayed no small self-sufficiency, along with a shameful want of knowledge of the fundamental facts they presume to write about: hence they have dishonoured the literature of this country by *Mosaic Geology*, *Scripture Geology*, and other works of cosmogony with kindred titles, wherein they have overlooked the aim and end of revelation, tortured the book of life out of its proper meaning, and wantonly contrived to bring about a collision between natural phenomena and the word of God. The Buggs and the Penns – the Nolans and the Formans – and some others of the same class, have committed the folly and the sin of dogmatizing on matters they have not personally examined, and, at the utmost, know only at second hand – of pretending to teach mankind on points where they themselves are uninstructed. Authors such as these ought to have first considered, that book learning (in whatever degree they may be gifted with it) is but a pitiful excuse for writing mischievous nonsense: and that to a divine or a man of letters, ignorance of the laws of nature and of material phenomena is then only disgraceful, when he quits his own ground and pretends to teach philosophy. Their learning (if perchance they possess it) has been but ill employed in following out the idle dreams of an irrational cosmogony: and they would be labouring at a task better fitted for their capacity, were they studying the simple and affecting lessons of Christianity, and trying to make its maxims of charity their rule of life [. . .] Men are still to be found, who not restrained by the wise and humane laws of their country, would try to stifle by personal violence, and crush by brute force, every truth not hatched among their own conceits, and confined within the narrow fences of their own ignorance.

[. . .]

Before a Geologist can condescend to reason with such men, they must first learn Geology. It is too much to call upon us to scatter our seed on a soil at once both barren and unreclaimed – it is folly to think, that we can in the same hour be stubbing up the thorns and reaping the harvest. All the writers of this school have not indeed sinned against plain sense to the same degree. With some of them, there is perhaps a perception of the light of natural truth which may lead them after a time to follow it in the right road: but the case of others is beyond all hope from the powers of rational argument [. . .]

A man of deep thought and great practical wisdom – one whose piety and benevolence have for many years been shining before the world, and of whose sincerity no scoffer (of whatever school) will dare to start a doubt – recorded his opinion in the great assembly of the men of science, who during the past year were gathered from every corner of the Empire within the walls of this University, *that Christianity had every thing to hope and nothing to fear from the advancement of philosophy*. These are golden words, and full of meaning to those who have wisdom to understand them [. . .]

Another indiscretion (far different however from the egregious follies I have just noticed) has been committed by some excellent Christian writers on the subject of Geology. They have not denied the facts established by this science, nor have they confounded the nature of physical and moral evidence: but they have prematurely (and therefore, without an adequate knowledge of all the facts essential to the argument) endeavoured to bring the natural history of the Earth into a literal accordance with the book of Genesis – first, by greatly extending the periods of time implied by the six days of creation (and whether this might be rightly done is a question only of criticism and not of philosophy) – and secondly, by endeavouring to shew, that, under this new interpretation of its words, the narrative of Moses may be supposed to comprehend, and to describe in order, the successive epochs of Geology. It is to be feared that truth may, in this way, receive a double injury; and I am certain the argument, just alluded to, has been unsuccessful. The impossibility of the task was however (as I know by my own experience) a lesson hard to learn: but it is not likely again to be attempted by any good Geologist. The only way to escape from all difficulties pressing on the questions of cosmogony has been already pointed out. We must consider the old strata of the earth as monuments of a date long anterior to the existence of man, and to the times contemplated in the moral records of his creation. In this view there is no collision between physical and moral truth. The Bible is left to rest on its appropriate evidences, and its interpretation is committed to learning and good sense of the critic and the commentator: while Geology is allowed to stand on its own basis, and the philosopher to follow the investigations of physical truth, wherever they may lead him, without any dread of evil consequences; and with the sure conviction that natural science, when pursued with a right spirit, will foster the reasoning powers, and teach us knowledge fitted, at once, to impress the imagination, to bear on the business of life, and to give us exalted views of the universal presence and unceasing power of God [. . .] In the furtherance of this object (though at the risk of being taxed with the fault of egotism and useless repetition) I will add one more passage, taken from an anniversary address to the Geological Society [. . .]

'No opinion can be heretical but that which is not true. Conflicting falsehoods we can comprehend; but truths can never war against each other. I affirm, therefore, that we have nothing to fear from the results of our enquiries, provided they be followed in the laborious but secure road of honest induction. In this way we may rest assured that we shall never arrive at conclusions opposed to any truth, either physical or moral, from whatsoever source that truth may be derived: nay rather (as in all truth there is a common essence), that new discoveries will ever lend support and illustration to things which are already known, by giving us a larger insight into the universal harmonies of nature'.

14

HENRY COLE, *POPULAR GEOLOGY SUBVERSIVE OF DIVINE REVELATION! A LETTER TO THE REV. ADAM SEDGWICK, WOODWARDIAN PROFESSOR OF GEOLOGY IN THE UNIVERSITY OF CAMBRIDGE, BEING A SCRIPTURAL REFUTATION OF THE GEOLOGICAL POSITIONS AND DOCTRINES PROMULGATED IN HIS LATELY PUBLISHED COMMENCEMENT SERMON, PREACHED IN THE UNIVERSITY OF CAMBRIDGE, 1832*

(London: Hatchard and Son, 1834)

Preface

A Commencement Sermon has been lately published, preached in the Chapel of Trinity College, Cambridge, by the Rev. Adam Sedgwick, Woodwardian Professor of Geology, in that University. The subject-matter of this Discourse is twofold. In its former part, it states, maintains, and propagates, the principles and doctrines of the popular 'new science' of *Geology*: which principles and doctrines and their promulgation, tend directly to falsify and subvert the inspired record of the world's Creation. The latter part of the Discourse treats of *moral philosophy*: the principles and doctrines of which stand equally opposed to the Word of God.

The publication of this Commencement Sermon was conspicuously announced in the leading Journal of the nation: and was ushered into public notice with the

DOI: 10.4324/9780429355653-17

most unbounded eulogies, and the most imposing recommendations. Copious extracts were made from its geological portion, and the nation was promised, that a like notice should be taken, on some future occasion, of the system of ethics which its latter division contained.

Deeply concerned, aggrieved, and astonished, that scientific principles and doctrines, so totally opposed to, and subversive of divine Revelation, should emanate from such a national source, and be thus publicly sounded forth to the notice of a Christian land: the Author could not resist the impulse he felt, to address the following letter to the Editor of the Journal alluded to: which letter is here subjoined, as, perhaps, the best explanation of the origin, nature, and design, of the present publication.

Professor Sedgwick's Commencement Sermon

To the Editor of 'The Times'

[. . .]

In your Journal of the 10th of January last, you gave rein to the most unbounded encomiums on that Discourse; characterizing it as one by which 'the Commencement-day of Trinity College, Cambridge, was signalized in December, 1832'. You lauded it as 'a work of great varied excellence, doing honour to the learning, the discriminative powers, the masculine spirit, and the integrity of Mr Sedgwick', and as being 'such a production as became the name and reputation of its author'.

Now, Sir, I trust your candour will permit your influential Journal to circulate, as widely as these praises have reached, a Declaration that the thus eulogized Positions which Mr Sedgwick has laid down in that Discourse, with reference to the popular science of Geology, stand in direct opposition to, and contradiction of, the eternal Truth of divine Revelation. Mr Sedgwick therein asserts, according to the revelation-subverting deductions of the 'new science', that 'the manifestations of God's power upon the earth have not been limited to the few thousand years of man's existence'; that 'the Geologist sees a long succession of monuments [anterior to the creation of man], each of which may have required a thousand years for its elaboration'; that 'the Geologist counts his time, not by celestial cycles, but by an index he has found in the solid framework of the globe itself' – 'Therefore', (says Professor Sedgwick,) '*We believe, independently of every written testimony,* that man has been but a few years [out of the numerous thousands of the globe's existence] a dweller upon earth'.

Thus, Sir, has Mr Sedgwick embodied and promulgated in his Sermon, in the national seat of our youth's Education (and which ought to be the national seminary of divine Truth also), that scripture-prostrating doctrine, to which the popular pursuit of the 'new science' is so widely and dangerously leading – namely, that the globe on which we live was created thousands of years before man was created upon it. Whereby the Word of God, the only foundation of truth in this and all divine matters, is denied and utterly superseded, and a number of little understood, undeterminable, and undatable (excepting as scripturally declared, determined, and dated) specimens of organic remains, exalted in its sacred stead. And this superseding exaltation, this impious preference, Mr Sedgwick himself admits, for he expressly avers, that himself and his fellow geological sophists, '*believe*' the world to have thus anteriorly existed, '*independently of every written testimony*'; counting their thousands of years of antecedent existence 'from the indexes found in the solid framework of the earth itself'.

Such, Sir, is the revelation-subverting philosophy of this 'new science', which is now spreading its perverting and baneful influence on all sides; and to which Mr Sedgwick's 'masculine spirit' has thus lamentably lent its aid. And though it would have been equally subversive of divine Revelation in the minds of students, had this been a mere lecture from the Professor in the geological schools, yet, when we consider that it was a Sermon in the public worship of God, preached by a member of the Church of England, on a specifically public occasion; and that too, in all probability, before the heads of the national seat of learning and religion; and, if not in the presence of, yet to the certain knowledge of, the Professor of Divinity himself; and, moreover, with the silent or expressed applause of them all; these momentous facts, Sir, make the matter, or ought to make it, a just cause of national grief and reflection; and no source of alarm at the manner in which the testimony and record of eternal Truth are thus set at nought and trampled under the academic feet of '*philosophy and vain deceit*'.

15

REV. WILLIAM BUCKLAND, *GEOLOGY AND MINEROLOGY WITH REFERENCE TO NATURAL THEOLOGY*, TREATISE 6, *THE BRIDGEWATER TREATISES*

(London: William Pickering, 1837 [1836])

Chapter II, Consistency of Geological Discoveries with Sacred History

It may seem just matter of surprise, that many learned and religious men should regard with jealousy and suspicion the study of any natural phenomena, which abound with proofs of some of the highest attributes of the Deity; and should receive with distrust, or total incredulity, the announcement of conclusions, which the geologist deduces from careful and patient investigation of the facts which it is his province to explore. These doubts and difficulties result from the disclosures made by geology, respecting the lapse of very long periods of time, before the creation of man. Minds which have been long accustomed to date the origin of the universe, as well as that of the human race, from an era of about six thousand years ago, receive reluctantly any information, which, if true, demands some new modification of their present ideas of cosmogony; and, as in this respect, Geology has shared the fate of other infant sciences, in being for a while considered hostile to revealed religion; so like them, when fully understood, it will be found a potent and consistent auxiliary to it, exalting our conviction of the Power and Wisdom, and Goodness of the Creator.

No reasonable man can doubt that all the phenomena of the natural world derive their origin from God; and no one who believes the Bible to be the world of God, has cause to fear any discrepancy between his word, and the results of any discoveries respecting the nature of his works; but the early and deliberative stages of scientific discovery are always those of perplexity and alarm, and during these stages the human mind is naturally circumspect, and slow to admit new conclusions in any department of knowledge. The prejudiced persecutors of Galileo apprehended dangers to religion, from the discoveries of a science, in which a Kepler, and a Newton found demonstration of the most sublime and glorious attributes of the

DOI: 10.4324/9780429355653-18

Creator. A Herschel has pronounced that 'Geology, in the magnitude and sublimity of the objects of which it treats, undoubtedly ranks in the scale of sciences next to astronomy'; and the history of the structure of our planet, when it shall be fully understood, must lead to the same great moral results that have followed the study of the mechanism of the heavens [. . .]

It must be candidly admitted that the season has not yet arrived, when a perfect theory of the whole earth can be fixedly and finally established, since we have not yet before us all the facts on which such a theory may eventually be founded; but in the mean while, we have abundant evidence of numerous and indisputable phenomena, each establishing important and undeniable conclusions; and the aggregate of these conclusions, as they gradually accumulate, will form the basis of future theories, each more and more nearly approximating to perfection [. . .] Admitting therefore, that we have yet much to learn, we contend that much sound knowledge has been already acquired; and we protest against the rejection of established parts, because the whole is not yet made perfect.

It was assuredly prudent, during the infancy of Geology, in the immature state of those physical sciences which form its only sure foundation, not to enter upon any comparison of the Mosaic account of creation with the structure of the earth, then almost totally unknown; the time was not then come when the knowledge of natural phenomena was sufficiently advanced to admit of any profitable investigation of this question; but the discoveries of the last half century have been so extensive in this department of natural knowledge, that, whether we will or not, the subject is now forced upon our consideration, and can no longer escape discussion. The truth is, that all observers, however various may be their speculations, respecting the secondary causes by which geological phenomena have been brought about, are now agreed in admitting the lapse of very long periods of time to have been an essential condition to the production of these phenomena.

It may therefore be proper, in this part of our enquiry, to consider how far the brief account of creation, contained in the Mosaic narrative, can be shown to accord with those natural phenomena, which will come under consideration in the course of the present essay. Indeed, some examination of this question seems indispensable at the very threshold of an investigation, the subject matter of which will be derived from a series of events, for the most part, long antecedent to the creation of the human species. I trust it may be shown, not only that there is no inconsistency between our interpretation of the phenomena of nature and of the Mosaic narrative, but that the results of geological enquiry throw important light on parts of this history, which are otherwise involved in much obscurity.

If the suggestions I shall venture to propose require some modification of the most commonly received and popular interpretation of the Mosaic narrative, this admission neither involves any impeachment of the authenticity of the text, nor of judgment of those who have formerly interpreted it otherwise, in the absence of information as to facts which have but recently been brought to light; and if, in this respect, geology should seem to require some little concession from the literal interpreter of scripture, it may fairly be held to afford ample compensation for this demand, by the large additions it has made to the evidence of natural religion, in a department where revelation was not designed to give information.

The disappointment of those who look for a detailed account of geological phenomena in the Bible, rests on a gratuitous expectation of finding therein historical information, respecting all the operations of the Creator in times and places with which the human race has no concern; as reasonably might we object that the Mosaic history is imperfect, because it makes no specific mention of the satellites of Jupiter, or the rings of Saturn, as feel disappointment at not finding in it the history of geological phenomena, the details of which may be fit matter for an encyclopedia of science, but are foreign to the objects of a volume intended only to be a guide of religious belief and moral conduct.

[. . .]

Several hypotheses have been proposed, with a view of reconciling the phenomena of Geology, with the brief account of creation which we find in the Mosaic narrative. Some have attempted to ascribe the formation of all the stratified rocks to the effects of the Mosaic Deluge; an opinion which is irreconcilable with the enormous thickness and almost infinite subdivisions of these strata, and with the numerous and regular successions which they contain of the remains of animals and vegetables, differing more and more widely from existing species, as the strata in which we find them are placed at greater depths. The fact that a large proportion of these remains belong to extinct genera, and almost all of them to extinct species, that lived and multiplied and died on or near the spots where they are now found, shows that the strata in which they occur were deposited slowly and gradually, during long periods of time, and at widely distant intervals. These extinct animals and vegetables could therefore have formed no part of the creation with which we are immediately concerned.

It has been supposed by others, that these strata were formed at the bottom of the sea, during the interval between the creation of man and the Mosaic deluge; and that, at the time of that deluge, portions of the globe which had been previously elevated above the level of the sea, and formed the antediluvian continents, were suddenly submerged; while the ancient bed of the ocean rose to supply their place. To this hypothesis also, the facts I shall subsequently advance offer insuperable objections.

A third opinion has been suggested, both by learned theologians and by geologists, and on the grounds independent of one another; viz. that the Days of the Mosaic creation need not be understood to imply the same length of time which is now occupied by a single revolution of the globe; but successive periods, each of great extent: and it has been asserted that the order of succession of the organic remains of a former world, accords with the order of creation recorded in Genesis. This assertion, though to a certain degree apparently correct, is not entirely supported by geological facts; since it appears that the most ancient marine animals occur in the same division of the lowest transition strata with the earliest remains of vegetables; so that the evidence of organic remains, as far as it goes, shows the origin of plants and animals to have been contemporaneous, if any creation of vegetables preceded that of animals, no evidence of such an event has yet been discovered by the researches of geology. Still there is, I believe, no sound critical, or theological, objection, to the interpretation of the word 'day', as meaning a long period; but there will be no necessity for such extension, in order to reconcile the text of Genesis with physical appearances, if it can be shown that the time indicated by the phenomena of Geology[1] may be found in the undefined interval, following the announcement of the first verse.

In my inaugural lecture, published at Oxford, 1820, pp. 31, 32, I have stated my opinion in favour of the hypothesis, 'which supposes the word "beginning", as applied by Moses in the first verse of the book of Genesis, to express an undefined period of time, which was antecedent to the last great change that affected the surface of the earth, and to the creation of its present animal and vegetable inhabitants; during which period a long series of operations and revolutions may have been going on; which as they are wholly unconnected with the history of the human race, are passed over in silence by the sacred historian, whose only concern with them was barely to state, that the matter of the universe is not eternal and self-existent, but was originally created by the power of the Almighty'.

[. . .]

The Mosaic narrative commences with a declaration, that 'In the beginning God created the heaven and the earth'. These few first words of Genesis may be fairly appealed to by the geologist, as containing a brief statement of the creation of the material elements, at a time distinctly preceding the operations of the first day: it is nowhere affirmed that God created the heaven and the earth on the *first day*, but in the *beginning*; this beginning may have been an epoch at an unmeasured distance, followed by periods of undefined duration, during which all the physical operations disclosed by Geology were going on [. . .] No information is given as to events which may have occurred upon this earth, unconnected with the history of man, between the creation of its component matter recorded in the first verse, and the era at which its history is resumed in the second verse; nor is any limit fixed to the time during which these intermediate events may have been going on; millions of millions of years may have been occupied the indefinite interval, between the beginning in which God created the heaven and the earth, and the evening or commencement of the first day of the Mosaic narrative.

The second verse may describe the condition of the earth on the evening of this first day; (for in the Jewish mode of computation used by Moses, each day is reckoned from the beginning of another evening). This first evening may be considered as the termination of the indefinite time which followed the primeval creation in the first verse, and as the commencement of the first of the six succeeding days, in which the earth was to be fitted up, and peopled in a manner fit for the reception of mankind. We have in this second verse, a distinct mention of earth and waters, as already existing, and involved in darkness; their condition also is described as a state of confusion and emptiness (*tohu bohu*), words which are usually interpreted by the vague and indefinite Greek term, 'chaos', and which may be geologically considered as designating the wreck and ruin of a former world.

At this intermediate point of time, the preceding undefined geological periods had terminated, a new series of events commenced, and the work of the first morning of this new creation was the calling forth of light from a temporary darkness, which had overspread the ruins of the ancient earth.[2]

We have further mention of this ancient earth and ancient sea in the ninth verse, in which the waters are commanded to be *gathered together* into one place, and the dry land to *appear*; this dry land being the same earth whose material creation had been announced in the first verse, and whose temporary submersion and temporary darkness are described in the second verse; the *appearance* of the land and the *gathering together* of the waters are the only facts affirmed respecting them in the ninth verse, but neither land nor waters are said to have been *created* on the third day.

[. . .]

It should be recollected that the question is not respecting the correctness of the Mosaic narrative, but of our interpretation of it; and still further, it should be borne in mind that the object of this account was, not to state *in what manner*, but *by whom*, the world was made. As the prevailing tendency of men in those days was to worship the most glorious objects of nature, namely, the sun and moon and stars; it should seem to have been one important point in the Mosaic account of creation, to guard the Israelites against the Polytheism and idolatry of the nations around them; by announcing that all those magnificent celestial bodies were no Gods, but the works of One Almighty Creator, to whom alone the worship of mankind is due.

Notes

1 A very interesting treatise on the Consistency of Geology with Sacred History has recently been published at Newhaven, 1833, by Professor Silliman, as a supplement to an American edition of Bakewell's Geology, 1833. The author contends that the period alluded to in the first verse of Genesis, 'In the beginning', is not necessarily connected with the first day, and that it may be regarded as standing by itself, and admitting of any extension backward in time which the facts may seem to require. He is further disposed to consider the six days of creation as periods of time of indefinite length, and that the word 'day' is not of necessity limited to twenty-four hours.

2 I learn from Professor Pusey that the words '*let there be light*', *yehi or*, Gen. i. 3, by no means necessarily imply, any more than the English words by which they are translated, that light had *never* existed before. They may speak only of the substitution of light for darkness upon the surface of this, our planet: whether light had existed before in other parts of God's creation, or had existed upon this earth, before the darkness described in v. 2, is foreign to the purpose of the narrative.

16

JOHN RUSKIN, *LETTERS ADDRESSED TO A COLLEGE FRIEND 1840–1845*

The Library Edition of John Ruskin's Works, 39 vols, Vol. 1: Early Prose Writings, 1903 [1894]

Letter XVI

January 8, 1843

1. MY DEAR C–, Many thanks for your kind letter and enclosure,[1] which I have read very carefully, and like exceedingly – especially the concluding part of it, which is very graceful and impressive; nor, on the whole, do I think you are at all wrong in taking advantage of the popular notion respecting the Fall, as it is too essential a part of most persons' faith to be lightly struck at, nor unless under very strong convictions of some necessary or important truth which it prevents the reception of. But when you are thinking of the subject yourself, for your own private edification and good, I wish you would tell me what is your notion of a *tree*.

You will most likely have a conception of a thing with leaves on it, and bringing forth flowers in its season. You cannot conceive a tree without leaves and flowers. Now what do you mean by a leaf and flower? You mean by the first, an instrument for depriving carbonic acid of its oxygen, and giving carbon to the plant. You can have no other meaning; for leaves are of all colours, and forms, appearances, and have nothing in common but this – this is the essence of a leaf. You mean by the second, a part of the plant which has in it organs of fructification. You can have no other meaning but this; for flowers have no common form, nor appearance, nor anything essential but this.

Therefore, you mean by the first, something which is perpetually giving to the plant that which it had not before; and by the second, a preparation for the production of another plant. You imply, therefore, growth – change of state – and preparation for a succeeding existence. Therefore, when you say 'a tree', you mean a growing, changing, and preparing thing.

Now it cannot grow for ever, for then there would not be nourishment for its substance. Whatever stops its growth must be a loss of energy in the vital functions – that is, incipient death. When you say a growing thing, therefore, you mean

DOI: 10.4324/9780429355653-19

something advancing to death. Neither can the new tree and the old tree exist together. One must perish to make room for the other. Therefore, every bud and blossom of the parent tree implies and necessitates its destruction.

Therefore, when you say a preparing thing, a fructifying thing, you mean a *dying* thing. Therefore, whenever you speak of a tree, you speak of death. That which has not in it the beginning and germ of death, is not a tree. Consequently, if there were trees in the Garden of Eden there was death; or, if there was not death, they could not have had leaves, nor flowers, nor any of those organs of growth or germination which now constitute the essence of a tree. People will look very grave at you, indeed, if you hint that there were no *flowers* in the Garden, and yet the very meaning of the word flower is – something to supply *death*.

But if you can suppose that Scripture tells you that there were trees in the Garden, and means in saying so something which had neither leaves nor flowers, nor any organs of a tree, you may give up your trust in the whole of it at once; for you can never tell, if there be such latitude of interpretation, what anything *means* throughout the book. Therefore, either Scripture is wholly to be distrusted, as meaning one thing when it says another – or there was death in Eden.

2. Again: what do you understand by the term 'lion'? Surely an animal with claws and sharp teeth. If it have not claws and teeth it is not a lion, it is some other animal – a different animal from any that we have any notion of, but not a lion. But if it have claws and teeth, do you suppose God gave it claws and teeth for nothing? The gift of an instrument supposes the appointment to a function. The claw is to catch with, the teeth are to tear with, and there is a particular juice in the stomach to digest meat with. Now to suppose that these were given without intention of being used, is the same thing as to suppose that your tongue was given to you without your being intended to talk or taste with it, and that it is by corruption of nature that you walk with your legs. A lion at peace with other animals is therefore a contradiction in terms – or at least it is the same thing as saying that God has adapted every muscle to a function which it was never intended to discharge. And though by special miracle the lion shall eat straw as the ox, that does not prove that it was made to eat straw, any more than the miracle of Elisha proves that iron was intended to be lighter than water – which, if it were, the whole economy of the world must be changed.

Hence, if these animals were at peace in Eden, they were either created with especial view to their *after* functions, and maintained for a short time at peace by especial miracle; or else they were different animals – not lions nor tigers, but things of which we have no conception, having different muscles, no claws, no digestive organs for meat, etc., etc. To the first of these positions, the naming by Adam gives the lie direct, for it implies knowledge of their nature; and how could Adam know their nature, when every one of their functions was miraculously suspended? The second position is more possible, partially implied by the speaking of the infant, but yet it supposes a *new creation* at the fall of Adam, which I cannot but think would have been at least indicated in some way or other in Scripture.

Further. By the institution of carnivora, one third more *happiness* is brought into existence. For the earth will only by its appointed constitution feed a certain number of herbivora; and by making them food to a higher series, one more step of existence is gained.

Further. There is not *one* text in Scripture, out of which you can squeeze the slightest evidence that death did not take place with the lower animals.

[. . .]

3. Further. All this evidence coming from the visible, present creation, and Scripture, we have, in addition, geological evidence of death extending for an infinite series of ages before man. Lyell has discovered the bones of the mastodon, the most recent of all fossils, in a bed *cut through* by the ancient course of the Niagara, three hundred feet above its present bed, and three miles and a half below the falls; in cutting back from this point, the river by the very lowest calculation must have been occupied 15,000 years. My own conviction is, therefore – it don't much matter what it is, but I believe it is most peoples who pay any regard whatsoever to modern science – that man in Eden was a growing and perfectible animal; that when perfected he was to have been translated or changed, and to leave the earth to his successors, without pain. In the doom of death he received what before was the lot of lower animals – corruption of the body – and, far worse, death of the soul. I believe the whole creation was in Eden what it is now, only so subjected to man as only to minister to him – never to hurt him. The words 'to dress it and keep it' speak volumes.

Ever yours affectionately, J. RUSKIN

Was there Death before Adam fell, in other parts of Creation?[2]

I. IT is always to be remembered that geologists, and, generally, the asserters of the existence of death previous to the Fall, appeal not to any text of Scripture for *proof* of their assertion – they affirm only that Scripture leaves the matter entirely undecided; and that therefore they are at liberty to follow out the conclusions to which they are led by *other* evidence. Hence, when it is allowed that such and such a text 'can neither prove nor disprove' anything relating to the question, they have all that they contend for.

[. . .]

II. The power of reproduction involves the necessity of death in many ways. First, because God never gave power without necessity for its use. If the trees first created on the earth were to be imperishable, there was no necessity for a power in them of creating others. The world would have been called into existence in perfection at once, as many trees and animals might have been created as would exist in perfection and happiness together, and all the complicated apparatus of fructification dispensed with. God never makes anything more complicated than is necessary, nor bestows a faculty without an object.

Secondly, the light little parenthesis of Miss C–, 'provided there be sufficient nourishment', begs the whole question. The farmer cannot grow wheat twice running in the same field, because *one* crop entirely exhausts the silicate of potash necessary to the existence of the plant. Nor will it grow again until the *death* either of the plant itself (as in straw used for manure), or of some other plant containing the same salt, has restored it to the soil. The sapling pine cannot rise to its full growth, nor, indeed, to *any* growth, until the *death* of its parent has restored to the soil its carbonate of potash. We may imagine a tree maintained for ever in full strength without demand upon the soil; but the moment we hear of its bearing seed, that moment we know that it must perish. Its seed implies that God has willed it to have a successor. Its successor cannot rise but out of its decay.

But it is not merely the death of *plants* which is implied by the *growth* of plants. They require in all cases an element for their growth, nitrogen, which they can only assimilate in one form, *ammonia*; for no chemical means, however powerful, can cause the combination of nitrogen with any other element but oxygen, unless it be presented in the form of ammonia.

It is accordingly found that *no* plants can grow unless supplied with ammonia; and they can be supplied with ammonia in one way only – by *animal* putrefaction. There is no ammonia in the soil; there is none in the decayed remnants of vegetable matter. It exists in the plant only in the crude and unripe juices; in the perfect plant, it exists separately as hydrogen and nitrogen, and cannot be assimilated by its successor. There is, therefore, only one source from which the plant can derive it, the atmosphere; but there is no ammonia in the atmosphere except what results from animal decay. All the nitrogen of animal matter is given off, on its decay, as ammonia. This ammonia combines in the atmosphere with the carbonic acid, which is the result of animal *breath*. The carbonate of ammonia so formed is dissolved in rain water, and presented in this form to the root of the plant.

We, again, require for our nourishment, not ammonia, but the nitrogenised substances, gluten, albumen, etc., of plants. Hence, each species of existence furnishes in its death food to the other, and the nourishment of one implies the simultaneous dying of the other.

Nor is it ammonia alone which the plant takes from the animal. Carbonic acid, also a product of decay, as well as of breath, is its staple nourishment – not more essential than ammonia, but required in far greater quantity. *We* are machines for turning carbon and oxygen into carbonic acid; the plant is a machine for turning carbonic acid into carbon and oxygen. Hence the plant is the supplement of the animal, and the animal of the plant.

Hence a balance must be kept between them; if either exceed its limit, it must perish for want of the other; and the inorganic constituents of the earth are left in a state of perpetual circulation from death to life, and *vice versa*.

Hence, whenever we talk of life, nourishment, or increase, we talk in the same breath of a supplementary death and diminution.

Nor were these laws otherwise in Eden. The green herb was to be for *meat*. This was destruction. Was it *less* destruction because violent and sudden? or did it less imply capability of decay, than if we had been told that the trees died themselves? We might as well say that the death of Abel did not imply capability of death in man.

And, finally, let us suppose for a moment that all these laws of nourishment and creation were suspended, and that there was sufficient matter for assimilation in the soil to supply *all* plants, multiply as they would, and sufficient nitrogen so prepared to nourish all animals, multiply as they would; and suppose death impossible. In two centuries after the creation the earth would have been packed tight with animals, and the only question remaining for determination would have been – which should be *uppermost*. Long before the flood the sea would have been one solid mass of potted fish, the air of wedged birds, and the earth of impenetrable foliage.

And let us not suppose for a moment that geology has opened to us worlds different in organisation or system from our own. It has but expanded before us the vast unity of system, the *one* great plan of progressive existence, of which we form, probably, the last link. The plants of past ages have the same organs, the same structure and development, as those growing now; none but the practised botanist can tell the leaves from each other. The animals played precisely the same part in relation to them; their organisation was the same as now, their ranks of destructive existence appointed in the same order. A few extraordinary (to us) creatures existed, peculiarly adapted for certain circumstances, but in no essential points, in nothing but outward form and strength, differing from their modern types. The digestion of the Ichthyosaurus is as regular and simple as that of any living aquatic beast of prey, and far more easily traceable. Even size is no unfailing characteristic. No fossil fish has been discovered fit to hold a candle to our modern sharks or whales, though the shark tribe was infinitely more numerous than it is now; but there were too many, and they kept each other thin. It is a curious fact, by-the-bye, though well known, respecting the beneficial influence of the carnivora even on the animals they prey upon, that if you stock a fish-pond with carp only, at the end of a year or two you will find all your fish miserably thin, and have no more weight of fish (if you drag the pond) than you put in. But if at first you put in with the carp a few pike, say one in four, you will, when you drag your pond, have twice the weight of carp, in good condition, and all your pike into the bargain.

I see that Miss C– objects that the growth of plants is not sufficient for animals as it is. Locally, it is not. Universally, it is far too great for them. Our farmers may raise the price of corn over a county, but the Great Forest stretches its uninhabitable growth over America, for the space of a thousand kingdoms. And even where vegetation is limited, this is simply because the plants are not fed by their own *death*; for though they have the animal *volatile* products of ammonia, etc., they have *not* the fixed salts except when they are laboriously restored in the form of manure.

With respect to the question respecting the naming of fish, I can only reply in the words of the questioner, that all such speculations lead us only into a labyrinth.

There are thousands of difficulties connected with the Mosaic account. What, for instance, does *Eden* include? For the *garden* was *in* Eden, and eastward in it. And was man, supposing he had stood, never to have left his primal and narrow nursery-and-seedsman sort of habitation? How, if so, could he 'replenish the earth and subdue it'?

The whole appears to me, but for the close geographical account of the Garden, very much like an Eastern allegory: was the same trial to be sustained by all? And how could it be sustained, unless gardens and trees of knowledge were multiplied over the earth as the population spread? etc., etc.[3] but however that may be, I think it is better always to read it without reference to matters of physical enquiry, to take the broad, simple statements of creation – innocence, disobedience, and guilt – and then to take in equal simplicity of heart such revelations as God may deign to give us of His former creations, and so to pass back through age before age of preparatory economy, without troubling ourselves about the little discrepancies which may appear to start up in things and statements which we cannot understand.

Creation may have been suspended in its functions for a moment – for the half-hour (divines seem to think it was little more) of man's probation. It matters not to us. What we are we know – and what we may be, we know; what we have been, God knows.

There is much of mystical in Scripture, which, doubtless, will one day be made manifest; but we do but waste our lives and peril our faith by trying to unravel it before its time. We shall not break the seal by dashing it against stones.

I have said, I see, that *no* ammonia exists in the atmosphere but what arises from the putrefaction of animals. This is not strictly true, for several mineral springs supply it in considerable quantity; not enough, however, in all the springs of Europe, to feed the vegetation of Lombardy for half a year.

Supplies of this kind are probably proportioned to the gradual increase of animal life, and consequent demand for more nitrogen. The immediate acting supply is deduced only from animal corruption. From every churchyard, from every perishing remnant of the life of the forest and the sea, rises the constant supply of carbonate of ammonia, which feeds the green leafage of spring, and expands the pulp of the bright fruit.

Liebig says that the source of this ammonia is sufficiently evident by its peculiar odour, if rain water be evaporated with a little sulphuric acid, and then tested with lime.

On the other hand, while the supply of ammonia is gradually, very slightly, but still certainly on the increase, that of carbonic acid is much diminished. Immense quantities of this acid existed formerly in the atmosphere, which fed the colossal vegetation of geological eras. By that vegetation it was gradually withdrawn; and, animal life not being sufficiently extended on the earth to feed on this vegetation, and so return the carbonic acid to the atmosphere, it was withdrawn for ever; its oxygen was restored by ordinary vegetable action, making the atmosphere purer for the abode of man, and its carbon deposited in the enormous coal-fields, which are now the source of all his vastest powers. Animal and vegetable life are now better balanced. The vegetable, having no extraordinary supply of carbonic acid, is diminished in growth; and the animal feeding on the air, and provides for the equal growth of the carbon to the air, and provides for the equal growth of the succeeding plant.

Notes

1 Editor's footnote: A sermon, no doubt, in which the questions discussed in this letter and in the essay following it were involved. (This extract is taken from volume 1 of Cook and Wedderburn's *Library Edition of John Ruskin's Works* (1903–12) and uses some of the editor's footnotes, which are prefaced with the initials CW).

2 CW: In the ed. of 1894 this essay preceded Letter xvi. It is clear, however, that it followed the Letter, being by way of rejoinder to criticisms made by 'C.' and his sister upon Ruskin's arguments in the Letter.

3 CW: Ruskin, it will be seen, had thought out for himself conclusions very similar to those which made so much stir when published twenty years later in *Essays and Reviews* (1860). His intercourse with Buckland would, however, have familiarised him with such speculations; passages from Buckland's *Bridgewater Treatises*, cited in the essay on 'Mosaic Cosmogony' in *Essays and Reviews*, adopt the same standpoint as Ruskin's. Later criticism has tended to interpret the story of Eden less as an allegory, than as a mythical tradition such as is found in the early sacred literature of other nations.

17

EDWARD HITCHCOCK, *THE RELIGION OF GEOLOGY AND ITS CONNECTED SCIENCES*

(Boston: Phillips, Samson, and Co., 1854 [1851])

TO MY BELOVED WIFE

Both gratitude and affection prompt me to dedicate these lectures to you. To your kindness and self-denying labors I have been mainly indebted for the ability and leisure to give any successful attention to scientific pursuits. Early should I have sunk under the pressure of feeble health, nervous despondency, poverty, and blighted hopes, had not your sympathies and cheering counsels sustained me. And during the last thirty years of professional labors, how little could I have done in the cause of science, had you not, in a great measure, relieved me of the cares of a numerous family! Furthermore, while I have described scientific facts with the pen only, how much more vividly have they been portrayed by your pencil! And it is peculiarly appropriate that your name should be associated with mine in any literary effort where the theme is geology; since your artistic skill has done more than my voice to render that science attractive to the young men whom I have instructed. I love especially to connect your name with an effort to defend and illustrate that religion which I am sure is dearer to you than every thing else. I know that you would forbid this public allusion to your labors and sacrifices, did I not send it forth to the world before it meets your eye. But I am unwilling to lose this opportunity of bearing a testimony which both justice and affection urge me to give. In a world where much is said of female deception and inconstancy, I desire to testify that one man at least has placed implicit confidence in woman, and has not been disappointed. Through many checkered scenes have we passed together, both on the land and the sea, at home and in foreign countries; and now the voyage of life is almost ended. The ties of earthly affection, which have so long united us in uninterrupted harmony and happiness, will soon be sundered. But there are ties which death cannot break; and we indulge the hope that by them we shall be linked together and to the throne of God through eternal ages.

DOI: 10.4324/9780429355653-20

In life and in death I abide

Your affectionate husband,

EDWARD HITCHCOCK.

Preface

[. . .] AFTER acknowledging that more than a quarter of a century has elapsed since this subject first engaged my attention, it may be useless for me to ask any indulgence from criticism. But really, I feel less prepared to write upon it than I did during the first five years in which I studied it. I have learnt that it is a most difficult subject. It requires, in order to master it, an acquaintance with three distinct branches of knowledge, not apt to go together. First, an acquaintance with geology in all its details, and with the general principles of zoology, botany, and comparative anatomy; secondly, a knowledge of sacred hermeneutics, or the principles of interpreting the Scriptures; thirdly, a clear conception of the principles of natural and revealed religion.

As examples of efforts made by men who were deficient in a knowledge of some of these branches, I am compelled to quote a large proportion of the works which, within the last thirty or forty years, have been written on the religion of geology; especially on its connection with revealed religion. I am happy to except such writers as Dr J. Pye Smith, Dr Chalmers, Dr Harris, Dr Buckland, Professor Sedgwick, Professor Whewell, Dr King, Dr Anderson, and Hugh Miller; for they, to a greater or less extent, acquainted themselves with all the subjects named above, before they undertook to write. But a still larger number of authors, although men of talents, and familiar, it may be, with the Bible and theology, had no accurate knowledge of geology. The results have been, first, that, by resorting to denunciation and charges of infidelity, to answer arguments from geology which they did not understand, they have excited unreasonable prejudices and alarm among common Christians respecting that science and its cultivators; secondly, they have awakened disgust, and even contempt, among scientific men, especially those of sceptical tendencies, who have inferred that a cause which resorts to such defences must be very weak. They have felt very much as a good Greek scholar would, who should read a severe critique upon the style of Isocrates, or Demosthenes, and, before he had finished the review, should discover internal evidence that the writer had never learnt the Greek alphabet.

On the other hand, prejudices and disgust equally strong have been produced in the mind of many a man well versed in theology and biblical exegesis by some productions of scientific men upon the religious bearings of geology, because they advanced principles which the merest tyro in divinity would know to be false and fatal to religion, and which they advocated only because they had never studied the Bible or theology.

And here I would remark that it does not follow, because a man is eminent in geology, that his opinion is of any value upon the religion of geology. For the two subjects are quite distinct, and a man may be a Coryphæus in the principles of geology, who is an ignoramus in its religious applications. Indeed, many of the ablest writers upon geology take the ground that its religious bearings do not belong to the science.

These statements, instead of pleading my apology for the following work, may only show my temerity and vanity. Nevertheless, they afford me an opportunity of calling the attention of the religious public to the great inadequacy of the means now possessed of acquiring a knowledge of the different branches of natural science. I refer especially to comparative anatomy, zoology, botany, and geology, in our literary and theological seminaries. The latter, so far as I know, do not pretend to give any instruction in these branches. And in our colleges that instruction is confined almost entirely to a few brief courses of lectures; often so few that the students scarcely find out how ignorant they are of the subjects; and hence those who are expecting to enter the sacred ministry vainly imagine that, at almost any period of their future course, they can, in a few weeks, become sufficiently acquainted with physical science to meet and refute the sceptic. In all our seminaries, however, abundant provision is made, as it ought to be, for the study of intellectual philosophy and biblical interpretation.

So well satisfied are two of the most enlightened and efficient Christian denominations in Great Britain – the Congregationalists and the Scottish Free Church – of the need of more extensive acquaintance with the natural sciences in ministers of the gospel, that they have attached a professorship of natural history to their theological seminaries. That in the New College in Edinburgh is filled by the venerable Dr Fleming; that in the New College in London by Dr Lankester. From a syllabus of Dr Fleming's course of lectures, which he put into my hands last summer, I perceive that it differs little from the instruction in natural science in the colleges of our country. This being the case, it strikes me that this is not exactly the professorship that is needed in the theological seminaries of our country. But they do need, it seems to me, professorships of natural theology, to be filled by men who are practically familiar with the natural sciences. If any such chairs exist in these seminaries, I do not know it. They are amply provided with instruction in the metaphysics of theology, hermeneutics, and ecclesiastical history; and I should be sorry to see these departments less amply provided for. But here is the wide field of natural theology, large enough for several professorships, which finds no place, save a nook in the chair of dogmatics. This might have answered well enough when the battle-field with scepticism lay in the region of metaphysics, or history, or biblical interpretation. But the enemy have, within a few years past, intrenched themselves within the dominions of natural science; and there, for a long time to come, must be the tug of the war. And since they have substituted skeletons, and trees, and stones, as weapons, in the place of abstractions, so must Christians do, if they would not be defeated. Let me refer to a few examples to show how inadequately furnished the minister must be for such a contest, who has used only the means of instruction provided in our existing seminaries, literary and theological.

Take the leading points discussed in the following lectures. How can a man who has heard only a brief and hurried course of thirty lectures on chemistry, twenty on anatomy and physiology, fifteen upon zoology, ten upon botany, ten upon mineralogy, and twenty upon geology, at the college, with no additional instruction at the theological seminary, – how can he judge correctly of points and reasoning difficult to be mastered by adepts in these sciences? How certain to be worsted in an argument with an accomplished naturalist who is a sceptic!

Suppose the sceptic takes the ground advocated by Oken and the author of the 'Vestiges'. Let the clergyman, whom I have supposed, read the works of Miller and Sedgwick in reply to the development hypothesis, and see whether he can even understand their arguments without a more careful study of the sciences on which they rest.

[. . .]

Lecture I. Revelation Illustrated by Science

THE leading object, which I propose in the course of lectures which I now commence, is to develop the relations between geology and religion. This cannot be done fully and fairly, however, without exhibiting also many of the religious bearings of several other sciences. I shall, therefore, feel justified in drawing illustrations and arguments from any department of human knowledge which may afford them. I place geology first and most conspicuous on the list, because I know of no other branch of physical science so prolific in its religious applications.

In treating of this subject, I shall first exhibit the relations between science and revealed religion, and afterwards between science and natural religion; though in a few cases these two great branches cannot be kept entirely distinct.

Geology is usually regarded as having only an unfavorable bearing upon revealed religion; and writers are generally satisfied if they can reconcile apparent discrepancies. But I regard this as an unfair representation; for if geology, or any other science, proves to us that we have not fairly understood the meaning of any passage of Scripture, it merely illustrates, but does not oppose, revelation.

A fundamental principle of Protestant Christianity is, that the Scriptures of the Old and New Testaments are the only infallible standard of religious truth; and I desire to hold up this principle prominently at the outset, as one to which I cordially subscribe. The mass of evidence in favor of the divine inspiration of the Bible is too great to be set aside by any thing short of scientific demonstration. Were the Scriptures to teach that the whole is not equal to its parts, the mind could not, indeed, believe it. But if it taught a truth which was only contrary to the probable deductions of science, science, I say, must yield to Scripture; for it would be more reasonable to doubt the probabilities of a single science, than the various and most satisfactory evidence on which revelation rests. I do not believe that even the probabilities of any science are in collision with Scripture. But the supposition is made to show how strong are my convictions of the evidence and paramount authority of the Bible.

But does it follow, from these positions, that science can throw no light upon the truths of Scripture? By no means; and it will be my leading object, in this lecture, to show how this may be done by science in general, and by geology in particular.

In discussing this subject, we ought to bear in mind the object of science, and the object of revelation. And by the term science I refer mainly to physical science. Its grand aim is, by an induction from facts, to discover the laws by which the material universe is governed. Those laws do, indeed, lead the mind almost necessarily to their divine Author. But this is rather the incidental than the direct result of scientific investigations, and belongs rather to natural theology than to natural science.

On the other hand, the exclusive object of revelation is of a moral character. It is a development of the divine character and the divine government; especially that part of it which discloses a plan for the reconciliation of a lost and wicked world to the favor of God by the death of his Son. Every other subject mentioned in Scripture is incidental, and would not have been noticed had it not some connection with the plan of salvation. The creation of the world and the Noachian deluge, for instance, are intimately related to the divine character and government, and therefore they are described; and the same is true of the various phenomena of nature which are touched upon in the Bible.

If these positions be correct, it follows, that as we ought not to expect to find the doctrines of religion in treatises on science, so it is unreasonable to look for the principles of philosophy in the Bible. Nay, we ought not to expect to find the terms used by the Sacred writers employed in their strict scientific sense, but in their popular acceptation. Indeed, as the Scriptures were generally addressed to men in the earliest and most simple states of society, with very limited views of the extent of creation, we ought to suppose that, in all cases where no new fact is revealed, the language was adapted to the narrow ideas which then prevailed. When, for instance, the sacred writers speak of the rising and setting of the sun, we cannot suppose they used language with astronomical correctness, but only according to appearances. Hence we ought not to be very confident, that when they employ the term *earth*, they meant that spherical, vast globe which astronomy proves the earth to be, but rather that part of it which was inhabited, which was all the idea that entered into the mind of a Jew. God might, indeed, have revealed new scientific as well as religious truth. But there is no evidence that in this way he has anticipated a single modern discovery. This would have been turning aside from the much more important object he had in view, viz., to teach the world religious truth. Such being the case, the language employed to describe natural phenomena must have been adapted to the state of knowledge among the people to whom the Scriptures were addressed.

Another inference from these premises is, that there may be an apparent contradiction between the statements of science and revelation. Revelation may describe phenomena according to apparent truth, as when it speaks of the rising and setting of the sun, and the immobility of the earth; but science describes the same according to the actual truth, as when it gives a real motion to the earth, and only an apparent motion to the heavens. Had the language of revelation been scientifically accurate, it would have defeated the object for which the Scriptures were given; for it must have anticipated scientific discovery, and therefore have been unintelligible to those ignorant of such discoveries. Or if these had been explained by inspiration, the Bible would have become a text-book in natural science, rather than a guide to eternal life.

The final conclusion from these principles is, that since science and revelation treat of the same subjects only incidentally, we ought only to expect that the facts of science, rightly understood, should not contradict the statements of revelation, correctly interpreted. Apparent discrepancies there may be; and it would not be strange, if for a time they should seem to be real; either because science has not fully and accurately disclosed the facts, or the Bible is not correctly interpreted; but if both records are from God, there can be no real contradiction between them. But, on the other hand, we have no reason to expect any remarkable coincidences, because the general subject and object of the two records are so unlike. Should such coincidences occur, however, they will render it less probable that any apparent disagreement is real.

If the positions taken in these preliminary remarks be correct, it will follow, that in judging of the agreement or disagreement between revelation and science, it is important, in the first place, that we rightly understand the Bible; and, in the second place, that we carefully ascertain what are the settled and demonstrated principles of science.

18

THOMAS EWBANK, *THE WORLD A WORKSHOP; OR, THE PHYSICAL RELATIONSHIP OF MAN TO THE EARTH*

(New York: D. Appleton and Company, 1855)

Preface

I am not aware of a single sentiment in the following pages to which the most devout mind can justly except, nor of a thought that is not in harmony with the deepest tone of admiration for the works of God, and with the purest feelings of love and reverence for Him; yet so it is, that kindred subjects are seldom brought forward without awakening opposition in persons who imagine the Ark of Truth endangered by the enunciation of speculations and deductions in science not included in their creeds, and who, on such occasions, eagerly put forth their hands to uphold it – as if it could be shaken or overthrown by error.

Truth, or rather the knowledge of it, is progressive. In nature there can be no end to its disclosures, for nothing is concealed. Upon every object, from an insect to a world, is written the purpose for which it is made. We may not always read aright, and no wonder, since we live in the infancy of systematic inquiry, and therefore cannot anticipate the results of its maturity; but our errors will be corrected by our successors, and theirs by those who succeed them.

That this mundane habitation was designed and literally fitted up for the cultivation and application of chemical and mechanical science as the basis of human development, will, I think, appear evident even from the imperfect examination here given it; and that it is essentially the same with other worlds, according to the condition of matter in them, and the physical constitution of their inhabitants, is all but an inevitable conclusion. To those who deny them to be centres of reasoning and active populations it is useless to reply till they can show for what other purposes they were made, and how this little earth, a mere atom among them, became so strange an exception. If we had had no knowledge of the existence of other orbs, it would have been unphilosophical to insist there were none besides our own; but now that we know they crowd every region of space, it would be positive folly to contend that all are barren of life and intelligence, of science and arts, except the one given to us.

DOI: 10.4324/9780429355653-21

It is preposterous to suppose the Divine Builder erects tenements for the purpose of keeping them empty. If they are not occupied, it is because they are not yet prepared to be so. It may be assumed that as soon as an orb is fitted for tenants they are put in possession of it; and then it is that another marvel is disclosed. Material natures require something to do as well as to reflect on; this is indispensable to their being – the purpose of it. Employment is, therefore, an element of existence, and hence, the industrial activities of the denizens of the universe; involving, as they must, infinities of modes and processes, and multiplied infinities of applications and results. The means by which this diversity is brought out might, on a first thought, be deemed inscrutable and incomprehensible, yet, like the effects of gravitation or of any universal law, it is very simply evolved. It depends on the diverse conditions of matter and the circumstances under which it exists, and as these cannot possibly be the same in any two worlds, much less in any two systems, neither can the occupations of those employed on it. These are, therefore, endless in numbers, because endless are the truths of which matter is the vehicle, and the applications of which it is capable.

[. . .]

Creation is not a medley of mingled purposes and disconnected things. The unity of design manifested in it is the theme of every philosopher, and not less observable and admirable is the fine chain of relationship that binds all the diverse forms and conditions of matter in one coherent whole. There are no violent transitions from series to series, but by almost imperceptible degrees differences open into species, species into genera, and genera into wider classes. And as with the contents of worlds so with worlds themselves; for they are merely larger divisions, and not the largest, since they merge into groups or systems, and systems, in all probability, into still more and more comprehensive departments. They are as intimately related to one another as are minerals, animals, or plants; and though we are not permitted to observe the alliance in their internal details, it is proclaimed externally to the utmost bounds of the heavens. There are no abrupt chasms in their outlines, dimensions, illumination, or movements, and by the strongest of analogies there can be none in their internal administrations.

In the latter respect the chain can be no more ruptured than in the former. The absence of the smallest link would break the continuity of the whole, and introduce disorder. On a matter so momentous, so overpowering in magnitude, as the interior economy of worlds, it would seem impossible that our orb should be the only one on which practical science is cultivated. There cannot be so wide a gap. There must be others more or less closely allied to it in this as in other respects; some in which the arts are prosecuted with higher and some with lower results. No truth is more patent than the unity of creation. There is nothing *sui generis* in it; nothing that stands solitary or alone – nothing that is not connected with and dependent on something else – not a boulder, a planet, or a sun, not an animal or the habits of one – not an order of intelligences or an occupation of intelligence.

[. . .]

But it is objected that physical industry and ingenuity are of too low and ephemeral a nature to enter into the sublime and everlasting plans of the Author of the Universe; that cultivation of mind must be the object of calling it into existence. True; but as matter is the agent on which God has printed his thoughts, may it not be the book which all minds are to read and to learn from? We know that he has made the elevation of human nature to depend on the study and application of principles impressed upon matter, and therefore it is consistent with his purposes and with his greatness to educate intelligences by means of it. And if one class why not two, or ten, or all? We know not that any are, or can be trained up without it; and as, wherever intelligences are, they are surrounded by it, and by displays of Divine wisdom shining forth in it, is it not reasonable to infer that it is a universal medium of mental and moral tuition for which purpose, instead of being collected into one inhabitable body, it has been gathered into an infinite number, every one different, and a theatre of different phenomena.

There must be something wrong in the general dislike to material labor, and to the association of it with other orbs. Few persons, lay or religious, feel comfortable about it, because they are not impressed with the cardinal truth that matter is the agent by which God everywhere proclaims himself; and that in it are sources of knowledge sufficient to exercise all orders of intelligences for ever.

[. . .]

It is to our own star that these pages are devoted. From careering among other worlds let us alight upon it, and scan from a stand-point that has seldom been occupied, the ceaseless labors of its living swarms. All are workers in and modifiers of matter. To man in common with the rest a task is given which, if fully understood, would place in a new and a better light this much abused orb of ours. It is the opinion of many that decay has seized its vitals, that its resources are approaching exhaustion, and the arts their climax; whereas in reality it is a spring of physical truths which man can never run dry. To suppose their current manifestations final is wrong, for before half of them can be found out and made use of new developments will have opened new tributaries. Chemistry and Physics, as the exponents of inorganic bodies, and Botany and Zoology of the organic, will pour, and continue to pour forth new elements, combinations, forms, forces, and motions. We have had pleasing illustrations of this in our day which, so far from inducing fear of the font failing, are prophetic of its fulness.

WASHINGTON, August, 1854

Section III

Chapter III, The Universe of Matter and of Mechanism

It is marvellous that any created being should be able to study its own organization, and reason on the causes and modes of its existence; that man, a piece of animated matter, should pry into his own structure, and, by dissecting the bodies of his fellows, find out the reasons that determined the forms and proportions of his own organs; and that he should then turn from himself and inquire into the nature and attributes of the Author of his being! The wonder is not greater than if balls of clay in the hands of a potter should ask 'What doest thou?' or if spinning-jennies and power-looms were to pause in their movements to inquire why they were made. Man is a tissue of marvels; his little seething brain, as if a part of the Godhead were located in it, spurns at boundaries to his thoughts. He neither confines them to the world he occupies nor to the visible heavens, but urges them through the invisible depths of space to learn, if possible, what is doing there. Nor is this all: not content with employing them on things of the present, he sends them into the future, and exercises them on the past. He is told, that in the beginning God created the heavens and the earth; but he longs to know how they were produced – by what principles and processes they were developed and are sustained.

That this amazing faculty is given for the great purposes of his education, it were a truism to assert; more than anything else it shows how illimitable are the soul's aspirations. As for imaginings of what was before the sidereal heavens appeared, they can hardly be carried further than a supposed condition of things, which may be illustrated by a discovery made some years ago of a subterranean structure, of unknown origin and antiquity. The proprietor entered with a light; his voice reverberated along the arches, and the dark and silent chambers were instantly charged with clouds of dancing atoms awakened into motion by his presence. So, it has been suggested, was the cold and boundless abyss first charged by the voice of God with the dust of which stars are made.

Wherever matter is, it proclaims the Builder of the universe, and discloses more or less of his plans. But why was it produced at all, since he requires for himself no visible or tangible manifestations of his ideas, nor any such means to record or peruse them? It arose then for some other purpose, and what was that? Surely, that it might be an arena for sensitive beings. From pure beneficence creation came; beneficence, whose essence is expansion; which ceases to be when it ceases to dilate, which cannot be confined in the breast even of the Deity. Through it the universe is working out the divinest of problems – the diffusion of the largest amount and diversity of enjoyment among the greatest variety of recipients. Hence, every form that matter here puts on, every condition and change of condition in it, is made an occasion to introduce fresh beings to disport themselves in the varying media. In this view, how admirable is the principle of limiting life, that legions enjoying it may be multiplied till they equal or surpass the atoms of which worlds are composed! Every particle of matter is the agent and symbol of enjoyment.

[. . .]

What the future destiny of the myriads of floating worlds and their populations is to be, no one can tell, nor why the Almighty adopted the existing plan of the universe. It is enough for us to know that he created matter, and everywhere appears as The Great Artificer in it; that he has ordained an essential intimacy between it and finite intelligences; commences their being with it; makes it the medium for the development, expansion, and employment of thought, by transfusing his wisdom through it, and making it their duty to seek that wisdom out – i.e. to discover the laws impressed on it, and by acting in accordance with them, to evolve from it all sciences and arts.

[. . .]

To conclude: Two lessons – one addressed to nations, the other to classes and individuals, may be learned from the preceding pages. First, that the stone of offence against which ancient nations stumbled and fell, was the rock upon which their prosperity and perpetuity should have been based. [. . .] The second is a corollary of the first; but enough has been advanced to induce artificers to hold their professions inferior to none, and to urge them to excel in them, remembering that in no character does the Creator so prominently, constantly, and universally appear as in that of THE MECHANICIAN. And if hereafter the parable of the talents is to be literally fulfilled, enlightened elaborators will be counted among the most profitable of servants, because of their activity in discovering and applying for the good of their kind, the great productive agencies located in the orbs on which they dwelt.

Part 3

GEOLOGY

Geology

Because of its wide-ranging significance to changing conceptualisations of environment and the directions science took in the first half of the nineteenth century, geology takes up much space in this volume. Interest in 'earth sciences' emerged in the Classical period and saw a renewed flowering in the Renaissance (with Andrea Cesalpino and Robert Hooke particularly important). Those wishing to explore geology's disciplinary 'prehistory' in more detail could do no better than starting with Chapter 3, Volume 1, of Charles Lyell's *Principles of Geology* (1830). Geology was consolidated as a discipline in the eighteenth century, when Edinburgh, Paris, and Freiburg were important centres, and the establishment of the Geological Society of London in 1807 created what would become the single most important network.

As we began to show in Part 2 of this volume, much of geology's significance to nineteenth-century science lay in the ways in which its analysis of earth history, via strata and fossils, led to collisions with a Biblical account of the earth that posited the earth's origins at around 6000 BC. The longer timespan opened up a chilly vision of 'deep time' – countless millennia of geological history from which humankind was seemingly absent and irrelevant. Geology also created better understanding of the earth as an enormous, dynamic system, the complexity and power of which dwarfed humanity. Geology thus participated in a number of ways in gradually reconceptualising the role and status of *Homo sapiens* in the world. By focusing on the fossilised remains of plants and animals, geology also meant investigating. the ways in which organisms interact with their environments; the manner in which extinctions occur; and, most provocatively, how new species come into existence. Without nineteenth-century geology, evolutionary theory would have been impossible, but it also influenced biogeography and ecology and a broader challenge to the anthropocentric visions of 'man and nature' underpinning Christianity. That geological ideas and discoveries creep into the sections in this volume on precursors, Natural Theology, comparative anatomy, evolutionary theory, and geography indicate the scope of its broader influence.

This section begins with James Hutton, the perceptive and inventive Edinburgh scientist whose conclusions about earth history set the pattern for geology's subsequent direction:

> The result, therefore, of our present enquiry is, that we find no vestige of a beginning. – no prospect of an end.

While elements of his work are anticipated by Andrea Cesalpino and Robert Hooke, and although Georges Buffon, Baron Cuvier, and Abraham Werner made substantial contributions to eighteenth-century geology, Hutton is justifiably regarded as its most significant founder. Amongst the first to address the subject in detail and to forward sustainable theories about the formation of the earth, he was pioneering in imagining a much-expanded conception of geological time, establishing uniformitarian ideas subsequently developed by James Playfair and Charles Lyell.

Two extracts from his work are included. The first, 'Concerning the System of the Earth, Its Duration, and Stability' (1785), is a précis of his address to the Royal Society of Edinburgh that year and is often referred to as Hutton's 'Abstract'. Expanded and revised, it became *Theory of the Earth* (1788) (our second extract). In the 'Abstract', Hutton establishes principles behind the formation of sedimentary rocks and suggests mechanisms by which orogeny and heat generate powerful processes of uplift capable of transporting deep ocean sedimentary rocks to terrestrial locations. It also addresses the issue of the time required for these processes. In Natural Theological mode, he claims to reveal God's design for creation, but his postulation of an expanded timescale for earth history set the infant science on a collision course with Biblical scholarship.

Theory of the Earth delves deeper into all of these matters, doubling down on earlier claims and drawing on a wider range of evidence. It opens with a Natural Theological, but also mechanistic, reading of the physical world. Confident that there need be no tension between a religious worldview and rational enquiry, he proposes a new arena of science, 'a subject on which hitherto opinion only, and not evidence, has decided'. Hutton initially describes the globe as a 'machine', but it is increasingly described as dynamic: he emphasises the importance of erosion in the creation of soils and the ways in which mountains are raised and denuded over time. The mechanistic view is also complicated in a counter-reading in which the earth is an 'organized body' capable of self-regulation in order to maintain overall harmony between formation and erosion.

In Part III, and following Hooke (see Part 1 of this volume), Hutton argues that the raising of sedimentary strata occurred as a result of heat from below and that this caused the fracturing, tilting, and displacement of strata into the disordered forms in which they are currently found. He posits a circular account of formation and erosion, in which continental lands ('a succession of worlds') are formed, raised, and then destroyed, to be replaced in their turn by new ones. Although subsequent geologists added to, adapted, or corrected Hutton's basic theory, particularly in understanding mechanisms of folding, faulting, tectonic movements, orogeny, and metamorphic action, and although he overestimates the proportion of the earth's strata that were formed in deep marine environments, the basic rudiments are breathtakingly prescient. It is impossible within the narrow confines of the extracts to give an adequate impression of the range and complexity of Hutton's evidence, reasoning, and insight, but Lyell's estimation is helpful:

> His application was unwearied, and he made frequent tours through different parts of England and Scotland, acquiring considerable skill as a mineralogist, and constantly arriving at grand and comprehensive views in geology. He communicated the results of his observations unreservedly, and with the fearless Spirit of one who was conscious that love of

> truth was the sole stimulus of all his exertions. When at length he had matured his views, he published, in 1788, his 'Theory of the Earth'.
>
> (*Principles Of Geology*, vol. 1, 61–2)

As we shall see, Hutton's work was crucial to Lyell's, itself probably the most important intervention in nineteenth-century geology.

The extract from Part III of the *Theory* builds on Part II (not included here), which is taken up with debates and theories concerning the ways in which sedimentary rocks are gradually formed either by aqueous solution or by heat and pressure, with Hutton inclining to see the latter as predominant. In the pages immediately preceding the extract from Part IV, Hutton, unable to find a means of demonstrating his claims about the slow rate of coastal erosion from geological evidence, turns to the writings of Classical authors.

The third extract is from Baron Georges Cuvier's *Essay on the Theory of the Earth* (1813). Primarily known as a key figure in comparative anatomy (see Part 4 of this volume), Cuvier brings his anatomical genius to palaeontology here. His work anticipates William Smith's (see extract later in this section) in seeing the identification of fossil species as a key process in dating strata. He also broadly follows Hutton in believing that strata at the bottom of oceans have been uplifted. Cuvier, however, forwards a catastrophist reading, in which the earth's strata can be explained by a series of brief, enormous upheavals. In this sense, he goes against the Huttonian (uniformitarian) direction geology would ultimately take: the extract includes his arguments for the suddenness of uplifted strata and his attempts (many of which rest on erroneous conclusions) to make his findings conformable with *Genesis*. Cuvier's 'Preliminary Remarks', implying that he might become for geology and anatomy what Sir Isaac Newton was for physics and astronomy, attest to his self-belief. His geology, he hoped, would prove so definitive as to represent 'an epoch in science'. While his Natural Theology would be gradually repudiated, Cuvier's importance to the establishment of palaeontology was considerable, and his claims to an elevated place in the annals of comparative anatomy have certainly been vindicated.

Our fourth extract is as interesting for what it conceals as for what it contains. Sir Everard Home's 'Some Account of the fossil Remains of an Animal more nearly allied to Fishes than any of the other Classes of Animals' appeared in the 1814 issue of *Philosophical Transactions of the Royal Society*, after Home's address to the Society that year. It is the earliest account of the extinct creature that would come to be known (and much celebrated) as *Ichthyosaurus*. While Home points out that 'this specimen was found upon an estate of Henry Host Henley, Esq. between Lyme and Charmouth, in Dorsetshire', no part of his discussion of its removal, analysis, and anatomy acknowledges its discovery by the brilliant teenage palaeontologist Mary Anning in 1812–13 (assisted by her brother, Joseph). It was the first of several complete skeletons that she would discover. In the sixth extract in this section, Reverend Conybeare's 'On the Discovery of an almost perfect Skeleton of the Plesiosaurus' (1823), Anning is again uncredited for her discovery, Conybeare referring only to 'the magnificent specimen recently discovered at Lyme' and lauding the work of a Mr Bullock in examining the bones at the Museum of Natural History. Although the first partial skeleton finds of plesiosaurs date back to 1605, Mary Anning excavated the first complete skeleton in 1823.

While Anning was much in demand by professional geologists and forged a financially successful career, the exclusively male scientific elites of the period never gave her full credit for her remarkable abilities and ground-breaking discoveries. The names of Home and Conybeare appear in the first accounts of the two Jurassic marine creatures, while the eleventh extract of this section, Mary Anning's letter to the *Magazine of Natural History* (1839), is the only example of her writing in print: incredibly brief, it meets a condescending and somewhat dismissive reply from the magazine's editor. Anning refers to a number of species in her brief letter: *Hybodus*, first described by Louis Agassiz, was a genus of Chondrichthyans of the Permian and Cretaceous periods. As well as having the two types of teeth described by Anning, they also featured a distinctive bony blade on their dorsal fin. The editor suggests that she in fact meant *Acrodus,* another extinct genus of cartilaginous fish from the Permian to Palaeocene periods from the Order Hybodontiformes. Despite the supercilious response to her letter, others recognised Anning's value: *Acrodus anningiae* was named by Louis Agassiz in honour of her pioneering contributions to palaeontology.

Taken together, these three extracts are illustrative of the generally marginal position of women in nineteenth-century science: at this point, and even later in the century, as the brilliant but unrecognised mycologist Beatrix Potter would discover, women were 'permitted' to become collectors and illustrators and to popularise science. As we shall see in Volume II, women also often created accounts of science mediated for women and children. To scale the bastions of professional science, and to achieve membership of the Royal Society, the Linnean, or other leading scientific organisations, was another matter altogether.

The extracts from Home and Conybeare are also of interest for what they reveal of the intimacy of geology and comparative anatomy – a repeated theme in forthcoming extracts. Insights such as Conybeare's about the anatomical significance of the shared number of neck vertebrae in all mammal species would eventually feed evolutionary theories about shared origins and stages of evolutionary development (see Parts 4 and 9 of this volume). That the reliance of palaeontology on the anatomical sciences was amongst the most fruitful scientific intersections of the period is nowhere more evident than in our fifth extract, from William Smith's *Strata Identified by Organized Fossils* (1816). Smith's contributions to geology cannot be overstated, although Simon Winchester's best-selling *The Map That Changed the World* (Further Reading) offers a partisan and somewhat contentious account of his career. As the extract makes clear, Smith's focus – in 'selling' his ideas, map, and book – was twofold: to persuade an often-reluctant geological community of the immense significance of his ideas and, with an eye to geology as an applied science, to appeal to the economic interests of farmers and landowners by demonstrating ways in which land could be exploited. His cartographic and stratigraphical work made it easier for rural landowners to judge the likelihood of particular resources being found at particular locations.

Smith's breakthrough lay in demonstrating that rocks could be dated relative to one another by observing their embedded fossil remains, but this now orthodox idea was not immediately accepted. Just as Anning was marginalised because of her gender, Smith was marginalised because of his humble birth. His achievement was belatedly recognised by the geological community in the 1830s, as professional members of the Geological Society began to predominate over its gentlemen-amateur founders. By 1839, Roderick Murchison, in *The Silurian System*, declared the central importance of Smith's 'fundamental principle of the identification of strata by their imbedded remains; the passage from one deposit to another being marked by a change in the animals which lived and died during the accumulation of each'. Hugh Miller also spoke warmly in *Testimony of the Rocks* (1857) of Smith's Herculean contributions, describing him as 'the father of English geology' (extracts from both works appear later in this section). In 1831, Smith was awarded the Wollaston Medal, the highest honour bestowed by the Geological Society, created expressly to reward him.

Smith's theory about a gradual and generally eastward tilting of strata first arose from his detailed observations of Somerset mines in the 1790s and his subsequent involvement as chief surveyor of the Somersetshire Coal Canal. That he went on to survey the entirety of British surface geology was partly a result of observations made while canal building across the country but also of a single-minded zeal that drove him on and left him bankrupt. It must be remembered that Smith's interest in fossils was not an obvious move at the time. For much of the seventeenth century, the term 'fossil' referred broadly to any unusual item found in rocks, while fossils proper were known as 'figured stones' and regarded as mysterious curios rather than the remains of creatures. The idea that fossils were animal remains was promulgated by Cesalpino and Hooke (see Part 1 of this volume), whose insistence on accurate scientific observation was given further impetus in the eighteenth century by developments in microscopy. By Smith's time, the idea was no longer controversial, but the subject of fossils remained deeply problematic, pointing as it did to the existence of extinct creatures prior to humankind and not recorded in the Mosaic narrative.

The work of Hutton, Anning, and Smith helped establish the theoretical underpinnings, practical work, and evidence base of modern geology, stratigraphy, and palaeontology. The following extracts turn to some of the major works that were built on these stellar foundations as Britain become the unquestioned leader in geology. Undoubtedly, the most important of these was Charles Lyell's three-volume *Principles of Geology* (1830–3), in which meticulous field work massively expanded the available evidence base and, most importantly, analysed it in ways that enabled him to argue convincingly for uniformitarianism as the basis of all studies of the earth's history, development, and current conditions. While

Hutton was the first uniformitarian, and Playfair the chief disseminator of Hutton's theory, Lyell's brilliance made it orthodox. Firmly opposed to catastrophism, Lyell contends that all geological evidence could be explained by the gradual elaboration of normal geological processes (formation, uplift, and erosion) occurring over a vastly expanded timescale. Rather than hypothesising different and more chaotic conditions in the geological past, Lyell argues that earth history can be understood by examining 'the evidence of those minute, but incessant mutations, which every part of the earth's surface is undergoing' in the present. He is particularly critical of Werner's 'Neptunist' theories about the marine formation of basalt. Lyell's work was certainly not accepted by everyone, and as it progressed, he responded to various criticisms.

There is also a step-change from Lyell's predecessors – a contrast between Hutton's emphasis on the centrality of 'mankind', for whom geological processes exist in order to create a 'habitable world', and Lyell's chillier, more remorseless view of deep time in which humankind seemed increasingly insignificant – absent, in large measure, from the earth's long history. However, he also sought to maintain Christian belief, arguing that through geological enquiries, 'we discover everywhere the clear proofs of a Creative Intelligence, and of His foresight, wisdom, and power'.

If Lyell's work represents the most important contribution to British geology in the period under consideration, Etheldred Benett's *A Catalogue of the Organic Remains of the County of Wilts* (1831) is a stellar achievement for different reasons, the first major work of someone often described as the first female geologist. An important scientific contribution in its own right, it adds considerably to palaeontological and stratigraphic knowledge. Benett's specialisation in Wiltshire paralleled the way that other major figures in geology focused on particular regions, and she recorded a number of significant discoveries. An authority on the middle Cretaceous as a result of her local knowledge and ability to finance detailed investigations of local quarries, her knowledge of marine fossils, particularly bivalves, sponges, and molluscs, was considerable. She made many discoveries, naming several fossil sponges under the now-defunct name *Polypothecia.* Species for which she is named include *Ventriculites Benettiæ*. The final part of the extract includes brief, illustrative examples of her methodical catalogue entries for these and related species, which she classified as sponges within the larger group Polyps, following Lamarck's zoological classification. The entries, in tabular form, list the species, the text or authority in which they had previously been cited, and the strata and location in which they appear.

Benett's financial independence meant she was able to go further than many aspiring female scientists of the period, but like Anning (and William Smith), she faced barriers that could not be overcome. Despite amassing a geological collection to rival any in the country – a collection consulted by numerous leading contemporary geologists – and despite becoming the correspondent and confidante of many key figures, she was ineligible for Geological Society membership because of her gender. The effect of a general atmosphere of exclusion and marginalisation are clear in the preface (the initial part of the extract), in which Benett is at pains to validate her work via the approval and assistance of male scientists. The fact that Benett produced only a limited run of a work she described as 'a thing of mere local interest' further indicates the discouraging environment in which she worked. Benett, it should be said, continues to be under-represented in histories of British geology.

The reference in her Preface to the failure, after fifteen years, of three male geologists to investigate and classify her collection of *Polypothecia* and her subsequent decision to undertake the task herself indicates both frustration and determination. A combination of deference to male authority and a desire to proclaim her own credentials is evident in the second part of the extract, her opening letter to influential Wiltshire landowner and writer Sir Richard Colt Hoare. This is also of interest in terms of references to 'Diluvian Detritus', indicative of adherence to catastrophism just as Lyell was challenging this orthodoxy. More generally, her survey of South Wiltshire geology is skilful, detailed, and precise, often referring to very recent scientific developments. In a parallel to botanical works by George Luxford and Anna Russell (Part 5 of this volume), Benett's work is part of a broader trend of nineteenth-century science of accumulating detailed regional data in order to enhance understanding of particular places and to enable regional comparisons. Like other geologists, Benett focuses particularly on quarries because they provide extensive exposure to successive strata: at Chicksgrove Quarry, Tisbury, for example, she investigated a 'fine section [showing] sixteen beds of this [Portland] series, singularly rich in Organic Remains'. While the rapid industrial exploitation of the British countryside in this period unleashed enormous environmental violence, quarries also yielded major scientific discoveries.

Benett's work demonstrates the impact of William Smith's stratigraphic principles in opening up opportunities for close investigation of fossils and of the excitement unleashed by Mary Anning's discoveries (to which Benett briefly alludes). The next extract, from Buckland's 1836 contribution to *The Bridgewater Treatises* (*Treatise 6: Geology and Minerology with Reference to Natural Theology*), is centred on early discoveries of the *Megalosaurus* and *Iguanodon*, which were opening up the dramatic realm of dinosaurs to the astonishment of scientists and the public. Buckland also alludes to the finds of ichthyosaurs and plesiosaurs covered in earlier extracts. That Buckland and his contemporaries initially thought megalosaurs were crocodile-like quadrupeds is a sign of the novelty of such finds: subsequently it became clear that they were bipedal and upright. Like his contemporaries, Buckland uses the term Saurians to collectively describe giant marine and terrestrial lizards. Only later (see Part 4 of this volume), did the term 'dinosaur' emerge through the work of Sir Richard Owen. Buckland's comments on the first discovered of the terrestrial dinosaurs, *Iguanodon*, relate to the groundbreaking discoveries of Gideon Mantell, to whom we will turn later. It is difficult to reflect the enormous shock and excitement that such discoveries elicited: the revelation that in the earth's distant past, long prior to evidence of *Homo sapiens*, the earth was dominated by gigantic, wildly exotic, and seemingly monstrous creatures held these early generations spellbound: the wider dissemination and popularisation of such discoveries is taken up in relevant sections of Volume II of this edition, but the contributions of palaeontology to a broader process of reconceptualising the earth's dynamic environments and processes are staggering.

One other note is worth making on this extract: Buckland's remarks on the perceived similarities between the teeth of *Iguanodon* and modern iguanas is one of many examples of the kinds of evidence that fed into the gradual realisation of evolutionary relationships between extinct and existing creatures. As Buckland commented, without offering an evolutionary explanation,

> we cannot but be impressed by the discovery of a resemblance, amounting almost to identity, between such characteristic organs as the teeth, in one of the most enormous among the extinct reptiles of the fossil world, and those of a genus whose largest species is comparatively so diminutive.

While Lyell's work addresses a particular theory, it also cast much light on the relatively recent Tertiary period. In the next extract, Sir Roderick Impey Murchison offered another significant contribution to Victorian geology. Sharing Lyell's commitment to evidence-led science, Murchison turns in *The Silurian System* (1839) to earlier periods. His book shows that great swathes of the earth's ancient history were still being discovered and theorised by the 'new science'. He also confounds the earlier orthodoxy that fossil remains were relatively scarce in the earth's older strata by demonstrating their wealth in the Silurian.

In the decade prior to *The Silurian System*, Murchison and Sedgwick were ranged against Sir Henry Thomas De la Beche and George Bellas Greenough (founders, respectively, of the Geological Survey of Great Britain and the Geological Society), in what was known as the Great Devonian Controversy. By 1840, this was settled in favour of Murchison and Sedgwick, whose periodisation was recognised by the establishment of the Devonian period – against the contention of their opponents that these strata belonged in the Carboniferous. *The Silurian System* also saw the naming of a new period. Immensely detailed, rigorous, and empirical in its approach, its expansion of geological knowledge is achieved by extensive field work in the borders between England and Wales. Murchison's decision to name his new era after the Silures, an early British tribe from the Welsh-English borders, marks the degree to which early nineteenth-century science had not yet become divorced from the arts and humanities. In 1879, the name of another ancient British tribe, the Ordovices, supplied the name of the period immediately prior to the Silurian: the establishment of the Ordovician (485–443 mya) resolved a dispute between later followers of Sedgwick and Murchison, who disagreed about whether rocks of these ages should be deemed Silurian or Cambrian.

In the second half of the extract, from Part 2, Chapter 43, Murchison's detailed review of Silurian fossiliferous remains provides evidence for, and anticipates, evolutionary theory, without at any point speculating in that direction. Murchison notes that the Silurian fossil species are less frequent than those of later geological periods; that the species become more frequent as the Silurian progresses; and that markedly different species occupied the beginning and end of the period, he refers to 'a true zoological transition' between older and newer forms. While Murchison draws no further conclusions, Darwin and others recognised the relevance of such evidences from the earth's deep past to the wider conditions governing species formation.

The tenth extract, from Gideon Mantell's *On the Pelorosaurus* (1850), continues our palaeontological coverage. Mantell's extraordinary record of early discoveries of 'sauropods' was mainly based on his explorations in his native Sussex, often centring on Tilgate Forest, and made him the most prolific of the early 'dinosaur hunters'. *Pelorosaurus* was the first species identified as a dinosaur but not the first to be discovered, this honour being taken by *Megalosaurus* in 1824 (*Iguanodon* followed in 1825). In 1841, Richard Owen discovered *Cetiosaurus* but erroneously viewed it as a marine reptile similar to crocodiles – an idea that Mantell contested. Mantell's identification of *Pelorosaurus* as a terrestrial dinosaur began the process of better understanding the physiology and habits of creatures whose existences could only be inferred by often-incomplete fossil remains.

Mantell was frequently embroiled in bad-tempered conflicts with Owen. Both leading exponents of applying comparative anatomy to palaeontology, their claims and counter-claims about the identification of fragmentary remains of *Pelorosaurus* and *Cetiosuarus* continued for some time. *On the Pelorosaurus* is an example of this, containing various submerged digs at his august colleague: Mantell slyly begins, for example, by offering his work as 'an explanation of certain discrepancies between some of my statements and those of other cultivators of this branch of comparative anatomy'. He suggests that 'calculations of the length and proportions of the original animal taken from a single bone, or from a few detached bones' (something for which Owen had become famous: see Part 4 of this volume) 'can afford but vague and unsatisfactory results'. In closing, he points to his own reputation by claiming to have been 'encouraged in my earliest researches by the illustrious founder of Palaeontology, Baron CUVIER', and alluding to having become the recipient of 'the highest award of the Geological Society': Mantell was the second recipient of the Wollaston Medal in 1835, three years ahead of Owen. Witheringly, he declares himself 'reluctant to discontinue researches which no other naturalist seemed disposed to undertake'. The disputes between the two men certainly involved very difficult matters of identification: clearer distinctions between *Pelorosaurus* and *Cetiosaurus* were not established until the twentieth century. Mantell was nonetheless correct in claiming that *Pelorosaurus* was terrestrial, although – like Buckland – he erroneously thought it to be crocodilian. In fact, it more closely resembled other large herbivores such as *Brontosaurus* and *Diplodocus*. His estimate of its length, however, was remarkably accurate. Martell also refers to the extinct flightless South Island giant moa (*Dinornis robustus*), famously first described by Owen in 1846.

The final two extracts, from the latter end of the period covered in this volume, return us to that central tension within Victorian geology between science and scripture. Philip Gosse's *Omphalos* and Hugh Miller's *Testimony of the Rocks* (both 1857) seek to reconcile the evidences of nature and the revealed wisdom of the Bible, but their contrasting methods are instructive.

Gosse begins by acknowledging the devoutness of many practicing geologists and is at pains not to impugn their integrity. He does not side with earlier scriptural geologists (see Part 2 of this volume), whom he describes as 'good men who merely *denounce* Geology and geologists', and specifically rejects claims found in Mellor Brown's *Reflections of Geology* (1838) and Granville Penn's *A Comparative Estimate of the Mineral and Mosaic Geologies* (1822). Gosse also rejects the 'gap creationist' interpretations of the 'days' of Genesis forwarded by Chalmers, Buckland, Sedgwick, Siliman, and Conybeare, quoting from Sedgwick's lectures and Conybeare's August 1834 article in *Christian Observer*. He is, moreover, deeply unimpressed by the evolutionary theories in *Vestiges of the Natural History of Creation* (1844) (see Part 9 of this volume). In rejecting scriptural approaches and mainstream geology, Gosse immediately signals a desire for a radical alternative and a belief that conflicts between science and scripture are apparent rather than real.

To achieve this alternative, he forwards the theory of prochronism – that while organic life is essentially circular in nature (individual organisms passing through *diachronic* life cycles which have neither beginning nor end because they generate new cycles of individual life), God's creative interventions are *prochronic* (i.e., they mark the moment of creation that initiates the various diachronic circularities of organic life). More controversially, Gosse asks whether it is conceivable 'that the strata of the surface of the earth, with their fossil floras and faunas, may possibly belong to a prochronic development of the mighty plan of the life-history of this world' without the fossil creatures ever having actually existed. *Omphalos* argues that God placed fossils within the strata of the earth to test the faith of his followers by seeming to provide evidence for the antiquity of the earth and of long periods of organic life, including extinctions, prior to the creation of *Homo sapiens*.

The reaction to Gosse's ideas was almost entirely negative, an indication of the degree to which opinion had been transformed by the accumulating evidences of geology and biology. One of Gosse's friends, the Reverend Charles Kingsley, novelist and proponent of Christian Socialism, informed Gosse that after reading *Omphalos*, 'it is not my reason, but my *conscience* that revolts' because he could not 'believe that God has written on the rocks one enormous and superfluous lie for all mankind'. That fossils '*pretend* to be the bones of dead animals' is for Kingsley preposterous and monstrous (Gosse, 1890, 280–1; and see Abberley, Further Reading). Gosse anticipated such objections, answering the charge that his theory meant that one must 'charge the Creator with forming objects whose sole purpose was to deceive us', but prochronism proved unconvincing to almost all of Gosse's readers and made *Omphalos* an interruption to what had been a brilliant career in popular science: Gosse's elevated status in the 1840s and early 1850s was due to his role in communicating the delights and fascinations of sea-shores and tidal pools and his pivotal role in the 'aquarium craze' of the period. After 1857, he resumed such works and never returned to prochronic controversies again. Although the least compelling of the section's extracts, it is amongst the most important precisely because of its status as a record of faith in crisis: it also contains a very useful summary of the various theories promoted in the preceding four decades.

Testimony of the Rocks was the final, posthumously published, work of the self-educated Scots geologist, folklorist, and Evangelical Hugh Miller, whose other geological works include *The Old Red Sandstone* (1841) and *The Footsteps of the Creator* (1850). Like most of his Geological Society colleagues, Miller accepted new evidence suggesting the great antiquity of the earth and that many species had existed (and become extinct) prior to the creation of humans. Although he argued that the succession of species demonstrated progress over time, he firmly rejected Lamarckian theories of evolutionary change, insisting on the direct role of a benign creator.

Most of the twelve lectures that make up *Testimony of the Rocks* were delivered to the Edinburgh Philosophical Institution (1852, 1855), the Young Men's Christian Association (1854), and the Geological Section of the British Association, Glasgow (1855). Miller accepts that fossil evidence made it necessary to eschew literal interpretations of *Genesis*: like so many of his believing brethren amongst Victorian geology, he thus tacitly acknowledges the primacy of science while seeking to maintain a basis for theistic belief in seeing the days of *Genesis* as 'prophetic' rather than literal: what he acknowledges here as a change in his view is attributed to his researches in the most recent periods of geological history, which he sees as coterminous with the 'final' creation described in the Old Testament. Despite his faith, his question is not how geology can be made to conform to scripture, but rather 'What are the facts scientifically determined which now demand a new scheme of reconciliation?' This is a subtle but important difference, in which priority is tacitly conceded to science. However, while Miller follows Sedgwick, Buckland, and others in conceiving the Biblical 'in the beginning' as allowing an extended earth history, he imagines many thousands of years rather than the millions posited in Lyell's *Principles of Geology*.

Miller focuses on the Pleistocene (2,580,000 to 11,700 years ago), an epoch following the Pliocene which featured repeated glaciations, including the most recent Ice Age. Pleistocene Drift refers to materials moved and deposited as a result of glaciation and related phenomena. In doing so, he speaks of a variety of existing and extinct species which may require some explanation. The Terebratulidae is a family of brachiopods. The species to which Miller refers is *Lobothyris punctate*. It was collected by William Smith as a child and known by him at the time as 'pundibs'. *Tellina proxima*, now *Macoma calcarea*, a bivalve of the Tellinidae family, is a still-existing clam species of the northern hemisphere. Another surviving bivalve, this time from the Astartidae family, *Astarte arctica* is native to the arctic and sub-arctic areas described by Miller.

Miller's extract is redolent with Biblical imagery, references, and cadences, an example of the power and beauty of Christian rhetoric in the service of Natural Theology and a sign of the confidence with which Miller, in the closing decade of his life and career, regarded the science to which he had devoted himself. It is as well, perhaps, that he did not live to see the decade that followed and the enormous rupture between science and religion that accompanied Darwinism. In this sense, Miller's work can be read as an elegy for a powerful force in terminal decline.

Taken together, the extracts included in this section record the rise of one of the most significant of the infant sciences of the nineteenth century, as well as its role in questioning previous religious orthodoxies and acting as a gateway for the progressively more secular visions of the second half of the century. In turning attention to issues of history, duration, process, disruption, and change, geology heralded in a new conception of time that proved profoundly significant. In turning attention to extinct animals, and deploying comparative anatomy, geology cast light on pressing biological issues of the present. The focus will turn, in the next section, to comparative anatomy itself before sections tracing developments in botany, zoology, global scientific exploration, geography, biogeography, proto-evolutionary thought, and agricultural sciences. Across all of these fields, geology left its traces.

Further reading

Abberley, Will, 'Deceptive Nature and Truthful Science in Charles Kingsley's Natural Theology', *Victorian Studies* 58:1 (2016), 34–56.

Barber, Lynn, *The Heyday of Natural History 1820–1870* (London: Cape, 1980).

Bowker, Geoff, 'In Defence of Geology: The Origins of Lyell's Uniformitarianism', *A History of Scientific Thought*, ed. Michel Serres (Oxford and Cambridge, Mass.: Blackwell, 1995).

Cadbury, Deborah, *The Dinosaur Hunters: A True Story of Scientific Rivalry and the Discovery of the Prehistoric World*, (London: Fourth Estate, 2001).

Dawson, Gowan, *Show Me the Bone: Reconstructing Prehistoric Monsters in Nineteenth-Century Britain and America* (Chicago; London: University of Chicago Press, 2016).

Emling, Shelley, *The Fossil Hunter: Dinosaurs, Evolution, and the Woman Whose Discoveries Changed the World* (Basingstoke: Palgrave Macmillan, 2009).

Guralnick, Stanley M., 'Geology and Religion Before Darwin: The Case of Edward Hitchcock, Theologian and Geologist (1793–1864)', *Isis* 63:4 (1972), 529–43.

Hawley, Duncan, 'Spotlight on William Smith's 1815 Geological Map: "A Delineation of the Strata of England and Wales with Part of Scotland . . ."', *Geography* 101:1 (2016), 35–41.

Jonsson, Fredrik Albritton, 'Abundance and Scarcity in Geological Time, 1784–1844', *Nature, Action and the Future: Political Thought and the Environment* (ed. Katrina Forrester and Sophie Smith) (Cambridge: Cambridge University Press, 2018), 70–93.

Gosse, Edmund (ed.), *The Life of Philip Henry Gosse* (London: Kegan Paul, 1890).

Lightman, Bernard V, *Victorian Science in Context* (Chicago: Chicago University Press, 1997).

Miller, Elizabeth Carolyn, *Extraction Ecologies and the Literature of the Long Exhaustion* (Princeton: Princeton University Press, 2021).

Moore, James R. 'Geologists and Interpreters of Genesis in the Nineteenth Century', *God and Nature: Historical Essays on the Encounter Between Christianity and Science*, ed. D.C. Lindberg and R.L. Numbers (Berkeley, Los Angeles and London: University of California Press, 1986), 322–50.

Nash, Sarah E., 'The Collections and Life History of Etheldred Benett (1776–1845)', *Wiltshire Archaeological and Natural History Magazine* 83 (1990), 163–9.

Pickford, Susan, '"I Have No Pleasure in Collecting for Myself Alone": Social Authorship, Networks of Knowledge, and Etheldred Benett's Catalogue of the Organic Remains of the County of Wiltshire (1831)', *Journal of Literature and Science* 8 (2015), 69–85.

Repcheck, Jack, *The Man Who Found Time: James Hutton and the Discovery of the Earth's Antiquity* (London: Pocket, 2004).

Rudwick, Martin J.S., 'The Shape and Meaning of Earth History', *God and Nature: Historical Essays on the Encounter Between Christianity and Science*, ed. D.C. Lindberg and R.L. Numbers (Berkeley, Los Angeles and London: University of California Press, 1986), 296–321.

———, Scenes from Deep *Time: Early Pictorial Representations of the Prehistoric World* (Chicago: The University of Chicago Press, 1992).

———, *Georges Cuvier, Fossil Bones, and Geological Catastrophes: New Translations & Interpretations of the Primary Texts* (Chicago: University of Chicago Press, 1997).

———, *Worlds Before Adam: The Reconstruction of Geohistory in the Age of Reform* (Chicago: University of Chicago Press, 2008).

Rupke, Nicolaas, *The Great Chain of History: William Buckland and the English School of Geology 1814–1850* (Oxford: Oxford University Press, 1983).

Stafford, Robert A., *Scientist of Empire: Sir Roderick Murchison, Scientific Exploration and Victorian Imperialism* (Cambridge: Cambridge University Press, 1989).

Thwaite, Ann, *Glimpses of the Wonderful: The Life of Philip Henry Gosse, 1810–1888* (London: Faber & Faber, 2002).

19

JAMES HUTTON, *ABSTRACT OF A DISSERTATION READ IN THE ROYAL SOCIETY OF EDINBURGH, UPON THE SEVENTH OF MARCH, AND FOURTH OF APRIL, MDCCLXXXV, CONCERNING THE SYSTEM OF THE EARTH, ITS DURATION, AND STABILITY*

(Edinburgh, 1785)

THE purpose of this Dissertation is to form some estimate with regard to the time the globe of the Earth has existed, as a world maintaining plants and animals; to reason with regard to the changes which the earth has undergone; and to see how far an end or termination to this system of things may be perceived, from the consideration of that which has already come to pass.

As it is not in human record, but in natural history, that we are to look for the means of ascertaining what has already been, it is here proposed to examine the appearances of the earth, in order to be informed of operations which have been transacted in times past. It is thus that, from principles of natural philosophy, we may arrive at some knowledge of order and system in the œconony of this globe, and may form a rational opinion with regard to the course of nature, or to events which are in time to happen.

The solid parts of the present land appear, in general, to have been composed of the productions of the sea, and of other materials similar to those now found upon the shores. Hence we find reason to conclude,

1*st*, That the land on which we rest is not simple and original, but that it is a composition, and has been formed by the operation of second causes.

2*dly*, That, before the present land was made, there had subsisted a world composed of sea and land, in which were tides and currents, with such operations at the bottom of the sea as now take place. And,

DOI: 10.4324/9780429355653-23

Lastly, That, while the present land was forming at the bottom of the ocean, the former land maintained plants and animals; at least, the sea was then inhabited by animals, in a similar manner as it is at present.

Hence we are led to conclude that the greater part of our land, if not the whole, had been produced by operations natural to this globe; but that, in order to make this land a permanent body, resisting the operations of the waters, two things had been required; 1*st*, The consolidation of masses formed by collections of loose or incoherent materials; 2*dly*, The elevation of those consolidated masses from the bottom of the sea, the place where they were collected, to the stations in which they now remain above the level of the ocean.

[. . .]

Thus the subject is considered as naturally divided into two branches, to be separately examined: *First*, By what natural operation strata of loose materials had been formed into solid masses; *Secondly*, By what power of nature the consolidated strata at the bottom of the sea had been transformed into land.

With regard to the *first* of these, the consolidation of strata, there are two ways in which this operation may be conceived to have been performed; first, by means of the solution of bodies in water, and the after concretion of these dissolved substances, when separated from their solvent; *secondly*, the fusion of bodies by means of heat, and the subsequent congelation of those consolidating substances.

[. . .]

With regard to the second branch, in considering by what power the consolidating strata had been transformed into land, or raised above the level of the sea, it is supposed that the same power of extreme heat, by which every different mineral substance had been brought into a melted state, might be capable of producing an expansive force, sufficient for elevating the land, from the bottom of the ocean, to the place it now occupies above the surface of the sea.

[. . .]

We are hence directed to look for the manifestation of this power and force, in the appearances of nature. It is here we find eruptions of ignited matter from the scattered volcano's [sic] of the globe; and these we conclude to be the effects of such a power precisely as that about which we now inquire. Volcano's [sic] are thus considered as the proper discharges of a superfluous or redundant power; not as things accidental in the course of nature, but as useful for the safety of mankind, and as forming a natural ingredient in the constitution of the globe.

[. . .]

A theory is thus formed, with regard to a mineral system. In this system, hard and solid bodies are to be formed from soft bodies, from loose or incoherent materials, collected together at the bottom of the sea; and the bottom of the ocean is to be made to change its place with relation to the centre of the earth, to be formed into land above the level of the sea, and to become a country fertile and inhabited.

That there is nothing visionary in this theory, appears from its having been rationally deduced from natural events, from things which have already happened; things which have left in the particular constitutions of bodies, proper traces of the manner of their production; and things which may be examined with all the accuracy, or reasoned upon with all the light, that science can afford. As it is only by employing science in this manner, that philosophy enlightens man with the knowledge of that wisdom or design which is to be found in nature, the system now proposed, from unquestionable principles, will claim the attention of scientific men, and may be admitted in our speculations with regard to the works of nature, notwithstanding many steps in the progress may remain unknown.

[. . .]

Having thus ascertained a regular system, in which the present land of the globe has been first formed at the bottom of the ocean, and then raised above the surface of the sea, a question naturally occurs with regard to time; what had been the space of time necessary for accomplishing this greater work?

20

JAMES HUTTON, 'THEORY OF THE EARTH; OR AN INVESTIGATION OF THE LAWS OBSERVABLE IN THE COMPOSITION, DISSOLUTION, AND RESTORATION OF LAND UPON THE GLOBE'

Transactions of the Royal Society of Edinburgh 1, 2, 209–304

Part 1: Prospect of the Subject to be Treated Of

When we trace the parts of which this terrestrial system is composed, and when we view the general connection of those several parts, the whole presents a machine of a peculiar construction by which it is adapted to a certain end. We perceive a fabric, erected in wisdom, to obtain a purpose worthy of the power that is apparent in the production of it.

[. . .]

The form and constitution of the mass are not more evidently calculated for the purpose of this earth as a habitable world, than are the various substances of which that complicated body is composed. Soft and hard parts variously combine, to form a medium consistence adapted to the use of plants and animals; wet and dry are properly mixed for nutrition, or the support of those growing bodies; and hot and cold produce a temperature or climate no less required than a soil. Insomuch, that there is not any particular, respecting either the qualities of the materials, or the construction of the machine, more obvious to our perception, than are the presence and efficacy of design and intelligence in the power that conducts the work.

[. . .]

Let us now confine our view, more particularly, to that part of the machine on which we dwell, that so we may consider the natural consequences of those operations which, being within our view, we are better qualified to examine.

DOI: 10.4324/9780429355653-24

This subject is important to the human race, to the possessor of this world, to the intelligent being Man, who foresees events to come, and who, in contemplating his future interest, is led to enquire concerning causes, in order that he may judge of events which otherwise he could not know.

If, in pursuing this object, we employ our skill in research, not in forming vain conjectures; and if *data* are to be found, on which Science may form just conclusions, we should not long remain in ignorance with respect to the natural history of this earth, a subject on which hitherto opinion only, and not evidence, has decided: For in no subject is there naturally less defect of evidence, although philosophers, led by prejudice, or misguided by false theory, have neglected to employ that light by which they should have seen the system of this world.

But to proceed in pursuing a little farther our general or preparatory ideas. A solid body of land could not have answered the purpose of a habitable world; for a soil is necessary to the growth of plants; and a soil is nothing but the materials collected from the destruction of the solid land. Therefore, the surface of this land, inhabited by man, and covered with plants and animals, is made by nature to decay, in dissolving from that hard and compact state in which it is found below the soil; and this soil is necessarily washed away, by the continual circulation of the water, running from the summits of the mountains towards the general receptacle of that fluid.

The heights of our land are thus levelled with the shores; our fertile plains are formed from the ruins of the mountains; and those travelling materials are still pursued by the moving water, and propelled along the inclined surface of the earth. These moveable materials, delivered into the sea, cannot, for a long continuance, rest upon the shore; for, by the agitation of the winds, the tides and currents, every moveable thing is carried farther and farther along the shelving bottom of the sea, towards the unfathomable regions of the ocean.

If the vegetable soil is thus constantly removed from the surface of the land, and if its place is thus to be supplied from the dissolution of the solid earth, as here represented, we may perceive an end to this beautiful machine; an end, arising from no error in its constitution as a world, but from that destructibility of its land which is so necessary in the system of the globe, in the œconomy of life and vegetation.

The immense time necessarily required for this total destruction of the land, must not be opposed to that view of future events, which is indicated by the surest facts and most approved principles. Time, which measures every thing in our idea, and is often deficient to our schemes, is to nature endless and as nothing; it cannot limit that by which alone it had existence; and as the natural course of time, which to us seems infinite, cannot be bounded by any operation that may have an end, the progress of things upon the globe, that is, the course of nature, cannot be limited by time, which must proceed in a continual succession. We are, therefore, to consider as inevitable, the destruction of our land, so far as effected by those operations which are necessary to the purpose of the globe [. . .]

But is this world to be considered this merely as a machine, to last no longer than its parts retain their present position, their proper forms and qualities? Or may it not be also considered as an organized body? Such as has a constitution in which the necessary decay of the machine is naturally repaired, in the exertion of those productive powers by which it had been formed.

This is the view in which we are now to examine the globe; to see if there be, in the constitution of this world, a reproductive operation, by which a ruined constitution may be again repaired, and a duration or stability thus procured to the machine, considered as a world sustaining plants and animals.

[. . .]

Here is an important question, therefore, with regard to the constitution of this globe; a question which, perhaps, it is in the power of man's sagacity to resolve; and a question which, if satisfactorily resolved, might add some lustre to science and the human intellect.

Animated with this great, this interesting view, let us strictly examine our principles, in order to avoid fallacy in our reasoning; and let us endeavour to support our attention, in developing a subject that is vast in its extent, as well as intricate in the relation of parts to be stated.

The globe of the earth is evidently made for man. He alone, of all beings which have life upon this body, enjoys the whole and every part; he alone is capable of knowing the nature of this world, which he thus possesses in virtue of his proper right; and he alone can make the knowledge of this system a source of pleasure and the means of happiness.

Man alone, of all the animated beings which enjoy the benefits of this earth, employs the knowledge which he there receives, in leading him to judge of the intention of things, as well as of the means by which they are brought about; and he alone is thus made to enjoy, in contemplation as well as sensual pleasure, all the good that may be observed in the constitution of this world; he, therefore, should be made the first subject of enquiry.

Now, if we are to take the written history of man for the rule by which we should judge of the time when the species first began, that period would be but little removed from the present state of things. The Mosaic history places this beginning of man at no great distance; and there has not been found, in natural history, any document, by which a high antiquity might be attributed to the human race. But this is not the case with regard to the inferior species of animals, particularly those which inhabit the ocean and its shores. We find in natural history monuments which prove that those animals had long existed; and we thus procure a measure for the computation of a period of time extremely remote, though far from being precisely ascertained.

[. . .]

In finding the relics of sea-animals of every kind in the solid body of our earth, a natural history of those animals is formed, which includes a certain portion of time; and for the ascertaining of this portion of time, we must again have recourse to the regular operations of this world. We shall thus arrive at facts which indicate a period to which no other species of chronology is able to remount.

In what follows, therefore, we are to examine the construction of the present earth, in order to understand the natural operations of time past; to acquire principles by which we may conclude with regard to the future course of things, or judge of those operations, by which a world, so wisely ordered, goes into decay; and to learn by what means such a decayed world may be renovated, or the waste of habitable land upon the globe repaired.

[. . .]

Part III: Investigation of the Natural Operations Employed in the Production of Land Above the Surface of the Sea

We seek to know that operation by means of which masses of loose materials, collected at the bottom of the sea, were raised above its surface, and transformed into solid land.

[. . .] We have found, that there had been operations, natural to the bowels of this earth, by which those loose and unconnected materials have been cemented together, and consolidated into masses of great strength and hardness; those bodies are thus enabled to resist the force of waves and currents, and to preserve themselves, for a sufficient time, in their proper shape and place, as land above the general surface of the ocean.

[. . .]

There is nothing so proper for the erection of land above the level of the ocean, as an expansive power of sufficient force, applied directly under materials in the bottom of the sea, under a mass that is proper for the formation of land when thus erected. The question is not, how such a power may be procured; such a power has probably been employed. If, therefore, such a power should be consistent with that which we found had actually been employed in preparing the erected mass; or, if such a power is to be reasonably concluded as accompanying those operations which we have found natural to the globe, and situated in the very place where this expansive power appears to be required, we should thus be led to perceive, in the natural operations of the globe, a power as efficacious for the elevation of what had been at the bottom of the sea in to the place of land, as it is perfect for the preparation of those materials to serve the purpose of their elevation.

[. . .]

The strata formed at the bottom of the ocean are necessarily horizontal in their position, or nearly so, and continuous in their horizontal direction or extent. They may change, and gradually assume the nature of each other, so far as concerns the materials of which they are formed, but there cannot be any sudden change, fracture or displacement naturally in the body of a stratum. But, if these strata are cemented by the heat of fusion, and erected with an expansive power acting below, we may expect to find every species of fracture, dislocation and contortion, in those bodies, and ever degree of departure from a horizontal towards a vertical position.

The strata of the globe are actually found in every possible position: For from horizontal, they are frequently found vertical; from continuous, they are broken and separated in every possible direction; and from a plane, they are bent and doubled [. . .]

Part IV: System of Decay and Renovation Observed in the Earth

[. . .]

To sum up the argument, we are certain, that all the coasts of the present continents are wasted by the sea, and constantly wearing away upon the whole; but this operation is so extremely slow, that we cannot find a measure of the quantity in order to form an estimate. Therefore, the present continents of the earth, which we consider as in a state of perfection, would, in the natural operations of the globe, require a time indefinite for their destruction.

But, in order to produce the present continents, the destruction of a former vegetable world was necessary; consequently, the production of our present continents must have required a time which is indefinite. In like manner, if the former continents were of the same nature as the present, it must have required another space of time, which also is indefinite, before they had come to their perfection as a vegetable world.

[. . .]

Let us suppose that the continent, which is to succeed our land, is at present beginning to appear above the water in the middle of the Pacific Ocean, it must be evident, that the materials of this great body, which is formed and ready to be brought forth, must have been collected from the destruction of an earth which does not now appear. Consequently, in this true statement of the case, there is necessarily required the destruction of an animal and vegetable earth prior to the former land; and the materials of that earth which is first in our account, must have been collected at the bottom of the ocean, and begin to be concocted for the production of the present earth, when the land immediately preceding the present had arrived at its full extent.

[. . .]

We have now got to the end of our reasoning; we have no data further to conclude immediately from that which actually is: But we have got enough; we have the satisfaction to find, that in nature there is wisdom, system, and consistency. For having, in the natural history of this earth, seen a succession of worlds, we may from this conclude that there is a system in nature; in like manner as, from seeing revolutions of the planets, it is concluded, that there is a system by which they are intended to continue those revolutions. But if the succession of worlds is established in the system of nature, it is in vain to look for any thing higher in the origin of the earth. The result, therefore, of our present enquiry is, that we find no vestige of a beginning. – no prospect of an end.

21

BARON GEORGES CUVIER, *ESSAY ON THE THEORY OF THE EARTH* (Paris, 1813)

Preface to the Fifth Edition

GEOLOGY, now deservedly one of the most popular and attractive of the physical sciences, was, not many years ago, held in little estimation; and even at present, there are not wanting some who do not hesitate to maintain, that it is a mere tissue of ill observed phenomena, and of hypotheses of boundless extravagance. The work of CUVIER now laid before the public, contains in itself not only a complete answer to these ignorant imputations, but also demonstrates the accuracy, extent, and importance of many of the facts and reasonings of this delightful branch of Natural History. Can it be maintained of a science, which requires for its successful prosecution an intimate acquaintance with Chemistry, Natural Philosophy and Astronomy,– with the details and views of Zoology, Botany, and Mineralogy, and which connects these different departments of knowledge in a most interesting and striking manner,– that it is of no value? Can it be maintained of Geology, which discloses to us the history of the first origin of organic beings, and traces their gradual development from the monade to man himself, – which enumerates and describes the changes that plants, animals, and minerals – the atmosphere, and the waters of the globe – have undergone from the earliest geological periods up to our own time, and which even instructs us in the earliest history of the human species,– that it offers no gratification to the philosopher? Can even those who estimate the value of science, not by intellectual desires, but by practical advantages, deny the importance of Geology, certainly one of the foundations of agriculture, and which enables us to search out materials for numberless important economical purposes?

Geology took its rise in the Academy of Freyberg, with the illustrious WERNER, to whom we owe its present interesting condition. This being the case, we ought not, (as is at present too much the practice), amidst the numerous discoveries in the mineral kingdom which have been made since the system of investigation of that great interpreter of nature was made known, forget the master, and arrogate all to ourselves. In this Island, Geology first took firm root in the north: in Edinburgh

DOI: 10.4324/9780429355653-25

the Wernerian geognostical views and method of investigation, combined with the theory of HUTTON, the experiments and speculations of HALL, the illustrations of PLAYFAIR, and the labours of the Royal and Wernerian Natural History Societies, excited a spirit of inquiry which rapidly spread throughout the Empire; and now Great Britain presents to the scientific world a scene of geological acuteness, activity, and enterprise, not surpassed in any other country.

On the Continent the writings of CUVIER, distinguished equally by purity and beauty of style, and profound learning, have proved eminently useful in aiding the progress of Geology. In this country CUVIER was first made known as a geologist by the publication of the present essay, which, from its unexampled popularity, has made his name as familiar to us as that of the most distinguished of our own writers.

ROBERT JAMESON
COLLEGE MUSEUM, EDINBURGH,
25th November 1826

Preliminary Observations

IN my work on Fossil Bones, the object which I proposed was to discover to what animals the osseous remains, with which the superficial strata of the globe are filled, may have belonged. In pursuing this object, I had to follow a path in which but little progress had hitherto been made. As an antiquary of a new order, I was obliged at once to learn the art of restoring these monuments of past revolutions to their original forms, and to discover their nature and relations; I had to collect and bring together in their original order, the fragments of which they consisted; to reconstruct, as it were, the ancient beings to which these fragments belonged; to reproduce them with all their proportions and characters; and, lastly, to compare them with those which now live at the surface of the globe:– an art almost unknown, and which presupposed a science whose first developments had scarcely yet been traced, that of the laws which regulate the co-existence of the forms of the different parts in organised beings. I had therefore to prepare myself for these inquiries, by others of a far more extensive kind, respecting the animals which still exist. Nothing, except an almost complete review of creation in its present state, could give a character of demonstration to the results of my investigation into its ancient state; but, from this review, I had at the same time to expect a great body of rules and affinities not less satisfactorily demonstrated; and it became obvious, that, in consequence of this essay upon a small portion of the theory of the earth, the whole animal kingdom would necessarily be in some measure subjected to new laws.

Thus I was encouraged in this twofold investigation, by the equal interest which it promised to possess, both with regard to the general science of anatomy, the essential basis of all those which treat of organised bodies, and with regard to the physical history of the globe, the foundation of mineralogy, geography, and even, it may be said, of the history of Man, and of all that it most concerns him to know with regard to himself.

If it be so interesting to us to follow, in the infancy of our species, the almost obliterated traces of extinct nations, why should it not also be so, to search, amid the darkness of the infancy of the Earth, for the traces of revolutions which have taken place anterior to the existence of all nations? We admire the power by which the human mind has measured the motions of the celestial bodies, which nature seemed to have concealed for ever from our view. Genius and science have burst the limits of space; and observations, explained by just reasoning, have unveiled the mechanism of the universe. Would it not also be glorious for man to burst the limits of time, and, by means of observations, to ascertain the history of this world, and the succession of events which preceded the birth of the human race? Astronomers have undoubtedly advanced more rapidly than naturalists; and the present period, with respect to the Theory of the Earth, bears some resemblance to that in which some philosophers fancied that the heavens were formed of polished stones, and that the moon was of the size of the Peloponnesus; but after ANAXAGORAS, came COPERNICUS and KEPLER, who pointed the way to NEWTON; and why should not natural history also one day have its Newton?

Plan of This Essay

WHAT I especially propose to present in this discourse, is the plan and the result of my labours regarding Fossil Bones. I shall also attempt to trace a rapid sketch of the efforts that have been made up to the present day, to restore the history of the revolutions of the globe. The facts which I have been enabled to discover, form, without doubt, only a small portion of those which would be necessary to complete this ancient history; but several of them lead to decisive consequences, and the rigorous manner in which I have proceeded in their determination, affords me reason to think that they will be regarded as points definitively fixed, and which in their aggregate will form an epoch in science. Lastly, I trust their novelty will be a sufficient excuse for me, if I claim for them the earnest attention of my readers.

My object will first be to shew by what relations the history of the fossil bones of terrestrial animals connects itself with the theory of the earth, and for what reasons a peculiar importance is to be attributed to it, with reference to this subject. I shall then unfold the principles upon which is founded the art of determining these bones, or, in other words, of recognizing a genus, and of distinguishing a species, by a single fragment of bone,– an art, on the certainty of which depends that of my whole work, I shall give a rapid account of the new species, and of genera previously unknown, which the application of these principles has led me to discover, as well as the different kinds of deposits in which they are contained. And as the difference between these species and those which exist at the present day is bounded by certain limits, I shall show that these limits much exceed those which now distinguish the varieties of the same species. I shall therefore make known to what extent these varieties may go, whether from the influence of time, or from that of climate, or, lastly, from that of domestication.

In this way I shall be enabled to conclude, and to induce my readers to conclude with me, that great events were necessary to produce the more considerable differences which I have discovered. I shall next mention the particular modifications which my researches must necessarily introduce into the hitherto received opinions regarding the revolutions of the globe; and, lastly, I shall inquire how far the civil and religious history of different nations corresponds with the results of observation with regard to the physical history of the Earth, and with the probabilities which these observations afford concerning the period at which societies of men may have found fixed places of abode, and fields susceptible of cultivation, and at which, therefore, they may have assumed a durable form.

First Appearance of the Earth

When the traveller passes over those fertile plains where gently flowing streams nourish in their course an abundant vegetation, and where the soil, inhabited by a numerous population, adorned with flourishing villages, opulent cities, and superb monuments, is never disturbed, except by the ravages of war, or by the oppression of the powerful, he is not led to suspect that Nature also has had her intestine wars, and that the surface of the globe has been broken up by revolutions and catastrophes. But his ideas change as soon as he digs into that soil which now presents so peaceful an aspect, or ascends to the hills which border the plain; his ideas are expanded, if I may use the expression, in proportion to the expansion of the view, and begin to embrace the full extent and grandeur of those ancient events, when he climbs the more elevated chains, whose base is skirted by these hills, or when, by following the beds of the torrents which descend from those chains, he penetrates, as it were, into their interior.

First Proofs of Revolutions on the Surface of the Globe

The lowest and most level parts of the earth, exhibit nothing, even when penetrated to a very great depth, but horizontal strata composed of substances more or less varied, and containing almost all of them innumerable marine productions. Similar strata, with the same kind of productions, compose the lesser hills to a considerable height. Sometimes the shells are so numerous as to constitute of themselves the entire mass of the rock; they rise to elevations superior to the level of every part of the ocean, and are found in places where no sea could have carried them at the present day, under any circumstances; they are not only enveloped in loose sand, but are often inclosed in the hardest rocks. Every part of the earth, every hemisphere, every continent, every island of any extent, exhibits the same phenomenon.

The times are past when ignorance could maintain, that these remains of organized bodies are mere sportings of nature, productions generated in the womb of the Earth, by its own creative powers; and the efforts made by some metaphysicians of the present day, will not probably succeed in bringing these exploded opinions again into repute. A scrupulous comparison of the forms of these remains, of their texture, and often even of their chemical composition, does not disclose the slightest difference between the fossil shells and those which still inhabit the sea: the preservation of the former is not less perfect than that of the latter; most commonly we neither observe detrition nor fracture in them, nothing, in short, that announces a violent removal from their original places; the smallest of them retain their sharpest ridges, and their most delicate spines. They have, therefore, not only lived in the sea, but they have also been deposited by it. It is the sea which has left them in the places where they are now found. But this sea has remained for a certain period in those places; it has covered them long enough, and with sufficient tranquillity to form those deposits, so regular, so thick, so extensive, and partly also so solid, which contain those remains of aquatic animals. The basin of the sea has therefore undergone one change at least, either in extent, or in situation. Such is the result of the very first search, and of the most superficial examination.

The traces of revolutions become still more apparent and decisive, when we ascend a little higher, and approach nearer to the foot of the great chains. There are still found many beds of shells; some of these are even thicker and more solid; the shells are quite as numerous, and as well preserved, but they are no longer of the same species. The strata which contain them are not so generally horizontal; they assume an oblique position, and are sometimes almost vertical. While in the plains and low hills it was necessary to dig deep, in order to discover the succession of the beds, we here discover it at once by their exposed edges, as we follow the valleys that have been produced by their disjunction. Great masses of debris form at the foot of the cliffs, rounded hills, the height of which is augmented by every thaw and tempest.

These inclined strata, which form the ridges of the secondary mountains, do not rest upon the horizontal strata of the hills which are situate at their base, and which form the first steps in approaching them; but, on the contrary, dip under them, while the hills in question rest upon their declivities. When we dig through the horizontal strata in the vicinity of mountains whose strata are inclined, we find these inclined strata re-appearing below; and even sometimes, when the inclined strata are not too elevated, their summit is crowned by horizontal ones. The inclined strata are therefore older than the horizontal strata; and as they must necessarily, at least the greater number of them, have been formed in a horizontal position, it is evident that they have been raised, and that this change in their direction has been effected before the others were superimposed upon them.

Thus the sea, previous to the deposition of the horizontal strata, had formed others, which, by the operation of problematical causes, were broken, raised, and overturned in a thousand ways; and, as several of those inclined strata which it had formed at more remote periods, rise higher than the horizontal strata which have succeeded them, and which surround them, the causes by which the inclination of these beds was effected, had also made them project above the level of the sea, and formed islands of them, or at least shoals and inequalities; and this must have happened, whether they had been raised by one extremity, or whether the depression of the opposite extremity had made the waters subside. This is the second result, not less clear, nor less satisfactorily demonstrated, than the first, to every one who will take the trouble of examining the monuments on which it is established.

[. . .]

Proofs That These Revolutions Have Been Sudden

It is of much importance to remark, that these repeated irruptions and retreats of the sea have neither all been slow nor gradual; on the contrary, most of the catastrophes which have occasioned them have been sudden; and this is especially easy to be proved, with regard to the last of these catastrophes, that which, by a two-fold motion, has inundated, and afterwards laid dry, our present continents, or at least a part of the land which forms them at the present day. In the northern regions, it has left the carcases of large quadrupeds which became enveloped in the ice, and have thus been preserved even to our own times, with their skin, their hair, and their flesh. If they had not been frozen as soon as killed, they would have been decomposed by putrefaction. And, on the other hand, this eternal frost could not previously have occupied the places in which they have been seized by it, for they could not have lived in such a temperature. It was, therefore, at one and the same moment that these animals were destroyed, and the country which they inhabited became covered with ice. This event has been sudden, instantaneous, without any gradation; and what is so clearly demonstrated with respect to this last catastrophe, is not less so with reference to those which have preceded it. The breaking to pieces, the raising up and overturning of the older strata, leave no doubt upon the mind that they have been reduced to the state in which we now see them, by the action of sudden and violent causes; and even the force of the motions excited in the mass of waters, is still attested by the heaps of debris and rounded pebbles which are in many places interposed between the solid strata. Life, therefore, has often been disturbed on this earth by terrible events. Numberless living beings have been the victims of these catastrophes; some, which inhabited the dry land, have been swallowed up by inundations; others, which peopled the waters, have been laid dry, from the bottom of the sea having been suddenly raised; their very races have been extinguished for ever, and have left no other memorial of their existence than some fragments, which the naturalist can scarcely recognize.

Such are the conclusions to which we are necessarily led by the objects that we meet with at every step, and which we can always verify, by examples drawn from almost every country. These great and terrible events are every where distinctly recorded, so as to be always legible by the eye skilled to decypher their history in the monuments which they have left behind.

But what is still more astonishing and not less certain, *life* has not always existed upon the globe; and it is easy for the observer to distinguish the point at which it has begun to deposit its productions.

Proofs That There Have Been Revolutions Anterior to the Existence of Living Beings

If we ascend to higher points of elevation, and advance towards the great ridges, the craggy summits of the mountain chains, we shall presently find those remains of marine animals, those innumerable shells, of which we have spoken, becoming more rare, and at length disappearing altogether. We arrive at strata of a different nature, which contain no vestiges of living beings. Nevertheless, their crystallization, and even their stratification, shew that they have been also in a liquid state at their formation; their inclined position, and the cliffs into which they are broken, shew that they also have been forcibly moved from their original places; the oblique manner in which they dip under the shelly strata, that they have been formed previously to these latter; and lastly, the height to which their rugged and bare peaks rise above all these shelly strata, that their summits had already emerged from the waters, when the shelly strata were forming.

Such are those celebrated Primitive Mountains which traverse our continents in different directions, raising themselves above the clouds, separating the basins of rivers from one another, affording, in their perennial snows, reservoirs which feed the springs, and forming, in some measure, the skeleton, and as it were the rough framework, of the Earth.

The eye perceives from afar, in the indentations with which their ridge has been marked, and in the sharp peaks with which it is bristled, indications of the violent manner in which they have been elevated. Their appearance, in this respect, is very different from that of those rounded mountains, and hills with long flat surfaces, whose less ancient masses have always remained in the situation in which they were quietly deposited by the waters of more recent seas.

These indications become more obvious as we approach. The valleys have no longer those gently-sloping sides, those salient and re-entering angles corresponding on either side to each other, which seem to denote the beds of ancient streams. They widen and they contract without any general rule; their waters, at one time, expand into lakes; at another, fall in torrents; and sometimes their rocks, suddenly approaching from each side, form transverse dikes, over which the waters tumble in cataracts. The dissevered strata, while they shew on one side their edges perpendicularly raised, on the other present large portions of their surface lying obliquely; they do not correspond in height, but those which, on one side, form the summit of the cliff, often dip underneath on the other, and are no longer visible.

Yet, amidst all this confusion, distinguished naturalists have been able to demonstrate, that there still reigns a certain order, and that those immense deposits, broken and overturned though they be, observe a regular succession with regard to each other, which is nearly the same in all the great mountain chains. According to them, Granite, of which the central ridges of the greater number of these chains consist, and which thus surmounts every other rock, is also the rock which is found deepest in the solid crust of the globe. It is the most ancient of those which we have found means of examining in the place assigned them by nature; and we inquire not at present, whether it owes its origin to a general fluid, which formerly held every thing in solution, or may have been the first consolidated by the cooling of a great mass in fusion, or even in a state of vapour. Foliated rocks rest upon its sides, and form the lateral ridges of these great chains; schists, porphyries, sandstones, and talcose rocks, intermingle with their strata; lastly, granular marbles, and other limestones destitute of shells, resting upon the schists, form the outer ridges, the lower steps as it were, the counterforts, of these chains, and are the last formations, by which this unknown fluid, this sea without inhabitants, would seem to have prepared materials for the mollusca and zoophytes, which were presently to deposit upon these foundations vast heaps of their shells and corals.

We even find the first productions of these mollusca and zoophytes appearing in small numbers, and scattered at greater or less distances, in the last strata of these primitive formations, or in that portion of the crust of the globe to which geologists have given the name of Transition rocks. Here and there we meet with beds containing shells, interposed between certain granites of later formation than the others, between schists of various kinds, and between some newer beds of granular marbles. Life, which was in the end to obtain entire possession of the globe, seems, in these primordial times, to have struggled with the inert nature which formerly predominated; and it was not until a considerable time after, that it obtained the ascendancy over it, and acquired for itself the exclusive right of continuing and elevating the solid envelope of the Earth.

Hence, it is impossible to deny, that the masses which now constitute our highest mountains, have been originally in a liquid state; and that they have for a long time been covered by waters in which no living beings existed. Thus, it has not been only since the appearance of life that changes have been operated in the nature of the matters which have been deposited; for the masses formed previous to that event, have varied, as well as those which have been formed since. They have also experienced violent changes in their position, and a part of these changes must have taken place at the period when these masses existed by themselves, and were not covered over by the shelly masses. The proof of this lies in the overturnings, the disruptions, and the fissures, which are observable in their strata, as well as in those of more recent formations, and which are in the ancient strata even in greater number and better defined.

But these primitive masses have also undergone other revolutions since the formation of the secondary strata, and have, perhaps, given rise to, or at least have partaken of, some of those changes which these strata themselves have experienced. There are actually considerable portions of the primitive formations uncovered, although placed in lower situations than many of the secondary formations; and we cannot conceive how it should have so happened, unless the primitive strata in those places had forced themselves into view, after the secondary strata had been formed. In certain countries, we find numerous large blocks of primitive substances scattered over the surface of secondary formations, and separated by deep valleys, or even by arms of the sea, from the peaks or ridges from which they must have been derived. We must necessarily conclude, therefore, either that these blocks have been ejected by eruptions, or that the valleys (which must have stopped their course) did not exist at the time of their being transported; or, lastly, that the motions of the waters by which they were transported, exceeded in violence any thing that we can imagine at the present day.

[. . .]

Older Systems of Geologists

DURING a long time, two events or epochs only, the Creation and the Deluge, were admitted as comprehending the changes which have been operated upon the globe; and all the efforts of geologists were directed to account for the present existing state of things, by imagining a certain original state, afterwards modified by the deluge, of which also, as to its causes, its operations, and its effects, each entertained his own theory.

Thus, according to one, the earth was at first invested with an uniform light crust, which covered the abyss of the sea; and which being broken up for the production of the deluge, formed the mountains by its fragments. According to another, the deluge was occasioned by a momentary suspension of cohesion among the particles of mineral bodies; the whole mass of the globe was dissolved, and the paste thus formed became penetrated with shells. According to a third, God raised up the mountains for the purpose of allowing the waters, which had produced the deluge, to run off; and selected those places in which there was the greatest quantity of rocks, without which the mountains could not have supported themselves. A fourth created the earth from the atmosphere of one comet, and deluged it by the tail of another: The heat which it retained from its origin, was what, in his opinion, excited the whole of the living beings upon it to sin; for which they were all drowned, excepting the fishes, whose passions were apparently less vehement.

It is evident, that, even while confined within the limits prescribed by the Book of Genesis, naturalists might still have a pretty wide range: they soon found themselves, however, in too narrow bounds; and when they had succeeded in converting the six days of creation into so many indefinite periods, the lapse of ages no longer forming an obstacle to their views, their systems took a flight proportioned to the periods which they could then dispose of at pleasure. Even the great Leibnitz amused himself, like Descartes, by conceiving the earth to be an extinguished sun, a vitrified globe, upon which the vapours falling down again, after it had cooled, formed seas, which afterwards deposited the limestone formations.

By Demaillet the whole globe was conceived to have been covered with water for many thousands of years. He supposed this water had gradually retired; that all the land animals were originally inhabitants of the sea; that man himself commenced his career as a fish; and he asserts, that it is not uncommon, even now, to meet with fishes in the ocean, which are still only half converted into men, but whose descendants will in time become perfect human beings.

The system of Buffon is merely an extension of that of Leibnitz, with the addition only of a comet, which, by a violent blow, struck off from the sun the liquefied mass of the earth, together with those of all the other planets at the same instant. From this supposition, he was enabled to assume positive dates, as, from the present temperature of the earth, it could be calculated how long it had taken to cool down so far; and, as all the other planets had come from the sun at the same time, it could also be calculated how many ages are still required for cooling the greater ones, and to what degree the smaller are already frozen.

[. . .]

Importance of Fossil Remains in Geology

[. . .] It is abundantly obvious, that it is to these fossil remains alone that we owe even the commencement of a theory of the earth, and that, without them, we should perhaps never have even suspected that there had existed any successive epochs, and a series of different operations, in the formation of the globe. By them alone we are, in fact, enabled to ascertain, that the globe has not always had the same external crust; because, we are thoroughly assured, that the plants and animals must have lived at the surface before they had thus come to be buried deep beneath it. It is only by analogy that we have been enabled to extend to the primitive formations, the conclusion which is furnished directly for the secondary by the organic remains which they contain; and if there had only existed formations in which no fossil remains were inclosed, it could never have been shewn that these formations had not all been of simultaneous origin.

It is also by means of the organic remains, slight as is the knowledge we have hitherto acquired of them, that we have been enabled to discover the little that we yet know respecting the nature of the revolutions of the globe. From them we have learned, that the strata in which they are buried have been quietly deposited in a fluid; that their variations have corresponded with those of the fluid in question; that their being laid bare has been occasioned by the transportation of this fluid to some other place; and that this circumstance must have befallen them more than once. Nothing of all this could have been known with certainty, had no fossil remains existed.

[. . .]

The organic remains, therefore, which have given rise to the theory of the earth, have, at the same time, furnished it with its principal illustrations;– the only ones, indeed, that have as yet been generally acknowledged.

It is this consideration which has encouraged us to investigate the subject. But the field is vast; and it is but a very small portion of it that could be cultivated by the labour of a single individual. It was necessary, therefore, to select a particular department; and the choice was soon made. The class of fossil remains which forms the subject of this work, engaged our attention at the very outset, because it appeared to us to be that which is the most fertile in precise results, and yet, at the same time, less known, and richer in new objects of research.

22

SIR EVERARD HOME, 'SOME ACCOUNT OF THE FOSSIL REMAINS OF AN ANIMAL MORE NEARLY ALLIED TO FISHES THAN ANY OF THE OTHER CLASSES OF ANIMALS'

(*Philosophical Transactions of the Royal Society*, 1814)

XXVIII. ***Some Account of the fossil Remains of an Animal more nearly allied to Fishes than any of the other Classes of Animals. By Sir*** **Everard Home,** ***Bart. F. R. S.***

Read June 23, 1814

THE study of comparative anatomy is not confined to the animals that at present inhabit the earth, but extends to the remains of such as existed in the most remote periods of antiquity; among these may be classed the specimen which forms the subject of the present Paper.

That the bones of the elephant, rhinoceros, hippopotamus, crocodile, and of many other animals should be met with in a fossil state in this island, in such numbers as to make it appear that at some distant period they were inhabitants of Great Britain, is perhaps one of the most wonderful circumstances that occurs in the history of the earth.

To discover the changes that have taken place in our globe, which can account for the remains of animals only fitted to live in warm climates being found in so northern a situation; and to explain the circumstance of human bones never having been met with in a fossil state, is the province of the geologist. To examine such fossil bones, and to determine the class to which the animals belonged, comes within the sphere of inquiry of the anatomist, and considerably increases its extent.

DOI: 10.4324/9780429355653-26

This branch of comparative anatomy not only brings to our knowledge races of animals very different from those with which we are acquainted, but supplies intermediate links in the gradation of structure, by means of which the different classes will probably be found so imperceptibly to run into one another, that they will no longer be accounted distinct, but only portions of one series, and show that the whole of the animal creation forms a regular and connected chain.

The fossil remains of animals are too frequently brought under our observation in a very mutilated state; or are so intimately connected with the substances in which they are deposited, that it is difficult to make out the figure of the bones. In the present instance, the pains that have been taken, and the skill which has been exerted in removing the surrounding stone, under the superintendance of Mr. Bullock in whose Museum of Natural History the specimen is preserved, have brought the parts distinctly into view.

This specimen was found upon an estate of Henry Host Henley, Esq. between Lyme and Charmouth, in Dorsetshire, in a cliff thirty or forty feet above the level of the sea-shore. It had been thrown down by the breaking off of a part of the cliff, and buried in the sand upon the shore, to the depth of nearly two feet. The skull was dug out in 1812, the other parts in the following year, at a distance of some feet.

The cliff is composed of that species of argillaceous limestone called blue lias, in which the fossil bones were deposited. Above the lias there is only a thin layer of black earth.

The figure and appearance of the fossil bones are so accurately shewn in the annexed drawings, (Plates xvii. xviii. xix. xx.) as to make a very particular description of them unnecessary. The head is four feet long, and all the parts of one side of the skull, and of the upper and under jaw, are very distinct: the vertebræ immediately behind the skull remain in their natural situation respecting the head; many of the other bones have been displaced in a greater or less degree, shewing that the skeleton, before the bones were rendered fossil, had been pressed upon by a considerable weight, which had broken many of them, and entirely destroyed others.

The lower jaw had been forced a little backwards, and the intermediate bone, by which it was attached to the skull, displaced; but a portion of it is seen projecting beyond the base of the lower jaw.

The bony sclerotic coat of the eye on one side is entire; that of the other is forced through the nose, and a part of it is seen in the opening of the nostril, which is enlarged by the bones in which it is situated being broken.

The vertebræ of the back have been twisted, and their spinous processes broken off; one of them in a detached state is preserved sufficiently entire to shew its shape, and the size of the canal which contained the spinal marrow.

The anterior surfaces of the dorsal vertebræ are exposed, and several of the ribs of the opposite side remain in their place, connected to the vertebræ, with their concave surfaces uppermost. Those of the other side are all forced down upon the vertebræ, and squeezed into a mass: the pressure has been so great, as to give many of them a fluted appearance, the middle line being more crushed than the edges; they are not only in close contact with the vertebræ, but are made to follow all the irregularities of their surface, so forcibly have they been beaten in upon them.

[. . .]

The jaws and scapulæ, both in shape and size, are more like those of the crocodile than of any fishes at present known, and the three small flat bones near the broken portion of the scapula, have a resemblance to those of the tarsus of a species of turtle.

These particulars, in which the bones of this animal differ from those of fishes, are sufficient to shew, that although the mode of its progressive motion has induced me to place it in that class, I by no means consider it as wholly a fish, when compared with other fishes, but rather view it in a similar light to those animals met with in New South Wales, which appear to be so many deviations from ordinary structure, for the purpose of making intermediate connecting links, to unite in the closest manner the classes of which the great chain of animated beings is composed.

23

WILLIAM SMITH, *STRATA IDENTIFIED BY ORGANIZED FOSSILS*

(London: W. Arding, 1816)

Introduction

THE present age is distinguished by many of the most extraordinary discoveries that were ever unfolded to the human mind; and amongst them the discoveries of Chemistry stand pre-eminent. The most extensively useful part of this science has, however, been long before the Public, and contributed greatly to the improvement of various branches of manufacture; but the benefits of Chemistry have not yet been extended to the soil.

Agriculture in this, as in most other instances, is the last to profit by any thing new. That easy analysis of the soil, which seemed to promise great advantages to the Farmer, by telling him correctly the component parts of the materials he has to work upon, has not been spread through the country, or even yet become an object of attention with many of the best informed Farmers, by whom the advantages of this science must be carried into effect; and while the theory is in the possession of one class of men, and the practice in another, who have little or no connexion, it is greatly to be feared, that the culture of land may long remain without its expected benefits from Chemistry.

In a similar way, also, the benefits resulting from the science of Botany, have been equally limited, and likely to remain so, until those who grow the grasses shall take the trouble to distinguish one from another, or until those who know them scientifically shall condescend to become the cultivators.

Nature furnishes the clue to each of these sciences, and to the most extensive application of their benefits.

She has also given the Farmer other more easy helps, to much of the useful knowledge he requires.

The method of knowing the Substrata from each other by their various substances imbedded, will consequently shew the difference in their soils. – All this is attainable by rules the most correct, and easily learnt, and also the simplest and most extensive that can well be devised; for by the help of organized Fossils alone, a science is established with characters on which all must agree, as to the extent

 DOI: 10.4324/9780429355653-27

of the Strata in which they are imbedded, those characters are universal; and a knowledge of them opens the most extensive sources of information, without the necessity of deep reading, or the previous acquirement of difficult arts.

The organized Fossils (which might be called the antiquities of Nature,) and their localities also, may be understood by all, even the most illiterate: for they are so fixed in the earth as not to be mistaken or misplaced; and may be as readily referred to in any part of the course of the Stratum which contains them, as in the cabinets of the curious; and, consequently, they furnish the best of all clues to a knowledge of the Soil and Substrata.

The practicality of thus distinguishing so great a variety of materials in the earth, as successively terminate at the surface being admitted; and their courses delineated in a large map of the Strata just published; I may now confidently proceed with a general account of those organized Fossils, which I found imbedded in each Stratum, and which first enabled me more particularly to distinguish one Stratum from another.

Fossil Shells had long been known amongst the curious, collected with care, and preserved in their cabinets, along with other rarities of nature, without any apparent use. That to which I have applied them is new, and my attention was first drawn to them, by a previous discovery of regularity in the direction and dip of the various Strata in the hills around Bath; for it was the nice distinction which those similar rocks required, which led me to the discovery of organic remains peculiar to each Stratum. Their perfect state of preservation, and most tender structure, raised a doubt respecting their diluvian origin, and a close attention to the Gravel Fossils, clearly provided *two distinct operations of water*.

The Fossils of the former deposit being all finely preserved, while those of the latter (which are chiefly superficial,) are all greatly rounded by attrition. Those of the first class are never found but in their respective sites in the Strata; – those of the latter, by their promiscuous mixture, superficial situation, and other circumstances, most strongly confirm the previous deposit and complete induration of the Strata which contain the former. Conceiving, therefore, the Gravel Fossils to be the most indubitable effects of a great body of water passing over the surface of the earth, with violence sufficient to tear up fragments of the Strata, round them by attrition, and drive them many miles from their regular beds to the promiscuous situations which they now occupy. These have been called *alluvial Fossils*, and the Gravel which contains them being thus clearly distinguished from the regular Strata beneath, much of the mystery in which Fossil Shells, and other materials of the earth were involved, seemed to be removed by this distinction [. . .]

Strata with Organized Fossils

THE eastern and south-eastern half of England, so far inland as a curved line from Exeter to Teesmouth, abounds with organized Fossils, regularly imbedded in the Strata. The vast expanse of red Marl and its Sandstone, has none of them, but they are very abundant in the Limestones which accompany it.

These, however, occupy but a small portion of the island, compared to the great extent of Strata before mentioned, and when it is considered that in the remainder of the Strata, Red and Dunstone, Killas and Granite, organized Fossils are not found, or very rare; they seem chiefly confined to the district before described, and to the Coalmeasures, the former nearly all animal, and the latter chiefly vegetable. The Muscles and Ammonites, found in Ironstone of the Coalmeasures, and the bituminized wood of blue clays, in the other district, being trifling exceptions to general rules so extensive.

The eastern side of the island is, therefore, best for the commencement of regular observations on the organized Fossils which are illustrative of its Geology. It is also necessary that the series of British Strata, for the simplification of science, should be considered in classes. The part above the chalk is one, and the principal divisions of which it is susceptible, are reducible to two – a great Sand and a great Clay, with a general parting of Crag; but each of these is subject to considerable variations.

The Sand lies next the Chalk, and the clay over that forms insular hills.

The great Sand is in many places interspersed with Clay, or Brickearth, and the Clay as frequently with Sand and Loam. Pebbles are common to both, but to what depth beneath the surface may be difficult to determine.

The chief partition Strata have not always the same appearance. The Crag being, in some parts of its course, composed of shells and sand, in some places of shells and clay, and in others of shells and coral, united in a soft stoney rock, which about Orford is used in building. In other places the shells are filled with, and imbedded in a hard blue grey Sandstone, and in some parts of their course they appear to be deficient, or found only thinly interspersed with a blue grey concreted loam, or indurated Brickearth. The alluvial Pebbles, Clay, and Sand spread over great breadths of the plains formed by the surface of these thin partition Strata, much increases the difficulty of tracing their outcrops.

[. . .]

In the present state of our knowledge of these Strata, and the shells they contain, any attempt at a minute division of them seems, therefore, more likely to perplex than instruct the reader.

The strong features only of the country, will therefore, first be noticed. The order of nature which is shown by my discoveries, suggests the outline of the work, and the different Strata serving like chapters for the principal divisions, the subject will be so treated; taking each of their outcrops in succession, from East to West. The figures of organized Fossils in each Stratum are printed on coloured paper, to correspond with the most general colour of the matter in which they are imbedded, and also with that by which their courses are represented on the Map; where otherwise, as in the Chalk, it will be particularly noticed under each head.

In England this class is separated into three portions, by vacancies on the heights of Hampshire, and in the sea by the Wash. The mouth of the Humber makes also a lesser division – but for these, the class might be said to extend from Dorsetshire to Yorkshire, for Pool Harbour is in one extremity, and Bridlington Bay in the other.

The northern-most of the three principal portions, North and South of the Humber, is small, long, and narrow, lying low, and as yet little noticed for the organized Fossils, except large bones washed out of the crumbly cliffs of Holderness, which correspond with those washed out of similar cliffs on the coast of East Norfolk, Suffolk, Essex, East Kent, and South Hants.

The middle and principal portion extends north-eastward from the Hampshire Hills to the coast of Norfolk; it flanks the Chalk through Surrey and Kent, on the south side of the Thames; the Buckinghamshire and Hertfordshire Chalk Hills on the north side. It embraces the whole Estuary of the Thames; spreads over nearly all Essex, three-fourths of Suffolk, and all the eastern half of Norfolk, except the Vales about Norwich and Aylesham.

The southern portion, chiefly in Hampshire and Dorsetshire, narrows both ways from its widest part about the new Forest, to its western extremity, near Dorchester, and its eastern, near Brighton. Its widest part is from Newport in the Isle of Wight, to the similar elevations of Chalk and down lands, between Salisbury and Winchester. Each of these districts is abundantly stored with organized Fossils. Large teeth and bones, greatly resembling those on the Continent, have been most frequently collected from the shores of the middle portion, and large vertebrae further inland, at Whitlingham, Leiston old Abbey, Diss, Hoxney, and Hawkedon.

24

REV. W.D. CONYBEARE, 'ON THE DISCOVERY OF AN ALMOST PERFECT SKELETON OF THE PLESIOSAURUS'

Transactions of the Geological Society of London, 1823

I am highly gratified in being able to lay before the Society an account of an almost perfect skeleton of Plesiosaurus, a new fossil genus, which, from the consideration of several fragments found only in a disjointed state, I felt myself authorized to propound in the year 1821, and which I described in the Geological Transactions for that and the following year. It is through the kind liberality of its possessor, the Duke of Buckingham, that this specimen has been placed for a time at the disposal of my friend Professor Buckland for the purpose of scientific investigation.

At the period of my former communications it was natural and even just that in the minds of many persons interested in such researches, much hesitation should be felt in admitting the conclusions of an observer who was avowedly inexperienced in comparative anatomy; and there might have then appeared reasonable ground for the suspicion that, like the painter in Horace, I had been led to constitute a fictitious animal from the juxtaposition of incongruous members, referable in truth to different species. But the magnificent specimen recently discovered at Lyme has confirmed the justice of my former conclusions in every essential point with the organization of the skeleton.

[. . .]

The specimen now exhibited presents other [particulars] of a most novel and interesting character, not to have been anticipated previously to the discovery of a skeleton the whole exterior portion of whose vertebral column was perfect. I particularly allude to the neck, which is fully equal in length to the body and tail united; and which surpassing in the number of its vertebrae that of the longest-necked birds, even the swan, deviates from the laws which were heretofore regarded as universal in quadruped animals, and the cetacea. I mention this circumstance thus early as forming the most prominent and interesting feature of the recent discovery, and that which in effect renders this animal one of the

DOI: 10.4324/9780429355653-28

most curious and important additions which geology has yet made to comparative anatomy.

I now proceed to the details in the usual order.

Head. – The present specimen, and another of this part only, in possession of Miss Philpot, confirm the restoration attempted from the distorted head figured in Plate XIX of the first volume of the second series of the Geological Transactions; and the latter extends our knowledge by exhibiting distinctly the occipital portion. We now also learn for the first time, that the head of this animal was remarkably small, forming less than the thirteenth part of the total length of the skeleton; while in the Ichthyosaurus its proportion is one-fourth. This proportional smallness of the head, and therefore of the teeth, must have rendered it a very unequal combatant against the latter animal, but the structure of its neck may perhaps be considered as compensating provision, supplying it with the means of security and of catching its prey.

Vertebrae. – The distinctions between the cervical and caudal vertebræ have been fully and correctly stated in my former communications; but I had not at that time observed more than twelve of the cervical, whereas the present specimen exhibits about thirty-five, or, including the anterior dorsal, which were placed before the humerus and bore only five ribs, forty-one. This great increase in the number of joints in the neck, is the more remarkable from the rigour with which nature appears, in most cases, to have enforced the law of a very limited number. In all quadrupedal animals, in all the mammalia (excepting only the tridactyl sloths which have nine), the series is exactly seven; and so strict is the adherence to this rule, that even the short and stiff neck of the whale, and the long and flexible neck of the camelopard, are formed out of the same elementary number; the vertebræ in the former instance being extremely thin and anchored together, and in the latter greatly elongated. [. . .]

25

CHARLES LYELL, *PRINCIPLES OF GEOLOGY, BEING AN ATTEMPT TO EXPLAIN THE FORMER CHANGES OF THE EARTH'S SURFACE, BY REFERENCE TO CAUSES NOW IN OPERATION*, 3 VOLS

(London: John Murray, 1830–3), Vol 1 (1830), Vol 2 (1832), Vol. 3 (1833)

Volume 1

Chapter 1

GEOLOGY is the science which investigates the successive changes that have taken place in the organic and inorganic kingdoms of nature; it enquires into the causes of these changes, and the influence which they have exerted in modifying the surface and external structure of our planet.

By these researches into the state of the earth and its inhabitants at former periods, we acquire a more perfect knowledge of its present condition, and more comprehensive views concerning the laws now governing its animate and inanimate productions. When we study history, we obtain a more profound insight into human nature, by instituting a comparison between the present and former states of society. We trace the long series of events which have gradually led to the actual posture of affairs; and by connecting effects with their causes, we are enabled to classify and retain in the memory a multitude of complicated relations – the various peculiarities of national character – the different degrees of moral and intellectual refinement, and numerous other circumstances, which, without historical associations, would be uninteresting or imperfectly understood. As the present condition of nations is the result of many antecedent changes, some extremely remote and others recent, some gradual, others sudden and violent, so the state of the natural world is the result of a long succession of events, and if we would

 DOI: 10.4324/9780429355653-29

enlarge our experience of the present economy of nature, we must investigate the effects of her operations in former epochs.

We often discover with surprise, on looking back into the chronicles of nations, how the fortune of some battle has influenced the fate of millions of our contemporaries, when it has long been forgotten by the mass of the population. With this remote event we may find inseparably connected the geographical boundaries of a great state, the language now spoken by the inhabitants, their peculiar manners, laws, and religious opinions. But far more astonishing and unexpected are the connexions brought to light, when we carry back our researches into the history of nature. The form of a coast, the configuration of the interior of a country, the existence and extent of lakes, valleys, and mountains, can often be traced to the former prevalence of earthquakes and volcanoes, in regions which have long been undisturbed. To these remote convulsions the present fertility of some districts, the sterile character of others, the elevation of land above the sea, the climate, and various peculiarities, may be distinctly referred. On the other hand, many distinguishing features of the surface may often be ascribed to the operation at a remote era of slow and tranquil causes – to the gradual deposition of sediment in a lake or in the ocean, or to the prolific growth in the same of corals and testacea. To select another example, we find in certain localities subterranean deposits of coal, consisting of vegetable matter, formerly drifted into seas and lakes. These seas and lakes have since been filled up, the lands whereon the forests grew have disappeared or changed their form, the rivers and currents which floated the vegetable masses can no longer be traced, and the plants belonged to species which for ages have passed away from the surface of our planet. Yet the commercial prosperity, and numerical strength of a nation, may now be mainly dependant on the local distribution of fuel determined by that ancient state of things.

Geology is intimately related to almost all the physical sciences, as is history to the moral. An historian should, if possible, be at once profoundly acquainted with ethics, politics, jurisprudence, the military art, theology; in a word, with all branches of knowledge, whereby any insight into human affairs, or into the moral and intellectual nature of man, can be obtained. It would be no less desirable that a geologist should be well versed in chemistry, natural philosophy, mineralogy, zoology, comparative anatomy, botany; in short, in every science relating to organic and inorganic nature. With these accomplishments the historian and geologist would rarely fail to draw correct and philosophical conclusions from the various monuments transmitted to them of former occurrences [. . .]

Volume 3

Chapter 1

Having considered, in the preceding volumes, the actual operation of the causes of change which affect the earth's surface and its inhabitants, we are now about to enter upon a new division of our inquiry, and shall therefore offer a few preliminary observations, to fix in the reader's mind the connexion between two distinct parts of our work, and to explain in what manner the plan pursued by us differs from that more usually followed by preceding writers on Geology.

All naturalists, who have carefully examined the arrangement of the mineral masses composing the earth's crust, and who have studied their internal structure and fossil contents, have recognized therein the signs of a great succession of former changes; and the causes of these changes have been the object of anxious inquiry. As the first theorists possessed but a scanty acquaintance with the present economy of the animate and inanimate world, and the vicissitudes to which these are subject, we find them in the situation of novices, who attempt to read a history written in a foreign language, doubting about the meaning of the most ordinary terms; disputing, for example, whether a shell was really a shell, – whether sand and pebbles were the result of aqueous trituration, – whether stratification was the effect of successive deposition from water; and a thousand other elementary questions which now appear to us so easy and simple, that we can hardly conceive them to have once afforded matter for warm and tedious controversy.

In the first volume we enumerated many prepossessions which biased the minds of the earlier inquirers, and checked an impartial desire of arriving at truth. But of all the causes to which we alluded, no one contributed so powerfully to give rise to a false method of philosophizing as the entire unconsciousness of the first geologists of the extent of their own ignorance respecting the operations of the existing agents of change.

They imagined themselves sufficiently acquainted with the mutations now in progress in the animate and inanimate world, to entitle them at once to affirm, whether the solution of certain problems in geology could ever be derived from the observation of the actual economy of nature, and having decided that they could not, they felt themselves at liberty to indulge their imaginations, in guessing at what might be, rather than in inquiring what is; in other words, they employed themselves in conjecturing what might have been the course of nature at a remote period, rather than in the investigation of what was the course of nature in their own times.

It appeared to them more philosophical to speculate on the possibilities of the past, than patiently to explore the realities of the present, and having invented theories under the influence of such maxims, they were consistently unwilling to test their validity by the criterion of their accordance with the ordinary operations of nature. On the contrary, the claims of each new hypothesis to credibility appeared enhanced by the great contrast of the causes or forces introduced to those now developed in our terrestrial system during a period, as it has been termed, of repose.

Never was there a dogma more calculated to foster indolence, and to blunt the keen edge of curiosity, than this assumption of the discordance between the former and the existing causes of change. It produced a state of mind unfavourable in the highest conceivable degree to the candid reception of the evidence of those minute, but incessant mutations, which every part of the earth's surface is undergoing, and by which the condition of its living inhabitants is continually made to vary. The student, instead of being encouraged with the hope of interpreting the enigmas presented to him in the earth's structure, – instead of being prompted to undertake laborious inquiries into the natural history of the organic world, and the complicated effects of the igneous and aqueous causes now in operation, was taught to despond from the first. Geology, it was affirmed, could never rise to the rank of an exact science, – the great number of phenomena must for ever remain inexplicable, or only be partially elucidated by ingenious conjectures. Even the mystery which invested the subject was said to constitute one of its principal charms, affording, as it did, full scope to the fancy to indulge in a boundless field of speculation.

The course directly opposed to these theoretical views consists in an earnest and patient endeavour to reconcile the former indications of change with the evidence of gradual mutations now in progress; restricting us, in the first instance, to known causes, and then speculating on those which may be in activity in regions inaccessible to us. It seeks an interpretation of geological monuments by comparing the changes of which they given evidence with the vicissitudes now in progress, or which may be in progress.

We shall give a few examples in illustration of the practical results already derived from the two distinct methods of theorizing, for we have now the advantage of being enabled to judge by experience of their respective merits, and by the relative value of the fruits which they have produced.

In our historical sketch of the progress of geology, the reader has seen that a controversy was maintained for more than a century, respecting the origin of fossil shells and bones – were they organic or inorganic substances? That the latter opinion should for a long time have prevailed, and that these bodies should have been supposed to be fashioned into their present form by a plastic virtue, or some other mysterious agency, may appear absurd; but it was, perhaps, as reasonable a conjecture as could be expected from those who did not appeal, in the first instance, to the analogy of the living creation, as affording the only source of authentic information. It was only by an accurate examination of living testacea, and by a comparison of the osteology of the existing vertebrated animals with the remains found entombed in ancient strata, that this favourite dogma was exploded, and all were, at length, persuaded that these substances were exclusively of organic origin.

In like manner, when a discussion had arisen as to the nature of basalt and other mineral masses, evidently constituting a particular class of rocks, the popular opinion inclined to a belief that they were of aqueous, not of igneous origin. These rocks, it was said, might have been precipitated from an aqueous solution, from a chaotic fluid, or an ocean which rose over the continents, charged with the requisite mineral ingredients. All are now agreed that it would have been impossible for human ingenuity to invent a theory more distant from the truth; yet we must cease to wonder, on that account, that it gained so many proselytes, when we remember that its claims to probability arose partly from its confirming the assumed want of all analogy between geological causes and those now in action.

By what train of investigation were all theorists brought round at length to an opposite opinion, and induced to assent to the igneous origin of these formations? By an examination of the structure of active volcanos, the mineral composition of their lavas and ejections, and by comparing the undoubted products of fire with the ancient rocks in question.

We shall conclude with one more example. When the organic origin of fossil shells had been conceded, their occurrence in strata forming some of the loftiest mountains in the world, was admitted as a proof of a great alteration of the relative level of sea and land, and doubts were then entertained whether this change might be accounted for by the partial drying up of the ocean, or by the elevation of the solid land. The former hypothesis, although afterwards abandoned by general consent, was at first embraced by a vast majority. A multitude of ingenious speculations were hazarded to show how the level of the ocean might have been depressed, and when these theories had all failed, the inquiry, as to what vicissitudes of this nature might now be taking place, was, as usual, resorted to in the last instance. The question was agitated, whether any changes in the level of sea and land had occurred during the historical period, and, by patient research, it was soon discovered that considerable tracts of land had been permanently elevated and depressed, while the level of the ocean remained unaltered. It was therefore necessary to reverse the doctrine which had acquired so much popularity, and the unexpected solution of a problem at first regarded as so enigmatical, gave perhaps the strongest stimulus ever yet afforded to investigate the ordinary operations of nature. For it must have appeared almost as improbable to the earlier geologists, that the laws of earthquakes should one day throw light on the origin of mountains, as it must to the first astronomers, that the fall of an apple should assist in explaining the motions of the moon.

Of late years the points of discussion in geology have been transferred to new questions, and those, for the most part, of a higher and more general nature; but, notwithstanding the repeated warnings of experience, the ancient method of philosophising has not been materially modified.

We are now, for the most part, agreed as to what rocks are of igneous, and what of aqueous origin, – in what manner fossil shells, whether of the sea or of lakes, have been imbedded in strata, – how sand may have been converted into sandstone, – and are unanimous as to other propositions which are not of a complicated nature; but when we ascend to those of a higher order, we find as little disposition, as formerly, to make a strenuous effort, in the first instance, to search out an explanation in the ordinary economy of Nature. If, for example, we seek for the causes why mineral masses are associated together in certain groups; why they are arranged in a certain order which is never inverted; why there are many breaks in the continuity of the series; why different organic remains are found in distinct sets of strata; why there is often an abrupt passage from an assemblage of species contained in one formation to that in another immediately superimposed, – when these and other topics of an equally extensive kind are discussed, we find the habit of indulging conjectures, respecting irregular and extraordinary causes, to be still in full force.

We hear of sudden and violent revolutions of the globe, of the instantaneous elevation of mountain chains, of paroxysms of volcanic energy, declining according to some, and according to others increasing in violence, from the earliest to the latest ages. We are also told of general catastrophes and a succession of deluges, of the alternation of periods of repose and disorder, of the refrigeration of the globe, of the sudden annihilation of whole races of animals and plants, and other hypotheses, in which we see the ancient spirit of speculation revived, and a desire manifested to cut, rather than patiently to untie, the Gordian knot.

In our attempt to unravel these difficult questions, we shall adopt a different course, restricting ourselves to the known or possible operations of existing causes; feeling assured that we have not yet exhausted the resources which the study of the present course of nature may provide, and therefore that we are not authorized, in the infancy of our science, to recur to extraordinary agents. We shall adhere to this plan, not only on the grounds explained in the first volume, but because, as we have above stated, history informs us that this method had always put geologists on the road that leads to truth, – suggesting views which, although imperfect at first, have been found capable of improvement, until at last adopted by universal consent. On the other hand, the opposite method, that of speculating on a former distinct state of things, has led invariably to a multitude of contradictory systems, which have been overthrown one after the other, – which have been found quite incapable of modification, – and which are often required to be precisely reversed.

[. . .]

Since in our attempt to solve geological problems, we shall be called upon to refer to the operation of aqueous and igneous causes, the geographical distribution of animals and plants, the real existence of species, their successive extinction, and so forth, we were under the necessity of collecting together a variety of facts, and of entering into long trains of reasoning, which could only be accomplished in preliminary treatises.

These topics we regard as constituting the alphabet and grammar of geology; not that we expect from such studies to obtain a key to the interpretation of all geological phenomena, but because they form the groundwork from which we must rise to the contemplation of more general questions relating to the complicated results to which, in an indefinite lapse of ages, the existing causes of change may give rise.

Chapter 26

[. . .]

Concluding Remarks. – In our history of the progress of geology, in the first volume, we stated that the opinion originally promulgated by Hutton, 'that the strata called primitive were mere altered sedimentary rocks', was vehemently opposed for a time, the main objection to the theory being its supposed tendency to promote a belief in the past eternity of our planet. Previously the absence of animal and vegetable remains in the so-called primitive strata, had been appealed to, as proving that there had been a period when the planet was uninhabited by living beings, and when, as was also inferred, it was uninhabitable, and, therefore, probably in a nascent state.

The opposite doctrine, that the oldest visible strata might be the monuments of an antecedent period, when the animate world was already in existence, was declared to be equivalent to the assumption, that there never was a beginning to the present order of things. The unfairness of this charge was clearly pointed out by Playfair, who observed, 'that it was one thing to declare that we had not yet discovered the traces of a beginning, and another to deny that the earth ever had a beginning'.

We regret, however, to find that the bearing of our arguments in the first volume has been misunderstood in a similar manner, for we have been charged with endeavouring to establish the proposition, that 'the existing causes of change have operated with absolute uniformity from all eternity'.

It is the more necessary to notice this misrepresentation of our views, as it has proceeded from a friendly critic whose theoretical opinions coincide in general with our own, but who has, in this instance, strangely misconceived the scope of our argument. With equal justice might an astronomer be accused of asserting, that the works of creation extend throughout infinite space, because he refuses to take for granted that the remotest stars now seen in the heavens are on the utmost verge of the material universe. Every improvement of the telescope has brought thousands of new worlds into view, and it would, therefore, be rash and unphilosophical to imagine that we already survey the whole extent of the vast scheme, or that it will ever be brought within the sphere of human observation.

But no argument can be drawn from such premises in favour of the infinity of the space that has been filled with worlds; and if the material universe has any limits, it then follows that it must occupy a minute and infinitesimal point in infinite space. So, if in tracing back the earth's history, we arrive at the monuments of events which may have happened millions of ages before our times, and if we still find no decided evidence of a commencement, yet the arguments from analogy in support of the probability of a beginning remain unshaken; and if the past duration of the earth be finite, then the aggregate of geological epochs, however numerous, must constitute a mere moment of the past, a mere infinitesimal portion of eternity.

It has been argued, that as the different states of earth's surface, and the different species by which it has been inhabited, have had each their origin, and many of them their termination, so the entire series may have commenced at a certain period. It has also been urged, that as we admit the creation of man to have occurred at a comparatively modern epoch – as we concede the astonishing fact of the first introduction of a moral and intellectual being, so also we may conceive the first creation of the planet itself.

We are far from denying the weight of this reasoning from analogy; but although it may strengthen our conviction, that the present system of change has not gone on from eternity, it cannot warrant us in presuming that we shall be permitted to behold the signs of the earth's origin, or the evidences of the first introduction into it of organic beings.

In vain do we aspire to assign limits to the works of creation in space, whether we examine the starry heavens, or that world of minute animalcules which is revealed to us by the microscope. We are prepared, therefore, to find that in time also, the confines of the universe lie beyond the reach of mortal ken. But in whatever direction we pursue our researches, whether in time or space, we discover everywhere the clear proofs of a Creative Intelligence, and of His foresight, wisdom, and power.

As geologists, we learn that it is not only the present condition of the globe that has been suited to the accommodation of myriads of living creatures, but that many former states also have been equally adapted to the organization and habits of prior races of beings. The disposition of the seas, continents, and islands, and the climates have varied; so it appears that the species have been changed, and yet they have all been so modeled, on types analogous to those of existing plants and animals, as to indicate throughout a perfect harmony of design and unity of purpose. To assume that the evidence of the beginning or end of so vast a scheme lies within the reach of our philosophical inquiries, or even of our speculations, appears to us inconsistent with a just estimate of the relations which subsist between the finite powers of man and the attributes of an Infinite and Eternal Being.

26

ETHELDRED BENETT, *A CATALOGUE OF THE ORGANIC REMAINS OF THE COUNTY OF WILTS*

(Warminster: J.L. Vardy, 1831)

Preface

SOME years since, at the request of Sir Richard Colt Hoare, I undertook to draw up the best account I could of the Geology of South Wiltshire: and I had proceeded as far as two hundred and fifty numbers in a catalogue of the fossils to accompany it; when my time was entirely engrossed by unforeseen circumstances, for such a length of time, as to make me almost despair of ever being able to fulfil my promise; and my subsequent ill health extinguished what little hope remained of my being able to accomplish it.

During the last summer, Sir R.C. Hoare again expressed a wish, that so interesting a portion of the history of South Wiltshire, should not be passed over in silence, and the following pages are my attempt to illustrate it. Those who know me best will be fully aware, that I have endeavoured to render the catalogue as correct as possible; and when I mention that it has been approved by Mr Greenough, it will run no risk of being despised in the Geological World.

If it should be objected to my new names in the genus Polypothecia, that they are all derived from external form; I beg to state, that three scientific gentlemen undertook, at different times, to describe and name this class of fossils, and to each I offered all the assistance which my very large collection of them afforded; that all have disappointed me; and having waited fifteen years, and the fossils being now, by the death of the late Mr J.S. Miller, again on my hands unnamed, I have done the best I could. Mr Miller did, however, publish a Prospectus of a work on them; to that I am indebted for the generic name 'Polypothecia'; and Mr D. Don obligingly gave me his valuable assistance in latinizing the characters I wished to express in their specific names.

When this catalogue was first thought of, my geological friends expressed a wish that it should be published separately; but considering it a thing of mere local interest, I have preferred printing a few copies only for the acceptance of my Friends.

[. . .]

DOI: 10.4324/9780429355653-30

TO SIR RICHARD COLT HOARE, BART.

Dear Sir,

In compliance with your request, I have drawn up the following slight Sketch of the Geology of South Wilts, and which I hope you may deem worthy of the place you have assigned to it in your work.

Our County and particularly the southern part of it, is exceedingly rich in Organic Remains; and is therefore not less interesting to the Geologist than to the Antiquary. Numerous Elephants' Teeth were dug up some years since at Fisherton Anger, near Salisbury, proving the Diluvian Detritus to exist there.

The London Clay is found at Clarendon Park, in a field on the road side leading to Romsey. The Plastic Clay occurs on Chittern Down, near Heytesbury; and the Beach Pebbles found there, form the pavement of the ladies' grottoes of the surrounding neighbourhood.

The downs are of great extent, on this side of the County; and the fossil content of those of Norton Bavent, Heytesbury, and their immediate vicinity, bear a close resemblance to those of Sussex; but those of Warminster, and Clay Hill, are essentially different, and much more sparing in their fossil contents: while, on the contrary, the Chalk of Pertwood, Chicklade, Berwick St Leonard, and Wiley, all near Hindon; and Ditchampton, near Wilton; is remarkable for the abundance of its Alcyonic Remains, chiefly in Flints, Echini, &c.; all of which vary materially from any of the other places specified.

The Chalk Marl, which is so local as to have been altogether unnoticed by Mr. Wm. Smith, is exceedingly well defined at Norton Bavent, at Bishopstrow, and at Stourton.

The town of Warminster stands on the Green Sand; and the remains of Alcyonia with which it abounds, more particularly on the west of the town, seem almost inexhaustible: a few remains of Testacea are sparingly scattered among them, but at Chute Farm, near Longleat, in a field called Brimsgrove, it would seem, said the late Mr Wm. Cunnington, as if a cabinet had been emptied of its contents, so numerous, and so various *were* the Organic Remains found there; now become scarce; but chiefly small species.

At Crockerton, south-west of Warminster, the Clay from below the Sand makes its appearance, with its accompanying fossils; and at the same bed occurs at Rudge, near Chilmark, Fossil Resin, similar to that at Highgate, is found at both places, but very sparingly, and at both the Clay is used for bricks and pottery.

I am not sufficiently acquainted with the late division of the Green Sand Formation into Upper and Lower Green Sand, to determine to which the Sand Hills belong, which rise at East Knoyle, and continue in a ridge to Fonthill, and on which Fonthill Abbey stands; but the Alcyonia of Warminster are not found there that I am aware of; while at Dinton, enough are seen to prove the identity of the Green Sand of Rudge, Dinton, and Barford, with that of Warminster. A bed of Gryphæa, more than a foot thick, is the peculiarity of the Dinton Sand Ridge, and they are plentiful at Rudge: these shells are siliceous casts at both these places; but at Dilton's Marsh, north-west of Warminster, where they are also numerous, they appear more like recent dead shells, chalky and brittle.

The Portland beds are in great strength at Tisbury; and Chicksgrove Quarry in that parish, is one of more than usual interest to the Geologist, on account of the fine section which it shows of sixteen beds of this series, singularly rich in Organic Remains: and the Purbeck Beds on Lady Down, also in the parish of Tisbury, have shewn that they contain the Icthyological Treasures of Dorsetshire. The siliceous Madrepore of Tisbury, is a subordinate bed in this series, and which has not yet been found elsewhere, with the exception of the agatized Madrepores of Antigua: they were first discovered by being turned up by the plough; but the sinking of a well at Burton's Cottage, near the Inn at Fonthill Gifford, has proved their geological position to be over the Portland Rock; they are extremely local.

The Kimmeridge Clay is seen near the Church at Tisbury, but I am unacquainted with its contents there: it appears again, with its characteristic fossils, at Binley Farm, also in the parish of Tisbury, to the west of Pythouse; and this appears to be the lowest stratum in this part of our County.

In North Wilts, the Coral Rag predominates at Blunsdon; and the fossils of the Kelloway Rock, and the beautiful Echini of Calne, have brought those beds of the Oolitic series into notice. Bradford is also indebted to the Pear Encrinite [. . .] for its celebrity in the Geological World.

From the above localities I have formed my Collection of Wiltshire Fossils: it is peculiarly rich in Alcyonia; probably not to be surpassed in those from the Green Sand Formation. I subjoin a Catalogue of the principal fossils named as far as they have come under my notice: those which are marked with an Asterisk are in my own Cabinet; the one marked n.g. is a new Genus; and those marked n.s are new Species.

E.B. 1st January, 1831

Since writing the above, I have found the following Memorandum:

Geological position of the Siliceous Madrepore. The sinking of a well in the field called Butcher's Knap, in the parish of Tisbury, the only place where the Coral Flint has been found, and which led to the discovery of the bed. – The usual rubble of the Portland beds, in Tisbury, ten feet. – Siliceous madrepore, one foot. – The usual succession of the Portland beds, in Tisbury, forty-two feet. – Water. – No sand between the beds.

A CATALOGUE OF WILTSHIRE ORGANIC REMAINS

[. . .]

POLYPI GLUTINUS

Order 7, Lamarck.

Spongiadæ			
Spongia botyroides	*Icones fossiles*, f. 32	Green Sand	Warminster
Spongia bullata		Ibid.	Chute Farm
Spongia undulata		Ibid.	Warminster
Spongia cucullate		Ibid	Ibid.
Spongus labyrinthicus	Geol. Suss. t.15, f. 7	Chalk Flints	Norton Bavent
Spongites cyathoides		Chalk Mark	Ibid.
Spongites urcaolatus		Ibid.	Ibid.
Spongites marginatus		Ibid.	Ibid.
Ventriculites radiatus	Geol. Suss. t. 11	Upper Chalk	Pertwood & Berwick St Leonard
Ventriculites alcyonoides	Org. Rem. ii, t.10, f.12	Ibid	Norton Bavent & Heytesbury
Ventriculites quadrangularis	Geol. Suss. t.15, f.6	Chalk Flints	Wiley
Ventriculites Benettiæ	Geol Suss. t.15, f.3	Ibid.	Heytesbury
Ventriculites reticulatus		Upper chalk	Pertwood
Ventriculites punctatus		Ibid.	Ibid.
Ventriculites notatus		Ibid.	Ibid.
Ventriculites quadratus		Upper Chalk & Chalk Flints	Ibid. St Leonard, Norton Bavent, & Heytesbury

[. . .]

Polypothecia clavellata	Miller's Prospectus	Chalk Flints	Pertwood, Wiley, & Ditchampton
Polypothecia fissa	Ibid.	Ibid.	Wiley
Polypothecia latissimi	Ibid.	Ibid.	Berwick St Leonard & Pertwood
Polypothecia maxima	Ibid.	Ibid.	Berwick St Leonard
Polypothecia (stems)	Ibid	Ibid.	Ibid., Wiley, & Pertwood
Polypothecia palmata	Ibid.	Chalk Flints & Green Sand	Wiley, Pertwood, & Warminster
Polypothecia infundibulum	Strata iden. Gn. Sand, f.1	Green Sand	Pertwood & Warminster

27

REV. WILLIAM BUCKLAND, *GEOLOGY AND MINEROLOGY WITH REFERENCE TO NATURAL THEOLOGY*, TREATISE 6, *THE BRIDGEWATER TREATISES*

(London: William Pickering, 1837 [1836])

Section IX

Megalosaurus[1]

THE Megalosaurus, as its name implies, was a Lizard, of great size, of which, although no skeleton has yet been found entire, so many perfect bones and teeth have been discovered in the same quarries, that we are nearly as well acquainted with the form and dimension of its limbs, as if they had been found together in a single block of stone.

From the size and proportions of these bones, as compared with existing Lizards, Cuvier concludes the Megalosaurus to have been an enormous reptile, measuring from forty to fifty feet in length, and partaking of the structure of the Crocodile and the Monitor.

As the femur and tibia measure nearly three feet each, the entire hind leg must have attained a length of nearly two yards: a metatarsal bone, thirteen inches long, indicates a corresponding length in the foot. The bones of the thigh and leg are not solid at the centre, as in Crocodiles, and other aquatic quadrupeds, but have large medullary cavities, like the bones of terrestrial animals. We learn from this circumstance, added to the character of the foot, that the Megalosaurus lived chiefly upon the land.

In the internal condition of these fossil bones, we see the same adaptation of the skeleton to its proper element, which now distinguishes the bones of terrestrial, from those of aquatic Saurians.[2] In the Ichthyosauri and Plesiosauri, whose paddles were calculated exclusively to move in water, even the largest bones of the arms and legs were solid throughout. Their weight would in no way have embarrassed their action in the fluid medium they inhabited; but in the huge Megalosaurus, and still more giant Iguanodon, which are shown by the character of their feet

 DOI: 10.4324/9780429355653-31

to have been fitted to move on land, the larger bones of the legs were diminished in weight by being internally hollow, and having their cavities filled with the light material of marrow, while their cylindrical form tended also to combine this lightness with strength.

The form of the teeth shows the Megalosaurus to have been in a high degree carnivorous: it probably fed on smaller reptiles, such as Crocodiles and Tortoises, whose remains abound in the same strata with its bones. It may also have taken to the water in pursuit of Plesiosauri and fishes.[3]

The most important part of the Megalosaurus yet found, consists of a fragment of the lower jaw, containing many teeth [. . .]. The form of this jaw shows that the head was terminated by a straight and narrow snout, compressed laterally like that of the Delphinus Gangeticus.

As in all animals, the jaws and teeth form the most characteristic parts, I shall limit my present observations to a few striking circumstances in the dentition of the Megalosaurus. From these we learn that the animal was a reptile, closely allied to some of our modern Lizards; and viewing the teeth as instruments for providing food to a carnivorous creature of enormous magnitude, they appear to have been admirably adapted to the destructive office for which they were designed [. . .]

In the structure of these teeth we find a combination of mechanical contrivances analogous to those which are adopted in the construction of the knife, the sabre, and the saw. When first protruded above the gum, the apex of each tooth presented a double cutting edge of serrated enamel. In this stage, its position and line of action were nearly vertical, and its form like that of the two-edged point of a sabre, cutting equally on each side. As the tooth advanced in growth, it became curved backwards in the form of a pruning knife, and the edge of serrated enamel was continued down to the base of the inner and cutting side of the tooth whilst, on the outer side, a similar edge descended, but to a short distance from the point, and the convex portion of the tooth became blunt and thick, as the back of a knife is made thick, for the purpose of producing strength.

[. . .]

Iguanodon[4]

As the reptiles hitherto considered appear from their teeth to have been carnivorous, so we find extinct species of the same great family, that assumed the character and office of herbivora. For our knowledge of this genus, we are indebted to the scientific researches of Mr Mantell. This indefatigable historian of the Wealden fresh-water formation, has not only found the remains of the Plesiosaurus, Megalosaurus, Hylæosaurus‡ and several species of Crocodiles and Tortoises in these deposits, of a period intermediate between the oolitic and cretaceous series, but has also discovered in Tilgate Forest, the remains of the Iguanodon, a reptile much more gigantic than the Megalosaurus, and which, from the character of its teeth, appears to have been herbivorous.[5] The teeth of the Iguanodon are so precisely similar, in the principles of their construction, to the teeth of the modern Iguana, as to leave no doubt of the near connection of this most gigantic extinct reptile with the Iguanas of our own time. When we consider that the largest living Iguana rarely exceeds five feet in length, whilst the congenerous fossil animal must have been nearly twelve times as long, we cannot but be impressed by the discovery of a resemblance, amounting almost to identity, between such characteristic organs as the teeth, in one of the most enormous among the extinct reptiles of the fossil world, and those of a genus whose largest species is comparatively so diminutive [. . .]

We know from the fragment of a femur, in the collection of Mr Mantell, that the thigh-bone of this reptile much exceeded in bulk that of the largest Elephant: this fragment presents a circumference of twenty-two inches in its smallest part, and the entire length must have been between four and five feet. Comparing the proportions of this monstrous bone with those of the fossil teeth with which it is associated, it appears that they bear to one another nearly the same ratio that the femur of the Iguana bears to the similarly constructed and peculiar teeth of that animal.

Notes

1 Buckland's footnote: 'This genus was established by the Author, in a Memoir, published in the Geol. Trans. of London, (Vol. I., N. S. Pt, 2, 1824), and was founded upon specimens discovered in the oolitic slate of Stonesfield, near Oxford, the place in which these bones have as yet chiefly occurred. Mr Mantell has discovered remains of the same animal in the Wealden fresh-water formation of Tilgate Forest; and from this circumstance we infer that it existed during the deposition of the entire series of oolitic strata. The author, in 1826, saw fragments of a jaw, containing teeth, and of some other bones of Megalosaurus, in the museum at Besançon, from the oolite of that neighbourhood'. In 1827, Mantell added an epithet in Buckland's honour, naming the creature *Megalosaurus bucklandii.*

2 'I learn from Mr Owen that the long bones of land Tortoises have a close cancellous internal structure, but not a medullary cavity'.

3 'Mr Broderip informs me that a living Iguana (I. Tuberculata), in the gardens of the Zoological Society of London, in the summer of 1834, was observed frequently to enter the water, and swim across a small pond, using its long tail as the instrument of progression, and keeping its fore feet motionless'. Although there is no evidence to disprove Buckland's theory, it is more likely that Megalosaurs were coastal but terrestrial predators and scavengers. Buckland refers to *Guide to the Gardens of the Zoological Society* (1829) by Broderip and Nicholas A. Vigors.

4 Buckland directs readers to Mantell's *Geology of Sussex.*

5 Buckland's footnote: 'The Iguanodon has hitherto been found only, with one exception, in the Wealden fresh-water formation of the south of England, intermediate between the marine oolitic deposits of the Portland stone and those of the greensand formation in the cretaceous series. The discovery, in 1834, (Phil. Mag. July 1834, p. 77), of a large proportion of the skeleton of one of these animals, in strata of the latter formation, in the quarries of Kentish Rag, near Maidstone, shews that the duration of this animal did not cease with the completion of the Wealden series. The individual from which this skeleton was derived had probably drifted to sea, as those which afforded the bones found in the fresh-water deposits subsequent to this marine formation, had been drifted into an estuary. This unique skeleton is now in the museum of Mr Mantell, and confirms nearly all his conjectures respecting the many insulated bones which he had referred to the Iguanodon'.

28

RODERICK MURCHISON, *THE SILURIAN SYSTEM*

(London: John Murray, 1839)

Part I: The Silurian System

Chapter I, Introduction

THE chief design of the present work is, to fill up an interval in geological history, by describing certain strata, which, although they occupy a considerable thickness in the crust of the globe, and connect the secondary deposits with the older slaty rocks, have never yet been adequately examined.

A few words will explain the previous condition of this subject.– When the materials of the earth's crust first became a subject of study, they were viewed principally with reference to their mineral characters; but the attention which was afterwards directed to the imbedded animals and plants, gradually produced a revolution in the science, and gave birth to what is now the largest and most important part of Geology. Since the period when Smith in England, and Cuvier and Brogniart in France, first identified strata by their fossils, a most rapid progress has been made in the application of this method of testing the age of rocks.

Sixteen years have now elapsed since Conybeare and Phillips, in their Outlines of the Geology of England and Wales, presented us with a connected view of the succession of the sedimentary British deposits, from the most superficial to those which support the carboniferous system; and, in the succeeding years, great and important additions have been contributed to our acquaintance with the youngest or tertiary deposits particularly by Mr Lyell.

But this method has not been extended as to carry the chronological succession *below* the Old Red Sandstone; partly, perhaps, from a preconceived opinion, that few organic remains were likely to be detected in these formations; partly from the belief, founded on just but inadequate observation, that the many mutations which these older rocks had undergone, must have nearly obliterated the evidences of their origin [. . .]

Though, undoubtedly, such reasons deterred many from grappling with this inquiry, it must not be supposed that the ancient strata have not been studied by enlightened observers. In Great Britain we may cite the names of MacCulloch, Greenough, and Sedgwick, as those of men prominently distinguished in throwing

DOI: 10.4324/9780429355653-32

light upon them [. . .] But these authors had not the peculiar advantages which have fallen to my lot; for in the regions which they studied, there is generally an abrupt boundary-line or break between the older and newer systems. [. . .] In fact, a perfect and unbroken series of links, connecting these older rocks with the younger deposits, does not occur in any portion of these islands which had previously been examined.

On the continent, where great attention had been bestowed upon the older and crystalline rocks, from the days of De Saussure and Werner to our own, the same belief, was imposed on the minds of geologists, that the great dislocations to which these ancient rocks had been subjected, had entirely dissevered them from those fossiliferous strata with which we were well acquainted. In short, there existed no foreign work in which rocks of this age were classified according to a law, founded upon superposition and characteristic organic remains.

But to proceed to facts connected with our own country. No one was aware of the existence below the Old Red Sandstone, of a regular series of deposits, containing peculiar organic remains. For example, although it was supposed, that the limestone of Dudley was of greater antiquity than the Old Red Sandstone, no one had observed that those deposits were connected by an intermediate formation of very considerable dimensions, full of organic remains. It is this formation, now termed the 'Ludlow Rocks', which seems to have most escaped attention, whilst from its position, as will appear in the sequel, it is the key which accurately reveals to us the relations of the inferior masses to the overlying strata with which we formerly were acquainted.

[. . .]

In this condition of the subject, I first explored the borders of England and Wales in 1831. The order of succession seen in the ridges on the left bank of the Wye, between Hay in Herefordshire and Builth in Brecknockshire, where the Old Red Sandstone is distinctly underlaid by a grey fossiliferous strata, first led me to suspect, that I had met with a district containing a good part of the evidence required to lead to a systematic study of our older formations; a surmise which was confirmed by following out these rocks upon their line of bearing to the neighbourhood of Ludlow and Wenlock, where I found them much expanded [. . .] Perceiving, however, that a subject so new and so large could not be really advanced, except through patient and repeated examination, I re-explored the same districts, in 1832, and submitted details of the new acquisitions to the Geological Society; accompanying my memoirs with a set of geologically coloured maps of the Ordnance Survey.

An effort to classify these deposits was then made; but it was not until the close of the summer of 1833, that I was enabled to publish a tolerably correct synopsis or table of the various formations in the extensive region, over which my observations had progressively extended. Seven years have since elapsed, during which my attention has been almost exclusively given to the development of this subject and its collateral branches.

During the summer in which my first observations were made, Professor Sedgwick commenced a general inquiry into the structure of North Wales, for which his previous acquaintance with the slaty and crystalline rocks of Cumberland eminently qualified him. He first endeavoured to connect the transition rocks of the age of Dudley with certain calcareous slates pointed out by Mr Greenough in North Wales; but finding no concordance between them, he was, to use his own expression, 'driven to a new base line', in other words, to work upwards from the central axis or oldest rocks of Wales. As soon as he perceived that I had observed the links which connect the Old Red Sandstone with some of the inferior masses of his region, he felt the importance of pushing the inquiry, and by his encouragement I was materially stimulated to do so. In speaking of the labours of my friend, I may truly say, that he not only shed an entirely new light on the crystalline arrangement or slaty cleavage of the North Welsh mountains, but also overcame what to most men would have proved insurmountable difficulties, in determining the order and relations of these very ancient strata amid scenes of vast dislocation [. . .]

I became convinced that, as this large and ancient group contained peculiar organic remains, and was marked by distinctness of physical features, lithological structure, and order of superposition, it was well entitled to be considered a separate *system*. [. . .]

A geographical term was finally adopted, derived from the Silures, whose power extended over the region where these rocks are best displayed, and the names of whose illustrious chief Caradoc (Caractacus), has been transmitted to us in a bold range of hills, composed of one of the most important formations of the system to be described. The term was no sooner proposed than sanctioned by geologists, both at home and abroad, as involving not theory, and as simply expressing the fact, that in the 'Silurian region', *a complete succession of fossiliferous strata is interpolated between the Old Red Sandstone and the oldest slaty rocks.*

[. . .]

Part 2: Organic Remains

Chapter XLIII, Introductory View of the Distribution of Organic Remains in the Older Formations

Geology reveals to us the extraordinary fact, and without its aid the fact never could have been known, that as the globe passed from one condition to another, whole races of animals perished, and were succeeded by others with organizations adapted to the altered state of our planet.

On this phenomenon is based the fundamental principle of the identification of strata by their imbedded remains; the passage from one deposit to another being marked by a change in the animals which lived and died during the accumulation of each. This, although the fossils of any one great series of beds possess a common character, yet those which are found in the lowest and highest strata of a great formation are for the most part dissimilar in species, and often in genera.

This principle, so little adverted to in the early days of the science, and yet so vitally essential to its advance, has hitherto been chiefly tested by the examination of the tertiary and secondary deposits, and its complete development required, that succeeding observers should laboriously work out its application to the numerous older strata which are contained in the crust of the globe [. . .]

Another remarkable fact illustrating this point of inquiry is, that although the older fossiliferous strata often contain *vast quantities* of organic remains, *the number of species is much smaller* than in the more recent deposits [. . .]

On comparing the fossils from different localities of each Silurian formation we find the closest agreement; certain forms being found in one formation, certain other forms in another; and by this means we have been able to determine the age of the rocks, and to verify the justness of the division by order of superposition.

Extending this mode of investigation to fossils obtained from other parts of the world, and comparing casts with casts, an identity of form as well as of muscular and other impressions will be at once admitted. It is on these grounds we assert, that while the Silurian System was in the course of deposition, similar *mollusks, crustaceans and corals*, lived in very distant parts of the globe, and where climates of very different temperatures now prevail.

[. . .]

Examining the strata with this view, we find that in the ascending geological series the quantity of species increases considerably as we approach the younger deposits, and that in proportion to this increase, their geographical distribution harmonizes more and more with that of existing nature. For example, in the older Tertiary strata (Eocene), there are scarcely any examples of shells identical, being found in opposite hemispheres, and in the younger Tertiary (Pliocene) absolutely none; though in these deposits the number of species is at least tenfold greater than in the ancient rocks under consideration.

Now if this position be admitted, it is manifest that it cannot be explained by reference to the present distribution of animal life. For not only has each latitude, in our days, its different products, but even in distant parts of the same latitude (and in the same hemisphere), the amount of variation of species is often very great.

[. . .]

The following pages and accompanying plates are intended to develop the amount of zoological mutation throughout a large portion of these ancient fossiliferous strata; and the inference we obtain is, that most of the groups which graduate into each other by lithological and geological characters, exhibit also a true zoological transition. This, in reviewing this ancient succession in ascending order, we perceive a certain number of species common to the Upper Formations of the Cambrian System and the Lower Silurian Rocks; and that while some of the Lower Silurian animals lived on to those days when the Upper Silurian beds were deposited, scarcely a single species which existed when the Silurian æra commenced, can be detected in the strata which mark its close.

[. . .]

The same order and change of species is apparent throughout the overlying formations. Thus, although there is a slight community of character between the upper members of the Carboniferous rocks and the lower strata of the New Red System, particularly in the plants; and although certain *mollusca* of the Magnesian Limestone differ little from those of the coal measures, we cannot examine even that formation, still less pass onwards through the overlying red sandstone and marl without perceiving, that in the Sauroids of those days, we are surrounded by entirely new types of animal life, which conduct us by almost imperceptible gradations into the period of the *Ammonites* and *Belemnites* [. . .] Such is the vast succession with which geology has made us acquainted.

29

MARY ANNING, LETTER TO *MAGAZINE OF NATURAL HISTORY* 3

(1839)

Extract of a Letter from Miss Anning, referring to the supposed frontal spine in the genus *Hybodus*.– 'In reply to your request I beg to say that the hooked tooth is by no means new; I believe that M. De la Beche described it fifteen years since in the Geological Transactions, I am not positive; but I know that I then discovered a specimen, with about a hundred palatal teeth, and four of the hooked teeth, as I have done several times with different specimens. I had a conversation with Agassiz on this subject; his remark was that they were the teeth by which the fish seized its prey,– milling it afterwards with its palatal teeth. I am only surprised that he has not mentioned it in his work. We generally find the *Ichthyodorulites* with them as well as cartaliginous bones'. *Mary Anning.– Lyme Regis, April 7*, 1839.

[As Miss Anning speaks of 100 *palatal* teeth, she probably refers to the genus *Acrodus*, which may very possibly be furnished with an organ similar to the one possessed by *Hybodus*, as the genera are closely allied. Mr. De la Beche makes no allusion to its existence in the Geological Transactions.– Ed.]

DOI: 10.4324/9780429355653-33

30

GIDEON MANTELL, *ON THE PELOROSAURUS; AN UNDESCRIBED GIGANTIC TERRESTRIAL REPTILE WHOSE REMAINS ARE ASSOCIATED WITH THOSE OF THE IGUANODON AND OTHER SAURIANS IN THE STRATA OF TILGATE FOREST, IN SUSSEX*

(London: R. and J. E. Taylor, 1850)

I had for a long while entertained the idea that among the fossil remains collected from the Wealden deposits of the South-East of England, there were indications of an enormous Lizard entirely distinct from the Iguanodon, Megolosaurs, Cetiosaurus, and other genera which have been named and more or less accurately determined; and I have at length obtained such evidence in support of my opinion, as induces me to submit to the Royal Society the data which appear to establish the existence of a terrestrial reptile contemporary with the Iguanodon, and which equalled, if not surpassed in magnitude, that colossal herbivorous Saurian.

[. . .]

The occurrence of very large isolated vertebrae, and portions of femora with medullary cavities, indicating animals of terrestrial habits, and of great size, and which, though assigned to the Cetiosaurus, could not properly be included in a genus of aquatic marine reptiles with solid bones like the Cetaceans, first suggested to me the probability of these having existed another genus of Saurians contemporary with those previously mentioned, and to which some of the supposed Cetiosaurian remains might belong. This idea, though vague, seemed to offer an explanation of certain discrepancies between some of my statements and those of other cultivators of this branch of comparative anatomy.

The stupendous humerus, or arm-bone of a terrestrial reptile from the strata of Tilgate Forest, in Sussex, which I have now the honour to place before the Royal Society, will, I believe, establish the correctness of that opinion.

DOI: 10.4324/9780429355653-34

This splendid fossil was obtained from the locality in which was situated the quarry represented in the frontispiece of my work on the Geology of the South-East of England, and that yielded to my early researches the teeth of the Iguanodon and numerous other highly interesting remains. The bone was imbedded in the fawn-coloured sandstone that prevails in the Wealden of that part of Sussex, at the depth of 25 feet beneath the surface of the soil. The distal parts of the bone, to the extent of 2 feet, was discovered in 1847, by MR PETER FULLER of Lewes, and some months afterwards the middle portion was found at a higher level [. . .] At length other fragments were discovered and extricated from the rock, and the whole replaced and cemented together in the admirable state in which the fossil now appears, by the intelligent collector in whose possession I had, a few weeks since, first the gratification of seeing this unique and stupendous relic from the Weald of my native country.

This bone is the right humerus, and bears a closer resemblance in its general character to the analogous element in the Crocodile, than to that of the Lacertians. It is 4½ feet long, and 32 inches in circumference at the distal extremity. It is in the ordinary state of mineralization of the bones from the Wealden sandstones, being of a rich umber colour, and very heavy from an impregnation of oxide of iron. The surface presents a smooth appearance, but upon close inspection is found to be finely striated: it evidently belonged to an animal arrived at maturity, but not aged.

[. . .]

The vertebrae which I would assign to the *Pelorosaurus* with but little hesitation are four anterior caudals of a very remarkable character, which I found many years since in the same stratum as the humerus above described, and at the distance of but a few yards. They were firmly embedded and lying in various positions in a large block of sandstone, and I succeeded, after much labour, in clearing them from the stone with their processes nearly entire [. . .]

When these splendid fossils were first discovered, I referred them to the Iguanodon; subsequently they were named by Professor OWEN *Cetiosaurus brevis*; and lastly, Dr MELVILLE and myself, in my Memoir on the Iguanodon, suggested the necessity of adopting a different specific appellation, and proposed that of 'Conybeari'; we were unwilling to remove them from the genus Cetiosaurus, till corroborative evidence was obtained to justify the change.

[. . .]

From Sandown Bay, in the Isle of Wight, I have the proximal end of a tibia of enormous size, the circumference of the head of the bone being 34 inches, a magnitude surpassing that required for a tibia to articulate with the largest known femur, and presenting such deviations in form from the tibiae of the Iguanodon, as to render it highly probable that this bone belonged to the *Pelorosaurus*.

Indications of the Pelorosaurus in the Oolitic strata.– The general accordance of the terrestrial fauna and flora of the Oolitic period, (as proved by the remains of land animals and plants imbedded in the fluvio-marine deposits of that formation.) with those of the Createaceous and Wealden, renders it probable that vestiges of most, if not all, of the genera and species of land reptiles that occur in the latter will be found in the former strata. Thus as the Iguanodon, Pterodactyles, with Clathrariae, and Dracaenae, are found in the Chalk, and the Megalosaurus, with Cycadeae and Coniferae, in the Wealden, traces of the *Pelorosaurus* may be expected to occur in the Oolite [. . .]

It may perhaps be expected that some estimate should be given of the probable magnitude of the reptile to which the humerus belonged; but calculations of the length and proportions of the original animal taken from a single bone, or from a few detached bones, can afford but vague and unsatisfactory results. With the view, however, of conveying some idea of the almost incredible bulk of the *Pelorosaurus*, it may be stated, that in the Gavial or Gangetic Crocodile, the length of the humerus is equal to one-eighteenth of the entire length of the animal from the snout end to the end of the tail [. . .] Computed by this standard the length of the Pelorosaurus would be 81 feet, and the girth of its body about 20 feet. But if we assume the length and number of the vertebrae as the scale, we should have a reptile of relatively very abbreviated proportions; but in either case, a Saurian far exceeding in size all living types, and equalling if not surpassing the magnitude the most colossal of the extinct forms.

From what has been advanced, we perceive that every addition to the zoology of the countries that flourished during the secondary geological ages, affords proof of the high development of the terrestrial reptiles, which appear to have enjoyed the same predominance in those ancient faunas, as the large Mammalia in those of the tertiary and human epochs. The trees and plants associated with the remains of the extinct Saurians, manifest by their affinity to existing forms, that countries in which they grew possessed as pure an atmosphere, as high a temperature, and as unclouded skies, as those of our tropical climes. There are, therefore, no legitimate grounds to support the hypothesis that during the 'Age of Reptiles' – the period when the reptilian class most prevailed – the earth was 'in the state of a half-finished planet', and its atmosphere too heavy from an excess of carbon, for the respiration of warm-blooded animals! Such an opinion can only have originated from an imperfect view of the phenomena which these problems embrace. There is as great a discrepancy between certain existing faunas and those of modern Europe, as that presented by the Wealden; for example, those of Australia, Tasmania, New Zealand, and the Galapagos Islands.

By a singular coincidence, on the same day that I obtained the humerus of the *Pelorosaurus* from Tilgate Forest, I received from my eldest son in New Zealand, the most interesting collection of the remains of the extinct giant birds of those islands that has reached this country: and I could but think, that had the respective localities and periods of these birds and reptiles,– both of a size far surpassing all other known types of their respective classes,– been interchanged, and the bones of the Dinornis of New Zealand referred to the Wealden epoch, what speculations would in all probability be hazarded to account for physical conditions assumed to be required by so marvellous a development in numbers and magnitude of the class Aves, to the almost entire exclusion of the Mammalia!

[. . .]

In conclusion I would beg to remark, that in adding another genus of terrestrial reptiles to those previously established as belonging to the Fauna of the Wealden, I am fully aware of the imperfect manner in which, from various unavoidable causes – especially the pressure of professional duties – my investigations have been carried out. But encouraged in my earliest researches by the illustrious founder of Palaeontology, Baron CUVIER, and honoured by the highest award of the Geological Society, I felt reluctant to discontinue researches which no other naturalist seemed disposed to undertake, lest some important additions to our knowledge of the ancient physical condition of the earth and its inhabitants should be uncrecorded and forgotten.

Chester Square, Pimlico,
November 1849

31

PHILIP GOSSE, *CREATION (OMPHALOS): AN ATTEMPT TO UNTIE THE GEOLOGICAL KNOT*

(London: J. Van Voorst, 1857)

I

The Cause

I hope I shall not be deemed censorious in stating my fear that those who cultivate the physical sciences are not always sufficiently mindful of the '*Humanum est errare*'. What we have investigated with no little labour and patience, what we have seen with our eyes many many times, in many aspects, and under many circumstances, we naturally believe firmly; and we are very prone to attach the same assurance of certainty to the inferences we have, *bonâ fide*, and with scrupulous care to eliminate error, deduced from our observations, as to the observations themselves; and we are apt to forget that some element of error may have crept into our actual investigations, and still more probably into our deductions. Even if our observations be so simple, so patent, so numerous, as *almost* to preclude the possibility of mistake in them, and our process of reasoning from them be without a flaw, still we may have overlooked a principle, which, though perhaps not very obvious, ought to enter into the investigation, and which, if recognised, would greatly modify our conclusions.

In this volume I venture to suggest such a principle to the consideration of geologists. It will not be denied that Geology is a science that stands peculiarly in need of being cultivated with that salutary self-distrust that I have above alluded to. Though a strong and healthy child, it is as yet but an infant. The objects on which its senses have been exercised [. . .] are indeed plain enough and numerous enough, when once discovered; but the inferences drawn from them [. . .] find their sphere in the most venerably remote antiquity,– an antiquity mensurable not by years or – centuries, but by *secula seculorum*. And the dicta, which its votaries rest on as certitudes, are at variance with the simple literal sense of the words of God.

I am not assuming here that the Inspired Word has been rightly read; I merely say that the plain straightforward meaning, the meaning that lies manifestly on the face of the passages in question, is in opposition with the conclusions which

DOI: 10.4324/9780429355653-35

geologists have formed, as to the antiquity and the genesis of the globe on which we live.

Perhaps the simple, superficial sense of the Word is not the correct one; but it is at least that which its readers, learned and unlearned, had been generally content with before; and which would, I suppose, scarcely have been questioned, but for what appeared the exigencies of geological facts.

Now while there are, unhappily, not a few infidels, professed or concealed, who eagerly seize on any apparent discrepancy between the works and the Word of God, in order that they may invalidate the truth of the latter, there are, especially in this country, many names of the highest rank in physical (and, among other branches, in geological) science, to whom the veracity of God is as dear as life. They cannot bear to see it impugned; they know that it cannot be overthrown; they are assured that He who gave the Word, and He who made the worlds, is One Jehovah, who cannot be inconsistent with Himself. But they cannot shut their eyes to the startling fact, that the records which *seem* legibly written on His created works do flatly contradict the statements which *seem* to be plainly expressed in His word.

Here is a dilemma. A most painful one to the reverent mind! And many reverent minds have laboured hard and long to escape from it. It is unfair and dishonest to class our men of science with the infidel and atheist. They did not rejoice in the dilemma; they saw it at first dimly, and hoped to avoid it.[1] At first they believed that the mighty processes which are recorded on the 'everlasting mountains' might not only be harmonized with, but might afford beautiful and convincing demonstrations of Holy Scripture. They thought that the deluge of Noah would explain the stratification, and the antediluvian era account for the organic fossils.

As the 'stone book' was further read, this mode of explanation appeared to many untenable; and they retracted their adherence to it. To a mind rightly constituted, Truth is above every thing: there is no such thing as a pious fraud; the very idea is an impious lie: God is light, and in Him is no darkness at all; and that religion which can be maintained only by dissembling or denying truth, cannot proceed from 'Him that is Holy, Him that is True', but from him who 'is a liar, and the father of it'.

Many upright and ardent cultivators of the young science felt that truth would be compromised by a persistence in those explanations which had hitherto passed current. The discrepancy between the readings in Science and the hitherto unchallenged readings in Scripture, became manifest. Partisans began to array themselves on either side; some, jealous for the honour of God, knew little of science, and rushed into the field ill-prepared for the conflict; some, jealous for science, but little conversant with Scripture, and caring less for it, were willing to throw overboard its authority altogether: others, who knew that the writings were from the same Hand, knew therefore that there must be some way of reconciling them, and set themselves to find it out.

Have they succeeded? If I thought so, I would not publish this book. Many, I doubt not, have been convinced by each of the schemes by which the discrepant statements have been sought to be harmonized. Each of them has had sufficient plausibility to convince its propounder; and, probably, others too. And some of them have attained a large measure of public confidence. Yet if any one of them is true, it certainly has not commanded universal assent. Let us examine how far they agree among themselves, who propose to reconcile Scripture and Science, 'the Mosaic and the Mineral Geologies'.

And first, it is, perhaps, right to represent the opinions of those who stand by the literal acceptation of the Divine Word. There have been some, indeed, who refuse to entertain the question of reconciliation, taking the high ground that, as the Word of God is and must be true, it is impious to set any evidence in competition with it. I cannot but say, my sympathies are far more with these than with those who, at the opposite pole of the argument, would make scientific deduction paramount, and make the Word go to the wall. But, then, we ought to be quite sure that we have got the very Word of God; and, so far from being impious, it seems highly proper and right, when conflicting evidence appears to flow out of what is indubitably God's *work*, to examine afresh the witnesses on both sides, that we may not make either testify what it does not.

Those good men who merely *denounce* Geology and geologists, I do not quote. There are the facts, 'written and engraven in stones', and that by the finger of God. How can they be accounted for?

Some have recourse to the assumption that the natural processes by which changes in the earth's surface are now going on, may have operated in antediluvian times with a rapidity and power of which we can form little conception from what we are cognisant of. The Rev J. Mellor Brown takes this ground, adducing the analogies of steam-power and electricity, as effecting in a few moments or hours, what formerly would have required several days or weeks to accomplish.

'God's most tremendous agencies may have been employed in the beginning of his works. If, for instance, it should be conceded that the granitic or basaltic strata were once in a state of fusion, there is no reason why we should not call in the aid of supposition to produce a *rapid* refrigeration. We may surround the globe with an atmosphere (not as yet warmed by the rays of the newly kindled sun) more intensely cold than that of Saturn. The degree of cold may have been such as to cool down the liquid granite and basalt in a few hours, and render it congenial to animal and vegetable life; while the gelid air around the globe may have been mollified by the abstracted caloric'.

[. . .]

Another class of this school of interpreters refers the stratification of the earth, either to the deluge alone, or to that convulsion conjoined with the one which is considered to have taken place on the third day of the Mosaic narrative. Perhaps the most eminent writer of this class is Mr. Granville Penn, whose opinions may be thus condensed.

He supposes that this globe has undergone only two revolutions. The first was the violent rupture and depression of the surface to become the bed of the sea, and the simultaneous elevation of the other portion to become dry land,– the theatre of terrestrial existence. This first revolution took place before the creation of any organized beings. The second revolution was at the Noachic Flood, when the former bed of the sea was elevated to become the dry land, with all its organic accumulations of sixteen centuries, while the former land was correspondingly depressed and overflowed. 'The earth must, therefore, necessarily exhibit manifest and universal evidences of the vast apparent ruin occasioned by its first violent disruption and depression; of the presence and operation of the marine fluid, during the long interval which succeeded; and of the action and effects of that fluid in its ultimate retreat'.

[. . .]

A totally different solution of the difficulty has been sought in the hypothesis, that the six 'days' of the Inspired Record signify six successive periods of immense though of undefined duration. This opinion is as old as the Fathers at least,[2] and not a few able maintainers of it belong to our own times.

[. . .]

A large and influential section of the students of Geology regard this hypothesis as untenable. Generally they may be described as holding that the history which is recorded in the igneous and fossiliferous strata does not come into the sacred narrative in any shape. As, however, that narrative commences with 'the beginning', and comes down to historic times, the facts so recorded must find their chronology within its bounds. Their place is accordingly fixed by this school of interpretation between the actual primordial creation (Gen. i. 1), and the chaotic state (ver. 2).

Let us hear an able and eloquent geologist, Professor Sedgwick, on the hypothesis just mentioned of the elongation of the six days:–

'They [certain excellent Christian writers on the subject of Geology] have not denied the facts established by this science, nor have they confounded the nature of physical and moral evidence; but they have prematurely (and, therefore, without an adequate knowledge of all the facts essential to the argument) endeavoured to bring the natural history of the earth into a literal accordance with the Book of Genesis; first, by greatly extending the periods of time implied by the six days of creation; and secondly, by endeavouring to show that under this new interpretation of its words, the narrative of Moses may be supposed to comprehend, and to describe in order, the successive epochs of Geology. It is to be feared that truth may, in this way, receive a double injury; and I am certain that the argument just alluded to has been unsuccessful'. – 'We must consider the old strata of the earth as monuments of a date long anterior to the existence of man, and to the times contemplated in the moral records of his creation'.

Many able theologians, who, though well acquainted with natural science, can scarcely be considered as geologists, have been satisfied with this solution of the problem. [. . .]

Probably the majority of our ablest geologists, men who have devoted their lives to the study and elucidation of geological phenomena, are to be found among those who advocate this scheme of reconciling those phenomena with the statements of the Holy Scriptures. Thus one of the earliest cultivators of the science, the Rev Dr Conybeare:–

> 'I regard Gen. i. 1 as an universal proposition, intended to contradict all the heathen systems which supposed the eternity of matter or polytheism; and ver. 2 I regard as proceeding to take up our planet in a state of ruin from a former condition, and describing a succession of phenomena effected in part by the laws of nature (which are no more than our expression of God's observed method of working), and in part by the immediate exercise of Divine power in directing and creating'.

[. . .]

The author of 'Vestiges of the Natural History of Creation' propounds a theory of organic origin much more worthy of God, than that 'mean view', which supposes Him 'to come in on frequent occasions with new fiats or special interferences'. Coolly bowing aside His authority, this writer has hatched a scheme, by which the immediate ancestor of Adam was a Chimpanzee, and his remote ancestor a Maggot!

In reviewing this array of opinions, is there not sufficient ground for regarding with caution the claim to certainty which has been boldly put forth for the conclusions of Geology? It cannot be denied that there is here room for a very considerable amplitude of choice among discordant hypotheses. All cannot be true [. . .] They vary widely as to their tenableness, and as to their prevalence. But if we leave out of view the fears of those who, from insufficient acquaintance with science, are not competent to adjudicate on its positions, and those who despise or decline Biblical authority altogether on this subject, we have still a somewhat wide range to choose from. [. . .]

I am not blaming, far less despising, the efforts that have been made for harmonizing the teachings of Scripture and science. I heartily sympathise with them. What else could good men do? They could not shut their eyes to the facts which Geology reveals: to have said they were not facts would have been simply absurd. Granting that the whole truth was before them – the whole evidence – they could not arrive at other conclusions than those just recorded; and, therefore, I do not blame their discrepancy *inter se*. *The true key has not as yet been applied to the wards*. Until it be, you may force the lock, but you cannot open it. Whether the key offered in the following pages will open the lock, remains to be seen.

VI

Laws

[. . .]

The course of nature is a circle. I do not mean the *plan* of nature; I am not speaking of a circular arrangement of species, genera, families, and classes [. . .] I decide nothing – concerning them; I am not alluding to any *plan* of nature, but to its *course*, *cursus*,– the way in which it *runs on*. This is a circle.

Here is in my garden a scarlet runner. It is a slender twining stem some three feet long, beset with leaves, with a growing bud at one end, and with the other inserted in the earth. What was it a month ago? A tiny shoot protruding from between two thick fleshy leaves scarcely raised above the ground. A month before that? The thick fleshy leaves were two oval cotyledons, closely appressed face to face, with the minute plumule between them, the whole enclosed in an unbroken, tightly-fitting, spotted, leathery coat. It was a bean, a seed.

Was this the commencement of its existence? O no! Six months earlier still it was snugly lying, with several others like itself, in a green fleshy pod, to the interior of which it was organically attached. A month before that, this same pod with its contents was the centre of a scarlet butterfly-like flower, the bottom of its pistil, within which, if you had split it open, you would have discerned the tiny beans, whose history we are tracing backwards, each imbedded in the soft green tissue, but no bigger than the eye of a cambric needle.

But where was this flower? It was one of many that glowed on my garden wall all through last summer; each cluster springing as a bud from a slender twining stem, which was the exact counterpart of that with which we commenced this little life-history.

And this earlier stem,– what of it? It too had been a shoot, a pair of cotyledons with a plumule, a seed, an integral part of a carpel, which was a part of an earlier flower, that expanded from an earlier bud, that grew out of an earlier stem, that had been a still earlier seed, that had been – and backward, *ad infinitum*, for aught that I can perceive.

The course, then, of a scarlet runner is a circle, without beginning or end:– that is, I mean, without a natural, a normal beginning or end. For at what point of its history can you put your finger, and say, 'Here is the commencement of this organism, before which there is a blank; here it began to exist?' There is no such point; no stage which does not look back to a previous stage, on which *this* stage is inevitably and absolutely dependent.

[. . .]

This sounding-winged Hawkmoth, which like a gigantic bee is buzzing around the jasmine in the deepening twilight, hovering ever and anon to probe the starry flowers that make the evening air almost palpable with fragrance,– this moth, what 'story of a life' can he tell? Nearly a year of existence he has spent as a helpless, almost motionless pupa, buried in the soft earth, from whence he has emerged but this evening. About a twelvemonth ago he was a great fat green caterpillar with an arching horn over his rump, working ever harder and harder at devouring poplar leaves, and growing ever fatter and fatter. But before that he had one day burst forth a little wriggling worm, from a globular egg glued to a leaf. Whence came the egg? It was developed within the ovary of a parent Hawkmoth, whose history is but an endless rotation of the same stages,– pupa, larva, egg, moth, pupa, larva, &c. &c.

[. . .]

Once more. The cow that peacefully ruminates under the grateful shadow of yonder spreading beech, was, a year or two ago, a gamesome heifer with budding horns. The year before, she was a bleating calf, which again had been a breathless foetus wrapped up in the womb of its mother. Earlier still it had been an unformed embryo; and yet earlier, an embryonic vesicle, a microscopically minute cell, formed out of one of the component cells of a still earlier structure,– the germinal vesicle of a fecundated ovum. But this ovum, which is the remotest point to which we can trace the history of our cow as an individual, was, before it assumed a distinct individuality, an undistinguishable constituent of a viscus,– the ovary,– of another cow, an essential part of *her* structure, a portion of the tissues of *her* body, to be traced back, therefore, through all the stages which I have enumerated above, to the tissues of another parent cow, thence to those of a former, and so on, through a vista of receding cows, as long as you choose to follow it.

This, then, is the order of all organic nature. When once we are in any portion of the course, we find ourselves running in a circular groove, as endless as the course of a blind horse in a mill. It is evident that there is no one point in the history of any single creature, which is a legitimate beginning of existence. And this is not the law of some particular species, but of all: it pervades all classes of animals, all classes of plants, from the queenly palm down to the protococcus, from the monad up to man: the life of every organic being is whirling in a ceaseless circle, to which one knows not how to assign any commencement,– I will not say any certain or even probable, but any *possible*, commencement. The cow is as inevitable a sequence of the embryo, as the embryo is of the cow. Looking only at nature, or looking at it only with the lights of experience and reason, I see not how it is possible to avoid one of these two theories, the development of all organic existence out of gaseous elements, or the eternity of matter in its present forms.

Creation, however, solves the dilemma. I have, in my postulates, begged the fact of creation, and I shall not, therefore, attempt to prove it. Creation, the sovereign fiat of Almighty Power, gives us the commencing point, which we in vain seek in nature. But what is creation? It is *the sudden bursting into a circle.* Since there is no one stage in the course of existence, which, more than any other affords a natural commencing point, whatever stage is selected by the arbitrary will of God, must be an unnatural, or rather a preternatural, commencing point.

The life-history of every organism commenced at some point or other of its circular course. It was created, called into being, in some definite stage. Possibly, various creatures differed in this respect; perhaps some began existence in one stage of development, some in another; but every separate organism had a distinct point at which it began to live. Before that point there was nothing; this particular organism had till then no existence; its history presents an absolute blank; *it was not.*

But the whole organisation of the creature thus newly called into existence, looks back to the course of an endless circle in the past. Its whole structure displays a series of developments, which as distinctly witness to former conditions as do those which are presented in the cow, the butterfly, and the fern, of the present day. But what former conditions? The conditions thus witnessed unto, as being necessarily implied in the present organisation, were non-existent; the history was a perfect blank till the moment of creation. The past conditions or stages of existence in question, can indeed be as triumphantly inferred by legitimate deduction from the present, as can those of our cow or butterfly; they rest on the very same evidences; they are identically the same in every respect, except in this one, that they were *unreal*. They exist only in their results; they are effects which never had causes.

Perhaps it may help to clear my argument if I divide the past developments of organic life, which are necessarily, or at least legitimately, inferrible from present phenomena, into two categories, separated by the violent act of creation. Those unreal developments whose apparent results are seen in the organism at the moment of its creation, I will call *prochronic*, because time was not an element in them; while those which have subsisted since creation, and which have had actual existence, I will distinguish as *diachronic*, as occurring during time.

Now, again I repeat, there is no imaginable difference to sense between the prochronic and the diachronic development. Every argument by which the physiologist can prove to demonstration that yonder cow was once a foetus in the uterus of its dam, will apply with exactly the same power to show that the newly created cow was an embryo, some years before its creation.

[. . .]

Permit me, therefore, to repeat, as having been proved, these two propositions:–
ALL ORGANIC NATURE MOVES IN A CIRCLE.
CREATION IS A VIOLENT IRRUPTION INTO THE CIRCLE OF NATURE.

XII

The Conclusion

[. . .]

Since creation and previous history are inconsistent with each other; as the very idea of the creation of an organism excludes the idea of pre-existence of that organism, or of any part of it; it follows, that such records are *false*, so far as they testify to time; that the developments and processes thus recorded have been produced without time, or are what I have called *prochronic*.

Nor is this conclusion in the least degree affected by the actual chronology of creation. The phenomena were equally eloquent, and equally false, whether any individual organism were created six thousand years ago, or innumerable ages; whether primitively, or after the successive creations and annihilations of former organisms.

The law of creation supersedes the law of nature; so far, at least, as the organic world is concerned. The law of nature, established by universal experience, is, that its phenomena depend upon certain natural antecedents: the law of creation is, that the same phenomena depend upon *no* antecedents. The philosopher who should infer the antecedents from the phenomena alone, without having considered the law of creation, would be liable to form totally false conclusions. In order to be secure from error, he must first assure himself that creation is eliminated from the category of facts which he is investigating; and this he could do only when the facts come within the sphere of personal observation, or of historic testimony. Up to such a period of antiquity as is covered by credible history, and within such a field of observation as history may be considered fairly cognisant of,– the inference of physical antecedents from physical phenomena, in the animal or vegetable world, is legitimate and true. But, beyond that period, I cannot safely deduce the same conclusion; because I cannot tell but that at any given moment included in my inquiry, creation may have occurred, and have been the absolute beginning of the circular series.

[. . .]

One of the familiar street-exhibitions in the metropolis is a tiny coach and horses of glittering metal; which, by means of simple machinery, course round and round the margin of a circular table. Let us suppose two youths of philosophical turn to come up during the process. They gaze for a while, and one asks his companion the following question.

'How long do you suppose that coach has been running round?'

'How long! for an indefinite period, for aught I know. I have counted twenty-two turns, and can see no change: nor can I suggest any point where the course could have begun'.

Here a shrewd lad, carrying a grocer's basket, breaks in.

'Oh no; there have been only six-and-twenty turns altogether. Four turns had been made when you came up. The whole began by the man taking the carriage out of a box; then he set it down out there, just opposite to us, and gave it a little push with his finger, and it has been running ever since. I saw him do it'.

Now perhaps you will say that a glance at the machinery beneath the table would show in a moment how many turns had been made, and how many could be made. Very true: but what if the tramp had locked up his clock-work, and would not let you look at it?

The only evidence worth a rush is that of the lad who saw the whirligig set a-going.

I wish it to be distinctly understood, that I am not proving the exact or approximate antiquity of the globe we inhabit. I am not attempting to show that it has existed for no more than six thousand years. I wish this to be distinctly stated, because I am sure I shall meet with many opponents unfair enough, or illogical enough, to misrepresent or misunderstand my argument, and sound the trumpet of victory, because I cannot demonstrate *that*. *All* I set myself to do, is to invalidate the testimony of the witness relied on for the indefinitely remote antiquity; to show that in a very large and important field of nature, evidence exactly analogous to that relied on would inevitably lead to a false conclusion, and must, therefore, be rejected, or received only contingently; received only as indicative of probability, and that only in the absence of any positive witness to the contrary.

Perhaps it may be objected, that there is no sufficient analogy between the phenomena from which the past history of a single organism is inferred, and those from which the past history of a world is inferred. Is there not?

Permit me to repeat an illustration I have already used. The geologist finds a fossil skeleton. His acquaintance with anatomy enables him to pronounce that the objects found are bones. He sees cylinders, condyles, cavities for the marrow, scars of attachment of muscles and tendons, foramina for the passage of nerves and blood-vessels; he finds the internal structure, no less than the form and surface, such as to leave not a doubt that these are real *bones*. Now universal experience has taught him that bones imply the existence of flesh; that flesh implies blood; that blood implies life; that life implies time. He therefore concludes unhesitatingly, that this skeleton was once alive, and that time passed over it in that living condition.

Is not this process of reasoning exactly parallel to that which he would have pursued if he had examined an animal the moment after its creation, (supposing this fact to be unknown to him,) and by which he would in like manner have inferred past time? And where is the vital difference between the two cases, which would operate to make a conclusion which is manifestly false in the one case, necessarily true in the other?

[. . .]

In order to perfect the analogy between an organism and the world, so as to show that the law which prevails in the one obtains also in the other, it would be necessary to prove that the development of the physical history of the world is circular, like that already shown to characterise the course of organic nature. And this I cannot prove. But neither, as I think, can the contrary be proved.

The life of *the individual* consists of a series of processes which are cyclical. In the tree this is shown by the successive growths and deaths of series of leaves: in the animal by the development and exuviation of nails, hair, epidermis, &c.

The life of *the species* consists of a series of processes which are cyclical. This has been sufficiently illustrated in the preceding pages, in the successive developments and deaths of generations of individuals.

We have reason to believe that species die out, and are replaced by other species, like the individuals which belong to the species, and the organs which belong to the individual. But is the life of *the species* a circle returning into itself? In other words, if we could take a sufficiently large view of the whole plan of nature, should we discern that the existence of species [. . .] necessarily involved the pre-existence of species [. . .], and must inevitably be followed by species?

I dare not say, we should; though I think it highly probable. But I think you will not dare to say, we should *not*.

It is certain that, when the Omnipotent God proposed to create a given organism, the course of that organism was present to his idea, as an ever-revolving circle, without beginning and without end. He created it at some point in the circle, and gave it thus an arbitrary beginning; but one which involved all previous rotations of the circle, though only as ideal, or, in other phrase, prochronic. Is it not possible – I do not ask for more – that, in like manner, the natural course of the world was projected in his idea as a perfect whole, and that He determined to create it at some point of that course, which act, however, should involve previous stages, though only ideal or prochronic?

All naturalists have speculated upon the great plan of Nature; a grand array of organic essences, in which every species should be related in like ratio to its fellow species, by certain affinities, without gaps and without redundancies; the whole constituting a beautiful and perfect unity, a harmonious scheme, worthy of the infinite Mind that conceived it. Such a perfect plan has never been presented by any existing fauna or flora; nor is it made up by uniting the fossil faunas and floras to the recent ones; *yet the discovery of the fossil world has made a very signal approach to the filling up of the great outline*; and the more minutely this has been investigated, the more have hiatuses been bridged over, which before yawned between species and species, and links of connexion have been supplied which before were lacking.

It is not necessary,– at least it does not seem so to me,– that all the members of this mighty commonwealth should have an actual, a diachronic existence; any-more than that, in the creation of a man, his foetal, infantile, and adolescent stages should have an actual, diachronic existence, though these are essential to his normal life-history. Nor would their diachronism be more certainly inferrible from the physical traces of them, in the one case than in the other. In the newly-created Man, the proofs of successive processes requiring time, in the skin, hairs, nails, bones, &c. could in no respect be distinguished from the like proofs in a Man of to-day; yet the developments to which they respectively testify are widely different from each other, so far as regards the element of time. Who will say that the suggestion, *that the strata of the surface of the earth, with their fossil floras and faunas, may possibly belong to a prochronic development of the mighty plan of the life-history of this world*,– who will dare to say that such a suggestion is a self-evident absurdity? If we had no example of such a procedure, we might be justified in dealing cavalierly with the hypothesis; but it has been shown that, without a solitary exception, the whole of the vast vegetable and animal kingdoms were created,– mark! I do not say *may* have been, but MUST have been created – on this principle of a prochronic development, with distinctly traceable records. It was *the law of organic creation.*

It may be objected, that, to assume the world to have been created with fossil skeletons in its crust,– skeletons of animals that never really existed,– is to charge the Creator with forming objects whose sole purpose was to deceive us. The reply is obvious. Were the concentric timber-rings of a created tree formed merely to deceive? Were the growth lines of a created shell intended to deceive? Was the navel of the created Man intended to deceive him into the persuasion that he had had a parent?

These peculiarities of structure were inseparable from the adult stage of these creatures respectively, without which they would not have been what they were [. . .]

If, then, the existence of retrospective marks, visible and tangible proofs of processes which were prochronic, was so necessary to organic essences, that they could not have been created without them,– is it absurd to suggest the *possibility* (I do no more) that the world itself was created under the influence of the same law, with visible tangible proofs of developments and processes, which yet were only prochronic?

Admit for a moment, as a hypothesis, that the Creator had before his mind a projection of the whole life-history of the globe, commencing with any point which the geologist may imagine to have been a fit commencing point, and ending with some unimaginable acme in the indefinitely distant future. He determines to call this idea into actual existence, not at the supposed commencing point, but at some stage or other of its course. It is clear, then, that at the selected stage it appears, exactly as it would have appeared at that moment of its history, if all the preceding eras of its history had been real. Just as the new-created Man was, at the first moment of his existence, a man of twenty, or five-and-twenty, or thirty years old; physically, palpably, visibly, so old, though not really, not diachronically. He appeared precisely what he would have appeared had he lived so many years.

[. . .]

I am endeavouring to show that a grand LAW exists, by which, in two great departments of nature at least, the analogues of the fossil skeletons were formed without pre-existence. An arbitrary acting, and an acting on fixed and general laws, have nothing in common with each other.

Finally, the acceptance of the principles presented in this volume, even in their fullest extent, would not, in the least degree, affect the study of scientific geology. The character and order of the strata; their disruptions and displacements and injections; the successive floras and faunas; and all the other phenomena, would be *facts* still. They would still be, as now, legitimate subjects of examination and inquiry. I do not know that a single conclusion, now accepted, would need to be given up, except that of actual chronology. And even in respect of this, it would be rather a modification than a relinquishment of what is at present held; we might still speak of the inconceivably long duration of the processes in question, provided we understand *ideal* instead of *actual* time;– that the duration was projected in the mind of God, and not really existent.

The zoologist would still use the fossil forms of non-existing animals, to illustrate the mutual analogies of species and groups. His recognition of their prochronism would in nowise interfere with his endeavours to assign to each its position in the scale of organic being. He would still legitimately treat it as an entity; an essential constituent of the great Plan of Nature; because he would recognise the Plan itself as an entity, though only an ideal entity, existing only in the Divine Conception. He would still use the stony skeletons for the inculcation of lessons on the skill and power of God in creation; and would find them a rich mine of instruction, affording some examples of the adaptation of structure to function, which are not yielded by any extant species. [. . .]

Here I close my labours. How far I have succeeded in accomplishing the task to which I bent myself, it is not for me to judge. Others will determine that; and I am quite sure it will be determined fairly, on the whole.

[. . .]

'IN SIX DAYS JEHOVAH MADE HEAVEN AND EARTH, THE SEA, AND ALL THAT IN THEM IS'.

Notes

1 Gosse's footnote: 'As Cuvier, Buckland, and many others. On the question whether the phenomena of Geology can be comprised within the short period formerly assigned to them, the Rev Samuel Charles Wilts long ago observed: "Buckland, Sedgwick, Faber, Chalmers, Conybeare, and many other Christian geologists, strove long with themselves to believe that they could: and they did not give up the hope, or seek for a new interpretation of the sacred text, till they considered themselves driven from their position by such facts as we have stated. If, *even now, a reasonable, or we might say* POSSIBLE *solution were offered, they would*, we feel persuaded, *gladly revert* to their original opinion"– *Christian Observer*, August, 1834'.

2 Gosse cites Christian scholars Origen of Alexandria (c.184–c.253CE) and St Augustine of Hippo (354–430CE).

32

HUGH MILLER, *TESTIMONY OF THE ROCKS OR GEOLOGY IN ITS BEARING ON THE TWO THEOLOGIES, NATURAL & REVEALED*

(Edinburgh: Thomas Constable & Co., 1857)

To the reader

[. . .] It will be seen that I adopt, in my Third and Fourth Lectures, that scheme of reconciliation between the Geologic and Mosaic Records which accepts the six days of creation as vastly extended periods; and I have been reminded by a somewhat captious critic that I once held a very different view, and twitted with what he terms inconsistency. I certainly did once believe with Chalmers and with Buckland that the six days were simply natural days of twenty-four hours each,– that they had compressed the entire work of the existing creation,– and that the latest of the geologic ages was separated by a great chaotic gap from our own. My labors at the time as a practical geologist had been very much restricted to the Palæozoic and Secondary rocks, more especially to the Old Red and Carboniferous Systems of the one division, and the Oolitic System of the other; and the long extinct organisms which I found in them certainly did not conflict with the view of Chalmers. All I found necessary at the time to the work of reconciliation was some scheme that would permit me to assign to the earth a high antiquity, and to regard it as the scene of many succeeding creations. During the last nine years, however, I have spent a few weeks every autumn in exploring the later formations, and acquainting myself with their peculiar organisms. I have traced them upwards from the raised beaches and old coast lines of the human period, to the brick clays, Clyde beds, and drift and boulder deposits of the Pleistocene era, and again from these, with the help of museums and collections, up through the mammaliferous crag of England, to its Red and its Coral crags. And the conclusion at which I have been compelled to arrive is, that for many long ages ere man was ushered into being, not a few of his humbler contemporaries of the fields and woods enjoyed life in their present haunts, and that for thousands of years anterior to even their appearance, many of the existing molluscs lived in our seas. That

 DOI: 10.4324/9780429355653-36

day during which the present creation came into being, and in which God, when he had made 'the beast of the earth after his kind, and the cattle after their kind', at length terminated the work by moulding a creature in his own image, to whom he gave dominion over them all, was not a brief period of a few hours' duration, but extended over mayhap millenniums of centuries. No blank chaotic gap of death and darkness separated the creation to which man belongs from that of the old extinct elephant, hippopotamus, and hyæna; for familiar animals such as the red deer, the roe, the fox, the wild cat, and the badger, lived throughout the period which connected their times with our own; and so I have been compelled to hold, that the days of creation were not natural, but prophetic days, and stretched far back into the bygone eternity. After in some degree committing myself to the other side, I have yielded to evidence which I found it impossible to resist; and such in this matter has been my inconsistency,– an inconsistency of which the world has furnished examples in all the sciences, and will, I trust, in its onward progress, continue to furnish many more.

Lecture Third

The Two Records, Mosaic and Geological

It is now exactly fifty years since a clergyman of the Scottish Church, engaged in lecturing at St Andrews, took occasion in enumerating the various earths of the chemist, to allude to the science, then in its infancy, that specially deals with the rocks and soils which these earths compose. 'There is a prejudice', he remarked, 'against the speculations of the geologist, which I am anxious to remove. It has been said that they nurture infidel propensities. It has been alleged that geology, by referring the origin of the globe to a higher antiquity than is assigned to it by the writings of Moses, undermines our faith in the inspiration of the Bible, and in all the animating prospects of the immortality which it unfolds. This is a false alarm. *The writings of Moses do not fix the antiquity of the globe*'.

The bold lecturer on this occasion,– for it needed no small courage in a divine of any Established Church to take up, at the beginning of the present century, a position so determined on the geologic side,– was at the time an obscure young man, characterized, in the small circle in which he moved, by the ardor of his temperament and the breadth and originality of his views; but not yet distinguished in the science or literature of his country, and of comparatively little weight in the theological field. [. . .] Time and experience have since impressed their stamp on these supposed eccentricities, and shown them to be the sagacious forecastings of a man who saw further and more clearly than his contemporaries; and fame has since blown his name very widely, as one of the most comprehensive and enlightened, and, withal, one of the most thoroughly earnest and sincere, of modern theologians. The bold lecturer of St Andrews was Dr Thomas Chalmers,– a divine whose writings are now known wherever the English language is spoken, and whose wonderful eloquence lives in memory as a vanished power, which even his extraordinary writings fail adequately to represent. And in the position which he took up at this early period with respect to geology and the Divine Record, we have yet another instance of the great sagacity of the man, and of his ability of correctly estimating the prevailing weight of the evidence with which, though but partially collected at the time, the geologist was preparing to establish the leading propositions of his science. Even in this late age, when the scientific standing of geology is all but universally recognized, and the vast periods of time which it demands fully conceded, neither geologist nor theologian, could, in any new scheme of reconciliation, shape his first proposition more skilfully than it was shaped by Chalmers a full half century ago [. . .]

[Chalmers] teaches, and teaches truly, that between the first act of creation, which evoked out of the previous nothing the *matter* of the heavens and earth, and the first act of the first day's work recorded in Genesis, periods of vast duration may have intervened; but further, it insists that the days themselves were but natural days of twenty-four hours each; and that, ere they began, the earth, though mayhap in the previous period a fair residence of life, had become void and formless, and the sun, moon, and stars, though mayhap they had before given light, had been, at least in relation to our planet, temporarily extinguished. In short, while it teaches that the successive creations of the geologist may all have found ample room in the period preceding that creation to which man belongs, it teaches also that the record in Genesis bears reference to but the existing creation, and that there lay between it and the preceding ones a chaotic period of death and darkness. [. . .]

William Smith, the 'Father of English Geology', as he has been well termed (a humble engineer and mineral surveyor, possessed of but the ordinary education of men of his class and profession), was born upon the English Oolite,– that system which, among the five prevailing divisions of the great Secondary class of rocks, holds exactly the middle place. The Triassic system and the Lias lie beneath it; the Cretaceous system and the Weald rest above. Smith, while yet a child, had his attention attracted by the Oolitic fossils; and it was observed, that while his youthful contemporaries had their garnered stores of marbles purchased at the toy shop, he had collected, instead, a hoard of spherical fossil terebratulæ, which served the purposes of the game equally well. The interest which he took in organic remains, and the deposits in which they occur, influenced him in the choice of a profession; and, when supporting himself in honest independence as a skilful mineral surveyor and engineer, he travelled over many thousand miles of country, taking as his starting point the city of Bath, which stands near what is termed the Great Oolite: and from that centre he carefully explored the various Secondary formations above and below. He ascertained that these always occur in a certain determinate order; that each contains fossils peculiar to itself; and that they run diagonally across the kingdom in nearly parallel lines from north-east to south-west. And, devoting every hour which he could snatch from his professional labors to the work, in about a quarter of a century, or rather more, he completed his great stratigraphical map of England. But, though a truly Herculean achievement, regarded as that of a single man unindebted to public support, and uncheered by even any very general sympathy in his labors, it was found to be chiefly valuable in its tracings of the Secondary deposits, and strictly exact in only that Oolitic centre from which his labors began. It was remarked at an early period that he ought to have restricted his publication to the formations which lie between the Chalk and the Red Marl inclusive; or, in other words, to the great Secondary division. The Coal Measures had, however, been previously better known, from their economic importance, and the number of the workings opened among them, than the deposits of any other system; and ere the publication of the map of Smith, Cuvier and Brogniart had rendered famous all over the world the older Tertiary formations of the age of the London Clay. But both ends of the geological scale, comprising those ancient systems older than the Coal, and representative of periods in which, so far as is yet known, life, animal and vegetable, first began upon our planet, and those systems of comparatively modern date, representative of the periods which immediately preceded the human epoch, were equally unknown. The light fell strongly on only that middle portion of the series on which the labors of Smith had been mainly concentrated. The vast geologic bridge, which, like that in the exquisite allegory of Addison, strode across a 'part of the great tide of eternity', 'had a black cloud hanging at each end of it'. And such was the state of geologic science when, in 1814, Dr Chalmers framed his scheme of reconciliation.

Since that time, however, a light not less strong than the one thrown by William Smith on the formations of the Lias and the Oolite has been cast on both the older and the newer fossiliferous systems. Two great gaps still remain to be filled up,– that which separates the Palæozoic from the Secondary division, and that which separates the Secondary from the Tertiary one. But they occur at neither end of the geological scale. Mainly through the labors of two distinguished geologists, who, finding the geologic school of their own country distracted by a fierce and fruitless controversy, attached themselves to the geologic school of England, and have since received the honor of knighthood in acknowledgment of their labours, both ends of the geologic scale have been completed. Sir Roderick Murchison addressed himself to the formations older than the Coal, more especially to the Upper and Lower Silurian systems, from the Ludlow rooks to the Llandeilo flags. The Old Red Sandstone too, a system which lies more immediately beneath the Coal, has also been explored, and its various deposits, with their peculiar organic remains, enumerated and described. And Sir Charles Lyell, setting himself to the other extremity of the scale, has wrought out the Tertiary formations, and separated them into the four great divisions which they are now recognized as forming. And of these, the very names indicate that certain proportions of their organisms still continue to exist. It is a great fact, now fully established in the course of geological discovery, that between the plants which in the present time cover the earth, and the animals which inhabit it, and the animals and plants of the later extinct creations, there occurred no break or blank, but that, on the contrary, many of the existing organisms were contemporary during the morning of their being, with many of the extinct ones during the evening of theirs. We know further, that not a few of the shells which now live on our coasts, and several of even the wild animals which continue to survive amid our tracts of hill and forest, were in existence many ages ere the human age began. Instead of dating their beginning only a single natural day, or at most two natural days, in advance of man, they must have preceded him by many thousands of years. In fine, in consequence of that comparatively recent extension of geologic fact in the direction of the later systems and formations, through which we are led to know that the present creation was not cut off abruptly from the preceding one, but that, on the contrary, it dovetailed into it at a thousand different points, we are led also to know, that any scheme of reconciliation which would separate between the recent and the extinct existences by a chaotic gulf of death and darkness, is a scheme which no longer meets the necessities of the case. Though perfectly adequate forty years ago, it has been greatly outgrown by the progress of geological discovery, and is, as I have said, adequate no longer; and it becomes a not unimportant matter to determine the special scheme that would bring into completest harmony the course of creation, as now ascertained by the geologist, and that brief but sublime narrative of its progress which forms a meet introduction in Holy Writ to the history of the human family. The first question to which we must address ourselves in any such inquiry is of course a very obvious one,– *What are the facts scientifically determined which now demand a new scheme of reconciliation?*

There runs around the shores of Great Britain and Ireland a flat terrace of unequal breadth, backed by an escarpment of varied height and character, which is known to geologists as the old coast-line. On this flat terrace most of the sea-port towns of the empire are built. The subsoil which underlies its covering of vegetable mould consists usually of stratified sands and gravels, arranged after the same fashion as on the neighboring beach, and interspersed in the same manner with sea shells. The escarpment behind, when formed of materials of no great coherency, such as gravel or clay, exists as a sloping, grass-covered bank,– at one place running out into promontories that encroach upon the terrace beneath,– at another receding into picturesque, bay-like recesses; and where composed, as in many localities, of rock of an enduring quality, we find it worn, as if by the action of the surf,– in some parts relieved into insulated stacks, in others hollowed into deep caverns,– in short, presenting all the appearance of a precipitous coast-line, subjected to the action of the waves. Now, no geologist can or does doubt that this escarpment was at one time the coast-line of the island,– the line against which the waves broke at high water in some distant age, when either the sea stood from twenty to thirty feet higher along our shores than it does now, or the land sat from twenty to thirty feet lower. Nor can the geologist doubt, that along the flat terrace beneath, with its stratified beds of sand and gravel, and its accumulations of sea shells, the tides must have risen and fallen twice every day, as they now rise and fall along the beach that at present girdles our country. But, in reference to at least human history, the age of the old coast-line and terrace must be a very remote one. Though geologically recent, it lies far beyond the reach of any written record. It has been shown by Mr. Smith of Jordanhill, one of our highest authorities on the subject, that the wall of Antoninus, erected by the Romans as a protection against the Northern Caledonians, was made to terminate at the Firths of Forth and Clyde, with relation, not to the level of the old coast-line, but to that of the existing one. And so we must infer that, ere the year A.D. 140 (the year during which, according to our antiquaries, the greater part of the wall was erected) the old coast-line had attained to its present elevation over the sea. Further, however, we know from the history of Diodorus the Sicilian, that at a period earlier by at least two hundred years, St Michael's Mount, in Cornwall, was connected with the mainland at low water, just as it is now, by a flat isthmus, across which, upon the falling of the tide, the ancient Cornish miners used to carry over their tin in carts. Had the relative levels of sea and land been those of the old coast-line at the time, St Michael's Mount, instead of being accessible at low ebb would have been separated from the shore by a strait from three to five fathoms in depth [. . .]

But even the incidental notice of Diodorus Siculus represents very inadequately the antiquity of the existing coast-line. Some of its caves, hollowed in hard rock in the line of faults and shifts by the attrition of the surf, are more than a hundred feet in depth; and it must have required many centuries to excavate tough trap or rigid gneiss to a depth so considerable, by a process so slow. And yet, however long the sea may have stood against the present coast-line, it must have stood for a considerably longer period against the ancient one. The latter presents generally marks of greater attrition than the modern line, and its wave-hollowed caves are of a depth considerably more profound. In determining, on an extensive tract of coast, the average profundity of both classes of caverns from a considerable number of each, I ascertained that the proportional average depth of the modern to the ancient is as two to three. For every two centuries, then, during which the waves have been scooping out the caves of the present coast-line, they must have been engaged for three centuries in scooping out those of the old one. But we know *historically*, that for at least twenty centuries the sea has been toiling in these modern caves; and who shall dare affirm that it has not been toiling in them for at least ten centuries more? But if the sea has stood for but even two thousand six hundred years against the present coast-line (and no geologist would dare fix his estimate lower), then must it have stood against the old line, ere it could have excavated caves one third deeper, three thousand nine hundred years. And both periods united (six thousand five hundred years) more than exhaust the Hebrew chronology. Yet what a mere beginning of geologic history does not the epoch of the old coast-line form! It is but a mere starting point from the recent period. Not a single shell seems to have become extinct during the last six thousand five hundred years! The shells which lie embedded in the subsoils beneath the old coast-line are exactly those which still live in our seas.

Above this ancient line of coast we find, at various heights, beds of shells of vastly older date than those of the low-lying terrace, and many of which are no longer to be found living around our shores. I spent some time last autumn in exploring one of these beds, once a sea bottom, but now raised two hundred and thirty feet over the sea, in which there occurred great numbers of shells now not

British, though found in many parts of Britain at heights varying from two hundred to nearly fourteen hundred feet over the existing sea level. But though no longer British shells, they are shells that still continue to live in high northern latitudes, as on the shores of Iceland and Spitzbergen; and the abundance in which they were developed on the submerged plains and hill-sides of what are now England and Scotland, during what is termed the Pleistocene period, shows of itself what a very protracted period that was. The prevailing tellina of the bed which I last explored,– a bed which occurs in some places six miles inland, in others elevated on the top of dizzy crags,– is a sub-arctic shell (*Tellina proxima*), of which only dead valves are now to be detected on our coasts, but which may be found living at the North Cape and in Greenland. The prevailing astarte, its contemporary, was *Astarte arctica*, now so rare as a British species, that many of our most sedulous collectors have never seen a native specimen, but which is comparatively common on the northern shores of Iceland, and on the eastern coasts of Norway, within the arctic circle. In this elevated Scottish bed of the Pleistocene period I laid these boreal shells open to the light by hundreds, on the spot evidently where the individuals had lived and died. Under the severe climatal conditions to which (probably from some change in the direction of the gulf stream) what is now Northern Europe had been brought, this tellina and astarte had increased and multiplied until they became prevailing shells of the British area; and this increase must have been the slow work of ages, during which the plains, and not a few of the table lands, of the country, were submerged in a sub-arctic sea, and Great Britain existed as but a scattered archipelago of wintry islands. But in a still earlier period, of which there exists unequivocal evidence in the buried forests of Happisburgh and Cromer, the country had not only its head above water, as now, but seems to have possessed over more than its present breadth of surface. During this ancient time,– more remote by many centuries than not only the times of the old coast-line, but than even those of the partial submergence of the island,– that northern mammoth lived in great abundance, of which the remains have been found by hundreds in England alone, together with the northern hippopotamus, and at least two northern species of rhinoceros. And though they have all ceased to exist, with their wild associates in the forests and jungles of the Pleistocene, the cave-hyæna, the cave-tiger, and the cave-bear, we know that the descendants of some of their feebler contemporaries, such as the badger, the fox, the wild cat, and the red deer, still live amid our hills and brakes. The trees, too, under which they roamed, and whose remains we find buried in the same deposits as theirs, were of species that still hold their place as aboriginal trees of the country, or of at least the more northerly provinces of the continent. The common Scotch fir, the common birch, and a continental species of conifer of the far north, the Norwegian spruce (*Abies excelsa*), have been found underlying the Pleistocene drift, and rooted in the mammiferous crag; and for many ages must the old extinct elephant have roamed amid these familiar trees. From one limited tract of sea bottom on the Norfolk coast the fishermen engaged in dredging oysters brought ashore, in the course of thirteen years (from 1820 to 1833), no fewer than two thousand elephants' grinders, besides great tusks and numerous portions of skeletons. It was calculated that these remains could not have belonged to fewer than five hundred individual mammoths of English growth; and, various in their states of keeping, and belonging to animals of which only a few at a time could have found sufficient food in a limited tract of country, the inference seems inevitable that they must have belonged, not to one or two, but to many succeeding generations. The further fact, that remains of this ancient elephant (*Elephas primigenius*) occur all round the globe in a broad belt, extending from the fortieth to near the seventieth degree of north latitude, leads to the same conclusion. It must have required many ages ere an animal that breeds so slowly as the elephant could have extended itself over an area so vast.

Many of the contemporaries of this northern mammoth, especially of its molluscan contemporaries, continue, as I have said, to live in their descendants. Of even a still more ancient period, represented by the Red Crag, seventy out of every hundred species of shells still exist; and of an older period still, represented by the Coraline Crag, there survive sixty out of every hundred. In the Red Crag, for instance, we find the first known ancestors of our common edible periwinkle and common edible mussel; and in the Coraline Crag, the first known ancestors of the common horse-mussel, the common whelk, the common oyster, and the great pecten. There then occurs a break in the geologic deposits of Britain, which, however, in other parts of Europe we find so filled up as to render it evident that no corresponding break took place in the chain of existence; but that, on the contrary, from the present time up to the times represented by the earliest Eocene formations of the Tertiary division, day has succeeded day, and season has followed season, and that no chasm or hiatus – no age of general chaos, darkness, and death – has occurred, to break the line of succession, or check the course of life. All the evidence runs counter to the supposition that immediately before the appearance of man upon earth, there existed a chaotic period which separated the previous from the present creation. Up till the commencement of the Eocene ages, if even then, there was no such chaotic period, in at least what is now Britain and the European continent: the persistency from a high antiquity of some of the existing races, of not only plants and shells, but of even some of the mammiferous animals, such as the badger, the goat, and the wild cat, prove there was not; and any scheme of reconciliation which takes such a period for granted must be deemed as unsuited to the present state of geologic knowledge, as any scheme would have been forty years ago which took it for granted that the writings of Moses *do* 'fix the antiquity of the globe'.

[. . .]

I come before you this evening, not as a philologist, but simply as a student of geological fact, who, believing his Bible, believes also, that though theologians have at various times striven hard to pledge it to false science, geographical, astronomical, and geological, it has been pledged by its Divine Author to no falsehood whatever. [. . .] Premising, then, that I make no pretensions to even the slightest skill in philology, I remark further, that it has been held by accomplished philologists, that the days of the Mosaic creation may be regarded, without doing violence to the genius of the Hebrew language, as successive periods of great extent. [. . .] I would, in any such case, at once, and without hesitation, cut the philological knot, by determining that that philology cannot be sound which would commit the Scriptures to a science that cannot be true. Waiving, however, the question as a philological one, and simply holding with Cuvier, Parkinson, and Silliman, that each of the *six* days of the Mosaic narrative in the first chapter were what is assuredly meant by the *day* referred to in the second,– not natural days, but lengthened periods,– I find myself called on, as a geologist, to account for but three of the six. Of the period during which light was created,– of the period during which a firmament was made to separate the waters from the waters,– or of the period during which the two great lights of the earth, with the other heavenly bodies, became visible from the earth's surface,– we need expect to find no record in the rocks. Let me, however, pause for a moment, to remark the peculiar character of the language in which we are first introduced in the Mosaic narrative to the heavenly bodies,– sun, moon, and stars. The moon, though absolutely one of the smallest lights of our system, is described as secondary and subordinate to only its greatest light, the sun. It is the apparent, then, not the actual, which we find in the passage,– what *seemed* to be, not what *was*; and as it was merely what appeared to be greatest that was described as greatest, on what grounds are we to hold that it may not also have been what *appeared* at the time to be made that has been described as made? The sun, moon, and stars may have been created long before, though it was not until this fourth period of creation that they became visible from the earth's surface.

[. . .]

If, taking the Mosaic days as equivalent to lengthened periods, we hold that, in giving their brief history, the inspired writer seized on but those salient points that, like the two great lights of the day and night, would have arrested most powerfully, during these periods, a human eye, we shall find the harmony of the two records complete. [. . .] Man came into being as the lastborn of creation, just ere the close of that sixth day – the third and terminal period of organic creation – to which the great mammals belong. Let me yet further remark, that in each of these three great periods we find, with respect to the classes of existences, vegetable or animal, by which they were most prominently characterized, certain well marked culminating points together, if I may so express myself,– twilight periods of morning dawn and evening decline. The plants of the earlier and terminal systems of the Palæozoic division are few and small: it was only during the protracted *eons* of the Carboniferous period that they received their amazing development, unequalled in any previous or succeeding time. In like manner, in the earlier or Triassic deposits of the Secondary division, the reptilian remains are comparatively inconsiderable; and they are almost equally so in its Cretaceous or later deposits. It was during those middle ages of the division, represented by its Liassic, Oolitic, and Wealden formations, that the class existed in that abundance which rendered it so peculiarly, above every other age, an age of creeping things and great sea monsters. And so also, in the Tertiary, regarded as but an early portion of the human division, there was a period of increase and diminution,– a morning and evening of mammalian life. The mammals of its early Eocene ages were comparatively small in bulk and low in standing; in its concluding ages, too, immediately ere the appearance of man, or just as he had appeared, they exhibited, both in size and number, a reduced and less imposing aspect. It was chiefly in its middle and latter, or Miocene, Pliocene, and Pleistocene ages, that the myriads of its huger giants,– its dinotheria, mastodons, and mammoths,– cumbered the soil. I, of course, restrict my remarks to the three periods of organic life, and have not inquired whether aught analogous to these mornings and evenings of increase and diminution need be sought after in any of the others.

Such are a few of the geological facts which lead me to believe that the *days* of the Mosaic account were great periods, not natural days; and be it remembered, that between the scheme of lengthened periods and the scheme of a merely local chaos, which existed no one knows how, and of a merely local creation, which had its scene no one knows where, geological science leaves us now no choice whatever. It has been urged, however, that this scheme of periods is irreconcilable with that Divine 'reason' for the institution of the Sabbath which he who appointed the day of old has, in his goodness, vouchsafed to man. I have failed to see any force in the objection. God the Creator, who wrought during six periods, rested during the seventh period; and as we have no evidence whatever that he recommenced his work of creation,– as, on the contrary, man seems to be the last formed of creatures,– God may be resting still. The presumption is strong that his Sabbath is an extended period, not a natural day, and that the work of Redemption is his Sabbath day's work. And so I cannot see that it in the least interferes with the integrity of the reason rendered to read it as follows:– Work during six periods, and rest on the seventh; for in six periods the Lord created the heavens and the earth, and on the seventh period *He* rested. The Divine periods may have been very great,– the human periods very small; just as a vast continent or the huge earth itself is very great, and a map or geographical globe very small. But if in the map or globe the proportions be faithfully maintained, and the scale, though a minute one, be true in all its parts and applications, we pronounce the map or globe, notwithstanding the smallness of its size, a faithful copy. Were man's Sabbaths to be kept as enjoined, and in the Divine proportions, it would scarcely interfere with the logic of the 'reason annexed to the fourth commandment', though in this matter, as in all others in which man can be an imitator of God, the imitation should be a miniature one.

[. . .]

One other remark ere I conclude. In the history of the earth which we inhabit, molluscs, fishes, reptiles, mammals, had each in succession their periods of vast duration; and then the human period began,– the period of a fellow worker with God, created in God's own image. What is to be the next advance? Is there to be merely a repetition of the past? – an introduction a second time of man made in the image of God? No. The geologist, in those tables of stone which form his records, finds no example of dynasties once passed away again returning. There has been no repetition of the dynasty of the fish, of the reptile, of the mammal.

The dynasty of the future is to have glorified man for its inhabitant; but it is to be the dynasty – 'the *kingdom*' – not of glorified man made in the image of God, but of God himself in the form of man. In the doctrine of the two conjoined natures, human and Divine, and in the further doctrine that the terminal dynasty is to be peculiarly the dynasty of HIM in whom the natures are united, we find that required progression beyond which progress cannot go. We find the point of elevation never to be exceeded meetly coincident with the final period never to be terminated,– the infinite in height harmoniously associated with the eternal in duration. Creation and the Creator meet at one point, and in one person. The long ascending line from dead matter to man has been a progress Godwards,– not an asymptotical progress, but destined from the beginning to furnish a point of union; and occupying that point as true God and true man,– as Creator and created,– we recognize the adorable Monarch of all the future!

Part 4

COMPARATIVE ANATOMY

Comparative Anatomy

As demonstrated in the previous section, comparative anatomy was a key enabler of the development of palaeontology, and some extracts in this section will reflect the overlap between these fields. This section also offers opportunities to understand the significant role comparative anatomists played in changing attitudes in biology. While anatomy might not seem the most obvious focus in an anthology relating to environments in the long nineteenth century, its insights were a crucial component in the development of a holistic, materialist understanding of organisms and their habitats. These new ways of analysing and conceiving the nexus comprising environments + organisms would ultimately enable Darwinism and ecology, even if this was emphatically not the intention of the Natural Theological authors in this section.

For Richard Owen, in *Lectures on Comparative Anatomy*, the science that gained him considerable celebrity offered scientific opportunities to perceive 'the power which the appreciation of the co-relations and interdependencies of the several parts of each organic machine gives us to interpret the nature of the whole from the observation of a part'. For Owen, this often involved reconstructing an extinct animal on the basis of single bones or limited remains, but, more crucially, anatomy meant that animals could only be understood by analysing their functionality within a given environment: the manner in which its eyes, teeth, stomach, limbs, or other features were adapted (or, for the devout, designed) to fulfil a particular role in the animal's total relations with the world in which it existed. While this focus on the relationship between organisms and environments was certainly part of earlier work (see Ray, Buffon, and Bruckner in Part 1 of this volume or Gilbert White in various later sections), in Natural Theology the key idea was that organisms existed in a pre-designed and guaranteed state of harmony with nature. As with geology (see Part 2), such assumptions were gradually interrogated: environments were seen to be changeable, often catastrophically, and relationships between organisms were increasingly regarded as competitive in ways that undermined belief in overall harmony.

A marker of the importance of this science within the period is that 1851 saw the first edition of *Gray's Anatomy*, a work of considerable interest and fame, but not included here because of its overwhelming focus on human anatomy and medical science – matters that are briefly touched upon in the following extracts but which largely lie beyond our scope.

The first extract, from Xavier Bichat's *General Anatomy, Applied to Physiology and Medicine* (1801), is a seminal early work that established many key principles by which comparative anatomy subsequently operated: in his review of previous landmarks in 'General Observations', Bichat draws a line between his own researches into the 'vital properties' of life and the imperfect understanding of earlier work by Georges Ernst Stahl, Albrecht von Haller, and Félix Vicq d'Azyr. Crucially, he defines anatomy as analysing 'the relations of living bodies with one another, and their relations with those that do not live', and he makes the fundamental claim that 'bodies have only functions relative to their properties': a clear statement of the requirement for materialist reading of organisms as functional systems.

To understand Bichat, it is helpful to compare him with Baron Georges Cuvier, whose geological work we have seen in Part 3 of this volume but whose contributions to anatomy are so significant as to overshadow Bichat. The second, third, and fourth extracts in this section are taken from Cuvier's *Lectures on Comparative Anatomy* (1802), *Essay on the Theory of the Earth* (1813), and *The Animal Kingdom* (1817). Bichat and Cuvier both reach for an understanding of organic life that anticipates understanding of the dynamic interconnections of organisms (with other organisms, with environments). For Bichat, the physical sciences operate in a realm governed by fixed and unchanging laws that determine the properties of physical forces and objects, while anatomy, a 'physiological' science, is different largely because of the complexity of organic life but also its status as living and in process. Working before the notion of entropy was understood, Bichat sees the physical universe as essentially unchanging and continuous, while organisms move ineluctably towards death. This process of decline, alongside the dynamic nature of organic life and its inherent complexity of relations, leads Bichat to conclude that 'all the vital functions are susceptible of numerous variations' and that 'they defy every kind of calculation'. In *The Animal Kingdom* (1817), Cuvier also emphasises the complexity of organic life and its rootedness in external phenomena:

> [Organic] Life, then, is a vortex (*tourbillon*), more or less rapid, more or less complicated, the direction of which is constant, and which always carries along molecules of the same kind, but into which individual molecules are continually entering, and from which they are constantly departing; so that the form of a living body is more essential to it than its matter.

Bichat promulgates proto-dynamic ideas when speaking of interchange between organisms and their environments: 'by nutrition the particles of the matter of inanimate bodies pass into living bodies, and vice versa'. Both authors, then, insist on the environmental interdependence of bodily systems: for Bichat, 'every thing is so connected and tied together in the living body, that no one part can be disordered in its functions, without being immediately perceived by the rest'. If we turn to the extracts from Cuvier, we discern in more detail the results of these ideas.

In *Lectures on Comparative Anatomy*, Cuvier underlines the materialist basis of his approach with an analogy concerning 'a female in the prime of youth and health'. He presents the living form – 'that elegant voluptuous form – that graceful flexibility of motion – that gentle warmth – those cheeks crimsoned with the roses of delight – those brilliant eyes, darting rays of love' – as an illusion that hides the physical reality of the material body in death: 'the body loses its heat; the muscles become flat, and the angular prominences of the bones appear; the lustre of the eye is gone; the cheeks and lips are livid'. The tension between these two states is powerful in Cuvier's work, for while he is often professionally occupied with examining fossilised remains or deceased corpses, his greatest insights into the body arise from regarding it as a living, functional organism.

Analogies are frequent in Cuvier's work and always important. At one point in the extract from *Lectures*, he refers to human beings as

> a kind of furnace, into which inert substances are successively thrown, which combine among themselves in various manners, maintain a certain place, and perform an action determined by the nature of the combinations they have formed, and at last fly off in order to become again subject to the laws of inanimate nature.

The analogy is mechanistic but also dynamically unpredictable. Significantly, in the second extract, from *Theory of the Earth*, Cuvier dispenses with references to animals as machines and elaborates on the idea of their systematic wholeness and the manner in which their anatomy is dependent on environment. In *Lectures*, the mechanistic exists in constant tension with the organic: 'Each animal', he argues at first, 'may be considered as a partial machine, co-operating with all the other machines, the whole of which form the universe: the organs of motion are the wheels and levers, in short, all the passive parts'. Immediately, however, he tempers this reading: 'the active principle, the spring which gives the impulse to every part, resides only in the sensitive faculty, without which the animal, plunged in a continual slumber, would be reduced to a state purely vegetative'. Thus, like Bichat, Cuvier insists on the importance of examining the 'vital functions' of creatures in order to account for them fully.

'Article IV. *Relations of the Organs*' in *Lectures* provides a clear example of Cuvier's methods in terms of locating an organism as a functional being within its environment. It demonstrates that the anatomical forms of animal mouths, stomachs, and limbs are mutually determined by their functional needs and status (herbivorous, omnivorous, or carnivorous), so that 'the system of digestive organs has also immediate relations with those of motion and sensation'. A carnivore, for example, 'should have a penetrating eye, a quick smell, a swift motion, address, and strength in the claws and in the jaws', as well as a particular digestive system, while herbivores must have a different anatomical system to fulfil their functional role. A powerful application of these ideas can be found in the John Ruskin extract in Part 2.

The second Cuvier extract (also featured in Part 3 of this volume), *Essay on the Theory of the Earth*, is chosen for the manner in which its palaeontology relies on, and generates further knowledge within, comparative anatomy. Turning to the difficulties of analysing fossilised quadrupeds when only fragmentary and incomplete remains are often available, Cuvier demonstrates that science overcomes such problems:

> comparative anatomy, when thoroughly understood, enables us to surmount all these difficulties, as a careful application of its principles instructs us in the correspondence and dissimilarity of the forms of organized bodies of different kinds, by which each may be rigorously ascertained, from almost every fragment of its various parts and organs.

Essentially, Cuvier applies the principles outlined in the previous extract concerning the necessary relationships between the various organs of an animal in order to make inferences about those parts of the organism that are not available for inspection. Because 'every organized individual forms an entire system of its own, all the parts of which mutually correspond, and concur to produce a certain definite purpose', it follows that 'in every one of their parts we discover distinct indications, not only of the classes and orders of animals, but also of their genera, and even of their species'. This revolutionary understanding of animals as systematic, functioning wholes is central to Cuvier's work and to the entire direction of comparative anatomy, biology, and related disciplines. The remainder of the extract offers examples of the 'astonishing results' arising from these methods. As Cuvier and others would demonstrate repeatedly in the decades that followed, 'the smallest fragment of bone' can 'by careful examination, assisted by analogy and exact comparison', help the anatomist 'to determine the species to which it once belonged, as certainly as if we had the entire animal before us' (on this, see Gowan Dawson, Further Reading).

The third extract from Cuvier, written a decade after the preceding two, represents a mature encapsulation of his work, as well as an enormous and meticulous anatomical study of fauna. Cuvier's introduction to *The Animal Kingdom* (1817) begins by recapitulating the distinction between the physical sciences (involved in 'isolating bodies, reducing them to their utmost simplicity, in bringing each of their properties separately into action, either mentally or by experiment, in observing or calculating the results, in short, in generalizing and correcting the laws of these properties for the purpose of establishing a body of doctrine, and, if possible, of referring the whole to one single law') and Natural History (which 'is confined to objects which do not allow of rigorous calculation, or of precise measurement in all their parts'). Positioning the natural sciences as necessarily more complex, he argues that one must study organisms 'in the different positions in which Nature places them, or in a comparison of different bodies together, until constant relations are recognized between their structures and the phenomena which they manifest' – to study them as functional wholes existing within specific environments.

Cuvier then argues that for the natural sciences to operate rigorously, they must have an adequate taxonomic system with which to classify organisms. He promotes the use of the 'natural system' or 'natural method', which emerged within botany but was also used in zoology. In botany, the 'natural method' arose in post-Revolutionary France as an alternative or addition to Linnaeus's 'sexual system' (which classified plants purely on floral anatomy). The French pioneered plant identification keys or manuals which combined step-by-step analyses of plant anatomy and appearance (size, leaf shape, habit, etc.) in order to focus on differences between plants and to cluster similar plants together in 'natural' ways. French botanical works frequently combined natural, Linnaean, and other methods, whilst in Britain, botanists were generally much slower to relinquish a Linnaean approach. This may have been due to both a post-Revolutionary suspicion of all things French and to a strong cultural association with Linnaeus (the Linnean Society of London being the most prestigious botanical society in the United Kingdom). Ultimately, the 'natural system' prevailed, and many of its findings were vindicated by genomic analysis.

The four extracts that follow are from the leading British comparative anatomist of the period, Sir Richard Owen. Taken together, they provide a useful overview of his contributions to palaeontology and comparative anatomy in ways that also contextualise the final extract in this section. For a useful study of Owen's work, see Nicolaas Rupke's work (Further Reading).

The first and briefest of the four extracts, Owen's 'Report on British fossil reptiles' for the 1841 *Report of the Eleventh Meeting of the British Association for the Advancement of Science*, was published anonymously, but the identity of the author would have been apparent to many. In it, Owen is the first to suggest the term 'dinosaur' to describe the growing number of prehistoric creatures whose remains were being found and examined (see Part 2 of this volume), and who were initially described as 'saurians' or 'sauropods'. Owen begins with an anatomical summary of the key features of the group he wishes to define, but his remarks on their character reveal the degree to which there was still uncertainty about their form – whether to regard them as either 'crocodilian' or 'lacertian' (lizard-like). In time, consensus would vindicate the latter view. While Owen's term '*Dinosauria*' to denote the terrestrial giant lizards was quickly accepted, his suggestion that their marine counterparts should be called '*Enaliosauria*' was never adopted.

Owen's interest in using comparative anatomy to understand extinct creatures is also exemplified in the second extract, from *Description of the Skeleton of an Extinct Gigantic Sloth* (1842). Although he refers to the species as *Mylodon robustus*, his earlier naming (from 1840), *Mylodon darwini*, is now accepted. The extract is remarkable for the light it sheds on the extent of palaeontological activity being undertaken by Europeans in South America during the first half of the nineteenth century and for the way in which Owen uses his anatomical knowledge to unpick the differences between the *Mylodon* and the *Megatherium*, another gigantic extinct species of South America which many believed was represented by the remains of the *Mylodon*. Drawing on Cuvierian principles about the lessons to be learnt by regarding animal remains as functional wholes existing within specific environments, Owen argues that the habits of the *Mylodon* become evident by the examination of the fossil remains. By painstaking investigations of skeletal and dental features, he concludes that *Mylodon* was equipped to scratch away the soil surrounding trees and then to upend them through violent force: 'extraordinary must have been the strength and proportions of that tree, which rocked to and fro, to right and left, in such an embrace, could long withstand the efforts of its ponderous assailant'.

As this extract suggests, Cuvier's remarks in the third extract about the anatomist being able to reconstruct an animal from one or few bones because of fixed interrelations between the parts of particular animals (such as herbivores and carnivores) is taken up by Owen, who achieved considerable fame because of his ability to do so. Perhaps most famously, in 1839 he inferred the existence of a hitherto-unknown giant flightless bird from New Zealand on the basis of a fragment of femur which he had acquired while Hunterian Professor of Comparative Anatomy and Physiology at the Royal College of Surgeons in London. That year, he announced to the Zoological Society of London that 'so far as a judgement can be formed of a single fragment . . . I am willing to risk the reputation for it on the statement that there has existed, if there does not now exist, in New Zealand, a Struthious bird nearly, if not quite, equal in size to the Ostrich'. Owen's claim was initially deemed too conjectural by the Zoological Society, who only reluctantly agreed to include it in their 1840 *Proceedings*. Owen was vindicated in spectacular fashion, however, when further bones were delivered to William Buckland, who sent them on to Owen. These remains enabled Owen to confirm his theory and to reconstruct a giant wingless bird, *Dinornis novae zealandiae*. The wider public excitement caused by Owen's feat indicates the growing status of science in the early Victorian period. It did much to establish Owen as a household name and raised the profile of comparative anatomy – much to the chagrin of rival palaeontologist and dinosaur-hunter Gideon Mantell, who continued to spar with Owen (on their disputes, see Part 3 of this volume).

The third extract, from *Lectures on the Comparative Anatomy and Physiology of the Vertebrate Animals* (1846), is from the period of Owen's established celebrity. They offer comprehensive coverage of the animal kingdom, and the extract from the section on fishes is chosen purely as representative of his general methods. In his opening remarks, Owen emphasises that enormous impacts of his field were key to advancements in medicine, surgery, zoology, geography, and geology, while also providing 'an insight into the true chronology and ancient geography of the globe'. Because of this, Owen claims, animal anatomy 'bids fair to take rank as the first of all sciences'.

Like 'the immortal Cuvier', Owen was an exponent of homology, the claim that different groups of animals (mammals, reptiles, amphibians, birds, fish) have essentially similar physiological structures (so that, for example, the forelimbs of mammals correspond to the wings of bats and birds and the pectoral fins of fish). Owen's work is also indebted to the stratigraphical fossil studies of William Smith. In terms of the geological debates outlined in Part 3 of this volume, Owen accepts an extended timeline of earth history and theories of successive extinctions. However, Owen's lectures were delivered in 1844, a year in which the controversial proto-evolutionary work *Vestiges of Creation* was published (see Part 9 of this volume), and Owen's stated fealty to the Natural Theological traditions of Ray and Paley, as well as his remarks at the end of this extract about 'the hypothesis of the development of a species' are of a piece with his wider promotion of a specifically Christian science. This is an irony given his suggestive statement that 'almost every part of the Human frame has its homologue in some inferior animal; and we at length begin to perceive that Man's organisation is a special modification of a more general type' and because so many of his homological and palaeontological findings provided materialist evidence for Darwin's work in *Origin of Species* (1859), where he argued that similarities of structures and functions of anatomical parts were the result of their descent from a common evolutionary ancestor and their subsequent adaptation to varying environments. Owen's rebuttal of *Vestiges* relies on positioning it as a further example of the discredited Lamarckian theories of development ('those once favourite and recently-revived hypotheses'), but he remained a fierce critic of Darwinism until his death in 1892.

The final Owen extract is taken from *The Nature of Limbs* (1849), which offers more detailed insights into his homology. Focusing on the 'the "arms and legs" in Man; the "fore-" and "hind-legs" of Beasts; the "wings" and "legs" of Bats and Birds; [and] the "pectoral fins" and "ventral fins" of Fishes', he outlines their homological relationship to one another – by which he means that these were essentially the same organ modified in different animals to fulfil a variety of functions. Owen's own comparisons – between the flipper of a dugong and the front limbs of a mole – invite evolutionary readings, as Owen points out the way in which the former is adapted for movement in water and the latter for the terrestrial habits of the burrowing mammal. Likewise, he shows the advantages of light wings for birds and heavy forelimbs for herbivorous quadrupeds. While proclaiming that 'the instrument must be equal to its office', he rejects the idea that homological diversification of forms indicates evolutionary development.

The final extract, from Louis Agassiz's *Twelve Lectures*, was also published in 1849 and extends the interest in embryology which Owen demonstrated in *Lectures*. Although his career commenced in Europe, by the late 1840s, Agassiz was firmly ensconced in American science and a public celebrity. Something of the excitement surrounding his public lectures at the Lowell Institute, Boston, is captured by the Preface, which announces the use of the relatively new system of Pitman Shorthand to capture the Professor's words.

Agassiz positions himself as an inheritor of Cuvier. Like Owen, he belongs in an earlier tradition than Darwin and his followers. However, as with Owen's emphasis on homology, Agassiz's advocacy of comparative embryology as a valuable new tool in zoological classification leads a writer with strong Natural Theological beliefs into territories fertile for the development of evolutionary ideas. Like Owen, Agassiz rejected Darwinism. In a less explicit way than Owen, Agassiz also resists the pre-Darwinian readings of embryonic development, such as *Vestiges of Creation*, that imply evolutionary change, and he explicitly argues that his own researches show 'the action of the intelligent Author of all these things'. Just as Owen's homology anticipated Darwinian thought in ways that would be unacceptable to him, so Agassiz's interest in embryology is shared by *Vestiges*, which rests much of its argument for the development of species on analysis on embryological analysis (see Part 9 of this volume). Indeed, Darwin's theories found some of their strongest evidence in the genetic discoveries (a form of embryology) of the Moravian mathematician Gregor Mendel in the 1860s (see Volume III of this anthology). Embryology has roots in Classical Greece, ancient India, and mediaeval Islamic traditions but proved of growing interest to scientists after 1700. Agassiz was amongst those sponsoring its rapid nineteenth-century development.

Commenting on the gradual progress of zoological taxonomy, Agassiz points to the role of Cuvierian anatomy in correcting many of the errors of previous naturalists before suggesting that embryology will supply the next step in perfecting nomenclatural systems. While anatomical 'structure is the principle upon which animals can be most satisfactorily classified', his proposed field of comparative embryology – the 'natural consequence of the natural progress and state of our science' – will provide 'the foundation of a natural system of Zoology'. Comparative embryology takes two related forms in Agassiz's proposals: the study of an embryo's development over time, as well as the study of life cycles such as those of frogs and insects and the comparison of different species via analysis of their embryonic forms and developments. Agassiz turns to 'the class of reptiles' to show the deficiencies of previous classifications and the various errors of earlier naturalists in distinguishing between frogs, toads, lizards, turtles, crocodiles, salamanders, and snakes. He demonstrates that while Cuvierian methods largely resolve such problems, comparative embryology offers greater taxonomic clarity and certainty.

The extracts gathered in this section are relevant not only in looking back to Parts 2 and 3 of this volume but also in looking forward, not least to Part 9, where much of the work of the comparative anatomists would yield unexpected evolutionary fruits. As we shall see in subsequent volumes, comparative anatomy continued to be of importance throughout the remainder of the period under study as it generated, and came into fruitful relationships with, new biological sciences such as genetics.

Further reading

Anderson, Thomas J., '*Aepyornis* as Moa: Giant Birds and Global Connections in Nineteenth Century Science, *British Journal for the History of Science* 46:4 (2013), 675–93.

Cadbury, Deborah, *The Dinosaur Hunters: A True Story of Scientific Rivalry and the Discovery of the Prehistoric World*, (London: Fourth Estate, 2001).

Cosans, Christopher, *Owen's Ape & Darwin's Bulldog: Beyond Darwinism and Creationism*. (Bloomington: Indiana University Press, 2009).

Dawson, Gowan, *Show Me the Bone: Reconstructing Prehistoric Monsters in Nineteenth-Century Britain and America* (Chicago: London: University of Chicago Press, 2016).
Feingold, Mordechai, *French Medical Culture in the Nineteenth Century* (Leiden: Brill, 2020).
Goodall, Jane, *Frankenstein's Science: Experimentation and Discovery in Romantic Culture, 1780–1830* (London: Taylor & Francis, 2016).
Haigh, Elizabeth, *Xavier Bichat and the Medical Theory of the Eighteenth Century* (London: Wellcome Institute, 1984).
Irmscher, Christoph, *Louis Agassiz: Creator of American Science* (Boston: Houghton Mifflin Harcourt, 2013).
Lightman, Bernard V, *Victorian Science in Context* (Chicago: Chicago University Press, 1997).
Lurie, Edward, 'Louis Agassiz and the Races of Man', *Isis* 45:3 (1954), 227–42.
———, *Louis Agassiz: A Life in Science* (Baltimore: Johns Hopkins University Press, 1988).
Rudwick, Martin J.S., *Georges Cuvier, Fossil Bones, and Geological Catastrophes: New Translations & Interpretations of the Primary Texts* (Chicago: University of Chicago Press, 1997).
Rupke, Nicolaas A., *Richard Owen: Victorian Naturalist* (New Haven: Yale University Press, 1994).
Winsor, Mary P., 'Louis Agassiz and the Species Question', *Studies in History of Biology* 3 (1979), 89–138.
———, *Reading the Shape of Nature: Comparative Zoology at the Agassiz Museum* (Chicago: University of Chicago Press, 1991).

33

XAVIER BICHAT, *GENERAL ANATOMY, APPLIED TO PHYSIOLOGY AND MEDICINE*, TRANS. GEORGE HAYWARD, 3 VOLS., VOL. 1

(Boston: Richardson and Lord, 1822 [first published in French, 1801])

General Observations

THERE are in nature two classes of beings, two classes of properties, and two classes of sciences. The beings are either organic or inorganic, the properties vital or non-vital, and the sciences physiological or physical. Animals and vegetables are organic – minerals are inorganic. Sensibility and contractility are vital properties; gravity, elasticity, affinity, &c. are non-vital properties. Animal and vegetable physiology, and medicine form the physiological sciences; astronomy, physics, chemistry, &c. are the physical sciences. These two classes of sciences have relation only to different phenomena; there are two other classes that correspond to these, which relate to the internal and external forms of bodies and their description. Botany, anatomy, and zoology, are the sciences of organic bodies; mineralogy, &c. of the inorganic. The first will occupy us, and we shall fix our attention especially upon the relations of living bodies with one another, and their relations with those that do not live.

I. General Remarks upon Physiological and Physical Sciences

THE differences between these sciences are derived essentially from those existing between the properties that preside over the phenomena, which are the object of each class of sciences. So immense is the influence of these properties, that they are the principle of all phenomena; whether we examine those of astronomy, of hydraulics, of dynamics, of optics, of acoustics, &c. we shall finally arrive by a connexion of causes to gravity, to elasticity, &c. as the end of our researches. So the vital property is the primum mobile to which we must ascend, whether

DOI: 10.4324/9780429355653-38

we consider the phenomena of respiration, of digestion, of secretion, circulation, inflammation, fevers, &c.– In giving existence to every body, nature has imprinted upon it a certain number of properties, that particularly characterize it, and by means of which it contributes in its own manner, to all the phenomena that are developed, succeed, and continually connect themselves in the universe. Cast your eyes upon that which surrounds you; turn them upon objects the most distant; whether, aided by the telescope, they examine those that swim in space, or, armed with the microscope, they enter the world of those, whose minuteness almost evades our view, every where you will find on one side physical properties, on the other vital properties, brought into action; every where you will see inert bodies gravitating upon each other, and reciprocally attracting; living bodies gravitate also, but above all they feel, and possess a motion which they owe only to themselves.

These properties are so inherent in bodies, that we cannot conceive of their existence without them. They constitute their essence and their attribute. To exist and to enjoy them are two things inseparable. Suppose that of a sudden they are deprived of them; instantly all the phenomena of nature cease, and matter alone exists. Chaos was only matter without properties; to create the universe, God endowed it with gravity, elasticity, affinity, &c. and to a part he gave sensibility and contractility.

This mode of considering the vital and physical properties, sufficiently shews, that we cannot ascend above them in our explanations, that they afford the principles, and that these explanations are to be deduced from them as consequences. The physical sciences, as well as the physiological, then, are composed of two things; 1st. the study of phenomena, which are effects; 2d. the research into the connexions that exist between them and the physical or vital properties, which are the causes.

For a long time these sciences have not been so considered; every fact that was observed, was made the subject of a particular hypothesis. Newton was the first to remark, that however variable the physical phenomena were, they could all be referred to a certain number of principles. He analyzed these principles and found that attraction enjoyed the most important place among them. Attracted by each other and by their sun, the planets describe their eternal courses; attracted to the centre of our system, the waters, air, stones, &c. move or tend to move towards it: it is truly a sublime idea, and one that serves as the basis to all the physical sciences. Let us render homage to Newton; he was the first who discovered the secret of the Creator, viz. a simplicity of causes reconciled with a multiplicity of effects.

The epoch of this great man was the most remarkable of human wisdom. Since that period, we have had principles from which we draw facts as consequences. This epoch, so advantageous to the physical sciences, was nothing to the physiological; what do I say? It retarded their progress. Mankind soon saw nothing but attraction and impulse in the vital phenomena.

[. . .]

Unknown to the ancients, the laws of life have begun to be understood during the last age only. Stahl had already remarked the tonic motions, but he did not generalize their influence. Haller was engaged particularly with sensibility and irritability; but in limiting one to the nervous system, and the other to the muscular, this great man did not consider them in the correct point of view; he made them almost insulated properties. Vicq d'Azyr changed them into functions in his physiological division and ranked them with ossification, digestion, &c. that is, he confounded the principle with the consequence. Thus you see, notwithstanding the labours of a crowd of learned men, how much the physiological sciences still differ from the physical. In these, the chemist refers all the phenomena that he observes to affinity: the natural philosopher, in his science, every where sees gravity, elasticity, &c. In the others, we have not as yet ascended, at least in a general manner, from the phenomena to the properties from which they are derived. Digestion, circulation, or the sensations, do not bring the idea of sensibility or contractility to the mind of the physiologist, as the movement of a watch proves to the mechanician that elasticity is the primum mobile of its motion; or as the wheel of a mill or of any machine, which running water sets in motion, proves to the natural philosopher that gravity is the cause. To place upon the same level in this respect these two classes of sciences, it is evidently necessary to form a just idea of vital properties. If their limits are not accurately assigned, we cannot with precision analyse their influence. I shall present here only general considerations on this point, which has been treated sufficiently in my Researches upon Life; what I shall add now will be but as a supplement to what has been explained in that work.

II. Of Vital Properties, and their Influence upon all the Phenomena of the Physiological Sciences

To assign the limits of these properties, we must follow them from bodies that are hardly developed, to those which are the most perfect. In the plants that seem to form the transition from vegetables to animals, you discover only an internal motion that is scarcely real; their growth is as much by the affinity of particles and consequently by juxtaposition, as by a true nutrition. But in ascending to vegetables better organized, you see them continually pervaded by fluids, that circulate in numerous capillary canals, which mount, descend, and run in a thousand different directions, according to the state of the forces that regulate them. This continual motion of fluids is foreign to the physical properties, the vital ones only direct it. Nature has endowed every portion of a vegetable with a faculty of feeling the impression of fluids, with which their fibres are in contact, and of reacting upon them in an insensible manner, to favour their course. The first of these faculties I call organic sensibility, the other, insensible organic contractility. This is very obscure in most vegetables; it is the same in the bones of animals. These two properties govern not only the vegetable circulations, which correspond in some measure to the capillary system of animals, but also the secretion, absorption, and exhalation of vegetables. Remark, in fine, that these bodies have only functions relative to their properties; that all the phenomena that animals derive from properties which they have more than vegetables, as the great circulation and digestion, for which there must be sensible organic contractility; as the sensations, for which there must be animal sensibility; and locomotion, the voice, &c. for which animal contractility is necessary; remark, I say, that these functions are essentially foreign to vegetables, since they have not vital properties to place them in action.

[. . .]

If we pass from vegetables to animals, we see the lowest of these, the zoophites, receive into a sac, which is alternately filled and emptied, the aliments that are to nourish them; we see them begin to unite sensible organic contractility or irritability to the properties which they have in common with vegetables, and consequently commence the performance of different functions, digestion in particular.

Thus far the organized bodies live wholly within themselves; they have no relation with that which surrounds them; animal life is wanting in them, or at least if it has commenced in these animo-vegetables, its rudiments are so obscure that we can hardly discover them. But this life begins to display itself in the superior classes, in worms, insects, mollusca, &c. On the one hand, the sensations, and on the other, locomotion, which is inseparable from them, are more or less fully developed. Then the vital properties necessary to the exercise of these new functions, are added to the preceding. Animal sensibility and contractility, obscure in the lower species, become more perfect, as we approach quadrupeds, and locomotion and the sensations become also more extensive. Sensible organic contractility then increases, and in proportion to that, digestion, circulation of the great vessels, &c. which are governed by it, receive a development which is constantly growing more perfect.

If we strictly examine the immense series of living bodies, we shall see the vital properties gradually augmenting in number and energy, from the lowest of plants to the first of animals, man; we shall see the lowest plants obedient to vital and physical properties; all plants are governed only by these, which, in them, consist of insensible contractility and organic sensibility; the lowest animals begin to add sensible organic contractility to these properties, afterwards animal sensibility and contractility. [. . .]

Man and the neighbouring species, which are the particular object of our researches, enjoy then evidently, all the vital properties, some of which belong to organic life, the others to animal life. 1st. Organic sensibility and insensible contractility have all the phenomena of the capillary circulation, of secretion, of absorption, exhalation, nutrition, &c. evidently dependant upon them in a state of health. In treating, therefore, of these functions, we must always ascend to these properties. In the state of disease, all the phenomena that suppose a disorder in these functions, are clearly derived from an injury of these properties [. . .]

Every where in the physiological sciences, in the physiology of vegetables and of animals, in pathology, in therapeutics, &c. there are vital laws, that govern the phenomena which are the object of these sciences; and that there is not one of these phenomena that does not flow from these essential and fundamental laws, as from its source.

If I should take a survey of all the divisions of physical sciences, you would see that the physical laws were ultimately the sole principle of all their phenomena; but this is so well known that it is not necessary to do it. I will consider an important subject, and one to which we are naturally led by the preceding observations. I mean, a parallel between physical and vital phenomena, and consequently between physical and physiological sciences.

III. Characteristics of the Vital Properties, Compared With Those of the Physical

When we consider, on one side, the phenomena which are the object of the physical sciences, and those that are the object of the physiological, we see how immense is the space that separates their nature and their essence. But this difference arises from that which exists between the laws of the one and the other.

Physical laws are constant and invariable; they are subject neither to augmentation or diminution. A stone does not gravitate towards the earth with more force at one time than another; in every case marble has the same elasticity, &c. On the other hand, at every instant, sensibility and contractility are increased, diminished, or altered; they are scarcely ever the same.

It follows, therefore, that the physical phenomena are never variable, that at all periods and under every influence they are the same; they can, consequently, be foreseen, predicted, and calculated. We calculate the fall of a heavy body, the motion of the planets, the course of a river, the ascension of a projectile, &c.: the rule being once found, it is only necessary to make the application to each particular case. Thus heavy bodies fall always in a series of odd numbers; attraction is in the inverse ratio of the square of the distances, &c. On the other hand all the vital functions are susceptible of numerous variations. They are frequently out of their natural state; they defy every kind of calculation, for it would be necessary to have as many rules as there are different cases. It is impossible to foresee, predict, or calculate, any thing with regard to their phenomena; we have only approximations towards them, and even these are often very uncertain.

There are two things in the phenomena of life, 1st. the state of health; 2d. that of disease; hence there are two distinct sciences; physiology considers the phenomena of the first state, pathology those of the second. The history of the phenomena in which the vital forces have their natural type, leads us to consider as a consequence, those phenomena that take place when these forces are altered. But in the physical sciences there is only the first history; the second is never found. Physiology is to the movements of living bodies, what astronomy, dynamics, hydraulics, hydrostatics, &c. are to those of inert ones; but these last have no such correspondent sciences as pathology. There is nothing in the physical sciences that corresponds to therapeutics in the physiological. For the same reason, every idea of medicament is absurd in the physical sciences. The object of a medicament is the restoration of properties to their natural type; but, the physical properties, never losing this type, have of course no need of restoration.

We see then that the peculiar instability of the character of the vital laws is the source of an immense series of phenomena, which form a peculiar order of sciences. What would become of the universe, if the physical laws were subject to the same commotions and the same variations as the vital? Much has been said of the revolutions of the globe, of the changes that the earth has undergone, of the overthrows that ages have gradually brought about, and upon which ages have accumulated without producing others: but you would see these overthrows and these general commotions in nature at every instant, if the physical properties had the same character as the vital.

[. . .]

Ordinary minds, in reading, stop at insulated facts that are presented; they do not embrace, at one view, the principles of the work. Oftentimes the author himself incautiously follows the impression given to a science in the age in which he writes. But the man of genius every where pauses at this impression, which should be henceforth entirely different in physical and physiological works. It is necessary to use a different language; for most of the words that are carried from the physical into the physiological sciences, continually refer to ideas that have no connexion with them. You see the living solids constantly undergoing composition and decomposition, every moment taking and rejecting new substances; on the other hand, inert bodies remain the same, and keep the same constituent principles, until friction and other causes destroy them. So in the elements of inert bodies, there is a constant uniformity, and an invariable identity in their principles, which is known when they have been once analyzed; whilst the principles of the living fluids are so continually changing, that it is necessary that many analyses should be made, under every possible circumstance [. . .]

It is the nature of vital properties to exhaust themselves; time wastes them. Elevated in the commencement of life, they remain stationary at the adult age, and afterwards are debilitated and become nothing. Prometheus, it is said, having formed some statues of men, snatched fire from heaven to animate them. This fire is the emblem of the vital properties; while it burns, life is supported; when it is extinguished, it ceases. It is, then, a part of the essence of these properties, to animate matter for a determinate time only; hence there are necessary limits to life. On the other hand, the physical properties, constantly inherent in matter, never abandon it; so that inert bodies have no limits to their existence, but what accident gives to them.

By nutrition the particles of the matter of inanimate bodies pass into living bodies, and vice versa; and we can evidently conceive that this matter has been endowed through an immense series of ages with physical properties. These properties are given to it at the creation, and will leave it only when the world shall end. This matter, in passing into living bodies, in the space that separates these two epochs, a space that immensity only can bound, this matter, I say, becomes possessed, at intervals, of vital properties, which are then united to physical properties. Here, then, is a great difference in matter, with regard to these two kinds of properties; one it enjoys by intermissions only, the other it possesses constantly.

34

BARON GEORGES CUVIER, *LECTURES ON COMPARATIVE ANATOMY, VOL. 1. ON THE ORGANS OF MOTION*, TRANS. WILLIAM ROSS

(London: T. N. Longman and O. Rees, 1802 [first published in French, 1800])

Lecture the First: Preliminary Observations upon the Animal Œconomy

Article I, General View of the Functions of Animal Bodies

THE idea of *Life* is one of those general and obscure ideas produced in us by observing a certain series of phænomena possessing mutual relations, and succeeding each other in a constant order. We know not indeed the nature of the link that unites these phænomena, but we are sensible that a connexion must exist; and this conviction is sufficient to induce us to give it a name, which the vulgar are apt to regard as the sign of a particular principle, though in fact that name can only indicate the totality of the phænomena which have occasioned its formation.

Thus, as the human body, and the bodies of several other animals resembling it, appear to resist, during a certain time, the laws which govern inanimate bodies, and even to act on all around them in a manner entirely contrary to those laws, we employ the terms *life* and *vital force* to designate what are at least apparent exceptions to general laws. It is, therefore, by determining exactly in what the exceptions consist, that we shall fix the meaning of those terms. For this purpose, let us consider the bodies I have mentioned, in their active and passive relations with the rest of nature.

For example, let us contemplate a female in the prime of youth and health. That elegant voluptuous form – that graceful flexibility of motion – that gentle warmth – those cheeks crimsoned with the roses of delight – those brilliant eyes, darting rays of love, or sparkling with the fire of genius – that countenance, enlivened by sallies of wit, or animated by the glow of passion, seem all united to form

DOI: 10.4324/9780429355653-39

a most fascinating being. A moment is sufficient to destroy this illusion. Motion and sense often cease without any apparent cause. The body loses its heat; the muscles become flat, and the angular prominences of the bones appear; the lustre of the eye is gone; the cheeks and lips are livid. These, however, are but preludes of changes still more horrible. The flesh becomes successively blue, green, and black. It attracts humidity, and while one portion evaporates in infectious emanations, another dissolves into a putrid sanies, which is also speedily dissipated. In a word, after a few short days there remains only a small number of earthy and saline principles. The other elements are dispersed in air, and in water, to enter again into new combinations.

It is evident that this separation is the natural effect of the action of the air, humidity and heat – in a word, of external matter upon the dead body; and that it has its cause in the elective attraction of those different agents for the elements of which the body is composed. That body, however, was equally surrounded by those agents while living, their affinities with its molecules were the same, and the latter would have yielded in the same manner during life, had not their cohesion been preserved by a power superior to that of those affinities, and which never ceased to act until the moment of death.

Of all the phænomena, the particular ideas of which enter into the general idea of life, this is what at first sight appears to constitute its essence, since we can form no conception of life without it, and since it evidently exists without interruption until the instant of dissolution.

But a further study of any living body convinces us, that the power which preserves the union of the moleculæ, notwithstanding the eternal forces which tend to separate them, does not confine its activity to this tranquil operation, and that the sphere of its action extends beyond the bounds of the living body itself. At least it does not appear that this power differs from that which attracts new moleculæ to deposit them between those that already exist: and this action of the living body, in attracting the surrounding moleculæ, is not less constant than that which its exercises in retaining its own; for besides that the absorption of the alimentary matter, its conversion into nutritive fluid, and its subsequent transmission to all the parts of the body, experiences no interruption, and continues from one repast to the other; another absorption constantly takes place at the external surface, and a third by the effect of respiration. The two latter are those only which exist in all living bodies which do not digest, that is to say, in all plants.

[. . .]

This consideration must modify the idea which we at first formed from the principal phænomena of life. Instead of a constant union in the moleculæ, we cannot avoid observing, that there is a continual circulation from the exterior to the interior, and from the interior to the exterior of bodies – a circulation which, though uniformly preserved, is notwithstanding fixed within certain limits. Thus living bodies may be considered as a kind of furnace, into which inert substances are successively thrown, which combine among themselves in various manners, maintain a certain place, and perform an action determined by the nature of the combinations they have formed, and at last fly off in order to become again subject to the laws of inanimate nature.

[. . .]

How much influence the faculties of sensation and motion, which animals possess in addition to those of vegetables, have over the organs of all the other faculties which are common to both these kinds of beings. The comparison which we shall hereafter make of the different orders of animals, will, in the same manner, demonstrate that the modifications of their principal functions exercise a similar influence on all the others:– such is the union and harmony which prevails in all the parts of living bodies.

[. . .]

We have thus described the principal functions which compose the animal œconomy. It is obvious that they may be divided into three distinct orders. There are some which, in constituting animals what they are, fit them for fulfilling the part that nature has assigned to them in the general arrangement of the universe – in a word, which would be sufficient for their existence, if that existence were only momentary. These are the faculties of sensation and motion: The latter enables them to execute certain actions, and the former determines their choice of the particular actions they are capable of performing. Each animal may be considered as a partial machine, co-operating with all the other machines, the whole of which form the universe: the organs of motion are the wheels and levers, in short, all the passive parts; but the active principle, the spring which gives the impulse to every part, resides only in the sensitive faculty, without which the animal, plunged in a continual slumber, would be reduced to a state purely vegetative; plants themselves, as Buffon has observed, may be called animals which sleep. These two functions form the first order, and are termed *animal functions*. But animal machines, unlike those we construct, possess an internal principle of preservation and reparation. This principle consists in the union of the different functions which serve to nourish the body, that is to say, *digestion*, *absorption*, *circulation*, *respiration*, *transpiration*, and the *excretions*. These form the second order, and are denominated *vital functions*.

[. . .]

Article IV, Relations of the Organs

[. . .]

In consequence of the connection that subsists between the organs of respiration and the modifications of several other functions, some of the latter have relations to one another which at first sight did not appear necessary. This is the reason why birds have in general an exceedingly strong stomach, and a very quick digestion. This also is the reason why their repasts are so frequently repeated; while reptiles, which among the red-blooded animals seem to be contrasted to them in every respect, astonish us by the little aliment they take and the length of time they abstain from food. These differences in the digestive powers do not depend on the nature of the organs in motion which characterise these two classes, but upon that of the organs of respiration, the modifications of which have an immediate relation with those of motion.

[. . .]

The system of digestive organs has also immediate relations with those of motion and sensation. The disposition of the alimentary canal determines, in a manner perfectly absolute, the kind of food by which an animal is nourished; but if the animal did not possess, in its senses and organs of motion, the means of distinguishing the kinds of aliments suited to its nature, it is obvious it could not exist.

An animal, therefore, which can only digest flesh, must, to preserve its species, have the power of discovering its prey, of pursuing it, of seizing it, of overcoming it, and tearing it in pieces. It is necessary, then, that this animal should have a penetrating eye, a quick smell, a swift motion, address, and strength in the claws and in the jaws. Agreeably to this necessity, a sharp tooth, fitted for cutting flesh, is never co-existent in the same species, with a foot covered with horn, which can only support the animal, but with which it cannot grasp anything; hence the law by which all hoofed animals are herbivorous; and also those still more detailed laws which are but corollaries of the first, that hoofs indicate dentes molares, with flat crowns, a very long alimentary canal, a capacious or multiplied stomach, and several other relations of the same kind.

Those laws which determine the relations of the organs belonging to the different functions, likewise exercise their powers on the different parts of the same system, and connect its variations with equal force. The application of these laws is particularly evident in the alimentary system, the parts of which are more numerous and distinct. The form of the teeth, the length, the convolutions, and the dilations of the alimentary canal, and the number and abundance of the dissolving liquors poured into it, have always an admirable relation to each other, and to the nature, the hardness, and the solubility of the substance the animal eats. This connection is so evident, that the skilful anatomist, upon knowing one of those parts, may easily conjecture most of the others, and may, agreeably to the preceding laws, even guess the extent of the other functions.

35

BARON GEORGES CUVIER, *ESSAY ON THE THEORY OF THE EARTH*

(Paris, 1813)

Of the Difficulty of Distinguishing the Fossil Bones of Quadrupeds

WHILE the study of the fossil remains of the greater quadrupeds is more satisfactory, by the clear results which it affords, than that of the remains of other animals found in a fossil state, it is also complicated with greater and more numerous difficulties. Fossil shells are usually found quite entire, and retaining all the characters requisite for comparing them with the specimens contained in collections of natural history, or represented in the works of naturalists. Even the skeletons of fishes are found more or less entire, so that the general forms of their bodies can, for the most part, be ascertained, and usually at least their generic and specific characters are de terminable, as these are mostly drawn from their solid parts. In quadrupeds, on the contrary, even when their entire skeletons are found, there is great difficulty in discovering their distinguishing characters, as these are chiefly founded upon their hair and colours, and other marks which have disappeared previous to their incrustation. It is also very rare to find any fossil skeletons of quadrupeds in any degree approaching to a complete state, as the strata for the most part only contain separate bones, scattered confusedly, and almost always broken and reduced to fragments, which are the only means left to naturalists for ascertaining the species or genera to which they have belonged.

It may be stated also, that most observers, alarmed by these formidable difficulties, have passed slightly over the fossil remains of quadrupeds, and have satisfied themselves with classing them vaguely, by means of slight resemblances, or have not even pretended to give them names. Hence this portion of the history of extraneous fossils, though the most important and most instructive, has been investigated with less care than any other.

Fortunately, comparative anatomy, when thoroughly understood, enables us to surmount all these difficulties, as a careful application of its principles instructs us in the correspondence and dissimilarity of the forms of organized bodies of

DOI: 10.4324/9780429355653-40

different kinds, by which each may be rigorously ascertained, from almost every fragment of its various parts and organs.

Every organized individual forms an entire system of its own, all the parts of which mutually correspond, and concur to produce a certain definite purpose, by reciprocal reaction, or by combining towards the same end. Hence none of these separate parts can change their forms without a corresponding change on the other parts of the same animal, and consequently each of these parts, taken separately, indicates all the other parts to which it has belonged. Thus, as I have elsewhere shewn, if the viscera of an animal are so organized as only to be fitted for the digestion of recent flesh, it is also requisite that the jaws should be so constructed as to fit them for devouring prey: the claws must be constructed for seizing and tearing it to pieces; the teeth for cutting and dividing its flesh; the entire system of the limbs, or organs of motion, for pursuing and overtaking it; and the organs of sense, for discovering it at a distance. Nature also must have endowed the brain of the animal with instincts sufficient for concealing itself, and for laying plans to catch its necessary victims.

Such are the universal conditions that are indispensable in the structure of carnivorous animals; and every individual of that description must necessarily possess them combined together, as the species could not otherwise subsist. Under this general rule, however, there are several particular modifications, depending upon the size, the manners, and the haunts of the prey for which each species of carnivorous animal is destined or fitted by nature; and, from each of these particular modifications, there result certain differences in the more minute conformations of particular parts, all, however, conformable to the general principles of structure already mentioned. Hence it follows, that in every one of their parts we discover distinct indications, not only of the classes and orders of animals, but also of their genera, and even of their species.

In fact, in order that the jaw may be well adapted for laying hold of objects, it is necessary that its condyle should have a certain form; that the resistance, the moving power, and the fulcrum, should have a certain relative position with respect to each other; and that the temporal muscles should be of a certain size: The hollow or depression, too, in which these muscles are lodged, must have a certain depth; and the zygomatic arch under which they pass must not only have a certain degree of convexity, but it must be sufficiently strong to support the action of the masseter.

To enable the animal to carry off its prey when seized, a corresponding force is requisite in the muscles which elevate the head, and this necessarily gives rise to a determinate form of the vertebræ to which these muscles are attached, and of the occiput into which they are inserted.

In order that the teeth of a carnivorous animal may be able to cut the flesh, they require to be sharp, more or less so in proportion to the greater or less quantity of flesh that they have to cut. It is requisite that their roots should be solid and strong, in proportion to the quantity and the size of the bones which they have to break in pieces. The whole of these circumstances must necessarily influence the development and form of all the parts which contribute to move the jaws.

To enable the claws of a carnivorous animal to seize its prey, a considerable degree of mobility is necessary in their paws and toes, and a considerable strength in the claws themselves. From these circumstances, there necessarily result certain determinate forms in all the bones of their paws, and in the distribution of the muscles and tendons by which they are moved. The fore-arm must possess a certain facility of moving in various directions, and consequently requires certain determinate forms in the bones of which it is composed. As the bones of the fore-arm are articulated with the arm-bone or humerus, no change can take place in the form and structure of the former without occasioning correspondent changes in the form of the latter. The shoulder blade also, or scapula, requires a correspondent degree of strength in all animals destined for catching prey, by which it likewise must necessarily have an appropriate form. The play and action of all these parts require certain proportions in the muscles which set them in motion, and the impressions formed by these muscles must still farther determine the forms of all these bones.

After these observations, it will be easily seen that similar conclusions may be drawn with respect to the hinder limbs of carnivorous animals, which require particular conformations to fit them for rapidity of motion in general; and that similar considerations must influence the forms and connections of the vertebræ and other bones constituting the trunk of the body, to fit them for flexibility and readiness of motion in all directions. The bones also of the nose, of the orbit, and of the ears, require certain forms and structures to fit them for giving perfection to the senses of smell, sight, and hearing, so necessary to animals of prey. In short, the shape and structure of the teeth regulate the forms of the condyle, of the shoulder-blade, and of the claws, in the same manner as the equation of a curve regulates all its other properties; and, as in regard to any particular curve, all its properties may be ascertained by assuming each separate property as the foundation of a particular equation; in the same manner, a claw, a shoulder blade, a condyle, a leg or arm bone, or any other bone separately considered, enables us to discover the description of teeth to which they have belonged; and so also reciprocally we may determine the forms of the other bones from the teeth. Thus, commencing our investigation by a careful survey of any one bone by itself, a person who is sufficiently master of the laws of organic structure, may, as it were, reconstruct the whole animal to which that bone had belonged.

[. . .]

By thus employing the method of observation, where theory is no longer able to direct our views, we procure astonishing results. The smallest fragment of bone, even the most apparently insignificant apophysis, possesses a fixed and determinate character, relative to the class, order, genus, and species of the animal to which it belonged; insomuch that, when we find merely the extremity of a well-preserved bone, we are able, by careful examination, assisted by analogy and exact comparison, to determine the species to which it once belonged, as certainly as if we had the entire animal before us. Before venturing to put entire confidence in this method of investigation, in regard to fossil bones, I have very frequently tried it with portions of bones belonging to well known animals, and always with such complete success, that I now entertain no doubt with regard to the results which it affords. I must acknowledge that I enjoy every kind of advantage for such investigations that could possibly be of use, by my fortunate situation in the Museum of Natural History; and, by assiduous researches for nearly fifteen years, I have collected skeletons of all the genera and sub-genera of quadrupeds, with those of many species in some of the genera, and even of several varieties of some species. With these aids, I have found it easy to multiply comparisons, and to verify, in every point of view, the application of the foregoing rules.

36

BARON GEORGES CUVIER, *THE ANIMAL KINGDOM, ARRANGED AFTER ITS ORGANISATION, FORMING A NATURAL HISTORY OF ANIMALS, AND AN INTRODUCTION TO COMPARATIVE ANATOMY, A NEW EDITION WITH ADDITIONS BY W.B. CARPENTER*

(London: M. S. Orr and Co, 1851 [first published in French, 1817])

Introduction. Of Natural History, and of Systems Generally

As few persons have a just idea of Natural History, it appears necessary to commence our work by carefully defining the proposed object of this science, and establishing rigorous limits between it and the contiguous sciences.

The word NATURE, in our language, and in most others, signifies – sometimes, the qualities which a being derives from birth, in opposition to those which it may owe to art; at other times, the aggregate of beings which compose the universe; and sometimes, again, the laws which govern these beings. It is particularly in this latter sense that it has become customary to personify Nature, and to employ the name, respectfully, for that of its Author.

Physics, or *Natural Philosophy*, treats of the nature of these three relations, and is either general or particular. *General Physics* examines, abstractedly, each of the properties of those moveable and extended beings which we call bodies. That department of them styled *Dynamics*, considers bodies in mass; and, proceeding from a very small number of experiments, determines mathematically the laws of equilibrium, and those of motion and of its communication. It comprehends in its different divisions the names of Statics, Mechanics, Hydrostatics,

DOI: 10.4324/9780429355653-41

Hydrodynamics, Pneumatics, &c., according to the nature of the bodies of which it examines the motions. Optics considers the particular motions of light; the phenomena of which, requiring experiments for their determination, are becoming more numerous.

Chemistry, another branch of General Physics, expounds the laws by which the elementary molecules of bodies act on each other when in close proximity, the combinations or separations which result from the general tendency of these molecules to unite, and the modifications which different circumstances, capable of separating or approximating them, produce on that tendency. It is a science almost wholly experimental, and which cannot be reduced to calculation.

The theory of Heat, and that of Electricity, belong almost equally to Dynamics or Chemistry, according to the point of view in which they are considered.

The method which prevails in all the branches of General Physics consists in isolating bodies, reducing them to their utmost simplicity, in bringing each of their properties separately into action, either mentally or by experiment, in observing or calculating the results, in short, in generalizing and correcting the laws of these properties for the purpose of establishing a body of doctrine, and, if possible, of referring the whole to one single law, under the universal expression of which all might be resolved.

Particular Physics, or *Natural History*,– for these terms are synonymous – has for its object to apply specially the laws recognized by the various branches of General Physics, to the numerous and varied beings which exist in nature, in order to explain the phenomena which they severally present.

In this extended sense, it would also include Astronomy; but that science, sufficiently elucidated by Mechanics, and completely subjected to its laws, employs methods too different from those required by ordinary Natural History, to permit of its cultivation by the students of the latter.

Natural History, then, is confined to objects which do not allow of rigorous calculation, or of precise measurement in all their parts. [. . .]

Natural History should, in strictness, employ the same modes of procedure as the general sciences; and it does so, in fact, whenever the objects of its study are so little complex as to permit of it. But this is very seldom the case.

An essential difference, in effect, between the general sciences and Natural History is, that, in the former, phenomena are examined, the conditions of which are all regulated by the examiner, in order, by their analysis, to arrive at general laws; while in the latter, they occur under circumstances beyond the control of him who studies them for the purpose of discovering, amid the complication, the effects of general laws already known. It is not permitted for him, as in the case of the experimenter, to subtract successively from each condition, and so reduce the problem to its elements: but he must take it entire, with all its conditions at once, and can analyze only in thought. Suppose, for example, we attempt to isolate the numerous phenomena which compose the life of an animal a little elevated in the scale; a single one being suppressed, the life is wholly annihilated.

Dynamics have thus become a science almost purely of calculation; Chemistry is still a science wholly (chiefly) of experiment; and Natural History will long remain, in a great number of its branches, one of pure observation.

[. . .]

Natural History has, moreover, a principle on which to reason, which is peculiar to it, and which it employs advantageously on many occasions; it is that of the conditions of existence, commonly termed final causes. As nothing can exist without the concurrence of those conditions which render its existence possible, the component parts of each must be so arranged as to render possible the whole living being, not only with regard to itself, but to its surrounding relations; and the analysis of these conditions frequently conducts to general laws, as demonstrable as those which are derived from calculation or experiment.

It is only when all the laws of general physics, and those which result from the conditions of existence, are exhausted, that we are reduced to the simple laws of observation.

The most effectual mode of observing is by comparison. This consists in successively studying the same bodies in the different positions in which Nature places them, or in a comparison of different bodies together, until constant relations are recognized between their structures and the phenomena which they manifest. These various bodies are kinds of experiments ready prepared by Nature, who adds to or subtracts from each of them different parts, just as we might wish to do in our laboratories, and shows us herself the results of such additions or retrenchments.

It is thus that we succeed in establishing certain laws, which govern these relations, and which are employed like those that have been determined by the general sciences.

The incorporation of these laws of observation with the general laws, either directly or by the principle of the conditions of existence, would complete the system of the natural sciences, in rendering sensible in all its parts the mutual influence of every being. [. . .]

All researches of this kind, however, presuppose means of distinguishing with certainty, and causing others to distinguish, the objects investigated; otherwise we should be incessantly liable to confound the innumerable beings which Nature presents. Natural History, then, should be based on what is called a System of Nature, or a great catalogue, in which all beings bear acknowledged names, may be recognized by distinctive characters, and distributed in divisions and subdivisions themselves named and characterized, in which they may be found.

[. . .]

There can only be one perfect method, which is the natural method. An arrangement is thus named in which beings of the same genus are placed nearer to each other than to those of all other genera; the genera of the same order nearer than to those of other orders, and so in succession. This method is the ideal to which Natural History should tend; for it is evident that, if we can attain it, we shall have the exact and complete expression of all nature. In fact, each being is determined by its resemblance to others, and its differences from them; and all these relations would be fully given by the arrangement which we have indicated. In a word, the natural method would be the whole science, and each step towards it tends to advance the science to perfection.

Of Living Beings, and of Organization in General

IF, in order to obtain a just idea of the essence of life, we consider it in those beings in which its effects are the most simple, we readily perceive that it consists in the faculty which certain corporeal combinations have, of enduring for a time, and under a determinate form, by incessantly attracting into their composition a part of surrounding substances, and rendering to the elements portions of their own proper substance.

Life, then, is a vortex (*tourbillon*), more or less rapid, more or less complicated, the direction of which is constant, and which always carries along molecules of the same kind, but into which individual molecules are continually entering, and from which they are constantly departing; so that the form of a living body is more essential to it than its matter.

As long as this movement subsists, the body in which it takes place is living it lives. When it is permanently arrested, the body dies. After death, the elements which compose it, abandoned to the ordinary chemical affinities, are not slow to separate, from which, more or less quickly, results the dissolution of the body that had been living. It was then by the vital motion that its dissolution was arrested, and that the elements of the body were temporarily combined.

All living bodies die after a time, the extreme limit of which is determined for each species; and death appears to be a necessary consequence of life, which, by its own action, insensibly alters the structure of the body wherein its functions are exercised, so as to render its continuance impossible.

In fact, the living body undergoes gradual but constant changes during the whole term of its existence. It increases first in dimensions, according to the proportions and within the limits fixed for each species, and for each of its several parts; then it augments in density, in most of its parts it is this second kind of change that appears to be the cause of natural death.

On examining the various living bodies more closely, a common structure is discerned, which a little reflection soon causes us to adjudge as essential to a vortex, such as the vital motion.

[. . .]

Life, then, in general, presupposes organization in general, and the life proper to each being presupposes the organization peculiar to that being, just as the movement of a clock presupposes the clock; and, accordingly, we behold life only in beings that are organized and formed to enjoy it; and all the efforts of philosophers have not yet been able to discover matter in the act of organization, either of itself or by any extrinsic cause. In fact, life exercising upon the elements which at every instant form part of the living body, and upon those which it attracts to it, an action contrary to that which would be produced without it by the usual chemical affinities, it is inconsistent to suppose that it can itself be produced by these affinities, and yet we know of no other power in nature capable of reuniting previously separated molecules.

37

[RICHARD OWEN], 'REPORT ON BRITISH FOSSIL REPTILES. PART II.', *REPORT OF THE ELEVENTH MEETING OF THE BRITISH ASSOCIATION FOR THE ADVANCEMENT OF SCIENCE; HELD AT PLYMOUTH IN JULY 1841*: 60–204

(1841)

THIS group, which includes at least three well-established genera of Saurians, is characterized by a large sacrum composed of five anchylosed vertebrae of unusual construction, by the height and breadth and outward sculpturing of the neural arch of the dorsal vertebrae, by the twofold articulation of the ribs to the vertebrae, viz. at the anterior part of the spine by a head and tubercle, and along the rest of the trunk by a tubercle attached to the transverse process only; by broad and sometimes complicated coracoids and long and slender clavicles, whereby Crocodilian characters of the vertebral column are combined with a Lacertian type of the pectoral arch; the dental organs also exhibit the same transitional or annectant characters in a greater of lesser degree. The bones of the extremities are of large proportional size, for Saurians; they are provided with large medullary cavities, and with well developed and unusual processes, and are terminated by metacarpal, metatarsal and phalangeal bones, which, with the exception of the ungual phalanges, more or less resemble those of the heavy pachydermal Mammals, and attest, with the hollow long-bones, the terrestrial habits of the species.

The combination of such characters, some, as the sacral ones, altogether peculiar among Reptiles, others borrowed, as it were, from groups now distinct from each other, and all manifested by creatures far surpassing in size the largest of existing reptiles, will, it is presumed, be deemed sufficient ground for establishing a distinct tribe or sub-order of Saurian Reptiles, for which I would propose the name of *Dinosauria*.

DOI: 10.4324/9780429355653-42

Of this tribe the principal and best established genera are the *Megalosaurus*, the *Hylaeosaurus*, and the *Iguanodon*; the gigantic Crocodile-lizards of the dry land, the peculiarities of the osteological structure of which distinguish them as clearly from the modern terrestrial *Sauria*, as the opposite modifications for an aquatic life characterize the extinct *Enaliosauria*, or Marine Lizards.

38

RICHARD OWEN, *DESCRIPTION OF THE SKELETON OF AN EXTINCT GIGANTIC SLOTH,* MYLODON ROBUSTUS, *WITH OBSERVATIONS ON THE OSTEOLOGY, NATURAL AFFINITIES, AND PROBABLE HABITS OF THE MEGATHERIOID QUADRUPEDS IN GENERAL*

(London: John Van Voorst, 1842)

Description of the Skeleton of the *Mylodon Robustus*

Introduction

THE Skeleton which is the subject of the present Memoir was discovered in the year 1841 by M. Pedro de Angelis, seven leagues north of the city of Buenos Ayres, in the fluvatile deposits constituting the extensive plain intersected by the great Rio Plata and its tributaries, and which has been raised during a recent geological epoch above the level of the sea.[1]

In this formation, and most probably anterior to its elevation, the animal must have been buried entire; and, if the present heat of the climate prevailed, soon after its death: for the parts of the skeleton were found little disturbed, and the very few bones that are wanting are such as would be likely to escape the search of the most diligent collector.

About the same time, and near the same place, a tesselated osseous carapace of some large quadruped, like an Armadillo, was exhumed; and information of this discovery having been communicated to the Royal College of Surgeons by Sir Woodbine Parish, late H.M. Chargé d'Affaires at Buenos Ayres, both this carapace and the above-mentioned skeleton were purchased by the College. They arrived in November 1841, in many pieces, fragile from the loss of the animal matter; but

DOI: 10.4324/9780429355653-43

after having been restored in some measure to their original tenacity, the parts of the carapace were reunited, the skeleton was articulated, and both are now placed in the Museum.

In receiving these rare and instructive evidences of the ancient zoology of South America, the College was expressly assured, by their discoverer, that the carapace and the skeleton belonged to two different animals. Already, indeed, the fossils discovered by Mr. Darwin in Patagonia and La Plata[2] and those by Dr Lund, the Danish naturalist, in Brazil,[3] had indicated that several species of large quadrupeds had become extinct in South America. But independently of these evidences, the complete state of the skeleton about to be described, establishes its essential relations to a family of the order *Bruta* of Linnæus (*Edentata*, Cuv.), distinct from the Armadillos, and throws much valuable light on the organization and affinities of other less known and more gigantic quadrupeds, its congeners and contemporaries in the ancient transatlantic world, and now alike removed from the scene of animated existence.

A few words on these extinct quadrupeds, and on the existing species of the Edentate order to which they are allied, will serve to show the peculiar interest and importance of the present accession to the Osteological Department of the Museum.

Of the three genera of living Edentata, which are peculiar to South America, viz. *Bradypus* (Sloth), *Dasypus* (Armadillo), and *Myrmecophaga* (Ant-eater), the last includes the largest species. The Great Ant-eater (*Myrmecophaga jubata*) equals in length, though not in height, a Newfoundland dog: the gigantic Armadillo[4] may attain to two-thirds of that bulk, but most of the species are of much smaller dimensions: the largest of the Sloths does not exceed two feet from the muzzle to the vent, but the anterior extremities are of disproportionate length.

The fossil quadruped, whose relations to the slow-paced *Bradypi* Cuvier first scientifically determined, thus proving that their peculiar type of organization had once been represented, on a gigantic scale, in the primæval forests of South America, was the Megatherium,[5] which in certain dimensions surpasses all known quadrupeds, existing or extinct.

A summary of the knowledge of the osteology of the Megatherium since acquired by Cuvier from a comparison of the descriptions and figures published by Garriga, Abildgaard, Pander and D'Alton,[6] is given in the posthumous edition of the 'Ossemens Fossiles', to which the able editor, M. Laurillard, has appended some valuable observations founded on the casts and Mr. Clift's description of the remains of the Megatherium presented by Sir Woodbine Parish to the Royal College of Surgeons in 1832.

In that summary the resemblance of the Megatherium to the Sloths in the descending process of the zygomatic arch, and in the angular process of the lower jaw, is pointed out; its supposed difference in both number and mode of implantation of the teeth is dwelt upon.[7]

Seven cervical, sixteen dorsal, and three lumbar vertebræ are assigned to the Megatherium by Cuvier, who regrets that the absence of the cartilages or sternal portions of the ribs, and also (with the exception of a single bone) of the sternum itself, prevents the form and capacity of the chest being recognized, or properly shown in the articulated skeleton at Madrid.

The sacrum is described to consist of five anchylosed vertebræ, whose spines form a continuous crest, indicating the existence of a tail of some length; which indication, the fossils brought to England by Sir Woodbine Parish confirmed.[8]

The difference between the Megatherium and Sloths in the proportions of the fore and hind extremities is pointed out; the existence of the peculiar characteristics of the scapula of the Sloth in the same bone of the Megatherium is dwelt on with emphasis equivalent to the value of such evidence of the real affinities of the fossil; and the marked degree in which the Megatherium differs from all other great quadrupeds with which it might be compared on account of its size, in possessing complete clavicles, is set forth. The humerus, ulna, and radius are described in some detail. The organization of the fore-foot is, however, involved in obscurity, from the faulty manner in which Cuvier believed the bones to have been articulated; and he regrets that M. Pander and D'Alton have not thrown further light on the subject. After a comparison of their figures with the bones of the fore-foot in existing Edentata, Cuvier concludes that the fore-feet of the Madrid skeleton are transposed, the right being on the left and the left on the right side; that the index, medius, and annular digits were the only ones provided with claws; that the thumb was clawless, and the little finger rudimental and concealed, in the living Megatherium, under the skin; the hand being thereby specially formed for cleaving the soil and digging, like that of the *Dasypus gigas*.

[. . .]

After mature deliberation on the skeleton of the Megatherium, Cuvier conceives himself permitted to form the following conjectures as to the nature and habits of the animal to which it belonged:– 'Its teeth prove that it lived on vegetables, and its robust fore-feet armed with sharp claws, make us believe that it was principally their roots which it attacked. Its magnitude and its talons must have given it sufficient means of defence. It was not of swift course, nor was this requisite, the animal needing neither to pursue, nor to escape'.

In these conclusions, the editor of the posthumous edition of the 'Ossemens Fossiles' coincides, but appends a warning against too hastily attributing to the Megatherium the fragments of the gigantic bony armour that had been found in the same formations of South America, suggested by his recognition of the remains of the foot of a great Armadillo among the fossils transmitted to England by Sir Woodbine Parish.

In a later memoir by Professor Weiss, on fragments of a gigantic osseous carapace discovered by the Prussian traveller Sellow, in the province of Monte Video, and attributed by the Professor to the Megatherium, this warning seems to have been disregarded.[9]

[. . .]

The collection of fossils brought to England from South America by Mr. Darwin has enabled me to add the following facts to the history of the Megatherium. The teeth of the Megatherium, for example, do not differ in number from those of the Sloths, as both Cuvier and M. de Blainville supposed; there being five, not four molars on each side of the upper jaw. Microscopic examination having demonstrated a marked difference in the intimate structure of the teeth of the Sloths and Armadillos, I have ascertained by this mode of investigation, that the teeth of the Megatherium have the same texture and composition as those of the Sloth. And if from identity of dental structure in two different animals we may predicate a similarity in their food, a glance at the bony frame-work of the Megatherium is sufficient to show that it must have resorted to other means of obtaining its leafy provender than that of climbing for it; whereby the necessity of inferring a proportionate magnitude of the trees which nourished the Megatherium is obviated.

[. . .]

In the absence, therefore, of a second skeleton of a Megatherium, more complete, and more authentic in regard to its discovery and mode of articulation than the existing one, further light on this subject can only be obtained, as it were, by reflexion from such a skeleton of some nearly allied species. The subject of the present memoir happily fulfils all the required conditions; and the desire to derive every possible advantage in the solution of this interesting problem in Comparative Anatomy from the singularly perfect skeleton of the *Mylodon robustus*, must plead as the excuse for any apparently undue detail in the following descriptions and comparisons.

[. . .]

Physiological Summary

[. . .]

Now we know that the Mylodon had a strong and powerful tail, but too short for prehensile purposes; its proportions being exactly such as to enable it to complete, with the two hind legs, a tripod strong enough to afford a firm foundation for the massive pelvis and adequate resistance to the forces acting upon and from that great osseous centre. The large and thick transverse, and upper and lower spinous processes, and especially the prolonged and capacious spinal canal, indicate the bulk and strength of the muscular masses which surrounded the tail and connected it with the pelvis: the natural co-adaptation of the articular surfaces shows that the ordinary inflection of the extremity of the tail was backwards, as in a 'cauda fulciens', not forwards, as in a 'cauda prehensilis'.

Viewing, then, the pelvis of the Mylodon as the fixed point towards which the fore-legs and anterior parts of the body were to be drawn in the gigantic leaf-eater's efforts to uprend the tree that bore its sustenance, the colossal proportions of the hind extremities and tail lose all their anomaly, and appear in just harmony with the robust claviculate and unguiculate fore-limbs, with which they combined their forces in the Herculean labour. The uncommon length of the sole of the foot, equalling in the Mylodon, perhaps surpassing in the Megatherium, the length of the femur; the prolongation of the os calcis which affords the strong posterior fulcrum, and the very powerful claw of the middle toe by which the opposite end of the foot would be kept fixed upon the ground,[10] become clearly intelligible, and their final purpose explicable on the above-developed idea of the exertions during which the well-based hind-limbs worked in steadying, and reacting upon, the trunk of the Megatherian wrestling with its passively but sturdily resisting source of subsistence. And thus the expanse of the pelvis, the superior bulk and strength of the hind-legs compared with the fore-legs, the peculiar length and organization of the hind-feet, the proportions and construction of the tail, all of which form a combination of characters common to the Megatherioids, and present in no other animals, yield the required explanation of the uses of the fore-extremities, which too nearly resemble those of other claviculate Edentata, to indicate, apart, their own proper functions in the Megatherioids.

If the foregoing physiological interpretation of the osseous frame-work of the gigantic extinct Sloths be the true one, they may be supposed to have commenced the process of prostrating the chosen tree by scratching away the soil from the roots; for which office we find in the Mylodon the modern scansorial fore-foot of the Sloth modified after the type of that of the partially fossorial Ant-eater. The compressed or subcompressed form of the claws, which detracts from their power as burrowing instruments, adds to their fitness for penetrating the inter spaces of roots, and for exposing and liberating them from the attached soil. This operation having been duly effected by the alternate action of the fore-feet aided probably by the unguiculate digits of the hind-feet, the long and curved fore-claws, which are habitually flexed and fettered in the movements of extension, would next be applied to the opposite sides of the loosened trunk of the tree: and now the Mylodon would derive the full advantage of those modifications of its fore-feet by which it resembles the Bradypus; the correspondence in the structure of the prehensile instruments of the existing and extinct Sloths, extending as far as was compatible with the different degrees of resistance to be overcome. In the small climbing Sloth the claws are long and slender, having only to bear the weight of the animal's light body, which is approximated by the action of the muscles towards the grasped branch, as to a fixed point. The stouter proportions of the prehensile hooks of the Mylodon accord with the harder task of overcoming the resistance of the part seized and bringing it down to the body. For the long and slender brachial and anti-brachial bones of the climbing Sloth we find substituted in its gigantic predecessor a humerus, radius and ulna of more robust proportions,– of such proportions, indeed, in the *Mylodon robustus*, as are unequalled in any other known existing or extinct animal. The tree being thus partly undermined and firmly grappled with, the muscles of the trunk, the pelvis and hind limbs, animated by the nervous influence of the unusually large spinal cord, would combine their forces with those of the anterior members in the efforts at prostration. And now let us picture to ourselves the massive frame of the Megatherium, convulsed with the mighty wrestling, every vibrating fibre reacting upon its bony attachment with a force which the sharp and strong crests and apophyses loudly bespeak:– extraordinary must have been the strength and proportions of that tree, which rocked to and fro, to right and left, in such an embrace, could long with stand the efforts of its ponderous assailant.

A few observations remain to be offered touching the most singular modifications of the feet of the Mylodon, viz. the thick and stunted outer toes, which were evidently enveloped in a kind of hoof. It would be difficult to conceive any modification of the Sloth's structure, whereby such a creature might, with dimensions rivalling those of the largest Pachyderms and unfitting it for an abode in trees, still continue to derive from them its sustenance, more simple and more effectual than those which the skeletons of the Mylodon and Megatherium have brought to light. Their power of standing and walking freely on the ground was gained by the addition of these strong hoofed digits to a tridactyle or didactyle unguiculate and prehensile foot. With this addition some minor modifications in the proportions of the claws were combined, and strength was given to other parts of the frame, to enable the animal to prostrate the trees it could not climb.

[. . .]

My theory of the Megatherian animals, in assigning to them the Herculean labour of uprooting and prostrating trees, for the acquisition of the food which is unequivocally indicated by their dental and maxillary organs, explains and requires all the other characteristics of their organization, and assumes no unknown condition of the vegetable world. Whoever is acquainted with the energy and rapidity of the growth of trees in the intertropical regions of the American continents, or with the enormous quantity of timber annually floated away by the great American rivers, can have little difficulty in conceiving that the interminable and by man unpenetrated forests of the primæval world would yield sustenance to many generations of huge quadrupeds, even though they might uproot the trees on the foliage of which they fed.

In fine, whatever be the value of such collateral evidence as may be deduced from the known conditions of the vegetable kingdom, a searching and impartial review of the anatomical facts and analogies detailed in the foregoing part of the present Memoir has led me to the conclusion, that:– All the characteristics which co-exist in the skeleton of the Mylodon and Megatherium conduce and concur to the production of the forces requisite for uprooting and prostrating trees; of which characteristics, if any one were wanting, the effect could not be produced; and that this hitherto unknown and most extraordinary mode of obtaining food, is the condition of the sum of such characteristics, and of the concourse of so great forces in one and the same animal [. . .]

Notes

1 See the Geological Introduction by Mr. Darwin to the Description of the Fossil Mammalia discovered in the Voyage of the Beagle, Part 1, 4to, 1838, p. 5.

2 These fossils were presented to the Royal College of Surgeons in 1836, and have been described in the Natural History of the Voyage of the Beagle, 'Fossil Mammalia', Parts I. to IV., 4to, 1838–40.

3 Comptes Rendus des Séances de l'Académie des Sciences, 1839, p. 570.
4 This existing species was called by Cuvier Dasypus gigas, before the remains of the ancient Glyptodons had been determined.
5 So named and described by Cuvier, in 1795, from impressions of the plates illustrating the work subsequently published by Bru and Garriga, entitled 'Descripcion del Esqueleto de un Quadrupedo muy corpulento y raro, que se conserva en el Real Gabinete de Historia Natural de Madrid, fol. 1796', and containing the description of the most perfect skeleton of the Megatherium yet obtained. This skeleton was discovered near the city of Buenos Ayres in 1789, and was transmitted to Madrid by the governor of the province, Don Hilario Sosa.
6 Das Riesen-Faulthier, Bradypus giganteus, abgebildet, beschrieben, und mit verwandten Ge schlechtern verglichen, von Dr. Chr. Pander und Dr. E. D'Alton. Bonn, fol. 1821.
7 The remains of the Megatherium transmitted to England by Sir Woodbine Parish showed that the teeth of the Megatherium are implanted as in the Sloths, and indeed in all other so called *Edentata*, by an undivided root of the same size as the crown; not, as Cuvier supposed, by a bifid fang.
8 The caudal vertebræ and bony dermal armour ascribed to the Megatherium by Don Damasio de Laranhaja (Bulletin de la Société Philomathique, 1823, p. 83), appertain to another great fossil animal, called by me *Glyptodon clavipes* – Geol. Trans., Second Series, vol. vi. p. 98.
9 Abhandlungen der Kön. Acad. der Wissenschaften zu Berlin, 1827' (*Treatises of the Royal Academy of Sciences, Berlin*).
10 This use of the large hind-claw was first pointed out by Dr Buckland (Bridgewater Treatise, i. p. 158); but the full value of the structure to the Megatherium could not be appreciated, when the hind-legs were supposed to merely aid in supporting the trunk, whilst one fore-leg was used to dig up roots.

39

RICHARD OWEN, *LECTURES ON THE COMPARATIVE ANATOMY AND PHYSIOLOGY OF THE VERTEBRATE ANIMALS, DELIVERED AT THE ROYAL COLLEGE OF SURGEONS OF ENGLAND IN 1844 AND 1846,* PART I. FISHES

(London: Longman, Brown, Green, and Longman, 1846)

Part I. Fishes

Introductory Lecture

FIRST, permit me to dwell a little on the inestimable privilege which we enjoy, in entering upon our Professional studies by the portal of Anatomy.

How vast and diversified a field of knowledge opens out before us as we gaze from that portal! Consider what it is that forms the subject of our essential introductory study; nothing less than the organic mechanism of the last and highest created product which has been introduced into this planet. [. . .]

Every new term which the Anatomical student has to commit to memory is associated with a recognisable object, with some part which may be vibrating, contracting, or pulsating, in his own frame.

First, we enter upon the study of Human Anatomy that we may know with what we have to deal as Operative Surgeons; and, as Physicians, may recognise the seat of disease. Then, that we may learn, by the structure and connections of the parts of the Human body, their office in the vital economy. We next test the physiological ideas, so acquired, by experiments on the lower animals, which we are thus led to dissect in order to find the amount of resemblance with the Human structure which must guide the operation, influence the judgment as to the result, and indicate the conditions for new experiments.

DOI: 10.4324/9780429355653-44

We cannot advance far into the lower region of Anatomy without appreciating the same admirable adjustment of means to ends which pervades the Human frame: thus the field of Physiology expands before us, and we are enabled to bear a part with a RAY or a PALEY in illustrating the doctrine of final causes, and demonstrating the 'Wisdom of God in the Creation'.

In extending our Anatomical comparisons, we cannot fail to be struck with the close general resemblance of the structure of the lower animals with that of Man: almost every part of the Human frame has its homologue in some inferior animal; and we at length begin to perceive that Man's organisation is a special modification of a more general type. From analysis, the philosophic mind is irresistibly led on to comparison and synthetic combination of the multitude of particulars observed. In grasping the abstract idea of the general type, we appreciate the precise nature of the characteristic modifications of the Human frame; and then only can we be said to know properly our own structure [. . .].

As such we begin to feel ourselves in possession of an instrument which can be brought to operate successfully in the solution of deep and difficult problems of more general interest in the commonwealth of knowledge, and which renders us indispensable auxiliaries in the advancement of Sciences which might at first appear to have but a remote relationship with Anatomy. I need not expatiate on the light which Anatomy lends to the Zoologist, in threading the intricate mazes of the natural affinities of animals: it is, by universal consent, admitted to be the essential basis of a sound system of classification. I need not dwell on the importance of the Comparative Anatomy of the minute and low organised Invertebrata in establishing true theories, and eradicating false notions, of the origin of living species; of which different hypothetical secondary causes have been from time to time offered for the acceptance or speculation of the thinking public.

But I would allude to the power which the appreciation of the co-relations and interdependencies of the several parts of each organic machine gives us to interpret the nature of the whole from the observation of a part.

By this principle its discoverer, the immortal CUVIER, and his successors in this application of Anatomy, have been enabled to restore and reconstruct many species that have been blotted out of the book of life. By this we determine from fossil bones or fragments, submitted to us by the Geologist, the species which are characteristic of different strata. By physiological deductions we can prove that such species, now extinct, have lived and died, generation after generation, through the period when those additions were made to the earth's crust which their remains characterise.

Thus, and thus only, can we obtain a clear idea of the lapse of time in which these formations have taken place. The order of superposition of strata indicates, indeed, their successive formation, but the determination of their organic remains proves that each formation was gradual and progressive.

One of the results of this application of Anatomy has been no less than the discovery of the law of succession of animal life on this planet, or the determination of the relative periods at which the different classes were successively called into being.

Another result may be expected, and is in progress, as a corollary of the preceding, viz. the determination of the true Chronology of the Earth.

We know that it has pleased God to grant us faculties, by the right use of which we may obtain a true knowledge of His works; and it seems part of His providence to permit certain parcels of knowledge to be thus introduced from time to time, to the dissipation of the erroneous notions which previously prevailed. [. . .] It has been reserved for the present generation to acquire more just ideas of the age of the world, and Anatomy has been, and must be, the chief and most essential means of establishing this important element in the earth's history.

But Anatomy aids not only the Geologist, but the Geographer: by comparing the local distribution of restored extinct species from coeval geological strata over all the earth, with the geographical distribution of existing animals, we obtain an insight into the past conditions of continents and islands; we determine that our own island, for example, once formed part of the continent, and obtain data for tracing out much greater mutations and alternations of land and sea.

Thus, upon Anatomy depends the safe and successful practice of Medicine and Surgery: the knowledge of the uses of parts, and of their essential nature in Man, viewed as modifications of a general type. Anatomy is the basis of right classification and philosophical Zoology: it unfolds the law of the introduction of animal life on this planet; it is essential to the right progress of Geology, and gives an insight into the true chronology and ancient geography of the globe.

Almost every day brings some new proof of the importance of the knowledge of Animal Organisation, which bids fair to take rank as the first of all sciences; and it is to Anatomy, that we have the high privilege to be introduced at the very outset of our professional studies.

[. . .]

The grand modification, by which a higher type of organisation is established, and one which becomes finally equal to all the contingencies, powers, and offices of animated beings, in relation to this planet, is the allocation of the mysterious albuminous electric pulp in a special cylindrical cavity, of which the firm walls rest upon a basal axis, forming the centre of support to the whole frame, and from which all the motive powers radiate, and this axial cylinder is called the 'Vertebral Column'; vertebral, as consisting of segments of the skeleton, which turn one upon the other, and as being the centre on which the whole body can bend and rotate; from the Latin '*verto, vertere*', to turn.

[. . .]

There are five special modifications of sensation in the vertebrated animals, three of which have special nerves, viz. smell, sight, and hearing. Taste appears to be less generally enjoyed by the Vertebrata, and its nerve is a large branch of an ordinary nerve, the fifth pair. Feeling, which, in its more exquisite degree, constitutes touch, seems a common property of all those nervous filaments, which, passing into the posterior columns of the central axis, are continued to the brain. Speaking generally, such are the attributes of the recipient or sensitive portion of the nervous axis in the Vertebrated animals. They can take cognisance of all the impressing powers which surround them; as the character and resistance of the surface which supports them, the flavour and fitness of the substances which nourish them, the purity of the atmosphere which they breathe, the delicate vibrations of that atmosphere which follow the mutual contact or percussion of sonorous bodies, and the finer vibrations of a more subtle æther, the appreciation of which produces the sense of sight.

With these means of perceiving, knowing, and investigating the world around them, the Vertebrated animals possess a proportionate power of acting upon and subduing it. Not any species is fixed to the earth; all can move, and every variety and power of animal locomotion is manifested in the vertebrated sub-kingdom. Yet some permanently retain the worm-like figure, which all primarily manifest in common with the embryos of the articulate series; but always with the grand difference of the dorsal nervous column. Such vermiform species glide by undulatory inflections of the entire body through the waters, or on the surface of the ground. But in most Vertebrata special instruments of locomotion are developed; some single from the median line, some in pairs; the latter never exceed four in number, two before or above, called arms, or pectoral extremities, and two below or behind, called legs, or pelvic extremities: thus, the vertebrated type is essentially tetrapodal. The solid mechanical supporting and resisting axis, framework, or leverage of these members is internal, vascular, and commonly ossified. It is covered, and, as it were, clothed by the muscles, which are attached to its outer surface [. . .]

In ascending to Man, we trace a very extensive and varied, but progressive course of development, through the great Vertebrated Series, which commences at a very low point.

It might, perhaps, be imagined that the lowest Vertebrated form began where the highest Invertebrated form ended, and made a direct step in advance in the scale of Animal Organisation. Such, indeed, ought necessarily to follow on the hypothesis of the development of species by progressive transmutation, and of the arrangement of animal life in a single and uninterrupted chain of being.

But truer views of the nature and direction of Zoological affinities, and a deeper insight into the laws of Development and of Unity of Organisation in the Animal Kingdom, concur to disprove those once favourite and recently-revived hypotheses. We have seen that the Invertebrata resemble each other only at the earliest and most transitory periods of their development, diverging thence, in special directions, to the manifestation of very distinct types of animal structure. So likewise we must look to the very beginning of the development of the Vertebrate animal before we shall discover that amount of concordance which will justify us in predicating 'Unity of Organisation' between it and any of the Invertebrated forms. And when, with infinite care and minutest scrutiny, availing ourselves of all the aids and appliances of optical art, we have arrived at clear and satisfactory demonstration of the greatest amount of resemblance, in constitution and properties, between the Vertebrate embryo and the Invertebrate adult, it is not with any of the higher forms of Invertebrata,– with neither the Cephalopod, the Arachnidan, nor the Insect,– that such organic correspondence is found to exist; but it is with the lowest forms and simplest beginnings of animal life,– with the infusorial monads. Only, in fact, during that period of the ovum-life of the Vertebrated being, in which the mysterious properties of the impregnated germ-vesicle are diffused and distributed by fissiparous multiplication amongst countless nucleated cells – the progeny of the primary germinal vesicle and coheirs of the seminal virtue – do we find such a form and such properties of the Vertebrated animal as justify us in affirming that there is 'Unity of Organisation' between it and an Invertebrate animal.

40

RICHARD OWEN, *ON THE NATURE OF LIMBS: A DISCOURSE DELIVERED ON FRIDAY, FEBRUARY 9 AT AN EVENING MEETING OF THE ROYAL INSTITUTION OF GREAT BRITAIN*

(London: John Van Voorst, 1849)

THE 'limbs' to which the limits of the present Discourse confine its application, are those of the Vertebrate Series of animals; they are the parts called the 'arms' and legs' in Man; the 'fore-' and 'hind-legs' of Beasts; the 'wings' and 'legs' of Bats and Birds; the 'pectoral fins' and 'ventral fins' of Fishes. I take for granted that it is generally known, as it is universally admitted by competent anatomists and naturalists, that these limbs or locomotive members, which, according to their speciality of form, have received the above special names, are answerable or 'homologous' parts: that the arm of the Man is the fore-leg of the Beast, the wing of the Bird, and the pectoral fin of the Fish. This special homology has been long discerned and accepted; but the general homology of the parts or their relation to the vertebrate Archetype, in short their 'Bedeutung' or essential nature, is not generally known. [. . .]

It must be owned, however, that the non-acceptance of these generalizations has been due more to indifference and to the non-appreciation of the value of the inquiries, than to a rigorous investigation of their merits. Very few of the exact conclusions as to the general homology of parts of the skeleton of animals have been admitted with thorough comprehension and fruition of the discovery by the actual cultivators of Natural History in this country; so that I can scarcely appeal to an example in illustration of my meaning with any hope that it will prove such to more than a very small portion of my hearers.

[. . .]

The parts to which I here refer, and to which alone the reasoning will apply that leads to the desired conclusion in the present lecture, are those in the Vertebrated

DOI: 10.4324/9780429355653-45

animals, serving chiefly for locomotion, but sometimes adapted to other offices. Many and multiform parts answering these purposes are present in the Invertebrated animals; but their framework is formed out of a distinct system of hard parts from those employed in the Vertebrata. Here it is the internal or endo-skeleton: in the Invertebrata it is the hardened skin, the dermo- or exo-skeleton.

The hard parts of the leg of a Crab or an Insect may be 'analogous' to the bones of the limb of a Quadruped, but they are not 'homologous' with them; and where there is no special homology, there can be no relations of a higher or more general homology between the parts.

The Vertebrated animals enjoy as extensive and diversified a sphere of active existence as the Invertebrated. They people the seas, and can move swiftly both beneath and upon the surface of water: they can course over the dry land, and traverse the substance of the earth: they can rise above that surface and soar in the lofty regions of aërial space.

The instruments for effecting these different kinds of locomotion – diving and swimming, burrowing and running, climbing and flying – are accordingly very different in their configuration and proportions. The simplest form of the locomotive member is that of the fin. The marine mammal called Dugong here offers us an example of such. It is a strong, stiff, short, broad, flat, and obtusely pointed paddle or oar; without other apparent joint than that which unites it to the body it has to propel: a joint permitting that degree of rotation with the oblique stroke that makes the movement of the oar most effective.

The instrument for burrowing, such as the Mole presents is not very different in form and character from the fin; but being destined to displace a denser element than water, it is shorter in proportion to its breadth, and much stronger: it resembles the fin in consisting, seemingly, of but one segment or joint, and being moveable as a whole only where it is set on to the trunk. The free border, however, instead of being smooth and thin, is notched, and armed with a row of hard, tooth-like, horny points, adapted for scraping and throwing back the soil. With such rapidity does the mole effect this purpose, that it may literally be said to 'swim through the earth'.

The third form of limb or locomotive member [. . .] offers a striking contrast to the burrowing trowel we were last contemplating. It is a thin, vastly expanded sheet of membrane, sustained, like an umbrella, by slender rays, and flapped by means of these to and fro in the air; and with such force and rapidity, as, combined with its extensive surface, to make it react upon the attenuated element more powerfully than gravitation can attract the weight to which the limb is attached, and consequently the body is raised aloft and moved swiftly through the air; in brief, the animal flies, and these instruments of its aërial course are called 'wings'.

When a quadruped has to move swiftly along the surface of the earth by reacting upon the hard ground, its limbs are as remarkable for their length and slenderness as those of the burrower or swimmer are for their shortness and breadth. In the racer the instruments of its rapid course are four long tapering columns, with joints permitting them to bend in opposite directions, and of the form familiar to all:[1]

each column rests upon a slightly expanded base encased by the hard horny sheath which we call the 'hoof'.

When the limbs are adapted for grasping as well as running, they are divided at their extremities into moveable appendages or digits, one of which can be opposed to the others and retain the object of their mutual pressure. Each of the four extremities is so organized in the ape and monkey, which are thus especially adapted for climbing and living in trees.

In Man the principle of special adaptation goes further; and, whilst one pair of limbs is expressly organized for locomotion and standing in the erect position, the other pair is left free to execute the manifold behests of his rational and inventive Will, and is exquisitely organized for delicate touch and prehension, emphatically called 'manipulation'.

Such are some of the more striking amongst the countless purposes to which the parts of animals called limbs are adapted, and such the consequent diversity of their outward shapes and proportions. We cannot be surprised at this; it could not be otherwise: the instrument must be equal to its office. And consider the various devices that human ingenuity has conceived and human skill and perseverance have put into practice in order to obtain corresponding results!

To break his ocean-bounds the islander fabricates his craft, and glides over the water by means of the oar, the sail, or the paddle-wheel. To quit the dull earth Man inflates the balloon, and soars aloft, and, perhaps, endeavours to steer or guide his course by the action of broad expanded sheets, like wings. With the arched shield and the spade or pick he bores the tunnel: and his modes of accelerating his speed in moving over the surface of the ground are many and various. But by whatever means or instruments Man aids, or supersedes, his natural locomotive organs, such instruments are adapted expressly and immediately to the end proposed. He does not fetter himself by the trammels of any common type of locomotive instrument, and increase his pains by having to adjust the parts and compensate their proportions, so as best to perform the end required without deviating from the pattern previously laid down for all. There is no community of plan or structure between the boat and the balloon, between Stephenson's locomotive engine and Brunel's tunnelling machinery: a very remote analogy, if any, can be traced between the instruments devised by man to travel in the air and on the sea, through the earth or along its surface.

Note

1 The passage relates to an illustration of a horse.

41

LOUIS AGASSIZ, *TWELVE LECTURES ON COMPARATIVE ANATOMY DELIVERED BEFORE THE LOWELL INSTITUTE IN BOSTON, DECEMBER AND JANUARY 1848–9*, ENLARGED EDITION

(Boston: Redding & Co., 1849)

Preface

WE feel both pleasure and pride in being able to present to the public the following Course of Lectures. It is the first enterprise of the kind in this city, and has therefore been attended with unusual trouble and expense.

Embryology has but recently become the subject of scientific investigation. Few persons have as yet entered upon it, and in this country it may be considered as entirely new; but it is destined to have a most important influence in the future progress of Zoology, and greatly to modify the present classification of animals. Prof. Agassiz has embodied in his Lectures all that has been hitherto done abroad, and has added numerous observations of his own, made in this country, and in a form at once highly scientific and so illustrated, as to be interesting to the common reader. The application here made of Embryology to the improvement of the classification of animals is peculiarly his own, as he has shown in his fourth Lecture.

The point of the Lectures is to demonstrate that a natural method of classifying the animal kingdom may be attained by a comparison of the changes which are passed through by different animals in the course of their development from the egg to the perfect state; the changes they undergo being considered as a scale to appreciate the relative position of the series.

The language has been retained almost precisely as delivered by the Professor, because, although in many instances it wears a foreign idiom, yet it is peculiarly expressive, and possesses a charm which would be lost in the attempt to reduce it to Saxon phrases.

DOI: 10.4324/9780429355653-46

In proof of the fullness and accuracy of Dr Stone's phonographic report, and also of the value of the phonographic system, we are enabled to state that several gentlemen had the curiosity to compare a portion of manuscript which the Professor had read, in one lecture, with the report of it; when it was found that every word appeared precisely as written, except that one word was missing, which the Professor stated he had purposely omitted in reading.

Boston, January, 1849.

Lecture I

THE time has passed when it was possible to doubt that there is order in Nature, when the existence of a general system regulating the whole creation could be questioned. However, it has been only step by step that man has acquired an insight into this plan. Knowledge was to be gained before this wonderful arrangement of nature could be understood. And it was not at once fully understood. Understanding has been acquired gradually, successively and with difficulties. However, now we have sufficient data to be able to satisfy ourselves that the various views which have been brought forward respecting the order of nature are not altogether fanciful, that they are not mere artificial means to assist us in our investigations.

We can be satisfied that they correspond more or less to nature. We have the positive hope that they will one day correspond entirely to the natural phenomena, when we see how the investigations which are carried on in different directions by different authors go on, converging gradually, assisting each other, and harmonising subjects which at first seemed entirely obscure, if not entirely inaccessible.

The first attempts to an illustration of the relations which exist among the natural phenomena – which exist in particular in the animal kingdom – were traced from external characters. It was from external appearances that scientific men in the beginning tried to combine animals, as it seemed to them they resembled each other most.

But the simple investigation of these external characters was not sufficient. Mistakes were made under the impression that the right thing had been found. Animals, for instance, like the whale, were placed among fishes; though now it is very well known that those animals have no relation to each other – do not even belong to the same class. Crocodiles and turtles were placed among the viviparous quadrupeds, because they have four legs. Barnacles were placed among shells – among oysters and clams – because they had a solid external covering; and other similar mistakes were made, which have been successively corrected.

The corrections of these mistakes have been made after a certain knowledge of the internal structure of animals had been obtained. And it was found so satisfactory to derive information from the investigation of their internal structure, that soon comparative anatomy and the knowledge of the internal structure of animals became the real foundation of the classifications of the animal kingdom.

It was the result of the brilliant investigations of Cuvier, to show that a natural arrangement of the animal kingdom could be based upon the structure of the beings which were to be classified. It was from such data that arrangements could be produced, according to which all the kinds of animals which were brought together were found to agree in the most essential peculiarities, even when they had not been previously investigated anatomically.

This is one of the promising results of those investigations of Cuvier which made internal structure the foundation of the natural system. But he found at the same time, that otherwise natural groups had the same structure; and that from a knowledge of a few individuals, a great many facts could be acquired. The knowledge of a few fish, enabled him to compare the whole class of fishes with reptiles; a knowledge of a few reptiles, enabled him to institute extensive comparisons between reptiles and birds; and again, between these and mammalia; and to find that all these animals agree in certain respects. And however many have been examined since, – and three or four times more have been examined than the number which Cuvier had known when he laid out his classification – however many have been studied since, they have all been found to agree in these essential particulars. So that it is now plain, that structure is the principle upon which animals can be most satisfactorily classified. And as I shall often have occasion to refer to this classification let me at once, in a few words, indicate which great divisions Cuvier introduced into his animal kingdom.

All the animals which I have mentioned. Fishes, Reptiles. Birds, and Mammalia, are combined together, because they have a series of backbones, called vertebrae, by anatomists; and hence the name of vertebrated animals. They agree in the general structure of their brain; they agree in the general arrangement of the fleshy parts, and in the general arrangement of the organs of life – as of the organs of respiration, the heart, the alimentary canal, and so on.

Another group, which was established on the same principle, is that to which we may refer worms, insects, crabs and lobsters – all animals whose bodies are divided into a series of moveable rings, – joints, which surround the body and enclose the soft parts; and which are provided with moveable legs, and in some, even in addition to these legs, also with wings. All these animals have a most remarkable arrangement of the nervous system; there being a series of swellings of nervous substance placed, one in each of the rings, and connected together by double threads, so that the nervous system is all contained in one cavity, not only the general arrangement of parts, but this most important organ of life is also different from that of vertebrates.

The next great group is that of Mollusca, containing cuttle-fish, snails, slugs, clams, and oysters, – all those animals which we generally call shell-fish – those which are provided with hard structures – the body being soft and generally surrounded by a great quantity of mucosity; the nervous system consisting simply of a circle surrounding the alimentary tube, with a swelling above the intestine, and another below, from which all the nervous threads arise, which are diffused into all parts of the body.

In these three groups of the animal kingdom, all parts are in pairs, placed on two sides of the longitudinal axis. In all of these there is an anterior and posterior part; two sides, a right side and a left side; and they have a back part and a lower part; they are, in fact, symmetrical.

But there is another group, in which there is a different arrangement. The mouth is in the centre of a circular body and from this mouth, the organs are placed like rays, diverging in all directions. Here we have no right, and no left side, no anterior and no posterior extremity. The body is star-shaped; and the nervous system has the same general structure, consisting of an horizontal ring around the entrance of the alimentary tube, and has no longer an upper and a lower swelling, as in Mollusca.

There have been a few modifications made in the details of this arrangement as proposed by Cuvier. Some of the animals placed among the mollusca, were found to belong to the group of articulate – Barnacles are one of this group; a very remarkable family, from the numerous shells around the body. Without knowing certainly, he had placed them among the mollusca; but on examination, it was found that their nervous system consisted of swellings, and that their bodies were divided into joints – and an additional evidence was obtained from a knowledge of their young, which were found to resemble, in the earlier stage, much more the Crustacea than the mollusca and indeed, that they were Crustacea, and assumed this covering only at a later epoch.

However important these anatomical researches have been, it is nevertheless my belief that in this line of investigation we have gained all the important information that we can gain; and that we have to run new tracks in order to improve our natural method, – that we must even give up this fundamental principle, as the ruling principle, if we will make further advance in this science. And my reason is this: The minute investigations which are now making in the anatomy of animals, are bringing forward such differences between them that we have no principle by which we can appreciate their value. And if we consider every difference in structure as sufficient to separate animals, the time would come when we should form as many groups – as many divisions – as there would be smaller groups in the animal kingdom, as it can be shown that even genera differ anatomically among themselves.

If I am not entirely mistaken, these new investigations, this new information, must be derived from embryological data. It is to the study of young animals – it is to the investigation of the formation of the germ within the egg, that we must appeal for a ruling principle to ascertain the real, natural position of the subdivisions of the minor groups in the animal kingdom. I acknowledge that the great divisions will always stand on the anatomical structure. But the subdivisions of the classes cannot rest upon anatomical investigation, and if I do not fail in my endeavors, I hope to show it to you satisfactorily. This new step is a natural consequence of the natural progress and state of our science.

Investigations have recently been carried on, more particularly than before, upon the growth of animals within the egg; and some facts have been brought to light which have their bearing on Zoology. Though how these facts have to be applied to the study of classification, has not yet been traced. Embryological investigations have been particularly made with reference to Physiology – that is, with reference to the mode of formation of the various organs which exist in animals, and not with reference to ascertaining their natural relation among themselves.

Another series of investigations which have modified considerably the views which were entertained of the structure of the animal kingdom, are those microscopical researches upon the intimate structure of the tissue of the mass of the body. Of what does the flesh, the bone, the nerve, the various masses of the body, consist? and how have they been gradually formed? has been the object of various microscopical investigations. And again, in this department facts have been brought to light of which we can avail ourselves in investigating the natural relation of animals. On introducing a series of Lectures on Embryology, my object is not to illustrate embryology in the same sense, in the same manner, in which it has generally been traced in the animal kingdom. My object is not merely Embryology; it is Comparative Embryology. And under Comparative Embryology, I mean the comparisons of those phenomena which have been traced in the growth of the different animals, and the different modifications which occur in individual species, throughout the different classes, in their natural gradation, when full grown.

[. . .]

Embryology traces all these changes from the first formation of an egg to the formation of the germ within the egg; but the germ is not yet formed. We have next to witness the formation of the animal; and afterwards we trace in the primitive egg, the successive changes of the first rudiments – we trace its transformations. We have first its formation in the egg. We trace afterward its transformation through changes of different forms. And it is important to distinguish between these two orders of phenomena – the formation of the germ, and the transformation of the animal into different outlines. The one would be the subject of embryology proper; the other is called the metamorphosis of an animal; and has been particularly studied among insects, where the new being passes through very different and quite distinct forms [. . .]

In other animals the metamorphoses are gradual. We see, for instance, the tadpole, from the singular form first seen [. . .] passing gradually into the form of a frog. But every metamorphosis takes place gradually, not seemingly from one animal to another, but by changes of the same animal to others and other forms.

Now, it will be the knowledge of this metamorphosis of animals which I intend to make the foundation of a natural system of Zoology. And how that is to be done, I will explain by an example, and refer to the class of reptiles; as I find it is in that class in which we have the most matured materials for such an investigation. I might have selected a more worthy subject than frogs and salamanders, and perhaps have alluded to the higher animals. But let me say, there is nothing unworthy of our attention in nature. And if we can trace the action of the creative power in these animals which we despise, let us consider that they were made by Him, and if they were worth making, they are worth considering by us.

The class of reptiles as it is now circumscribed, is a very natural one, though it was not always so in the works of natural history [. . .]

Laurenti, an Austrian naturalist, was the first who described these most carefully, bringing together frogs, lizards, turtle, salamanders, toads, and combining in one natural division all the principal animals which we now refer to it. – But his classification was not much better on that account. He placed in one and the same division, salamanders, and lizards, and crocodiles, which we now know to be widely different; and he did not place in that class another group of animals, which we refer to it the Cæcilia. [. . .]

Linnaeus followed the same example. He brought together turtles, crocodiles, lizards, snakes, frogs and salamanders, but unfortunately left in the same class some fishes, which he combined with the reptiles, owing to some peculiarities of their solid frame. Linnaeus also left the salamanders with the lizards, because they had four legs.

[. . .] Brongniart, the celebrated geologist of Paris, studied these animals, and happily threw great light upon the subject, when he showed that reptiles could be divided into four groups – the turtles being one, the lizards another, the snakes a third and the Batrachians, as he called the frogs and salamanders, the fourth and last group. And in this, for the first time, we see salamanders separated from lizards and brought into connexion with frogs and toads. He had noticed that these animals undergo similar changes – that they are equally naked – that they have not the scales which characterize higher reptiles, and he therefore brought them together, but he left out an animal which really belonged to that class. A naked snake called Caecilia by naturalists, was left out and included among the snakes.

I shall use the term Batrachia to designate all those animals which are allied to frogs and salamanders. We have a great variety of these animals. After the publication of the works of Brongniart, Oppel, Dumeril, etc (who also introduced new views on the subject), they were extensively studied, so that in the museums these animals became more numerous, and it became necessary to introduce some subdivisions among them. Now let me show what sort of animals are referred to this order of Batrachia. And in the first place we have the type of frogs [. . .] Animals which have four fingers in the anterior leg, and five behind. There is no tail to those belonging to this group – we refer to the frog and the treetoad. There is a web in the finger of the frog; but in the treetoad there is a kind of web, and it is floating. But in the toad the fingers are entirely free.

In the salamanders there is a tail. There are four fingers at the termination of the anterior extremity and five at the termination of the posterior extremity. Without the tails, salamanders would be compared with frogs and toads. If their body was somewhat more contracted they would resemble each other very strongly. And indeed, their internal structure is similar. On account of the presence or absence of a tail, these have been divided into two groups – without a tail and with the tail. The tail is shorter and thicker and the whole body is more contracted [. . .] Here are gills which do not exist in any other of this group, gills which exist in the whole life only with fishes; but which here exist simultaneously with lungs in the body. [. . .] There is another type which is not figured, in which there is no tail, no legs, and only a transient and temporary gill. It is the Cæcilia – the so-called naked snake.

[. . .]

From want of a principle, all these details differ in the various authors. No one is ruled by anything but his impression – his feeling about it. And I think that we can substitute a principle, and we can show that this principle has nothing arbitrary, and is given to us by nature.

Let us trace the metamorphoses of frogs, and there we have the key. What are the changes which frogs and salamanders undergo? In the beginning, for instance, salamanders are animals without legs at all, with a long tail and large gills on the side of the head. A change takes place; the gills remaining and growing larger, when an anterior pair of legs appears, and in another stage the gills are reduced when the second pair of legs appears. Here the anterior pair has four fingers, but here is a further change of the same animal, when it loses its gills entirely, and the posterior pair of legs assumes an additional finger, the animal having four fingers forward and five backwards.

What changes does the frog have? Hatched, he is an animal without legs and without gills. The salamander is hatched with gills, but there is an epoch when it is without gills, and without tail, and without head, and only a fissure on the sides to indicate where the gills will be formed, but not yet external gills. The frog has not yet gills, and not yet a tail distinct from the body. But next, the tail makes its appearance, when the head separates more distinctly from the mass of the body, and the tail grows longer, and [. . .] the tail grows still larger. But in addition to that we have a pair of anterior legs, and the gills have disappeared. Then we have the same growth in the posterior legs coming out, though not yet as large as they are here. You see that the size of the tail in proportion to the main mass is reduced, and finally the tail disappears entirely, and we have a frog.

Here, in these facts we have not only the history of the transformation of salamanders and frogs, but we have a natural system of batrachians, and there is no longer any arbitrary arrangement in our system possible. Every thing is indicated in the metamorphoses of the animals.

[. . .] The web fingers are observed in all these early stages of growth, and those which have distinct fingers, when fully grown, have them webbed when young. Therefore, we shall see that the frogs are not to be placed higher. And frogs must be lowest, next treetoads and then toads the highest, because their fingers are finally entirely separated.

And in conclusion, I will say, that in studying the metamorphoses of animals, we may find in the transformations – in the different formations through which they pass, from the first formation up to the full grown condition, a natural scale by which we can measure and estimate the position to ascribe to any animal belonging to this family.

And, undoubtedly, the various genera of this family which I have mentioned, will find their places as soon as all the different metamorphoses of these different animals are known. At present, we know only the transformations of frogs and of salamanders, through the researches of European Naturalists. The metamorphoses of the numerous species of that family which occurs in the United States not having been investigated.

But this agreement of transformation is most remarkable. Nevertheless, we must acknowledge that these perfect animals which occur in different parts of the world in our day, are not copies from metamorphoses from the different stages of the growth of frogs; but they are animals of a peculiar kind, produced in various parts of the world, showing proof that there is one and the same plan ever producing the formation of this whole class, as well in the developement of the young from the beginning of their growth to their full grown stage, as in the formation of the different animals which inhabit different parts of the globe.

There is a freedom in the developement of this plan, a freedom in which we can see the action of the intelligent Author of all these things.

We read here the intelligent action of the Creator in the production of these animals; and we read more than the intelligent invention of his creation. We read the omnipresence of his action, as his action is developed on all parts of the globe, in the United States, in Europe, in Japan, in South America, and in all the portions of the globe. – And when developed in that way in its actual condition, we see that every one of them, when reproducing its species, passes through these different changes – the higher one, through more of the changes; the lower one, undergoing only the earlier modifications.

Part 5

BOTANY

Botany

THIS section offers an overview of the state of botany from the late eighteenth century to the mid-Victorian period, reflecting the production of standard botanical works of plant classification, the staggering expansion of plant knowledge as European nations explored the globe, efforts to pursue the economic potential of botanical knowledge, and works devoted to the aesthetic qualities of flora. It is a story of change and development, both of botany's status as a science and of changing attitudes to environment.

One of the oldest of the natural sciences, with roots in Classical Europe (Aristotle, Theophrastus, Pliny the Elder), ancient India (Parāśara), and mediaeval Arabic scholarship (Al-Asma'I, Abu Hanifa Dinawari, Ibn Juljul, Ibn al-Baytar), the study of plants became a major concern of the European Renaissance, when botany, astrology, medicine, herbalism, and arcane belief systems collided in fascinating, unstable ways. As Richard H. Grove (1995, 73–90) has demonstrated, Arabic and southern Indian botanical expertise shaped the genesis of an emerging scientific botany from the sixteenth century onwards. Grove's *Green Imperialism* is also a valuable guide to the impacts of imperial activity on the establishment of biological sciences, the formation of botanical gardens that registered and physically represented knowledge of global flora, and the emergence of conservationism in response to the environmental crises inflicted by colonialism.

Traceable in the long nineteenth century is a gradual movement away from Renaissance traditions as a result of Enlightenment commitment to empirical enquiry but also a growing interest in landscapes and environments as a result of the work of 'gentlemen amateurs' in natural history; the growth in gardening, market gardening, and landscape gardening; the influence of Romanticism; and the impact of continental authorities such as Linnaeus, Buffon, and de Candolle. We see growing professionalisation in botany and its firm establishment as a science preoccupied with morphology and taxonomy. This urge to observe, classify, and name can partly be read as a broader urge to control and exploit environments, and it is certainly the case that many of the extracts in this section are interested in the economic potential of plants, a need sharpened by rising populations, changing demographics, and periodic harvest problems (particularly in the 'Hungry Forties') and shaped by agricultural policies and Corn Law debates about imports of foreign produce. In a historical context overshadowed by Malthusian anxieties about population growth and food supplies, botanists often positioned their work as a patriotic duty to be aided by advancements in agricultural chemistry, plant and animal breeding, and post-Enclosures land management. Part 10 of this volume will turn to agriculture, but overlaps with botany are clear.

While botany often intersected with economics, it sometimes also helped in the process of forging an altogether different conceptualisation of environment as a holistic, intersecting whole made up of mutually connected elements. Botany's role in the establishment of proto-ecological ideas is evident in some extracts, while others retain traditional Natural Theology assumptions about human sovereignty over environment. The degree to which authors viewed plants as non-human others worthy of protection wildly varies. Some included here operated firmly within a nexus of environmental violence that extended far beyond the United Kingdom.

As well as generating scientific work, botany gained increasing popularity amongst the leisured middle classes, who participated enthusiastically in the growing body of literature available in books, periodicals, and magazines and in the various botanical crazes – particularly for orchids and ferns (subjects taken up in Volume II). It should also be noted that botany was often gendered in ways that made it more accessible to women, although the Linnaean focus on plant sexuality was sometimes deemed problematic, and the general trend for professionalisation also involved demarcating scientific botany as exclusively male. The fact that only two of our authors are female indicates the barriers women faced in entering botany's higher echelons. While women were incredibly active, often as popularisers, illustrators, collectors, and members of local botanical societies, their work was more often than not unheralded. Issues of gender within botany will be covered in more detail in Volume II, but readers wishing to pursue this important topic further can consult Ann Shteir (see Further Reading). Botany also offered opportunities for self-advancement for elements of the educated working classes, for whom botany represented a means of escape from urban environments as well as opportunities for advancement and recognition for those able to make contributions to the growing body of botanical knowledge, but like women, they faced largely insuperable barriers to elite science.

Many of the extracts exemplify a desire to establish reliable botanical primers that students and enthusiasts could use to reliably identify plants. This meant providing brief but precise descriptions of plant anatomy, drawing on an increasingly specialised vocabulary that permitted authors and readers to understand the size, shape, and position of petals, calices, stems, leaves, and others. Although the descriptions to be found in standard Floras of the period may appear dry and even forbidding, they encouraged generations of amateur botanists to go out into the countryside in search of plants and to engage with the countryside in ways that often promoted belief in the importance of preserving its beauties.

The first extract, from the first volume of William Curtis's *Flora Londinensis* (1777–91), reflects botany at the opening of our period. Alongside Benjamin Stillingfleet's *A Calendar of Flora* (1755), James Sowerby's *English Botany* (1791–1814), and James Edward Smith's *Flora Britannica* (1800–4), it is a major work of pre-Victorian plant science. It is a departure from works like William Hudson's *Flora Anglica* (1761), written purely in Latin. Curtis was also influenced by a vast Enlightenment-era project, *Flora Danica*, an atlas of botany proposed in 1753 by G.C. Oeder, professor of botany at Copenhagen's Botanic Garden, started in 1761, and only completed in 1883, under the cumulative stewardship of eight editors. Curtis follows these models in establishing a rigorous basis for botany via exhaustive cataloguing of species existing in a particular area, clear identification, and rigorous classification ('the ascertaining and fixing of the plants') while also scrutinising botany's economic advantages ('numberless improvements in Medicine, Agriculture, &c').

The story of botany between the sixteenth and eighteenth centuries is essentially a movement from Herbals (works on the medicinal qualities of plants) to Floras (scientifically organised catalogues of plants in a specific area). Regional and national Floras were pioneered in the 1500s by Ulisse Aldrovandi and Charles de l'Écluse, but the shift from Herbals was gradual. Curtis largely aims to distinguish his work from earlier English Herbals (by Nicolas Culpepper, John Gerard, and John Parkinson), with their primary medicinal focus and classification systems based on arbitrary or astrological systems. Although Curtis cautions against the unscientific nature of this tradition, he does not reject it entirely. Unlike later, more single-mindedly scientific botanies, Curtis offers a lively mixture of science, gardening, medicine, agriculture, aesthetics, natural history, and personal observation – making his work a transition between the Herbal and the Flora. His advocacy of the beauty of many native wild plants, evident in his remarks on foxgloves, is also a sign of some degree of conservationist consciousness, and there is perhaps a Romantic and environmentalist note in his lament that 'we are too apt to treat with neglect the beautiful plants of our own country, merely because they are common and easily obtained'. As well as wandering in search of plants, Curtis collected seeds and created a garden to observe plants throughout their life cycles. His London focus makes him a pioneer of urban nature study during a period when the capital's growth was markedly accelerating. For modern readers, references to Charlton Wood (probably Hanging Wood in the Borough of Greenwich, almost entirely swallowed up by development) are elegiac, an indicator of just how much the city has expanded in the past 150 years.

The Curtis extracts include his Preface and three plant entries, sadly without the charming accompanying illustrations: see Gill Saunders (Further Reading) for more on the subject of botanical illustration. To fully indicate his methodology, the first plant entry includes the (Linnaean) anatomical information and the prose description. The second and third entries include only the descriptions. In the first plant extract (chickweed), Curtis follows Linnaeus's listing of floral and fruit anatomy: chickweed has five calices, five equal petals, and a trilocular capsule with one valve. He then lists various botanical names attributed to chickweed, including Linnaeus's initial classification in *Genera Plantarum* (1737); the same author's *Flora Suecica* (1745) and *Systema Vegetabilium* (1774); and (using initials or shortened forms) to works by Gerard, Hudson, Albrecht von Haller, John Ray, Giovanni Antonio Scopoli, and others. Curtis's Latin anatomical descriptions have not been included. Extracts from Curtis's *The Botanical Magazine* (1787–present) feature in Volume II.

The second extract is from the most famous work of British eighteenth-century naturalism, Gilbert White's *Natural History of Selborne and Its Antiquities* (1789), of such importance and variety that it is variously anthologised throughout the first two volumes of this anthology. *Selborne* was written up from letters White addressed to Thomas Pennant and Daines Barrington, key figures in the amateur-gentleman tradition of Natural History. *Selborne* is more preoccupied with zoology (birds especially) than botany, but White aspired to write a *Flora Selborniensis* (working on this in 1766), and his *The Garden Kalendar* (horticultural notes kept between 1751 and 1771) indicates his considerable enthusiasm. There are also moments in *Selborne* when the curate-naturalist of the Hampshire village that gives the work its title turns his attention to plants. Two have been included here, both from his Barrington correspondence.

Initial remarks in letter XL offer a fascinating glimpse of botany's status at the time. White comments that 'the standing objection to botany has always been, that it is a pursuit that amuses the fancy and exercises the memory, without improving the mind or advancing any real knowledge'. He concedes the charge 'is but too true' wherever 'the science is carried no farther than a mere systematic classification'. In what can be read as an anticipation of the taxonomic direction which many botanists would pursue, White insists that alternatives are advisable: 'the botanist that is desirous of wiping off this aspersion should be by no means content with a list of names', but instead, 'should study plants philosophically, should investigate the laws of vegetation, should examine the Powers and virtues of efficacious herbs, should promote their cultivation; and graft the gardener, the planter, and the husbandman, on the phytologist'. This comprehensive gaze constitutes White's particular genius, but it is a model followed not so much by mainstream professional botanists but by figures like John Ruskin (see this section and subsequent volumes), Richard Jefferies, W.H. Hudson, and Edward Thomas (see Volume IV of this anthology). In this sense, White is both a significant naturalist and a founder of a British nature-writing tradition centred on alert, sympathetic figures describing and recording their countryside observations.

It is unclear whether White knew William Curtis or *Flora Londinensis*, but there is a remarkable similarity between Curtis's practice and White's recommendation for botanists. Like Curtis, White advocates the economic advantages of botany: speaking particularly of the cultivation of grass, and quoting from 'A Voyage to Brobdingnag' in Jonathan Swift's *Gulliver's Travels* (1726), he advocates cultivating 'two blades of grass where one alone was seen before'. White is also the fountainhead of localised natural history. In the second letter, he provides Barrington with a list of rarer plants seen within the compass of his daily walks – but he is uninterested in the detailed analysis of floral anatomy and the apparatus of botanical terminology that characterise subsequent botanists. Instead, he meticulously describes plant locales and habitats. Many of his botanical names have now changed. Unsurprisingly, some of the plants are rarer than in White's time or absent altogether, but many can be found by retracing White's movements around the village. Although White's world was famously local – largely confined to a twelve-mile radius around his home – his observations show awareness of global networks of social, scientific, and economic exchange. Noting that 'the productions of vegetation have had a vast influence on the commerce of nations, and have been the great promoters of navigation', he nods to the enormous importance in the period of 'sugar, tea, tobacco, opium, ginseng, betel, paper, &c'. Offering a naïve view of the 'mutual intercourse' underpinning international trade, he reminds Barrington that 'without the knowledge of plants and their culture, we must have been content with our hips and haws, without enjoying the delicate fruits of *India* and the salutiferous drugs of *Peru*'. Such issues are taken up in many subsequent extracts.

While Johann Wolfgang von Goethe is best known as a leading German Romantic poet, playwright, novelist, philosopher, and commentator, he was also a confidante of the geographer Alexander von Humboldt (see Parts 7 and 8 of this volume) and deeply interested in science. Goethe's botanical work provided cutting-edge research into plant behaviour. The third extract in this section is from *The Metamorphosis of Plants* (1790), but also of interest is his work on the spiral tendency in vegetation (see Further Reading). *Metamorphosis* employs a rigorous method of close observation of annual plants and scientific reasoning to theorise that all plant organs are modifications of the same basic material – essentially a new (and valid) claim in botany.

In demonstrating his central argument, Goethe exemplifies the close attention to morphology that dominated botany from this point forwards. The use of specialist anatomical terms parallels comparative anatomists and zoologists. Goethe also displays wider interest in the impact on plant growth, appearance, and behaviour of environmental conditions – for example, in the ways that individuals of the same species of plant modify their organs when growing at higher or lower altitudes – in ways that are reminiscent of Curtis's discussion of chickweed but also of Darwin's observations of South American flora in a later extract in this section. Scrutiny of such behaviours and modifications fed the fields of biogeography, evolutionary science, and ecology, because focusing on the interaction of plants with specific conditions leads to consideration of a) their ability to exist in different climates, soils, and topographies; b) the possibility that environmentally induced modifications to organisms might be reproducible and lead to species change; and c) the wider study of whole environments via the science of ecology. Goethe's work is one of countless examples in this anthology, and in long nineteenth-century natural sciences more generally, of an incremental shift towards dynamically materialist readings of environment and of the manner in which Enlightenment science birthed attitudes that led to the proliferation, specialisation, and professionalisation of sciences.

The fourth extract is from Volume 2 of an 1864 edition of *English Botany*. Quickly known as 'Sowerby's Botany' because of the leading role played by James Sowerby, the complexities of its authorship offer fascinating insights into the class basis of the natural sciences at this period. The work, running to 36 volumes between 1791 and 1814, was conceived, published, edited, and illustrated by Sowerby, a botanical illustrator and naturalist. The brief, formal descriptions of plant anatomy were primarily written by the founder of the Linnean Society, Sir James Edward Smith, who at first refused to have his name associated with the work because he believed that professional association with Sowerby – a member of the artisan class – would prejudice his scientific standing. Once the work met public and professional acclaim, however, Smith abruptly insisted that from the fourth volume onwards his name be included on the title page. With some justice, though, it has remained 'Sowerby's Botany' in the eyes of most readers.

As well as indicating both the dominance of elite gentlemen in eighteenth- and early nineteenth-century natural history, and a growing desire amongst others to find a space within its circles, Sowerby's work offers fascinating glimpses of the growing status of botany. It is also inordinately pleasing in its ambition and tone. Its volumes gradually provided near-comprehensive coverage of known English plant species, and its formal botanical descriptions (thanks to Smith) were systematic, rigorous, and reliable. The volume's charm, however, lies in Sowerby's digressive and lyrical discussions of plant cultivation, habits, forms, beauty, and cultural associations. Sowerby's love of plants is the central motive power, but as a publisher with an eye for the market, he astutely judged the appetites of an expanding middle-class readership enthused by Romantic landscape feeling, British and overseas travel, and the delights of gardening. Environments were being reconceived through botanical works such as these, which could inspire individuals to wander the countryside in search of Sowerby's wild flowers (and, unfortunately, to collect specimens) or to create their own idealised environments in suburban gardens.

Above all, Sowerby's own exquisitely rendered colour plates – comprising 2,592 hand-coloured copper-plate engravings, including three fold-outs – appealed to an enormous public enthusiasm for the visual arts. Sowerby adeptly appealed to different classes and pockets: in 1814, one could purchase all 36 volumes for £55 7*s*, but individual coloured plates were available for 1*s* each in quarto or 6*d* in octavo. The extract from Volume 2 (1791) offers examples from the coverage of a single family – the violets (*Violaceœ*) – and a single species, the sweet violet (*Viola odorata*). Smith's formal, taxonomic, and morphological descriptions are broadly similar to Curtis's, but the later sections demonstrate Sowerby's magical approach – invoking a range of poetry here, as he does throughout his floral studies. These include a number of sources which are not referenced: Perdita's monologue from Shakespeare's *The Winter's Tale*, Act IV, Scene VI, and a speech by Laertes in *Hamlet*, Act V, Scene I; Stanza 3 of P.B. Shelley's posthumously published 'Remembrance' (1824); Homer's *The Odyssey*, Book V, and Fragment 64 of Pindar's lyrical poetry, in which Athens is described as 'City of light, with thy violet crown, beloved of the poets'; William Hunnis's 'A Nosegay Always Sweet' (1573); and John Milton's *Comus* (1634). Sowerby includes the lines 'And the violets gleam like amethysts, from the dewy moss beneath', which appear in 'The Sicilian Captive' by Felicia Hemans. However, Hemans was not born until 1793, two years after Sowerby's work first appeared. The poem was published in 1825 in *The New Monthly Magazine and Literary Journal*, and in Hemans's 1828 collection, *Records of Woman*. Given Sowerby's death in 1822, this apparent anachronism can only be explained either by Hemans having plagiarised the line (from a source which I cannot trace, but which Sowerby knew – a ridiculous hypothesis) or by the 1864 editors amending the edition to suit their own poetical proclivities (an odd but plausible theory).

Like Curtis, Sowerby continues to embed medicine within botany, outlining remedies associated with violets and tracing the plant's historical associations and cultural meaning. Like White, Sowerby's references are also impressively global. He refers to 'Oriental Sherbet' (Sharbat), a sweet cordial made from fruits or flowers, popular throughout the Middle East. He also invokes Islam: according to one of his earliest biographers, Muhammad had much to say in praise of violets: 'The oil of violets (al-banafsaj) is the chief oil'; 'The most excellent oil is violet. Anoint yourselves with it, for its excellence over the rest of the oils is like our excellence over men'; 'Oil of violets among the oils is as the believer among men. . . . It is hot in the winter and cold in the summer. The rest of the oils do not have this merit. . . . If people knew what is in the violet, it would be valued in dinars'; 'Use oil of violets, for the excellence of violet over all the oils is as the excellence of the Ahl al-Bayt over people' (Ibn Ishaq, *Life of God's Messenger*, c. 767 CE).

The next extract moves forward forty years (the intervening years are well represented under Popular Botany in Volume II of this anthology) to William Jackson Hooker's *British Flora* (1830). Doing so permits appreciation of the ways in which botany had developed during this interval, in particular its growing preoccupation with nomenclature. While British botany had long staunchly supported Linnaeus's methods – the sexual system through which plants were classified according to a largely numerical analysis of floral anatomy – French botany had from the late eighteenth century followed the 'natural system', first promoted by the French-Swiss botanist Augustin de Candolle. The 'natural system' asked a wider range of questions about a plant's appearance, habits, and physiology in order to create 'natural' groupings. Hooker adopts and defends the Linnaean sexual system as the best tool of classification but does not attack the Natural System, even going so far as to cite Lindley's 'natural system' work (published in the same year and included in this section) as a valuable guide.

Hooker's intention was to produce a work 'serviceable in advancing the cause of Botanical science in this country'. He somewhat follows Sowerby in including 'short notices of the uses and properties, or some little historical remarks relative to the species' in order to avoid a common tendency to see botany 'as a dry and profitless employment, a system of hard words, destitute of any real utility to mankind'. All the same, Hooker focuses much more on nomenclature and anatomy than Sowerby, uses highly technical language, and largely abbreviates the wider cultural history of plants. His work is particularly interesting, though, for its sustained focus on plant habitats: by emphasising the habitats suitable to particular plants, Hooker nods to the biogeographical impulses of Humboldt and others (see Part 8 of this volume). Hooker's prominence within Victorian botany led to his stewardship of Kew Gardens from 1831, but he was also a botanical explorer, and Hooker used his role at Kew to sponsor and support plant collectors, including his close friend, the botanist and scientific patron Sir Joseph Banks (see Part 7 of this volume), and oversaw a massive expansion of its herbarium. In speaking of the 'anticipations of the pleasure we may have to bestow on kindred minds with our own, when sharing with them our discoveries and our acquisitions', Hooker quotes from Smith's *An Introduction to Physiological and Systematical Botany* (1807) and also refers to Sowerby's *English Botany* and his own *Flora Scotica* (1821) from his period in charge of Glasgow Botanical Gardens.

The extracts included are a somewhat random selection, including two non-native species introduced to the United Kingdom: valerian, established since Roman times, and, perhaps more surprisingly, coriander. Hooker suggests the latter had escaped from cultivation in East Anglia, and it was regarded at the time as primarily medicinal rather than culinary. The final selection reflects Hooker's at-the-time cutting-edge interest in ferns, a group that would soon become massively popular in middle-class England, establishing a craze that dominated the 1850s (see Volume II of this anthology). Here, his desire to exert primacy in the field against the error of others can be glimpsed between the extremely dry anatomical descriptions.

The following extract is from John Lindley's *Outline of the First Principles of Botany* (1830). Lindley, a significant and prolific botanist, is best known for advocating his own modification of the 'natural system'. The anatomical focus and logical framework of the brief propositions that make up the extract share the aspirations of comparative anatomy to consider the relationship between form and function, but to a much lesser extent. Lindley's approach was antithetical to John Ruskin, who despaired when finding Lindley of no help as he sought to answer what he described as 'the child's question' of how a tree grows its trunk, in 'Of Leaf Beauty' (1860): to understand Ruskin's very different methods, and his disgust at what he regards as Lindley's narrow approach, see Volume IV of this anthology. Like the herbarium tradition of the eighteenth century of which it is a logical product, Lindley's approach consists in isolating an individual plant specimen for the purposes of analysing its anatomical features. Ecology and biogeography, by contrast, involved understanding a plant as a functioning part of its environment. For Lindley, the preparation of dry specimens for preservation and analysis was the aim. For ecologists and biogeographers, the examination of a living plant was key.

The following extract, by Dublin botanist Lady Katherine Sophia Kane, is another example of a Flora designed to aid students of botany in the task of fieldwork identification. *The Irish Flora; Comprising the Phænogamous Plants and Ferns* (1833) is the first scientific study of Ireland's wild plants, and it was widely lauded for its precision, accuracy, and usefulness – it quickly became a primer for Irish university students of botany. The extract includes Kane's preface, often amusing in its dismissiveness of earlier studies, and some of her descriptions of *Trifolium* (that quintessentially Irish group, shamrocks). The latter, following the Linnaean system, also refer to Sowerby's *English Botany* and Hooker's *British Flora*, but as the preface attests, Kane was well connected, and able to draw on widespread expertise within Irish botanical circles. At a time when many talented female botanists faced insuperable difficulties in achieving publication, often instead providing materials or illustrations to male-authored works or seeking informal involvement in botanical societies, Kane's achievement is striking.

Lamentably little work has been produced on Kane, but readers can consult Donal Synnott and E. Charles Nelson (see Further Reading).

The next two extracts show the influence of Gilbert White and William Curtis in the creation of highly localised Floras. George Luxford's *A Flora of the Neighbourhood of Reigate, Surrey* (1838), drawing on his intimate knowledge of a very specific area, is one of many examples from the period. While Luxford deems his work 'trifling [from a] scientific point of view', Floras such as his were vital in beginning to gain a clearer picture of U.K. floral distribution and in answering a range of other questions: how and why certain species appeared in some regions and not in others, how and why certain species were common and others rare, whether they were thriving or declining, and which species were associated with one another and why this might be. Works like Luxford's are in this respect parochial cousins of the grand biogeographical enterprises of von Humboldt and other 'new world' explorers (see Parts 7 and 8 of this volume, and Volume II). A selection of entries have been taken from across Luxford's study to give a sense of his methodology, but the work as a whole is valuable in ways that would not have been anticipated by the author: for those seeking to understand the degree to which biodiversity has been lost over time, historical studies of particular locations offer a means to compare past and present populations.

Luxford's *Flora* should also be situated in relation to a Romantic tradition in which personal engagement with particular places was linked to transcendent experience, or simply to greater sympathy with nature. Reflecting on the 'delightful occupation' he had set himself, Luxford speaks in poetic and pastoral language of its consolations:

> my thoughts were often wafted far away from the smoke and noise of such a situation, to the pure air and quiet of the 'lanes and alleys green, dingles and bushy dells', so bountifully decked with these 'wildings of nature'.

Like Sowerby, Luxford quotes from Milton's *Comus*, but his cultural references are far more abbreviated than in Sowerby or Curtis. Luxford's reference to botanising near 'a busy manufacturing town' needs explanation. The preface is signed off 'London, April 17th, 1838', but his botanical knowledge up to this point must also have been based on his experiences of the Midlands, where he ran printing companies in Birmingham until relocating to London in 1837. Presumably he also made frequent visits to the Reigate region during this time.

The next extract, from Anna Russell's *Catalogue of Plants Found in the Neighbourhood of Newbury*, was published one year after Luxford's. Perhaps the most important female botanist of her generation, Russell became a member of the Botanical Society of London and contributed papers on fungi to *Journal of Botany*, as well as producing hundreds of scientific illustrations – a remarkable achievement given the constraints on women at this time. Her Newbury *Catalogue* is even more local and intimate than Luxford's work, providing incredibly specific locations for plants – copses, orchards, woods, and bogs – and drawing on the expertise of collaborators who are listed only by initials. The brief extract is an exemplar of her methods, which involve focusing on habitats rather than taxonomy or morphology. Like Luxford, her work is extremely valuable to later botanists and ecologists seeking to understand changes to biodiversity in particular locations. Like Kane, Russell is little covered in studies of Victorian botany but does appear in Ogilvie and Harvey's excellent study (see Further Reading).

While Luxford and Russell give a flavour of botany at a local English level, the following extract illustrates the global reach of Victorian science. Charles Darwin's *Journal of Researches into the Geology and Natural History of the Various Countries Visited by H.M.S. Beagle* (1839), often referred to as *Voyage of the Beagle*, is amongst the most significant texts of Victorian natural history, and extracts from it have been included in Parts 6–8 of this volume. Although Darwin's primary interests are geological, zoological, topographical, and anthropological, he also collected and analysed plants and always treats flora as part of broader enquiries into the dynamics of particular locations. This biogeographical (and proto-ecological) perspective is often strikingly modern. In the first part of the extract, taken from Darwin's extensive journeys across the Argentinean pampas, he is struck by the deleterious effects of the introduction of European plants and animals on both native species and whole landscapes and habitats. Extracts from Darwin's momentous time in the Galápagos archipelago likewise situate botanical subjects in broader contexts: he begins with geology, topography, and climate, which he uses to explain the relative paucity of plants he is able to collect, and the atypical character of the floral suite: 'such wretched-looking little weeds', he notes, 'would have better become an arctic, than an equatorial Flora'. Acutely aware of the unusual nature of the flora and fauna of the Galápagos, but at this time unable to make the leap to evolutionary explanations, Darwin nonetheless brilliantly focuses on relationships between environment and organic form, while his collecting work also contributed to the expansion of global botanical knowledge.

While some might question the inclusion here of an extract from John Ruskin's celebrated work of art criticism *Modern Painters I* (1843), it has been selected for good reasons. The preceding extracts, to varying degrees, are all committed to orthodox botanical investigation, founded on nomenclature and external observation of parts, or specific investigations of particular locales. Ruskin's very different approach is intriguing for its combination of empirical investigation – into the rules of growth and appearance of trees – and art. Ruskin aims to test the truth or falsity of landscape artists against the forms of nature, essentially asking whether J.M.W. Turner or the widely lauded Old Masters of seventeenth-century landscape art are most faithful to nature in their representations of trees. In setting out the series of rules that structure the chapter, Ruskin is uninterested in describing anatomical features as Lindley and others do, arguing that 'the disciplined eye and the life in the woods are worth more than all botanical knowledge'. The radicalism of Ruskin's approach lies in what he shares with Darwin: a proto-ecological insistence in examining the ongoing dynamic lives of organisms and the manner in which their forms are the result of complex interactions with environment. In this respect, Ruskin's method is the product of genuine engagements with nineteenth-century science (as exemplified in the extract from his work in Part 2 of this volume) but also, and just as importantly, his commitment to Romanticism's call to experience and venerate nature. Likewise, the impact of his Natural Theological convictions led him to believe that paying attention to nature cultivates individual humans and improves human nature. Ruskin's claims about tree appearance and growth are verifiably true, but his tree observations are not simply about demonstrating 'correct' forms (of leaves, boughs, and trunks) but also about his conviction that the overpowering energy and infinite variety of environment are sources of joy, inspiration, and consolation. Above all, what is evident throughout this extract – and what is strikingly absent in Lindley – is wonder at nature and a devotion to tracing those 'truths' of plant growth that have no place within standard botanies.

What irritates Ruskin most about the faulty tree studies of the 'Old Masters' is their lack of interest in, and attention to, nature. For Ruskin, their attempts to compose landscapes from imagination, in the studio and far from the scenes supposedly depicted, is a failure to reverence and learn from divine Creation. By contrast, modern painters such as Turner and J.D. Harding exemplify the Romantic call to experience nature directly and create landscape art directly inspired by real landscapes. Ruskin's extensive experiences of landscape travel – through France, Italy, the Alps, and the wilder parts of Britain – formed part of his own training in landscape and art, following paths trodden by the Romantic poets and artists, a training similar to many other middle- and upper-class men of his generation but wildly different in the degree of influence on the formation of his ideas. Other extracts from Ruskin in subsequent volumes will further demonstrate the uniqueness and the importance of his approach to natural history. Ruskin's titles for paintings and painters in the extract do not always correspond to their modern names. See Woof, Garside, and Haslam (Further Reading) for full details of the works and artists mentioned in this chapter.

The following two extracts resume our focus on the intersections of botany and international exploration undertaken by the European nations. Written respectively by the son and father of the botanically prestigious Hooker family, *Flora Antarctica* (1844) and *Flora Niger* (1849) demonstrate the ambitious nature of early to mid-nineteenth-century botanical voyages while also revealing the often-colonial attitudes of explorers to the 'undiscovered' lands of the 'new world'. These are just two of several such voyages undertaken by the Royal Navy in these years. Such journeys – including Fitzroy's and Darwin's earlier *HMS Beagle* adventures, underlined Britain's technological, military, and political dominance but also had clear scientific aims and profoundly shaped and influenced Victorian attitudes to environment. *Flora Antarctica* is the account of Joseph Dalton Hooker's time as botanist on Captain Ross's four-year voyage to Australasia Antarctica, South Africa, and South America. In an unexcerpted section of the opening chapter, Hooker outlines an extraordinary four-year itinerary comprising Prince Edward Islands, Crozet Islands, Kerguelen, Tasmania, Campbell's Island, Antarctica, Franklin Island, Australia, New Zealand, the Falkland Islands, Cape Horn, Hermite Islands, Cockburn Island, and Cape of Good Hope. Two of the ships involved in the voyage, *HMS Erebus* and *HMS Terror*, became famous for their role in the ill-fated Franklin Expedition, which set off a year after the publication of *Flora Antarctica.*

The extract includes parts of Hooker's 'Summary of the Voyage' and an example of his time in Tierra del Fuego (where he followed in Darwin's footsteps). While Ross's voyage was primarily focused on investigating terrestrial magnetism, it followed customary practice in including a naturalist on board in order to 'use every exertion to collect the various objects of Natural History which the many heretofore unexplored countries about to be visited would afford'. The extract underlines the manner in which botanists collaborated across nations in the formation of new knowledge: as well as acknowledging the work of Darwin, Hooker cites several explorers and authorities involved in explorations of the southern hemisphere during the previous century. Such voyages also generated significant biological collections, which further whetted elite and popular appetites for the 'exotic' – subjects taken up in more detail in Parts 7 and 8 of this volume, and in Volume II.

While the examples of Hooker's plant descriptions – of canelos and three species of barberry (plants now part of British horticulture) – follow patterns laid down by the preceding standard botanical works in this section, they are primarily of interest for the broadly biogeographical insights they disclose. Hooker's engagements in debates about the taxonomic status of canelo are particularly interesting because questions about how much a species varied over a geographical range, and whether these variations should be regarded as sub-species, varieties, or new species, would become a central ground on which evolutionary theory was constructed. The appetite for exploration, knowledge, and specimens also symbolises Victorian belief in the unquestionable sovereignty of 'civilised' western cultures – their entitlement to visit, occupy, and exploit lands for resources and to impose themselves in various ways on the inhabitants of the 'new lands' that they visited.

Further details of the exploration described by *Flora Niger* and its aims under the auspices of the African Civilization Society, as well as Sir William Jackson Hooker's role as a Kew-based editor, can be found in another extract from this work in Part 7 of this volume. The journal entries included here are from the voyage's German botanist, Dr Theodore Vogel, who lost his life to malaria during the journey – a fate shared by large numbers of the crew. Much more information on Vogel, and his assistant, Mr Ansell of the Horticultural Society of London, is included in the Part 7 extract, which also alludes to the fact that because both men suffered ill health during the journey, their botanical collections were not 'extensive or well-preserved'. Despite this, *Flora Niger* makes extensive use of Vogel's journal for his descriptions of flora.

The selection is principally from Vogel's time in present-day Liberia, which was established as a colony in 1822 by the American Colonization Society. Compared to Darwin's account of the limited flora of the Galápagos, the most striking feature is the proliferation of species and the extreme fertility of the soil. Vogel records dozens of species, both wild and cultivated. In his exploration of the area around Cape Palmas, Vogel is particularly interested in attempts by American colonists to establish farms, a phenomenon dating back to 1831, when Maryland legislators funded a project to relocate ex-slaves from the United States to Africa, overseen on the ground by Episcopal missionaries. Three years later, the Maryland State Colonization Society (a branch of the national society) established a settlement (Maryland in Africa). At the time of Vogel's visit, this was an independent colony, later folded within the Republic of Liberia. While Liberia's agricultural economy would later be unhelpfully dominated by rubber plantation cash crops, the picture painted by Vogel is of a much broader range of cultivated flora. In general, and unsurprisingly given the nature of the voyage as part of a project to 'improve' Africa, Vogel focuses attention on economic botany and agricultural crops, and his approach is more utilitarian and less theoretical than Darwin's or J.D. Hooker's. Throughout the journal, however, as in the selection here from his journeys within present-day Ghana, he focuses on wild plants, including the complex floral suites associated with mangroves. Unlike Darwin, Vogel was rarely able to make extensive journeys inland, and many of his observations are limited to coastal perspectives, an indicator of the difficulties (practical, climactic, topographical, and medical) of European exploration into the interior of the supposedly 'dark continent'.

The final extract, from Thomas Ewbank's peculiar 1855 work, *The World a Workshop*, is a yet more pronounced example of economic botany. It more explicitly exemplifies a feature of many of the other entries in this section – from Sowerby and Curtis to *Flora Niger* – that much of the interest in the vegetable kingdom was not scientific per se, nor particularly interested in its aesthetic qualities: when Ewbank exclaims 'how much there is to elicit and exhaust admiration' in the world of trees, he speaks not with a Ruskinian devotion to natural beauty but of the range and diversity of items producible from these 'wonderful compounds of matter'. His attitude is perhaps best encapsulated in his definition of vegetables as 'literal mechanisms for elaborating matter' and in his argument that 'to improve their products, they themselves must be improved or changed, just as artificial contrivances are, when required to turn out better goods'. Herein, amplified and underscored, is the logic of human agriculture, pastoralism, and environmental sovereignty: that nature is a material made to be shaped by humans for the exclusive benefit of humans.

Ewbank's central focus on nature's capacity to supply humankind with its wants leads to an often-statistical approach: a veritable goldmine of eye-catching figures illustrates the fecundity of plant life, its financial value, and the efforts involved in creating products from natural sources. As a tribute to human ingenuity and progress at a time of enormous confidence in the possibilities of industrial technology, these celebrations may strike modern readers as illustrative of the sharp acceleration of environmental exploitation during the period. Ewbank's highly international survey of vegetable productions also reflects the reach of modern capitalist infrastructures: multiple references to global plants and products reinforce the impression of a shrinking world and of the excitement that this expansion of experience generated. While his work is fed by knowledge from international travel and exploration, Ewbank is keen to point out that vast areas of the earth's surface remained to be explored, and exploited. It is partly for this reason that Ewbank is supremely confident not simply of the right of humans to exploit all that nature can give but of the inexhaustibility of its provisions. In the light of our current biodiversity and resource crises, his belief in the power of environments to endlessly replenish themselves is seriously misplaced. Always looking ahead, Ewbank experiences few Malthusian anxieties as he speculates a future population of 'three, and perhaps to five thousand millions' (a fivefold increase on the world figure for the 1850s but a paltry figure glimpsed from our beleaguered factory planet). Calculating the potential of the vegetable kingdom to feed, clothe, shelter, and assist humankind, he is sure that 'a perpetual supply is secured'. Anxieties about resource depletion – particularly in relation to coal – had already emerged and would come to dominate scientific and public discourse, but they are almost entirely absent from Ewbank's vision.

Ewbank's analyses indicate both the ways in which flora had proved economically valuable and how it could be further exploited. While his language and approach feel modern in figuring the world as a factory made up of different 'departments' (geology, zoology, and botany) to provide different products for the prime consumer, *Homo sapiens*, this is also Natural Theology, rooted in assumptions that the earth was designed with 'singular wisdom and economy' and illustrated with simplistic and unconvincing arguments. This unstable combination of Natural Theology and nineteenth-century industrial logic makes the work as chilling as it is gripping. Remorselessly anthropocentric in reducing everything non-human to the level of exploitable product, Ewbank's attitudes are rooted in religious cosmogonies which place humankind at the apex of creation and grant authority over every part of an environment that is thus rendered unquestioningly other. Further extracts from *The World a Workshop* in subsequent sections of this volume underline the degree to which Ewbank's vision largely runs counter to the emerging readings of environment which begin to displace *Homo sapiens* from its unquestioned position of primacy while auguring an ecological vision of connection, mutualism, and intersectionality. These issues will be further explored in subsequent sections.

Further reading

Allan, Mea, *The Hookers of Kew 1795–1911* (London: M. Joseph, 1967).

Armstrong, Patrick, *The English Parson-Naturalist: A Companionship Between Science and Religion* (Leominster: Gracewing, 2000).

Barber, Lynn, *The Heyday of Natural History 1820–1870* (London: Cape, 1980).

Clifford, David, Elizabeth Wadge, Alexandra Warwick and Martin Willis (eds.). *Repositioning Victorian Sciences: Shifting Centres in Nineteenth Century Scientific Thinking* (London: Anthem Press, 2006).

Desmond, Ray, *Dictionary of British and Irish Botanists: Including Plant Collectors, Flower Painters and Garden Designers* (London: Taylor & Francis, 1994).

Goethe, Johann Wolfgang von. *The Metamorphosis of Plants* (Cambridge: MIT Press, 2009).

Henderson, Paul, *James Sowerby; the Enlightenment's Natural Historian* (London: Kew Publishing/Natural History Museum, 2015).

Holland, Jocelyn, *The Procreative Poetics of Goethe, Novalis, and Ritter: German Romanticism and Science* (London: Taylor & Francis, 2009).

Mabey, Richard, *Gilbert White: A Biography of the Author of The Natural History of Selborne* (London: Century Hutchinson, 1986).

Nelson, E. Charles, 'Katherine Sophia Baily (Lady Kane) and The Irish Flora (1833)', *Archives of Natural History* 46:1 (2019), 44–57.

Ogilvie, Marilyn and Joy Harvey, *The Biographical Dictionary of Women in Science: Pioneering Lives From Ancient Times to the Mid-20th Century* (Abingdon: Routledge, 2003).

Richards, R.J., *The Romantic Conception of Life: Science and Philosophy in the Age of Goethe* (Chicago: Chicago University Press, 2002).

Saunders, Gill, *Picturing Plants: An Analytical History of Botanical Illustration.* (Berkeley: University of California Press, 1995).

Shteir, Ann. B. *Cultivating Women, Cultivating Science: Flora's Daughters and Botany in England, 1760–1860* (Baltimore: Johns Hopkins University Press, 1996).

Synnott, Donal, 'Botany in Ireland, *Nature in Ireland: A Scientific and Cultural History*, ed. John Wilson Foster and C.G. Helena (Dublin: The Lilliput Press, 1997).

Walls, Laura Dassow, 'Natural History in the Anthropocene', *A Global History of Literature and the Environment*, ed. John Parham and Louise Westling (Cambridge: Cambridge University Press, 2017), 187–200.

Woof, Lawrence, Roger Garside, and Ray Haslam (eds.), *The Electronic Edition of John Ruskin's Modern Painters I* (Lancaster University, 2000): www.lancaster.ac.uk/fass/ruskin/empi/

42

WILLIAM CURTIS, *FLORA LONDINENSIS*, 6 VOLS, VOL. 1

(London: B. White, 1777)

Preface

ALTHOUGH the Author does not here mean to give a Preface at large, reserving that until the first volume, containing thirty-six numbers or two hundred and sixteen plants, shall be completed; yet he presumes it will be satisfactory to his subscribers, and the public, to be informed a little more fully of the nature and design of the work; as it will also give him an opportunity of answering some few objections that have been made to the plan of it.

The primary design of it then, is to facilitate, a knowledge of the plants of our own country, and establish each species and variety on a firm basis: this the Author considers as the grand desideratum at present; this arduous task once accomplished, a way will be opened, and a foundation laid for numberless improvements in Medicine, Agriculture, &c.

To be enabled to do this, he means to take the greatest pains in the examination of those plants which he figures; to have them drawn from living specimens most expressive of the general habit or appearance of the plant as it grows wild; to place each plant, as much as is consistent, in the most pleasing point of view; and to be very particular in the delineation of the several parts of the flower and fruit, more especially where they characterize the plant.

And in order that he may obtain a more perfect knowledge of each plant; that he may see it in every stage of its growth, from the germination to the maturity of its seed; that he may compare and contrast the several species together; that he may make experiments to elucidate the nature of such as are obscure, or bring into more general use those which bid fair to be of advantage to the public; he is now cultivating each of them in a garden near the city, into which, by the kind assistance of his friends, he has already introduced, in the course of one, year, about five hundred different species, including sixty of that most valuable tribe of plants, the grasses.

Although the ascertaining and fixing of the plants will be his principal object, yet to make the work more useful to the public, as well as instructive and entertaining to the young botanist, his utmost endeavours will be used to lay before them whatever may be found *useful* in old botanic writers; and here they must not be

DOI: 10.4324/9780429355653-48

surprized to find many of the numerous and imaginary virtues, which they attributed to every plant, purposely omitted: the discoveries made by modern authors, particularly relative to *Agriculture* and *Rural Oeconomy*, will be carefully attended to; as here seems to be a field just opening to view, from whence the pubic is likely to draw great and lasting advantages: and as a knowledge of the plants themselves is first necessary, and for want of which, indeed, the experimental farmer cannot effectually communicate his improvements, he finds himself peculiarly happy in contributing his share to the public good.

He is nevertheless sensible how inadequate his abilities, or indeed the abilities of any *one* person are, to render a work of this kind any way compleat; he therefore respectfully solicits the assistance of those, who wish well to the improvement of *English Botany* and *English Agriculture*: any information that they shall be pleased to communicate, shall with those favours he has already received from divers of his friends, be gratefully acknowledged [. . .]

It now remains to obviate some few objections which have been made to the plan of this work; and first, it has been suggested to the Author, that it would have been better received if, instead of pursuing the present plan, he had published those plants only which were not figured in the *Flora Danica*, a work now carrying on in *Denmark* under the auspices of the *King*; but a few moments reflection, must he presumes be sufficient, to convince every unprejudiced person, how inadequate such a partial publication would have been to the making a knowledge of the plants of our country more general among ourselves – at best such a work could only answer the purpose of those few individuals who are in possession of that part of the *Flora Danica* already published; and as that is still going on, there is no doubt but the same plants would be published by both Authors.

[. . .]

Alsine media. Common Chickweed

ALSINE *Linnaei. Gen. Pl.* PENTANDRIA TRIGYNA.	
	Cal. 5-phyllus. *Petala* 5 -aequalia. *Caps.* 1-locularis, 3-valvis.
	Raii Syn. Gen. 24. HERBÆ PENTAPETALÆ VASCULIFERÆ
ALSINE	*media. Linnaei Syst. Vegetab.* p. 246. *Flora Suecic. p.* 37.
ALSINE	foliis petiolaris, ovata laceolatis, petalis bipartitis. *Haller hist. helv. n.* 880.
ALSINE	*media. Scopoli Fl. Carniol. n.* 376.
ALSINE	*media. Baubin pin. p.* 250.
ALSINE	*media* seu minor. *Gerard emac.* 611. *Raii Syn. p.* 347, Common Chickweed. *Hudson Fl. Angl. p.* 113. *Oeder Fl. Dan.* 525, 438.
ROOT	annual, fibrous capillary.

STALKS	numerous, tender, round, striking root here and there, branched, jointed and stringy, *hairy on one side only*, growing thicker towards the top.
LEAVES	of a pointed oval shape, smooth, slightly hairy at the edges, the lowermost standing on footstalks, the uppermost sessile, connate.
FOOT-STALKS	of the leaves broadest at bottom and hairy.
FOOT-STALKS	of the flowers, each sustaining one flower, proceeding from the bosoms of the leaves, hairy, when the flowering is over hanging down, finally becoming upright.
CALYX	a PERIANTHUM of five leaves, each of which is lanceolate, concave, keel-shaped at bottom, with a margin at the edge, hairy, *and longer than the Petals*.
STAMINA	five white FILAMENTS, placed betwixt the Petals, furnished at bottom with a little Gland; ANTHERÆ roundish, of a purplish colour.
PISTILLUM	GERMEN somewhat oval; STYLES three, filiform; STIGMATA, simple.
SEED-VESSEL	a CAPSULE of one cavity, splitting into six valves.
SEEDS	from eight to fifteen, somewhat kidney-shaped, of a brownish orange colour, with a rough surface, connected to the receptacle by little foot-stalks.

CHICKWEED being a plant which will grow in almost any situation is consequently liable to assume many different appearances: when it grows in a rich soil, and shady situation, it will frequently become so large as to resemble the *Cerastium aquaticum*; while at other times, on a dry barren wall, its leaves and stalks will be so minute, as to make the young botanist take it for some species different from the common Chickweed: happily however it affords marks which if attended to, will readily distinguish it from the *Cerastium*, and every other plant: exclusive of its differing from the *Cerastium* in its generic character, its Petals are shorter than the leaves of its Calyx; while in the *Cerastium* they are longer; hence a considerable difference will be observable at first sight in the size of the flowers of these two plants: and from all other plants related to it, it may be distinguished by the singular appearance of its stalk, which is *alternately hairy on one side only*.

The most common number of its *Stamina* with us is five; yet I have often seen it with less, and sometimes with more; and this inconstancy in the number of its Stamina has been noticed by most botanic writers [. . .]

Of annual plants there are few more troublesome: it sows itself plentifully in the summer, and remains green throughout the winter, flowering during the whole time, if the weather be mild: but its chief season for flowering is in the Spring. In rich garden mould, where the ground is highly cultivated, and in the fields about town, it does a deal of mischief: by the quickness of its growth and the great number of its shoots, it covers and choaks many young plants; hence it should be carefully weeded from dunghills.

The seeds are very beautiful, and have the greatest affinity to those of the *Cerastium aquaticum*.

When the flowers first open, the foot-stalks which support them are upright; as the flowers go off they hang down; and when the seeds become ripe, they again become erected.

LINNAEUS has observed that the flowers open from nine in the morning till noon, unless rain falls on the same day, in which case they do not open: from what little observations I have made on this plant, it is not subject to be affected precisely in the same manner here, having seen in the month of March, the blossoms continue rather widely expanded after repeated showers of rain.

It is considered as a wholesome food for Chicken and small Birds, whence, as RAY observes, it has obtained its name: boiled it resembles Spinach so exactly as scarcely to be distinguished from it, and is equally wholesome, being a plant which may be procured almost any where very early in the spring, it may be no bad substitute where Spinach or other greens are not to be had in plenty, and much preferable to Nettle-tops and other plants which the lower sort of people seek after in the Spring with so much avidity. Swine are very fond it, and prefer it to Turnep-tops. It is eaten by many Insects, particularly by the Caterpillar of the *Phalæna Villica* or *Cream Spot Tyger Moth*, and other hairy Caterpillars of the Tyger kind.

As a medicine it contains no active principle; but is frequently applied to hot, painful, and inflammatory swellings, either by itself, bruised, or mixed with poultices, with good success.

[. . .]

Antirrhinum linaria. Common Yellow Toad Flax

[. . .]

Mr Ray in his *Historia Plantarum*, has collected the Authorities of several writers who speak highly of the medical virtues of this Plant. At the same time that we by no means believe in all the Virtues which are attributed to many plants by the old Authors, we would be carefull of rejecting all their accounts, particularly when there is some reason to think they may be founded in Truth, the mention of them may at least serve to excite such of the Faculty as have proper opportunities to give them a fair trial, and either reject them entirely, or bring them more generally into practice.

According to some it opperates both by Stool and Urine, and so much by the latter, as to acquire among the *Germans* the name of *Harnkrout*. A small Glass of the distilled Water mixed with a drachm of the bark of the *Ebulus* or *Water Elder* in powder, powerfully provokes Urine, and is recommended in Dropsical Cases. The distilled water or juice of the Plant put in the Eyes, takes away the redness and inflammation of them, as TRAGUS asserts from his own long observation and experience. Made into an Ointment with lard and mixed with the yolk of Egg, it takes away the violent pain arising from the Piles.

The flowers of this plant are frequently found double with two or more Spurs, and a singular variety of it which LINNAEUS called *Peloria*, is said by Mr HUDSON to grow in *Clapham* in Surry, this rare monstrosity we shall not fail to figure.

In its common state, the Toad Flax grows very common on banks by the road sides, which it decorates not a little by its singular and beautiful Flowers. It may with the greatest ease be cultivated in Gardens, and raised either from Seeds or Roots; the Seed is ripe at the latter end of September.

Digitalis purpurea. Fox-glove

[. . .]

Was it not that we are too apt to treat with neglect the beautiful plants of our own country, merely because they are common and easily obtained, the stately and elegant *Fox-glove* would much oftener be the pride of our gardens than it is at present; for it is not only peculiarly striking at a distance, but its flowers and their several parts become beautiful in proportion to the nearness of our view: How singularly and how regularly do the blossoms hang over one another! How delicate are the little spots which ornament the inside of the flower! And like the wings of some of our small Butterflies smile at every attempt of the Painter to do them justice: how pleasing is it to behold the nestling Bee hide itself in its pendulous blossoms! while extracting its sweets which furnish our tables with honey, and our manufacturers with wax: nor are the more interior parts of the flower less worthy of our admiration, or less adapted to the improvement of the young Botanist: here all the parts of the fructification being large, he will readily obtain a distinct idea of them; but more particularly of the form of the Antheræ, and the alteration which takes place in them, previous to and after the discharge of the Pollen [. . .]

The flowers of this plant are in general of a fine purple colour, and like all other purple flowers are liable to variations; sometimes we find the blossoms of a milk white or cream colour, and some other varieties of it are mentioned by RAY, but the white is the most common. Such as would wish to cultivate it, may raise it either from seed, which is very small for the size of the plant, or from young plants. It grows naturally in a dry and gravelly soil, and in such situations is common enough over most parts of England; about *Charlton-Wood* it is very plentiful, and flowers in July and August.

According to the testimony of many writers, the juice or decoction of this plant taken inwardly, acts as an emetic and purgative, and that too with considerable violence; hence Mr RAY properly advises it to be given to such only as have robust constitutions. PARKINSON affirms that it is very efficacious in the cure of the Epilepsy; but he unites with it in his prescription *Polypody of the* Oak, so that there is no knowing to which of the plants the merit of curing this stubborn disease is due.

The flowers or herb either bruised or made into an ointment are strongly recommended in Schrophulous tumours and ulcers; and so great an opinion have the Italians of its virtues as a vulnerary, that they have the following proverb concerning it. '*Aralda tute le piaghe* salda'. Fox-glove cures all wounds. *Raii Hist. Plant.*

43

GILBERT WHITE, *THE NATURAL HISTORY AND ANTIQUITIES OF SELBORNE, IN THE COUNTY OF SOUTHAMPTON: WITH ENGRAVINGS AND AN APPENDIX*

(London: T. Bensley, 1789)

LETTER XL. [TO DAINES BARRINGTON]

DEAR SIR, SELBORNE, June 2, 1778.

The standing objection to botany has always been, that it is a pursuit that amuses the fancy and exercises the memory, without improving the mind or advancing any real knowledge: and, where the science is carried no farther than a mere systematic classification, the charge is but too true. But the botanist that is desirous of wiping off this aspersion should be by no means content with a list of names; he should study plants philosophically, should investigate the laws of vegetation, should examine the Powers and virtues of efficacious herbs, should promote their cultivation; and graft the gardener, the planter, and the husband man, on the phytologist. Not that system is by any means to be thrown aside; without system the field of Nature would be a pathless wilderness; but system should be subservient to, not the main object of, pursuit.

Vegetation is highly worthy of our attention; and in itself is of the utmost consequence to mankind, and productive of many of the greatest comforts and elegancies of life. To plants we owe timber, bread, beer, honey, wine, oil, linen, cotton, &c. what not only strengthens our hearts, and exhilerates our spirits, but what secures us from inclemencies of weather and adorns our persons. Man, in his true state of nature, seems to be subsisted by spontaneous vegetation; in middle climes, where grasses prevail, he mixes some animal food with the produce of the field and garden: and it is towards the polar extremes only that, like his kindred bears and wolves, he gorges himself with flesh alone, and is driven, to what hunger has never been known to compel the very beasts, to prey on his own species.

The productions of vegetation have had a vast influence on the commerce of nations, and have been the great promoters of navigation, as may be seen in the

DOI: 10.4324/9780429355653-49

articles of sugar, tea, tobacco, opium, ginseng, betel, paper, &c. As every climate has its peculiar produce, our natural wants bring on a mutual intercourse; so that by means of trade each distant part is supplied with the growth of every latitude. But, without the knowledge of plants and their culture, we must have been content with our hips and haws, without enjoying the delicate fruits of *India* and the salutiferous drugs of *Peru*.

Instead of examining the minute distinctions of every various species of each obscure genus, the botanist should endeavour to make himself acquainted with those that are useful. You shall see a man readily ascertain every herb of the field, yet hardly know wheat from barley, or at least one sort of wheat or barley from another.

But of all sorts of vegetation the grasses seem to be most neglected; neither the farmer nor the grazier seem to distinguish the annual from the perennial, the hardy from the tender, nor the succulent and nutritive from the dry and juiceless.

The study of grasses would be of great consequence to a northerly, and grazing kingdom. The botanist that could improve the swerd of the district where he lived would be an useful member of society: to raise a thick turf on a naked soil would be worth volumes of systematic knowledge; and he would be the best commonwealth's man that could occasion the growth of 'two blades of grass where one alone was seen before'. I am, &c.

LETTER XLI. TO THE SAME.

DEAR SIR, SELBORNE, July 3, 1778

In a district so diversified with such a variety of hill and dale, aspects, and soils, it is no wonder that great choice of plants should be found. Chalks, clays, sands, sheep-walks and downs, bogs, heaths, woodlands, and champaign fields, cannot but furnish an ample Flora. The deep rocky lanes abound with filices, and the pastures and moist woods with fungi. If in any branch of botany we may seem to be wanting, it must be in the large aquatic plants, which are not to be expected on a spot far removed from rivers, and lying up amidst the hill country at the spring heads. To enumerate all the plants that have been discovered within our limits would be a needless work; but a short list of the more rare, and the spots where they are to be found, may be neither unacceptable nor unentertaining:

Helleborus fœtidus, stinking hellebore, bear's foot, or setterworth, all over the High-wood and Coney-croft-hanger: this continues a great branching plant the winter through, blossoming about January, and is very ornamental in shady walks and shrubberies. The good women give the leaves powdered to children troubled with worms; but it is a violent remedy, and ought to be administered with caution.

Helleborus viridis, green hellebore, – in the deep stony lane on the left hand just before the turning to Norton-farm, and at the top of Middle Dorton under the hedge: this plant dies down to the ground early in autumn, and springs again about February, flowering almost as soon as it appears above ground.

Vaccinium oxycoccos, creeping bilberries, or cranberries, – in the bogs of Bin's-pond;

Vaccinium myrtillus, whortle, or bleaberries on the dry hillocks of Woolmer-forest;

Drosera rotundifolia, round-leaved sundew. *Drosera longifolia*, long-leaved ditto. In the bogs of Bin's-pond.

Comarum palustre, purple comarum, or marsh cinque foil, – in the bogs of Bin's-pond;

Hypericum androfœmum, Tutsan, St. John's wort in the stony, hollow lanes;

Vinca minor, less periwinkle, – in Selborne-hanger and Shrub wood;

Monotropa hypopithys, yellow monotropa, or birds' nest, – in Selborne-hanger under the shady beeches, to whose roots it seems to be parasitical – at the north-west end of the Hanger;

Chlora perfoliata, Blackstonia perfoliata, Hudsoni, perfoliated yellow-wort, – on the banks in the King's-field;

Paris quadrifolia, herb Paris, true-love, or one-berry, – in the Church Litten coppice;

Chrysosplenium oppositifolium, opposite golden saxifrage, – in the dark and rocky hollow lanes;

Gentiana amarella, autumnal gentian, or fellwort, – on the Zig-zag and Hanger;

Lathrea Squammaria, tooth-wort, – in the Church Litten coppice under some hazels near the foot-bridge, in Trimming's garden hedge, and on the dry wall opposite Grange-yard;

Dipsacus pilosus, small teasel, – in the Short and Long Lith.

Lathyrus sylvestris, narrow-leaved, or wild lathyrus, – in the bushes at the foot of the Short Lith, near the path;

Ophrys spiralis, ladies' traces, – in the Long Lith, and towards the south-corner of the common;

Ophrys nidus avis, birds' nest ophrys, – in the Long Lith under the shady beeches among the dead leaves; in Great Dorton among the bushes, and on the Hanger plentifully;

Serapias latifolia, helleborine, – in the High-wood under the shady beeches;

Daphne laureola, spurge laurel, – in Selborne Hanger and the High wood;

Daphne mezereum, the mezereon, – in Selborne Hanger among the shrubs at the south-east end above the cottages.

Lycoperdon tuber, truffles, – in the Hanger and High-wood.

Sambucus ebulus, dwarf elder, walwort, or danewort, – among the rubbish and ruined foundations of the Priory.

44

JOHANN WOLFGANG VON GOETHE, *GOETHE'S ESSAY ON THE METAMORPHOSIS OF PLANTS, TRANSLATED BY EMILY M. COX; WITH EXPLANATORY NOTES BY MAXWELL T. MASTERS*

(reprinted from the *Journal of Botany*, December 1863) (np: J. E. Taylor, 1863 [1790])

Introduction

1 No one who has paid any attention to the growth of plants, can have failed to observe that some of their external organs occasionally undergo a change; and assume, sometimes entirely, or in a greater or less degree, the appearance of the organ situated next in order.

2 Thus, for example, a single flower is changed into a double one, petals being developed in the place of stamens, either bearing a perfect resemblance in form and colour to the other petals of the corolla, or still retaining visible signs of their origin.

3 If we reflect that the plant has in this way the power of making an actual retrograde step, and of reversing the order of growth, we shall get more insight into nature's ordinary method of proceeding, and shall learn to understand those laws of transformation by which she produces one part from another, and exhibits the most different forms by the modification of a single organ.

4 The secret relation subsisting between the different external organs of plants, such as leaves, calyx, corolla, and stamens (which are developed in succession, and, as it were, out of one another), has long been acknowledged by naturalists in a general way; indeed, much attention has been bestowed upon it; and the title 'Metamorphosis of Plants', has been given to the operation by which one and the same organ presents itself to us under various disguises.

5 This metamorphosis is of three kinds, – regular, irregular, and accidental.

DOI: 10.4324/9780429355653-50

6 Regular metamorphosis may be equally well styled progressive; for it may be observed constantly and gradually at work from the first seed-leaves to the mature fruit; mounting upwards through a series of transformations, as by an imaginary ladder, to that crowning aim of nature, the propagation of the plant by the male and female organs. I have been attentively observing this process for several years, and it is for the purpose of explaining it that I propose to write this Essay. I shall treat of annual plants only, and the manner in which they progress from the seed to the fruit.

7 Irregular metamorphosis might be equally well styled retrogressive. For as in the former case nature hastens forward to her great object, she here takes one or more steps backward. In the former instance, with irresistible impulse and powerful effort she forms the flowers and fits them for their office; in the latter she seems, as it were, to relax, and irresolutely leaves her work in an unfinished, weakly condition, pleasing often to the eye, but intrinsically powerless and inactive. By means of practical observations made upon this kind of metamorphosis, we shall unveil that which in the ordinary way of development is concealed from us, and here shall see clearly what there we dare only infer. We may thus hope to attain, with the greatest certainty, the purpose we have in view.

8 We will not take into consideration the third kind of metamorphosis, which is produced accidentally and by external causes (especially through the operation of insects), as it might lead us out of our way, and interfere with our object [. . .]

1. Of the Seed-leaves

10 Having undertaken to observe the successive steps in the growth of a plant, let us first direct our attention to it when it begins to germinate. We can at this stage easily and exactly distinguish its component parts. Its coverings (which we will not now stay to examine) remain more or less concealed in the soil; and (in many instances) the root becomes established, before the plant exhibits those first organs of its upward growth, which were previously hidden in the seed.

11 These organs are called cotyledons; also seed-lobes, seed-leaves, etc., from their different forms.

12 They are often unshapely, charged as it were with a crude substance, and very thick in proportion to their breadth; their vessels are not recognizable, and can scarcely be distinguished from the general mass; they have moreover very little resemblance to leaves, and we are in danger of being led to regard them, erroneously, as distinct organs.

13 Yet in many plants they nearly approach the form of a leaf; they become flatter, and on being exposed to light and air they assume a deeper green; the vessels become recognizable and more like the veins of a leaf.

14 At length they assume the appearance of true leaves; their vessels are perfectly developed, and their similarity to the leaves subsequently produced, shows that they are not distinct organs, but simply the first leaves of the stem.

15 Now as we cannot realize the idea of a leaf apart from the node out of which it springs, or of a node without a bud, we may venture to infer that the point at which the cotyledons are attached, is the first true node of the plant. This view is confirmed by those plants which emit buds from the axils of their cotyledons, and develope perfect branches from these first nodes; as the common Bean (*Vicia Faba*).

16 The cotyledons are generally two in number; and here we have a remark to make, the importance of which will appear by-and-by. The leaves of this first node often appear in pairs, whilst the subsequent leaves of the stem are placed alternately; an approximation and connection being thus shown between parts, which nature subsequently separates, and places at a distance. The case is still more remarkable when the cotyledons appear like a number of little leaves round a common axis; whilst upon the stem which rises from the centre, the subsequent leaves are developed singly; this may be observed in the different kinds of Pine; the cotyledons of which are a crown of needle shaped leaves [. . .].

18 Let us, however, pause to remark that even those cotyledons which most resemble leaves, when compared with the subsequent stem leaves, are always imperfectly formed. Their margin is entire, with as few traces of incisions in it, as of hairs on the surface, or of any of those vessels which are to be observed in perfect leaves.[1]

II. On the Formation of the Stem-leaves at the Successive Nodes of the Stem

19 We are now able to observe with accuracy the successive formation of the leaves; as the progressive operations of nature all take place before our eyes. Some, or many, of the leaves which now appear, often exist previously in the seed, enclosed between the cotyledons; and are then called the plumule. Their shape, relatively to that of the cotyledons and of the future leaves, varies in different plants; but they differ most from the cotyledons in being flat, and of a delicate texture; and especially in being formed like true leaves, in being perfectly green, and in being situated on a visible node. Their connection with the future stem-leaves can no longer be denied; they are nevertheless inferior to them in the imperfect state of their margin.

20 At each successive node the form of the leaf attains greater perfection; the midrib lengthens, and the side-ribs which arise from it extend more or less towards the margin. The different relations of the ribs to each other are the principal cause of the various shapes we observe in leaves; which are notched, deeply incised, or are formed of many leaflets, and thus resemble

little branches. The Date Palm affords a striking instance of the most simple form of leaf becoming gradually but deeply divided. As the leaves succeed each other the midrib lengthens; till at last it tears asunder the numerous compartments of the simple leaf, and an extremely compound, branch-like leaf is formed.

21 The development of the leaf-stalk keeps pace with that of the leaf; the stalk being either closely coherent with the leaf, or so formed as ultimately to be easily severed from it.

22 We see in different kinds of plants that this independent leaf stalk has a tendency to assume the form of a leaf; as in the Orange. Its structure, which for the present we pass over, will afford us matter for future consideration.

[. . .]

24 Whilst the leaves principally derive their first nourishment from the more or less modified fluids which they draw from the stem; it is to the light and air that they are indebted for their increased perfection in form, and for the delicacy of their tissue. The cotyledons which are produced beneath the covering of the seed, are charged as it were with nothing but a crude kind of sap, are scarcely at all, or but rudely organized, and undefined; in the same way the leaves of plants which grow under water are more rudely organized than others which are exposed to the air; nay, even the same kind of plant will develope smoother and more imperfectly formed leaves when growing in low, damp situations, than it will if transplanted to a higher region; where, on the contrary, the leaves will be rough, hairy, and more delicately finished.

[. . .]

27 We observe in many plants that one node arises from another. In the jointed stems of the cereals, grasses and reeds, this is obvious; but it is not so obvious in plants whose centre is either hollow throughout, or filled with pith, or cellular tissue. The supposed important functions of the pith being now, on good ground, called in question; and the impulsive and productive power once claimed for it unhesitatingly attributed to the inner side of the second bark (the so-called pulp), we can more easily understand that whilst an upper node arises from the previous one, and receives the sap by means of it (receives it too in a more elaborated condition from the intervening operation of the leaves), it must not only attain to a more perfect state itself, but must consequently transmit a more elaborated sap to its own leaves and buds.

28 Whilst, therefore, the less pure fluids are got rid of, purer ones are introduced; and the plant having been gradually brought into a more perfect condition, attains the end prescribed to it by nature. We see the leaves at length perfectly developed in size and form, and soon become aware of a fresh phenomenon, which tells us that the period we have been observing has reached its termination, and that a new one is approaching; that, namely, of the Flower.

III. Transition to the Flowering Period

29 The transition to the period at which the flower appears, takes place with greater or less rapidity. In the latter case the stem-leaves generally become gradually smaller and less divided, whilst increasing more or less in width at their base; at the same time the space between the nodes of the stem, if not perceptibly lengthened, becomes at least more slender and more delicately formed.

30 It has been observed that if a plant is supplied with copious nourishment, the flowering-period is delayed; but that moderate or even scanty nourishment accelerates it. The function of the stem leaves is thus clearly shown. As long as there are crude juices to be carried off, the plant must be provided with organs competent to effect the task. If superfluous nourishment is forced on the plant, this task must be continued, and flowering becomes almost impossible. But, on the other hand, if nourishment is withheld, that operation of nature is facilitated and hastened; the organs of the nodes (leaves) become more refined in texture, the action of the purified juices becomes stronger, and the transformation of parts having now become possible, takes place without delay.

IV. On the Formation of the Calyx

31 This transformation often takes place rapidly; the stem at once becomes tapering and delicately-formed, and shoots upwards from the node at which the last perfect leaf was developed, terminating in a whorl of leaves collected round an axis.

32 It appears to us a fact capable of the clearest proof, that the leaves of the calyx are the same organs as those whose formation we have hitherto been observing as stem-leaves; though now often in a very altered condition, and collected round a common centre.

[. . .]

34 We further see, in many flowers, unaltered stem-leaves collected together so as to form a kind of calyx immediately below the inflorescence. That they are stem-leaves we need only appeal to the normal appearance still retained, and to botanical terminology, which designates them by the name of *Folia floraliu* (bracts).

35 We must now observe the case in which the transition to the flowering-period proceeds slowly; the stem-leaves gradually diminish in size, become altered in appearance, and gently insinuate themselves into the calyx; as may be very easily seen in the common calyx (involucrum) of Composite flowers; especially in Sunflowers and Marigolds.[2]

36 Nature's power of collecting a number of leaves round a common axis is seen to produce even a closer union, so as to render these clustered and modified leaves still more difficult to recognize; that is to say, it unites their edges one

with the other, often entirely, but frequently only in part. The crowded and closely-pressed leaves are brought into the nearest contact with each other while yet in a tender state, an anastomosis is effected by the operation of the elaborated juices which the plant now contains, and they thus form a bell-shaped, or so-called monosepalous calyx; which betrays its compound origin by the manner in which its border is more or less incised or divided. We may find evidence of this by comparing a number of deeply-divided calyces with polysepalous ones; especially if we attentively consider the common calyces (involucres) of many Composite flowers. Thus, we shall find that the calyx of a Marigold, which is defined in systematic descriptions as simple and much divided, consists both of attached and imbricated leaves; amongst which, as we said above, diminished stem-leaves have, as it were, insinuated themselves.

[. . .]

38 Thus has nature formed the calyx; by uniting together around a common centre, generally in a certain definite number and order, many leaves, and consequently many nodes, which she had previously produced in succession, and at some distance from each other. Should, however, the flowering-period have been checked by an excessive and superfluous degree of nourishment, they would have remained separate from each other, and would still have retained their original form. Nature, therefore, forms no new organ in the calyx, but simply unites and modifies those organs with which we are already acquainted, and advances by this means a step nearer to her object.

V. On the Formation of the Corolla

[. . .]

40 The transition of the calyx into the corolla is exhibited in various ways; for although the general colour of the calyx usually remains green, like that of the stem-leaves, it often shows a change in one part or another, at the tips, the edges, or at the back, or over the whole of the inner surface, while the outer surface remains green; and whenever this change of colour occurs, we see it combined with an increased refinement of texture. In this manner an ambiguous kind of calyx is produced, which might with equal propriety be called a corolla (perianth of Linnæus).

41 We remarked that from the seed-leaves upwards a great development takes place both in the size and form of the leaves, especially in their margins; and that a subsequent diminution of their size occurs in the calyx; we have now to observe a second act of expansion, by which the corolla is produced. The flower-leaves (petals) are usually larger than the calyx-leaves (sepals); and it is to be remarked that as a contraction of the organs occurs in the calyx, so (having been in a high degree refined by means of a farther filtration of the fluids in passing through the calyx) they again expand in the form of petals, and assume the appearance of entirely new and distinct organs. Their delicate

organization, their colour, and their scent would make it impossible to recognize their origin, if we had not frequent opportunities of stealthily observing nature when departing from her general rule.

42 Thus, for instance, within the calyx (epicalyx) of a Pink, a second calyx is often found, which being partly green was to all appearance originally designed for a monosepalous notched calyx; but its jagged tips and edges, transformed into incipient and spreading petals, betray both by their colour and texture the relationship that exists between the corolla and the calyx.

43 The relationship of the corolla to the stem-leaves is also shown in different ways; for stem-leaves already more or less coloured may be seen on many plants, far below the inflorescence; those nearest to it being coloured throughout.

44 Those instances also in which nature, as it were, altogether omits the calyx, afford additional opportunities of observing the transformation of the stem-leaves into petals. On the stalks of tulips, for example, a coloured petal almost perfect in form may often be seen. The case is even more remarkable when a leaf, half green and half coloured, remains attached to the stem by the green part, as more properly belonging to it, whilst the coloured portion is carried up with the corolla; so that the leaf is literally torn asunder.

[. . .]

VI. On the Formation of the Stamens

[. . .]

47 Some plants normally produce their petals in a transitional state; as *Canna*, and other plants of the same family. In this instance a true petal, but slightly changed, is contracted at the upper part, and exhibits an anther; in relation to which the rest of the petal stands in the place of the filament.

48 In those flowers whose habit it is to become double, we may trace this transition through all its different stages. In Roses, among perfect coloured petals, others may often be seen which are contracted both in the middle and at the side. This is occasioned by a little protuberance more or less resembling a perfect anther; and in the same proportion the whole petal assumes the form of a stamen. In the case of many double Poppies, some of the petals of the very double corolla are little changed, and tipped with perfectly-developed anthers; whilst others are more or less contracted by anther-like protuberances.

49 When all the stamens are changed into petals, the flower produces no seed; but if any of the stamens are developed whilst the process by which the flower becomes double is going forward, fertilization may take place.

50 A stamen, then, is produced by the re-appearance of the self-same organ diminished and refined, which we just before saw expanded as a petal. The truth of the proposition put forward above is hereby again confirmed; and our

attention becomes still more closely riveted on this operation of alternate contraction and expansion, by means of which nature at length attains her object.

[. . .]

XVIII. Recapitulation

112 It is my wish that this attempt to explain the metamorphosis of plants, may not only contribute something towards the solution of this problem, but may give occasion to further investigations and results. The observations on which it is grounded, which were made at different times, have been collected and arranged by Batsch in his 'An leitung zur Kenntniss und Geschichte der Pflanzen' and it will soon appear whether the step we have taken has brought us any nearer to the truth. Let us now review as briefly as possible the leading points in the foregoing essay.

113 When we consider the indications of vital powers existing in plants, we find them manifesting themselves in two different ways; first, by growth during the development of the stem and leaves; secondly, by reproduction effected in the flower and fruit. When we narrowly watch the growth of a plant, we see that as it mounts upwards from node to node, and from leaf to leaf, a kind of reproduction is going forward, differing from the sudden reproduction effected in the flower and fruit, inasmuch as it is a series of successive and distinct developments. This power of gradual growth by the production of buds, is most closely related to that which effects reproduction at once. Under different circumstances a plant may, on the one hand, be forced continuously to produce leaf-buds, or, on the other, to develope the flower. The former result is produced by an accumulation of crude juices; the latter by the preponderance of the subtile powers latent in the plant.

114 The manner in which the two different kinds of reproduction take place, has been indicated by the application of the term successive to reproduction by leaf-buds, whilst we spoke of reproduction by the flower and fruit as sudden. A plant, whilst it is producing leaf buds, increases more or less in size, it developes a stalk or stem, the nodes are generally separated by perceptible intervals, and leaves expand in all directions. But, on the contrary, when a plant produces the flower, all the parts become contracted; increase in height and breadth has ceased, and all the organs, now in an extremely contracted state, are developed in close proximity.

115 But whether a plant produces leaf-buds, flower, or fruit, it is still the selfsame organ which is carrying nature's laws into effect, though performing different offices, and disguised under different forms. The same organ which on the stem expands as the leaf, exhibiting every variety of form, is contracted in the calyx, again expands in the petal, and is once more contracted in the stamens and pistils, to expand for the last time in the fruit.

116 This operation of nature is combined with another; by means of which different organs are assembled round a common centre, in a definite number and order; subject however to variation in many flowers, and under certain circumstances. [. . .]

119 Now in the same way as we have endeavoured to deduce all the apparently different organs of a plant, whether producing buds or flowers, from one and the same organ, – namely, the leaf, which is usually developed at the nodes; we have farther ventured to refer to the same origin, the fruit (seed-vessel), within which the seeds lie safely enclosed.

120 It was obviously necessary to adopt some general term, by which to indicate the one organ which we see metamorphosed under so many different forms; and which we could also employ in comparing these variations with each other. The thing to be now aimed at is to keep habitually in view the two contrary directions, if we may so speak, in which these variations are developed. For we may say with equal truth that a stamen is a diminished petal, or that a petal is an expanded stamen; that a sepal is a diminished stem-leaf in a more refined condition, or that a stem-leaf is a sepal in a state of expansion occasioned by crude juices.

121 Thus also it is immaterial whether we speak of the stem, as the flower and fruit in a state of extension, or whether, as above, we regard the flower and fruit as a shortened stem.

Notes

1 'Occasionally, however, the cotyledons are lobed or notched at their margins, as in the Geranium; while at other times they possess hairs on their surface, as in Gossypium; or little vesicular glands, as in Myrtles, etc. These instances do but afford further proofs of the identity between the cotyledons and the leaves. For a full account of the homologies of these organs, see De Candolle, 'Organographie Végétale', vol. ii. p. 97'.

2 'Similar instances of the close similarity that exists between the leaves, and the bracts constituting an involucre, may be seen in many Umbelliferous plants; as the Carrot, the Anemone, etc. A remarkable instance is figured in the "Gardeners' Chronicle", Sept. 11, 1852, of a Dahlia, in which the bracts, or scales of the involucre, and the paleæ (scales) of the receptacle, instead of retaining their usual membranous state, have all assumed the texture, colour, and veins of leaves; even narrowing their bases into footstalks. So we have seen the bracts of the Plantain, *Plantago major*, presenting in all respects the form and size of the ordinary leaves; and we have observed similar changes in the scales of the strobile of the Hop, and in those of the Larch, *Cryptomeria*, etc'.

45

JAMES SOWERBY [AND JAMES EDWARD SMITH], *ENGLISH BOTANY; OR, COLOURED FIGURES OF BRITISH PLANTS*, VOL. 2: RESEDACAE TO SAPINDACEAE, 3RD ED., ED. JOHN T. BOSWELL SYME

(London: Robert Hardwicke, 1864 [1791])

Order VIII. – Violaceæ

HERBS or shrubs with alternate (rarely opposite) leaves, generally entire or crenate, more rarely laciniate. Stipules leaf-like or scale-like, usually deciduous in the shrubby species. Flowers often solitary, with 2 bracteoles on the pedicels, or arranged in cymes, racemes, or panicles; generally perfect but sometimes polygamous, irregular or regular. Calyx generally persistent, of 5 imbricated sepals. Petals 5, hypogynous or slightly adhering to the calyx. Perfect stamens 5, hypogynous or slightly perigynous; anthers sessile or sub-sessile, disposed in a ring and frequently united; connective often dilated and forming a membranous scale-like appendage beyond the anther-cells, which open by a longitudinal cleft, or very rarely by an apical pore. Staminodes present only in the sub-order Sauvagesieæ. Ovary free, sessile, 1-celled, with parietal placentas, generally 3 in number. Style simple; sometimes thickened or incurved at the apex, with a stigma on the underside; sometimes subulate, with a terminal stigma; more rarely cleft at the apex, or absent, so that the stigmas become sessile. Ovules on each placenta numerous, rarely 1 or 2, anatropous. Fruit a l-celled capsule, opening by as many valves as there are placentas, rarely indehiscent. Seeds with a very short funiculus, and most commonly a hard or leathery testa; albumen fleshy, usually plentiful; embryo in the axis of the albumen generally straight. Cotyledons flat; radicle next the hilum.

Genus I. – Viola. Linn.

SEPALS 5, prolonged backwards beyond the point of insertion, persistent. Petals 5, unequal, spreading, with short claws; the lower one generally larger, and furnished

DOI: 10.4324/9780429355653-51

with a spur at the base; the lateral ones often with a patch of hairs at the base of the lamina. Petals absent in the later flowers of some species. Anthers sub-sessile, forming a ring round the ovary; connective produced into a membranous scale at the apex; the two lowest anthers often spurred at the base, the spurs included in that of the lower petal. Style thickened at the apex, curved downwards, with the stigma on the underside, or dilated into a hollow knob, obliquely truncate at the apex, with the stigmatic portion on the rim on the underside. Capsule opening loculicidally with a spring into 3 valves. Seeds roundish-ovoid, with a hard and generally shining testa, and a short dilated strophiole-like funiculus.

Annual or perennial herbs, or sometimes undershrubs, with alternate leaves and small or foliaceous stipules. Peduncles axillary, 1- (very rarely 2-) flowered. Flowers inclined, blue, purple, white, yellow, or variegated with these colours.

The word Violet is derived from a Greek word ἴου (ion). The ancients held a legend that Violets were the first food of the cow Io, one of Jupiter's mistresses.

Sub-Genus 1. – *Nominium.* Gr. & Godr.

LATERAL petals forming a less angle with the lowest petal than with the upper ones. Style usually clavate and curved, with the stigma on the inner side, or (more rarely) dilated and excavated so as to form an oblique disk at the apex.

Herbs with or without distinct stems. Stipules scarcely leaf like. Petalous flowers produced in spring or early summer, and often barren; the seed being produced from apetalous flowers, which appear later in the season.

[. . .]

Species II. – *Viola odorata.* Linn.

Reich. Ic. Fl. Germ. et Helv. Vol. III. *Viol.* Tab. VIII. Fig. 4498.

Rootstock short, branched, scarred, producing radical leaves and leafless peduncles; scions always present, elongated, resembling runners. Leaves roundish-ovate, deeply cordate at the base, with an obtuse or somewhat acute angle at the apex, crenate, with short hairs on the veins, edges, and petiole. Stipules sub-membranous, lanceolate, with glandular hair-like processes on the margin. Bracts about or above the middle of the peduncles. Flowers fragrant. Style very little enlarged upwards, and slightly hooked at the apex. Capsule globose, downy.

In thickets and shady places. Not uncommon, but probably often of garden origin. It, however, appears to be undoubtedly wild in the South of England.

England, [Scotland,] Ireland. Perennial. Spring and (apetalous flowers) Summer.

Rootstock somewhat fleshy, with ring-like scars left by former leafstalks; from above these scars the scions are produced during the summer. These scions elongate, take root at the extremity (much as in the strawberry plant), and become independent plants, though still connected with the parent; after the second year, however, the intermediate part of the connecting stem decays, the terminal portion becoming the rootstock of the young plant, which thus commences a separate existence. Scapes produced from the axils of the leaves, 2 to 4 inches high, with two small nearly opposite lanceolate denticulated bracts about, or above the middle. Flowers drooping, ½ to ⅝ inch across, bluish purple, lilac, or white. In the white variety the lateral petals are often destitute of the little hairy tuft, and in this state it is the *Viola imberbis* of Leighton, but it seems to be rather a state than a true variety. In the purple flowers the lower petal has darker lines at the base, but in the white these are not present. Spur of the lower petal very short, blunt. Capsule about ⅜ inch in diameter, roundish, almost truncate or retuse at the apex. Plant rather deep green; the petioles, margins and veins of the leaves, the peduncles, the edges of the sepals, and the capsule, with short scattered hairs.

Sweet Violet

FRENCH, *Violette Odorante*. German, *Wohlriechendes Veilchen*

To describe the Violet, or to write of the qualities and useful properties of a flower so surrounded with the atmosphere of poetry and sentiment, is a somewhat difficult task; yet this pretty flower is not only valued for its beauty and delicious scent, but has its reputation in the practice of the healing art even at the present day. A syrup is made from the petals which is a favourite remedy for infantile disorders, and is certainly less dangerous than many which are administered by those who believe in doses. The root is a powerful emetic, and is frequently used to adulterate ipecacuanha. A dose of from forty to fifty grains of the powdered root acts powerfully. M. Boullay has discovered the presence of a principle called violine in all parts of the plant, analagous in external characteristics to the emeta of ipecacuanha, and possessing the same emetic properties. It is an alkaline substance, and forms salts by its union with acids; it is soluble in alcohol, but hardly so in water. The flowers of the Violet yield their purple colour to water, and form a good test for the presence of acids in the same way as litmus is used. In olden times they were used as remedies in many disorders, and were supposed to be especially serviceable to the eyes and in ague. Vitruvius tells us that the flowers were not only used to adulterate or counterfeit the celebrated blue of Athens, but were also employed to 'moderate anger, to cure ague and inflammation of the lungs, to allay thirst, procure sleep, and comfort and strengthen the heart, as well as for cooling plaisters', besides being worn in garlands as a charm against the 'falling sickness' and headaches. Pliny gives a long list of their virtues, affirming they are good for inflammation, cooling to weak eyes, quinsey, swellings, &c., and recommends the blossoms to be worn as garlands for the preservation of the head. The seeds were formerly believed to counteract the effect of a scorpion's sting. 'Violets', says Gerarde, 'have a great prerogative above others, not only because the mind conceiveth a certain pleasure and recreation by smelling and handling those most odoriferous flowers, but also for that very many by these Violets receive ornament and comely grace; for there be made of them garlands for the head, nosegaies, and posies which are delightful to look on and pleasant to smell to, speaking nothing of their appropriate vertues. Yea, gardens themselves receive by these the greatest ornament of all, chiefest beauty and most excellent grace; and the recreation of the mind which is taken hereby cannot be but very good and honest; for they admonish and stir up a man to that which is comely and honest,– for flowers, through their beauty, variety of colour, and exquisite forme, do bring to a liberal and gentle manly mind the remembrance of honestie, comliness, and all kinds of vertues; for it would be an unseemly and filthy thing (as a certain wise man saith) for him that doth look upon and handle faire and beautiful things to have his mind not faire, but filthy and deformed'. He goes on to enumerate the many excellencies of his favourite flower as a medicine: 'Syrup of Violets', says he, 'is good against inflammation of the lungs and brest, against pleurisie and cough'. This belief has not shared the fate of most of our good friend's remedies; it is still given, and in the country is a favourite medicine for coughs and hoarseness. The French make great use of Violets in their confitures and household remedies; and we have seen and partaken of a delicate sweetmeat composed simply of the Violet flower prepared with sugar, yet retaining its delicious perfume. In the neighbourhood of Stratford-on-Avon Violets are largely grown for the purposes of perfume and as a colouring agent. The syrup forms a principal ingredient in the Oriental sherbet; and with this in view, probably, Mahomed asserts that the Violet is as superior to other flowers as he himself claimed to be over the rest of mankind. The association of the Violet with female beauty is of very ancient date; for, long before we read of 'violet-like eyelids', we are told that the Britons used them as a cosmetic; for in a Celtic poem extant they are recommended to be employed, steeped in goat's milk, as a certain mode of increasing female beauty, perhaps by giving the blue tinge of woad to the complexion, then so much admired. Thus we see that it is not alone the external attractions of scent and beauty which have given its charm to the Violet, but a certain notion of its value as a useful plant. Shakespeare alludes to the Violet frequently and variously:

'Violets dim,
But sweeter than the lids of Juno's eyes'.

And again

'Lay her i' the earth;
And from her fair and unpolluted flesh
May Violets spring'.

Violets find a very constant place in churchyards and on the resting-places of the dead, placed there by the hands of those who love to associate the ideas of purity and beauty with departed loved ones. Shelley says,

'Lilies for a bridal bed,
Roses for the matron's head,
Violets for a maiden dead'.

The Violet was a great favourite with the Greeks. Homer, as translated by Cowper, says, –

'Everywhere appeared
Meadows of softest verdure purpled o'er
With Violets: it was a scene to fill
A god from heaven with wonder and delight'.

Athens was noted for its love of Violets, – 'ancient Violet-crowned Athens'. The same epithet was applied to the Muses; and Homer even calls Venus Ιοστφυνον, 'crowned with Violets'. Plutarch says: 'Its exhalations greatly assist in removing affections of the head caused by wine'. The Violet was the appropriate May-day prize bestowed on the troubadour, or the minnesinger of olden times. It was afterwards replaced by a golden Violet; and at Toulouse a society was instituted, which afterwards became the Academy of Floral Games, in which this prize was striven for. The 'true blue' of the Violet has ever been associated with fidelity; as in the old sonnet we have,

'Violet is for faithfulnesse,
Which in me shall abide,
Hoping likewise that from your heart
You will not let it slide'.

The sweet Violet and the sweet bird of song are associated not alone in poetic fancy, for it is a curious fact that where the Violet grows there may be heard the nightingale. Milton says:

'In the Violet-embroidered vale, The love-lorn nightingale Nightly her sad song mourneth well'. And who amongst us will not welcome the first Violets of the early spring, 'Gleaming like amethysts in the dewy moss'?

46

WILLIAM JACKSON HOOKER, *THE BRITISH FLORA; COMPRISING THE PHÆNOGAMOUS, OR FLOWERING PLANTS, AND THE FERNS*

(London: Longman, Rees, Orme, Brown, & Green, 1830)

Introduction

THE object which the Author proposed to himself, in preparing a new Flora of the British Empire, was of a twofold nature: 1stly, to provide the young Student with a description of our native plants, arranged according to the simplest method; and 2dly, to afford the more experienced Botanist, a manual, that should be useful in the field as well as in the closet. In regard to the first object, the experience of nearly an hundred years has proved to every unprejudiced mind, that no system has appeared which can be compared to that of the immortal Swede, for the facility with which it enables any one, hitherto unpractised in Botany, to arrive at a knowledge of the Genus and species of a plant. The Linnæan method is, therefore, here adopted.

It has been the opinion of the author, and of many of his friends, that, in most of the Floras hitherto published, however excellent in other respects, either too much or too little space has been devoted to the generic and specific descriptions and synonyms; in the one case, swelling the book to a size, which induces both expense to the purchaser, and difficulty in consulting the several volumes; in the other, reducing the technical characters to the shortest possible compass, so that they can scarcely be made available, except to those who are already partially acquainted with the plant under examination, or with some of its near allies. Between these extremes, the author has attempted to steer a middle course, by giving diagnostic remarks where, and where only, they have appeared to him necessary; confining the synonyms, with few exceptions, to those of the writer who first described the

DOI: 10.4324/9780429355653-52

plant, to a good figure, and a reference to a single Flora of England and of Scotland; and by adopting such an arrangement of the subject matter as would best occupy every portion of the page, without rendering it obscure to the reader. How far his endeavours have proved successful, must be left to the experience and judgment of those for whose use the work is particularly intended. Should it prove serviceable in advancing the cause of Botanical science in this country, the end which was fondly anticipated at the commencement of the undertaking will be fully accomplished. During the progress of the labour, it occurred to the Author that he could give an additional interest to the volume by subjoining short notices of the uses and properties, or some little historical remarks relative to the species, the origin of the generic names, &c.: thereby recommending the pursuit of which it treats, to the attention of the many, who are still apt to look upon Botany, as a dry and profitless employment, a system of hard words, destitute of any real utility to mankind.

[. . .]

Nor let it be supposed that the author is advocating the cause of an Artificial System, to the exclusion of a natural one; for if anyone can be more alive than another to the real advantage derivable from a knowledge of the characters of plants, when naturally combined, it is assuredly he, whose duty it is to teach the Science to those who are destined for the profession of medicine. The former method will soon enable the student to ascertain the Foxglove, the Cinchonas, the Squill, and innumerable other plants of which he would be ashamed to be ignorant: but the study of the latter alone will put it in his power to extend his inquiries, and with a prospect of success, to analyse other plants of the same Natural Order, among which he may expect to find the same or more powerful principles than what are hitherto known to us. This alone lays open a wide field of usefulness to the Botanist and the Physician; and with the view to so desirable an object, the name of the Natural Order to which each Genus belongs is mentioned in the following pages; and in the Appendix will be found a complete list of those Orders as far as British Botany is concerned, together with an enumeration of the Genera belonging to them. That the remarks upon the Natural Orders could not, owing to the limited nature of the present work, be further extended, is the less to be regretted, now that Mr. Lindley has published his Synopsis of the British Flora, arranged according to the Natural Orders [. . .]

The labour of compiling the Flora of a country by a careful examination and comparison of specimens themselves, whether in a living or dried state, can only be appreciated by those who have been engaged in an employment of the same kind. The collecting of materials, indeed, in their native hills and vallies, upon the sea shore, in the woods, and among the majestic alpine scenery with which the northern parts of our island, eminently abound, generally in the society of friends of a congenial taste, or students full of ardour and enthusiasm, has been a very delightful occupation, especially when taken in conjunction with 'anticipations of the pleasure we may have to bestow on kindred minds with our own, when sharing with them our discoveries and our acquisitions'. And the task of describing them has, in the present instance, been considerably lightened by the able assistance afforded by many of the most able Botanists of our country.

[. . .]

III. TRIANDRIA. 3 *Stamens.*
ORD. I, MONOGYNIA. 1 *Pistil.*
Flowers superior. *Gen. 1–5.*
1. VALERIÁNA. Linn. Valerian.

Cal. a thickened margin to the top of the germen, at length unfolding into a feathery *pappus*. *Cor.* monopetalous, 5-cleft, gibbous or spurred at the base. *Fruit* l-seeded, crowned with the feathery *pappus.– Nat. Ord.* VALERIANEÆ. *DeC.* – Name from *Valeo*, to be powerful, on account of the medicinal effects.

1 V. *rúbra*, Linn. (red Valerian); corolla with a long spur, *stam.* l, *leaves* ovato-lanceolate. *Engl. Bot.* t. 1532. *Hook. Scot.* i. p. 14. *Engl. Flora*, v. i. p. 42.

Chalk pits and old walls: but probably originally the outcast of gardens. Chalk-pits in Kent apparently wild, and certainly very abundant. Its native country is the south of Europe. Fl. June–Sept. One foot or more high, glabrous, somewhat glaucous, entire or slightly toothed. *Leaves*, as in all the species of this and the following genus, opposite. *Flowers* fine deep rose-colour, arranged in numerous uni-lateral corymbose spikes. This constitutes Centranthus of *DeC.* and assuredly is a good genus, which I have retained here on account of its affinity with the Valerians. For, if separated from them, it must be referred to another Class, MONANDRIA.

[. . .]

2 V. *officinalis*, Linn. (*great wild Valerian*) *corolla* gibbous at the base, *leaves* all pinnated, *leaflets* lanceolate nearly uniform serrated, *Engl. Bot.* t. 698. *Hook.* Scot. i. p. 15. *Engl. Fl.* v. i. p. 43.

Ditches, sides of rivers and moist woods, abundant. *Fl.* June, July. *Roots* tuberous, warm, aromatic and employed in medicine, as those of the φου of Dioscorides, *V. Dioscoridis Sm.* which is not the *V. Phu* of Linn. Cats are very fond of these roots, and the scent attracts rats. The leaves are much used by the poor as an application to fresh wounds; hence the plant has received the name of *all-heal*. Whole plant 2–4 feet high; *Stems* striated. Lower *leaves* on long footstalks. *Flowers* pale flesh-colour.

[. . .]

(III. CORIANDRUM *TRIBE.*)
49. CORIANDRUM. Linn. Coriander.

Cal. of 5 teeth. *Pet.* obcordate, point inflexed, outer ones radiant, bifid. *Fruit* globose. *Carpels* with 5 primary ridges depressed and wavy, of which the 2 lateral ones are placed in front of an accessory margin to the inner face: the 4 secondary ridges more prominent and carinated. *Interstices* without *vitlæ*, the inner face of the carpel having 2 *vitlæ*. *Seed* hollowed in front, covered by a loose membrane – Universal *involucre* 0. Partial on one side. *Carpels* cohering, separated with difficulty. Name from χορις, *a Bug*, in allusion to the intolerably fetid smell of the bruised foliage. Sir J.E. Smith retains in this genus the *Biforis* of Spreng. which has a fruit of 2 lobes.

1. C. *satívum*, Linn. (common Coriander). *Engl. Bot.* t. 67. *Engl. Fl.* v. ii. p. 67.

   Fields and waste places, about Ipswich and in Essex, in the neighbourhood of which it had formerly been cultivated. Fl. June:– This is the only true species of the genus, and is well known as a medicinal plant. The *seeds* are highly aromatic, and sold enveloped in sugar as *Coriander comfits*. *Stem* erect, leafy. *Lower leaves* bipinnate, the *pinnæ* pinnatifid with broad, wedge-shaped, toothed segments: the upper leaves gradually more compound, with the segments very narrow and linear, those of the uppermost leaves almost setaceous. *Fruit* very curious: each *carpel* is hemisphærical; on its inner and flat side having a projecting margin, which combines with the opposite one so as to leave no line or furrow between the two, and they form a complete little ball or globe; having, however, when quite ripe, 10 obscure elevated lines or ribs.

[. . .]

CLASS XXIV. CRYPTOGAMIA (*part of*).
Stamens and pistils not visible.
ORD. I. FILICES. *Ferns.*

Fructification only of one kind upon the same species. *Çapsules* spiked or racemed, or generally collected into *clusters* of various shapes (*sori*) mostly upon the back or margin of the *frond*, naked or covered with an *involucre*; with or without an elastic *ring*. *Seeds* minute. Perennial *plants*, in perfection during the greater part of the year, especially in the summer months.

[. . .]

5. CISTÓPTERIS. *Bernhardi*. Bladder-fern. (Cystea, Sm.)

Sori roundish. *Involucre* inserted, by its broad cucullate base, at the underside of the *sorus*, opening by a lengthened free extremity, which points towards the apex of the frond.– Name; χιστή, a *little box*, and πτεϑις, a *Fern*.– I have taken a different view of the structure of the *Involucre* from Sir J.E. Smith, and I trust a correct one. Its texture is thin and delicate and altogether widely different from *Aspidium*. Species with the above character exist in North and South America, as well as in Europe.

1 C. dentáta, (toothed Bladder-Fern); *fronds* bipinnate, *pinnæ* lanceolate, *pinnules* ovate obtuse bluntly and unequally toothed rarely subpinnatifid, *rachis* winged.

a *fronds* oblongo-lanceolate. Cystea dentata, Engl. Fl. v.iv.p. 300. Aspidium dentatum, Sw. Hook. Scot. ii. p. 155. Cyathea dentata, Engl. Bot. t. 1588. Polypodium dentatum, Dicks.

b *fronds* oblongo-ovate. Cystea angustata, Engl. Fl. v. iv. p. 301. Polypodium rhæticum, Dicks. Cyathea fragilis, B. Sm.

North of England and Wales, abundant. Scotland, Mr. Dickson. Ben Lawers. This is certainly the most common species of the Genus in Wales, where it seems to hold the place that C. fragilis does in Scotland, and from which I think it is distinct. I possess specimens of Cystea dentata and C. angustata from Mr. Dickson, and I can find no difference, except that the latter is a little broader in the frond than the former, and perhaps the pinnules are rather more divided, so as to approach nearer to the following species. This is the same as the Aspidium tenue of American Botanists.

2 C.frágilis, Bernh. (brittle Bladder-Fern); *fronds* bipinnate, *pinna* lanceolate, *pinnules* ovato-lanceolate deeply pinnatifid, *segments* ovate or lanceolate toothed, *rachis* winged. Cystea fragilis, Engl. Fl. v. iv. p. 298. Aspidium fragile, Sw. Hook. Scot. i. p. 155. Cyathea fragilis, Engl. Bot. t. 1587.

Rocks and walls, in the mountainous parts of Great Britain. Cheddar, Somersetshire. Rev. Mr. Berkeley. Most abundant in Scotland. It will be seen that this differs from the preceding, principally in its more pinna and narrower segments.

3 C. alpina, Desv. (laciniated Bladder-Fern); *fronds* tripinnate, *pinnules* confluent ovato-oblong pinnatifid rather spreading, the *segments* broadly and shortly linear obtuse, with 2 or 3 blunt erect teeth, *rachis* winged. Aspidium alpinum, Sw. Willd. Polypodium alpinum, Jacq. Ic. v. iii. t. 642. (excellent). Cystea regia, Engl. Fl. v. iv. p. 302. Cistopteris regia, Desv. Cyathea regia, Forst. Fl. Br. p. 1140. Cyathea incisa, Engl. Bot. t. 163. Aspidium regium, Hook. Scot. ii. p. 155. Polypodium regium, Linn?

On a wall (since destroyed) at Low Layton, Essex, plentifully; Mr. T. F. Forster. Having received authentic specimens of the Layton plant, from Mr. E. Forster, and compared them with continental ones, and with figures and descriptions of Aspidium alpinum, especially the plates of Jacquin and

Schkuhr, I can without hesitation, pronounce them to be identical. But I dare not introduce the Welsh, nor the Scotch station; believing, as I do, that the C. fragilis has there been mistaken for it. The species is most distinct, the fronds being more divided even than the last species, the divisions linear, with few and very blunt teeth. The fructification is exactly that of a Cistopteris.

47

JOHN LINDLEY, *AN OUTLINE OF THE FIRST PRINCIPLES OF BOTANY*

(London: Longman & Co., 1830)

Preface

THE want of some English work on Botany, at once of a mere elementary character, and comprehending all the more important points of the science, has given rise to the publication of the following pages. The propositions which they contain are such as it is of the most indispensable importance for a student to understand; and they all appear to be strictly deducible either from the facts recorded by observers worthy of confidence, or from the experience of the author. They form the basis of the Lectures delivered by him in the University of London, and are purposely divested of illustrative or explanatory matter; his only object having been to reduce the first principles of Botany to their simplest form.

[. . .]

An Outline of the First Principles of Botany

1 PLANTS are not separable from animals by any absolute character; the simplest individuals of either kingdom not being distinguishable by our senses.

2 Animals are for the most part incapable of multiplying by mechanical or spontaneous division of their trunk.

3 Plants are for the most part congeries of individuals, multiplying by spontaneous or artificial division of their trunk or axis.

4 Generally speaking, the latter are fixed to some substance from which they grow, are destitute of locomotion, and are nourished by absorption through their cuticle (38).

5 Plants consist of a membranous transparent tissue, formed by a combination of oxygen, hydrogen, and carbon, to which azote is occasionally superadded.

6 Their tissue appears under four forms, viz. cellular tissue, woody fibre, spiral vessels, and ducts. These are called elementary organs.

DOI: 10.4324/9780429355653-53

I. Elementary Organs

7 Of these CELLULAR Tissue (Tela cellulosa, *Lat.*; Tissu cellulaire, *Fr.*; Pulp and Parenchyma, *of old writers*; Zellgewebe, *Germ.*) is the only form universally found in plants; the other forms are often either partially or entirely wanting.

8 Cellular tissue is composed of transparent vesicles, the sides of which are not perforated by visible pores (17).

9 Each vesicle is a distinct individual, cohering with the vesicles with which it is in juxtaposition.

10 Therefore, the apparently simple membrane that divides two contiguous cells is in fact double.

11 If the adhesion of the contiguous cells be imperfect, spaces will exist between them. Such spaces are called *intercellular passages*.

12 The vesicles of cellular tissue, when separate, are round or oblong; when slightly and equally pressed together, they acquire an hexagonal appearance; stretched lengthwise, they become prismatical, cylindrical, or fusiform.

13 Cellular tissue, the vesicles of which fit together by their plane faces, is called *parenchyma*.

14 Cellular tissue, the vesicles of which are elongated and overlie each other at the extremities, is called *prosenchyma*.

15 *Parenchyma* constitutes all the pulpy parts of the medulla or pith (82), the medullary rays (113), a portion of the bark (102), and all that is interposed between the veins of the leaves and of other appendages of the axis.

16 *Prosenchyma* is confined to the bark and wood, in which it is mixed with woody fibre (19).

17 The function of the cellular tissue is to transmit fluids in all directions; the membrane of which it is composed is, therefore, permeable, although not furnished with visible pores (8).

18 It has been supposed that the cellular tissue is self-productive, one vesicle giving birth to many others.

19 WOODY FIBRE (Vasa fibrosa, *Lat.*; Tissu cellulaire allongé, *Fr.*; Clostres, *Fr.*; Baströhren, *Germ.*) is tissue consisting of elongated tubes tapering to each end, and, like the vesicles of cellular tissue, imperforate to the eye.

20 It may be considered a form of the cellular tissue itself, to which it is frequently referred.

[. . .]

38 The CUTICLE is an external layer of parenchyma, the cells of which are compressed, and in a firm state of cohesion.

39 The spaces seen upon the cuticle, when examined by a microscope, represent these cells.

40 It is, therefore, not a peculiar membrane, but a form of cellular tissue.

41 It is spread over all parts of plants, except the stigma (345).

42 The mass of cellular tissue lying beneath the cuticle of the bark is called the epidermis.

43 The cuticle is often furnished with stomata.
44 STOMATA are oval spaces lying between the sides of the cells, opening into intercellular cavities in the subjacent tissue, and bordered by a rim, the nature of which is not well known.
45 It is not improbable that this appearance of a rim is due to the juxtaposition of two elastic vesicles, closing up or opening the aperture on which they lie, according to circumstances.
46 Stomata are found abundantly upon leaves, particularly on the lower surface of those organs; occasionally upon all parts that are modifications of leaves, especially such as are of a leafy texture; and on the stem.
47 Stomata have not been found upon the roots, nor on colourless parasitical plants, nor the submersed parts of plants, nor on cellular plants destitute of ducts; they are rare, or altogether absent from succulent fruits, and from all parts in a state of anamorphosis.
48 Any part in which there is an unusual degree of cellular developement, is said to be in a state of *anamorphosis*,
49 The function of stomata is to facilitate evaporation.

II. Compound Organs

50 From peculiar combinations of the elementary organs are formed the compound organs.
51 The compound organs are the axis (52) and its appendages (158).
52 The Axis may be compared to the vertebral column of animals.
53 It is formed by the developement of an embryo, or of a leaf-bud.
54 An embryo is a young plant, produced by the agency of sexes, and developed within a seed.
55 A leaf-bud is a young plant, produced without the agency of sexes, and either enclosed within rudimentary leaves called scales, or naked.
56 Seeds propagate the species.
57 Leaf-buds propagate the individual.
58 All the phenomena connected with the growth of plants are caused by an inherent vital action.

48

LADY KATHERINE SOPHIA KANE, *THE IRISH FLORA; COMPRISING THE PHÆNOGAMOUS PLANTS AND FERNS*

(Dublin: Hodges and Smith, and London: Longman, Rees and Co, 1833)

Preface

It is a matter of surprise that Ireland should have remained so long without any descriptive catalogue of its plants, worthy of notice. It has been remarked, that when England and France had their provincial Floras, the Botany of this island was as much unknown as that of an island in the Pacific; although its peasantry possessed a very considerable knowledge of plants, which is proved by the fact that all the Irish herbals which were published before the beginning of the 18th century, were furnished with an Irish name for almost every plant they contained; but among its enlightened inhabitants it has remained almost a sealed book, while men of science have been occupied investigating other countries not possessing half its richness in vegetable productions [. . .]

The same want of knowledge particularly distinguishes the early attempts that were made to throw light on our Flora; it is only through the medium of English works, that the Irish could have been instructed in the Botany of their country, 'which nurtures on its mountains the *Andromeda daboecii*, the *Dryas octopetala*, and the *Saxifraga umbrosa* of the Alps; and on the borders of its enchanting lakes, the *Arbutus unedo* of Greece'. The first attempt that was made towards an Irish Flora, was by Caleb Threlkeld, an English Dissenting Minister and Physician, who, in 1727, published his *Synopsis Stirpium Hibernicarum*, or a Catalogue of indigenous Plants, growing in the environs of Dublin; it contains 535 plants with their Latin, English, and Irish names, and a minute account of their economical and medicinal virtues, interspersed with quaint moral and political reflections, which should have no place in a work of the kind; but without any scientific characters to distinguish the plants, except what the long nomenclature he adopts affords. In 1735, Dr K'Eogh published a similar treatise under the following pompous title,

DOI: 10.4324/9780429355653-54

Botanologia Universalis Hibernica, or 'A General Irish Herbal, calculated for this kingdom, giving an account of the herbs, shrubs, and trees, naturally produced therein, in English, Irish, and Latin; with a true description of them, and their medicinal virtues and qualities'. He enumerated about 600 plants, arranged alphabetically, each plant has its English, Irish, and old Latin name, after which follow its numerous virtues; seldom accompanied by any botanical description. The book is useless and now almost unknown. The next notice of our Flora appeared in a more respectable manner: Dr Rutty in his History of Dublin, published in 1772, describes 377 plants, found in the county, classed into those which are used for food, in domestic economy, in dying, and those that are poisonous. But still there was no guide to the student in the fields – no Irish work to enable the botanist, by short determinate characters, to discover the name of an unknown plant, to which purpose a Flora alone can be applied.

[. . .]

For the publication of this volume little apology is necessary; that an Irish Flora has long been a desideratum is true, and it is equally true that the desideratum has now been very imperfectly supplied; but it has at least one value, that of affording the learner a scientific description of the plants of our island without the confusion of additional genera and species, the study of which through the British Floras will always retard his progress. The determinate generic and specific characters are, in many instances, copied verbatim from Smith, and the secondary are selected from the long and minute descriptions of the English Flora, and from Hooker's British Flora. The habitats which the work contains are almost all afforded by Mr White of the Glasnevin Botanic Garden, and their authenticity is placed beyond all doubt by his well known accurate and penetrating researches.

[. . .]

This little work is presented to the public without much fear of its failure, or expectation of its success; a book that can neither boast of originality of invention nor elegance of style, will pass equally unnoticed by censure or applause; but if it could clear away a single difficulty in the student's path, or if, by affording facility for practice, it would induce any one to commence a pursuit which might possibly soften down some of the asperities of life, the author would feel every wish fulfilled, every object for which it was undertaken accomplished.

[. . .]

289.TRIFOLIUM

1 T. *officinale. Melitot.* Clusters unilateral; legume prominent, acute, transversely wrinkled, hairy, with 2 seeds; stem erect; stipules awl-shaped. Eng. Flor. v. 3, p. 295. Eng. Bot. v. 19, t. 1340.

Stem 2 or 3 feet high, branched, leafy; leaves obovate, narrow; clusters rather long, in axillary stalks; flowers yellow, small, numerous.

Hab. Hedges; Finglas quarries; on the lands of Abbyville, Baldoyle, and Kilbarrack; Feltrum hill, and Rush.

2 T. *ornithopoides. Bird's-foot Trefoil.* Flowers about 3 together; legume prominent, 8 seeded, twice as long as the calyx; stems reclining. Eng. Flo. v. 3, p. 298.

Stems lying on the ground, generally simple; leaflets inversely heart-shaped; flowers about 3, on axillary stalks, reddish, long, and slender.

Hab. Gravelly pastures.

3 T. *repens. White Trefoil.* Heads of flowers globose; flowers somewhat stalked; legume within the calyx, 4-seeded; stems creeping, solid. Eng. Flo. v. 3, p. 299. Eng. Bot. v. 25, t. 1769.

Stems prostrate, branched, leafy; leaflets roundish, with a whitish curved stripe in the middle; flowers white, numerous, small, in a dense head.

Hab. Pastures and road side, common.

4 T. *pratense. Honey-suckle Trefoil.* Heads dense; stems ascending; petals unequal; calyx hairy, 4 of its teeth equal; stipules ovate, bristle pointed. Eng. Flo, v. 3, p. 302. Eng. Bot. v. 25, t. 1770.

Stem about a foot high; leaflets rather large, elliptical, with a pale spot; flowers small, numerous, purple, in dense axillary heads.

Hab. Meadows, common.

5 T. *medium. Zigzag Trefoil.* Heads lax; stems zigzag and branching; pet. Nearly equal; stipules tapering, converging; 2 upper calyx teeth rather the shortest. Eng. Flo. v. 3, p. 302. Eng. Bot. v. 3, t. 190.

Habit very like the last, the stipules are linear, the heads rather larger, and the cal. less hairy.

Hab. Elevated dry pastures; land above the Little Dargle, and about Roebuck.

6 T. *maritimum. Teasel-headed Trefoil.* Spikes or heads ovate, somewhat hairy; stipules lanceolate, erect; calyx teeth after flowering, dilated, leafy, and spreading; leaflets obovate, oblong. Eng. Flo. v. 3, p. 303. Eng. Bot. v. 4, t. 220.

Stems spreading or recumbent, branched; leaflets dark green, rather narrow, hairy on both sides; stipules very long and narrow, flowers pale purple.

Hab. Salt marshes; in the island of Lambay.

49

GEORGE LUXFORD, *A FLORA OF THE NEIGHBOURHOOD OF REIGATE, SURREY, CONTAINING THE FLOWERING PLANTS AND FERNS*

(London: John Van Voorst, 1838)

Preface

It has been my aim in the following pages, to exhibit the botanical productions of a portion of the county of Surrey, equally interesting from the beauty of its scenery and its numerous historical associations. I have also endeavoured to render the work a guide to the localities of the plants growing in the district of which it treats; and the only merit it can claim is that of being a faithful record of my own researches in the neighbourhood of Reigate: so that however trifling it may be considered in a scientific point of view, I trust a certain value will attach to it, as being,– not a mere compilation from the labours of others,– but the result of actual observation. I believe I have not admitted a single plant which I have not seen growing in the recorded localities, with the exception of the very few collected by friends of undoubted veracity, whose names are invariably given as authorities. I possess specimens of all the plants so collected; and have, whenever it was practicable, verified the observations of my friends.

[. . .]

Sir W.J. Hooker has truly observed that the collection of materials for a Flora among their native hills and valleys, is a very delightful occupation; for this I was kindly permitted to use opportunities such as fall to the lot of but few persons, placed in situations similar to that which I then held. The arrangement of the materials then collected, I afterwards found to be a most agreeable relaxation from the hurry and bustle incident to business in a crowded manufacturing town. During the progress of this pleasing employment, on meeting with plants collected in well-known spots in my former 'daily walks and ancient neighbourhood', my thoughts were often wafted far away from the smoke and noise of such a situation, to the pure air and quiet of the 'lanes and alleys green, dingles and bushy dells',

DOI: 10.4324/9780429355653-55

so bountifully decked with these 'wildings of nature', and to the 'summers of old' when they were collected.

[. . .]

MELICA. Melic-grass.

M. *uniflora.* Wood Melic-grass.

Shady hedge-banks, as in the lanes about the Park.

M. *cærulea.* Purple Melic-grass.

Boggy parts of Reigate Heath. Most abundant in the bogs on Tilgate Forest, Sussex. The Rev. H. Davies observes', It is remarkable that this plant has perfectly withstood the sulphureous fumes of the copper-works at Amlwch, where every other vegetable, within a certain distance, even the crustaceous lichens, have been destroyed; and from its extreme toughness, it seems designed for some uses, which we do not know to have been hitherto made of it. It is likewise most easily propagated, as it is naturally produced in cold steril soil'. *Welsh Botanology*, p. 9.

[. . .]

OXALIS. Wood Sorrel. OXALIDEÆ, De Candolle.

Acetosella. Common Wood Sorrel.

Hedge-banks, in the Park and elsewhere. Sir J. Smith observes, 'Few of our wild flowers are more elegant. The flowers are solitary, drooping, bell-shaped, white or purplish, always streaked with fine branching purple veins'. It is not improbable that this plant is the original Irish Shamrock, as Mr. Bicheno has ingeniously argued, in a paper read before the Linnean Society a few years ago.

[. . .]

AGROSTEMMA. Cockle.

A. *Githago.* Corn Cockle.

Corn-fields. A very showy plant, but one of the pests of the farmer: it should be pulled up by hand, before flowering.

[. . .]

NYMPHÆA. White Water- Lily.

N. *alba.* Great White Water- Lily.

My specimens were collected in the mill-pond, at Ifield, Sussex, by The Rev. C.T. Smith; in whose company I first saw this magnificent plant, in a secluded pool on Furnace Farm, Worth, Sussex, where it was growing in the greatest plenty. In the English Flora the flowers are said to be without scent; I have found them, on the contrary, to give out a powerful and exceedingly disagreeable odour.

[. . .]

CARDAMINE. Lady's Smock Bitter Cress.

C. *amara*. Bitter Lady's Smock. 1000. p. 4–6.

Abundant in an alder copse at Littleton. The violet coloured anthers will serve to distinguish this species.

C. *pratensis*. Meadow Lady's Smock. Cuckoo Flower. 776. p. 5.

Damp meadows. Sir J.E. Smith says of the flowers of this plant, 'They come with the Cuckoo, whence one of their English, as well as Latin, names; and they cover the meadows as with linen bleaching, which is supposed to be the origin of the other, now extended to the whole genus. They are associated with pleasant ideas of spring; and join with the White Saxifrage, the Cowslip, Primrose and Harebell, to compose many a rustic nosegay'.

50

ANNA WORSLEY RUSSELL, *CATALOGUE OF PLANTS FOUND IN THE NEIGHBOURHOOD OF NEWBURY*

(npl: np, 1839)

THE neighbourhood of Newbury presents unusual attractions for the lover of Botany, whose researches are sure to be repaid by a large proportion of the rare, as well as beautiful, among nature's flowery treasures. The great diversity of soil – the combination of chalk, bog, sand, and gravel, together with what may be called the almost mountainous character of some of its features, afford a remarkable variety of plants, among which the interesting family of Orchideæ holds a conspicuous station. The following list has been drawn up after a very short acquaintance with the neighbourhood, and must necessarily therefore be imperfect: it will, nevertheless, be found to bear ample testimony to the truth of the foregoing remarks.

[. . .]

Hexandria, Class VI

Peplis Portula (Water Purslane)

North Heath.

Galanthus nivalis (Snowdrop)

Wild and abundant in a hedge-row near Eling, Hampstead Norris; on the bank in Richard Barns' orchard, and a few other places – J.L. Lane leading to Enborne; meadow near the Bell, Boxford – J.B.

Narcissus Pseudo-narcissus (common Daffodil)

In an orchard near the churchyard, Hampstead; in Barns' orchard; in Mrs Watts' orchard, Blewbury, and many other places, but not plentiful – J.L. Meadows at Highclere and East Woodhay – J.B.

Convallaria majalis (Lily of the Valley)

Enborne, Hampstead, and Pea woods – J.B.

DOI: 10.4324/9780429355653-56

Allium vineale (Crow garlic)

In the corn-fields about Aston, amongst wheat; in the wet lands between Moreton and Mackney. Quite a nuisance and very injurious – J.L. Cornfields at High Woods and North Heath.

Allium ursinum (broad-leaved Garlic or Ramsons)
Allium schœnoprasum (Chive Garlic)

In a meadow called Horse Croft, at Blewbury, and in some other moist meadows – J.L.

Ornithogalum pyrenaicum (spiked Star of Bethlehem)

Plentiful in Ashridge wood, Ilsley; in the woods between Compton and Hampstead Norris; in Beech wood – J.L. Near Langley – Dr Lamb.

Hyacinthus non-scriptus (wild Hyacinth or Hare-bell)
Muscari racemosum (Starch Grape-Hyacinth)

Watson's Botanist's Guide.

Narthecium ossifragum (Lancashire Bog-Asphodel)

Snelsmore and Woodhay commons – J.B.

Fritillaria Meleagris (common Fritillary)

In a meadow called Thorn Croft, Blewbury; in a meadow by Burghfield bridge. A beautiful and rare plant – J.L.

Juncus conglomeratus (common Rush)
Juncus acutiflorus (sharp-flowered jointed Rush)
Juncus lampocarpus (shining-fruited jointed Rush)
Juncus obtusiflorus (blunt-flowered jointed Rush)

Cold-ash common.

Juncus uliginosus (lesser Bog jointed Rush)

Snelsmore common.

Juncus bofunius (Toad Rush)
Juncus squarrosus (Heath Rush)
Rumex Hydrolapathum (great Water Dock)
Rumex crispus (curled Dock)

Common in damp shady places about Blewbury and Hagbourn – J.L.

Rumex sanguineus (bloody-veined and green-veined Dock)

Before some of the door-ways and in the gardens at Blewbury – J.L.

Rumex obtusifolius (broad-leaved Dock)

Common about Blewbury and Hampstead Norris, in moist places – J.L.

Rumex acetosa (common Sorrel)
Rumex acetosella (Sheep's sorrel)
Triglochin palustre (Marsh Arrow-grass)

Bagnor Marsh.

Colchicum autumnale (common Meadow-Saffron)

In Ashridge wood, East Ilsley; in a hedge-row near Hatch Gate, Hampstead Norris. Not common – J.L. In a meadow near Burghclere parsonage, very plentiful – J.B.

51

CHARLES DARWIN, *JOURNAL OF RESEARCHES INTO THE GEOLOGY AND NATURAL HISTORY OF THE VARIOUS COUNTRIES VISITED BY H.M.S.* BEAGLE *UNDER THE COMMAND OF CAPTAIN FITZROY, R.N. FROM 1832 TO 1836*

(London: Henry Colburn, 1839)

Chapter 6, September 1833, Bahia Blanca to Buenos Ayres

NEAR the Guardia we find the southern limit of two European plants, now become excessively common. The Fennel in great profusion covers the ditch banks in the neighbourhood of Buenos Ayres, Monte Video, and other towns. But the cardoon (*Cynara cardunculus*)[1] has a far wider range; it occurs in these latitudes on both sides of the Cordillera, across the continent. I saw it in unfrequented spots in Chile, Entre Rios, and Banda Oriental. In the latter country alone, very many (probably several hundred) square miles are covered by one mass of these prickly plants, and are impenetrable by man or beast. Over the undulating plains where these great beds occur, nothing else can live. Before their introduction, however, I apprehend the surface supported as in other parts a rank herbage. I doubt whether any case is on record, of an invasion on so grand a scale of one plant over the aborigines. As I have already said, I nowhere saw the cardoon south of the Salado; but it is probably that in proportion as that country becomes inhabited, the cardoon will extend its limits. The case is different with the giant thistle (with variegated leaves) of the Pampas, for I met with it in the valley of the Sauce. According to the principles so well laid down by Mr Lyell, few countries have undergone more remarkable changes, since the year 1535, when the first colonist of La Plata landed with seventy-two horses. The countless herds of horses, cattle, and sheep, not only have altered the whole aspect of the vegetation, but they have almost banished the guanaco, deer, and ostrich. Numberless other changes must likewise have taken

DOI: 10.4324/9780429355653-57

place; the wild pig in some parts probably replaces the peccari; packs of wild dogs may be heard howling on the wooded banks of the less frequented streams; and the common cat, altered into a large and fierce animal, inhabits rocky hills. I have alluded to the invasion of the cardoon: in a like manner, the islands near the mouth of the Parana, are thickly clothed with peaches and orange-trees, springing from seeds carried there by the waters of the river.

Chapter 15, November 1834, Chiloe and Chonos Islands

[. . .]

November 10th

THE Beagle sailed from Valparaiso to the southward, for the purpose of surveying the southern part of Chile, the island of Chiloe, and the broken land called the Chonos Archipelago, as far south as the Peninsula of Tres Montes. On the 21st we anchored in the bay of S. Carlos, the capital of Chiloe.

This island is about ninety miles long, with a breadth of rather less than thirty. The land is hilly, but not mountainous, and is every where covered by one great forest, excepting a few scattered green patches, which have been cleared round the thatched cottages. From a distance the view somewhat resembles Tierra del Fuego; but the woods, when seen nearer, are incomparably more beautiful. Many kinds of fine evergreen trees, and plants with a tropical character, here take the place of the gloomy beech of the southern shores. In winter the climate is detestable, and in summer it is only a little better. I should think there are few parts of the world, within the temperate regions, where so much rain falls. The winds are very boisterous, and the sky almost always clouded; to have a week of fine weather is something wonderful. It is even difficult to get a single glimpse of the Cordillera; during our first visit only one opportunity occurred, and that was before sunrise, when the Volcano of Osorno stood out in bold relief; and it was curious to watch, as the sun rose, the outline gradually fading away in the glare of the eastern sky.

[. . .]

December 1st

DURING the four succeeding days we continued sailing southward. The general features of the country remained the same, but it was much less thickly inhabited. On the large island of Tanqui there was scarcely one cleared spot; the trees on every side extending their branches over the sea-beach. I one day noticed some very fine plants of the panke (*Gunnera scabra*), which somewhat resembles the rhubarb on a gigantic scale, growing on the sandstone cliffs. The inhabitants eat the stalks, which are subacid, and tan leather with the roots, and prepare a black dye from them. The leaf is nearly circular, but deeply indented on its margin: I measured one which had a diameter of nearly eight feet, and therefore a circumference of no less than twenty-four! The stalk is rather more than a yard high, and each plant sent out four or five of these enormous leaves – presenting together a very noble appearance.

Chapter 29, September to October 1835, Galapagos Archipelago

September 15th

THE Beagle arrived at the southernmost of the Galapagos islands. This archipelago consists of ten principal islands, of which five much exceed the others in size. They are situated under the equatorial line, and between five and six hundred miles to the westward of the coast of America. The constitution of the whole is volcanic. With the exception of some ejected fragments of granite, which have been most curiously glazed and altered by the heat, every part consists of lava, or of sandstone resulting from the attrition of such materials. The higher islands (which attain an elevation of three, and even four thousand feet) generally have one or more principal craters towards their centre, and on their flanks smaller orifices. I have no exact data from which to calculate, but I do not hesitate to affirm, that there must be, in all the islands of the archipelago, at least two thousand craters. [. . .]

Considering that these islands are placed directly under the equator, the climate is far from being excessively hot; a circumstance which, perhaps, is chiefly owing to the singularly low temperature of the surrounding sea. Excepting during one short season, very little rain falls, and even then it is not regular: but the clouds generally hang low. From these circumstances the lower parts of the islands are extremely arid, whilst the summits, at an elevation of a thousand feet or more, possess a tolerably luxuriant vegetation. This is especially the case on the windward side, which first receives and condenses the moisture from the atmosphere.

In the morning (17th), we landed on Chatham Island, which, like the others, rises with a tame and rounded outline, interrupted only here and there by scattered hillocks – the remains of former craters. Nothing could be less inviting than the first appearance. A broken field of black basaltic lava is everywhere covered by a stunted brushwood, which shows little signs of life. The dry and parched surface, having been heated by the noonday sun, gave the air a close and sultry feeling, like that from a stove: we fancied even the bushes smelt unpleasantly. Although I diligently tried to collect as many plants as possible, I succeeded in getting only ten kinds; and such wretched-looking little weeds would have better become an arctic, than an equatorial Flora.

The thin woods, which cover the lower parts of all the islands, excepting where the lava has recently flowed, appear from a short distance quite leafless, like the deciduous trees of the northern hemisphere in winter. It was some time before I discovered, that not only almost every plant was in full leaf, but that the greater number were now in flower. After the period of heavy rains, the islands are said to appear for a short time partially green. The only other country, in which I have seen a vegetation with a character at all approaching this, is at the volcanic island of Fernando Noronha, placed in many respects under similar conditions.

The natural history of this archipelago is very remarkable: it seems to be a little world within itself; the greater number of its inhabitants, both vegetable and animal, being found nowhere else [. . .]

October 3rd

[. . .]

I will now offer a few general observations on the natural history of these islands. I endeavoured to make as nearly a perfect collection in every branch as time permitted. The plants have not yet been examined, but Professor Henslow, who has kindly undertaken the description of them, informs me that there are probably many new species, and perhaps even some new genera. They all have an extremely weedy character, and it would scarcely have been supposed, that they had grown at an inconsiderable elevation directly under the equator. In the lower and sterile parts, the bush, which from its minute brown leaves chiefly gives the leafless appearance to the brushwood, is one of the Euphorbiaceæ. In the same region an acacia and a cactus (*Opuntia Galapageia*),[2] with large oval compressed articulations, springing from a cylindrical stem, are in some parts common. These are the only trees which in that part afford any shade. Near the summits of the different islands, the vegetation has a very different character; ferns and coarse grasses are abundant; and the commonest tree is one of the Compositæ. Tree-ferns are not present. One of the most singular characters of the Flora, considering the position of this archipelago, is the absence of every member of the palm family. Cocos Island, on the other hand, which is the nearest point of land, takes its name from the great number of cocoa-nut trees on it. From the presence of the Opuntias and some other plants, the vegetation partakes more of the character of that of America than of any other country.

Notes

1 D'Orbigny (vol. i., p. 474), says that the cardoon and artichoke are both found wild. Dr Hooker (Botanical Magazine, vol. iv., p. 2862), has described a variety of the Cynara from this part of South America under the name of *inermis*. He states that botanists are now generally agreed that the cardoon and the artichoke are varieties of one plant. I may add, that an intelligent farmer assured me, he had observed in a deserted garden, some artichokes changing into the common cardoon. Dr Hooker believes that Head's vivid description of the thistle of the Pampas applies to the cardoon; but this is a mistake. Captain Head referred to the plant, which I have mentioned a few lines lower down, under the title of giant thistle. Whether it is a true thistle I do not know; but it is quite different from the cardoon, and more like a thistle properly so called.

2 Magazine of Zoology and Botany, vol. i., p. 466.

52

JOHN RUSKIN, 'OF TRUTH OF VEGETATION', *THE LIBRARY EDITION OF JOHN RUSKIN'S WORKS*, 39 VOLS, VOL. 3 *MODERN PAINTERS I*, 1903

(London: George Allen, 1903–12)

Section VI, Chapter I, Of Truth of Vegetation

§ 1. Frequent occurrence of foliage in the works of the old masters

WE have now arrived at the consideration of what was, with the old masters, the subject of most serious and perpetual study. If they do not give us truth here, they cannot have the faculty of truth in them: for foliage is the chief component part of all their pictures, and is finished by them with a care and labour which, if bestowed without attaining truth, must prove either their total bluntness of perception, or total powerlessness of hand. With the Italian school, I can scarcely recollect a single instance in which foliage does not form the greater part of the picture; in fact, they are rather painters of tree-portrait than landscape painters; for rocks, and sky, and architecture are usually mere accessories and backgrounds to the dark masses of laborious foliage, of which the composition principally consists. Yet we shall be less detained by the examination of foliage than by our former subjects; since where specific form is organized and complete, and the occurrence of the object universal, it is easy, without requiring any laborious attention in the reader, to demonstrate to him quite as much of the truth or falsehood of various representations of it, as may serve to determine the character and rank of the painter.

§ 2. Laws common to all forest trees. Their branches do not taper, but only divide

IT will be best to begin as nature does, with the stems and branches, and then to put the leaves on. And in speaking of trees generally, be it observed, when I say

DOI: 10.4324/9780429355653-58

all trees, I mean only those ordinary forest or copse trees of Europe, which are the chief subjects of the landscape painter. I do not mean to include every kind of foliage which by any accident can find its way into a picture, but the ordinary trees of Europe: oak, elm, ash, hazel, willow, birch, beech, poplar, chestnut, pine, mulberry, olive, ilex, carob, and such others. I do not purpose to examine the characteristics of each tree; it will be enough to observe the laws common to all. First, then, neither the stems nor the boughs of any of the above trees *taper*, except where they fork. Wherever a stem sends off a branch, or a branch a lesser bough, or a lesser bough a bud, the stem of the branch is, on the instant, less in diameter by the exact quantity of the branch or the bough they have sent off, and they remain of the same diameter; or if there be any change, rather increase than diminish until they send off another branch or bough. This law is imperative and without exception; no bough, nor stem, nor twig, ever tapering or becoming narrower towards its extremity by a hair's-breadth, save where it parts with some portion of its substance at a fork or bud, so that if all the twigs and sprays at the top and sides of the tree, which are, and *have been*, could be united without loss of space, they would form a round log of at least the diameter of the trunk from which they spring.

§ 3. Appearance of tapering caused by frequent buds

But as the trunks of most trees send off twigs and sprays of light under-foliage, of which every individual fibre takes precisely its own thickness of wood from the parent stem, and as many of these drop off, leaving nothing but a small excrescence to record their existence, there is frequently a slight and delicate appearance of tapering caused in the trunk itself; while the same operation takes place much more extensively in the branches; it being natural to almost all trees to send out from their young limbs more wood than they can support; which, as the stem increases, gets contracted at the point of insertion, so as to check the flow of the sap, and then dies and drops off, leaving all along the bough, first on one side, then on another, a series of small excrescences sufficient to account for a degree of tapering, which is yet so very slight that if we select a portion of a branch with no real fork or living bough to divide it or diminish it, the tapering is scarcely to be detected by the eye; and if we select a portion without such evidences of past ramification, there will be found none whatsoever.

§ 4. And care of nature to conceal the parallelism

But nature takes great care and pains to conceal this uniformity in her boughs. They are perpetually parting with little sprays here and there, which steal away their substance cautiously and where the eye does not perceive the theft, until, a little way above, it feels the loss; and in the upper parts of the tree, the ramifications take place so constantly and delicately, that the effect upon the eye is precisely the same as if the boughs actually tapered, except here and there, where some avaricious one, greedy of substance, runs on for two or three yards without parting with anything, and becomes ungraceful in so doing.

§ 5. The degree of tapering which may be represented as continuous

HENCE we see that although boughs may and must be represented as actually tapering, they must only be so when they are sending off foliage and sprays, and when they are at such a distance that the particular forks and divisions cannot be evident to the eye; and farther, even in such circumstances, the tapering never can be sudden or rapid. No bough ever, with appearance of smooth tapering, loses more than one tenth of its diameter in a length of ten diameters. Any greater diminution than this must be accounted for by visible ramification, and must take place by steps, at each fork.

§ 6. The trees of Gaspar Poussin

AND therefore we see at once that the stem of Gaspar Poussin's tall tree, on the right of the La Riccia, in the National Gallery, is a painting of a carrot or a parsnip, not of the trunk of a tree. For, being so near that every individual leaf is visible, we should not have seen, in nature, one branch or stem actually tapering. We should have received an *impression* of graceful diminution; but we should have been able, on examination, to trace it joint by joint, fork by fork, into the thousand minor supports of the leaves. Gaspar Poussin's stem, on the contrary, only sends off four or five minor branches altogether, and both it and they taper violently, and without showing why or wherefore; without parting with a single twig, without showing one vestige of roughness or excrescence; and leaving, therefore, their unfortunate leaves to hold on as best they may. The latter, however, are clever leaves, and support themselves as swarming bees do, hanging on by each other.

§ 7. And of the Italian school generally, defy this law

BUT even this piece of work is a jest to the perpetration of the bough at the left-hand upper corner of the picture opposite to it, the View near Albano. This latter is a representation of an ornamental group of elephants' tusks, with feathers tied to the ends of them. Not the wildest imagination could ever conjure up in it the remotest resemblance to the bough of a tree. It might be the claws of a witch, the talons of an eagle, the horns of a fiend; but it is a full assemblage of every conceivable falsehood which can be told respecting foliage, a piece of work so barbarous in every way, that one glance at it ought to prove the complete charlatanism and trickery of the whole system of the old landscape painters. For I will depart for once from my usual plan, of abstaining from all assertion of a thing's being beautiful or otherwise; I will say here, at once, that such drawing as this is as ugly as it is childish, and as painful as it is false; and that the man who could tolerate, much more, who could deliberately set down such a thing on his canvas, had neither eye nor feeling for one single attribute or excellence of God's works. He might have drawn the other stem in excusable ignorance, or under some false impression of being able to improve upon nature; but *this* is conclusive and unpardonable. Again, take the stem of the chief tree in Claude's Narcissus. It is a very faithful portrait of a large boa constrictor, with a handsome tail; the kind of trunk which young ladies at fashionable boarding-schools represent with nosegays at the top of them by way of forest scenery.

§ 8. The truth, as it is given by J. D. Harding

LET us refresh ourselves for a moment, by looking at the truth. We need not go to Turner, we will go to the man who next to him is unquestionably the greatest master of foliage in Europe, J.D. Harding. Take the trunk of the largest stone-pine, plate 25 in 'The Park and the Forest'. For the first nine or ten feet from the ground it does not lose one hair's-breadth of its diameter. But the shoot broken off just under the crossing part of the distant tree is followed by an instant diminution of the trunk, perfectly appreciable both by the eye and the compasses. Again, the stem maintains undiminished thickness up to the two shoots on the left, from the loss of which it suffers again perceptibly. On the right, immediately above, is the stump of a very large bough, whose loss reduces the trunk suddenly to about two thirds of what it was at the root. Diminished again, less considerably, by the minor branch close to this stump, it now retains its diameter up to the three branches broken off just under the head, where it once more loses in diameter; and finally branches into the multitude of head-boughs, of which not one will be found tapering in any part, but losing itself gradually by division among its off-shoots and spray. This is nature, and beauty too.

§ 9. Boughs, in consequence of this law, must diminish where they divide. those of the old masters often do not

BUT the old masters are not satisfied with drawing carrots for boughs. Nature can be violated in more ways than one, and the industry with which they seek out and adopt every conceivable mode of contradicting her is matter of no small interest. It is evident from what we have above stated of the structure of all trees, that as no boughs diminish where they do not fork, so they cannot fork without diminishing. It is impossible that the smallest shoot can be sent out of the bough without a diminution of the diameter above it; and wherever a branch goes off it must not only be less in diameter than the bough from which it springs, but the bough beyond the fork must be less by precisely the quantity of the branch it has sent off.[1] Now observe the bough underneath the first bend of the great stem in Claude's Narcissus; it sends off four branches like the ribs of a leaf. The two lowest of these are both quite as thick as the parent stem, and the stem itself is much thicker after it has sent off the first one than it was before. The top boughs of the central tree, in the Marriage of Isaac and Rebecca, ramify in the same scientific way.

§ 10. Boughs must multiply as they diminish. Those of the old masters do not

BUT there are farther conclusions to be drawn from this great principle in trees. As they only diminish where they divide, their increase of number is in precise proportion to their diminution of size; so that whenever we come to the extremities of boughs, we must have a multitude of sprays sufficient to make up, if they were united, the bulk of that from which they spring. Precision in representing this is neither desirable nor possible. All that is required is just so much observance of the general principle as may make the eye feel satisfied that there is something like the same quantity of wood in the sprays which there is in the stem. But to do this there must be, what there always is in nature, an exceeding complexity of the outer sprays. This complexity gradually increases towards their extremities, of course exactly in proportion to the slenderness of the twigs. The slenderer they become, the more there are of them, until at last, at the extremities of the tree, they form a mass of intricacy, which in winter, when it can be seen, is scarcely distinguishable from fine herbage, and is beyond all power of definite representation; it can only be expressed by a mass of involved strokes.

§ 11. Boughdrawing of Salvator

ALSO, as they shoot out in every direction, some are nearer, some more distant; some distinct, some faint; and their intersections and relations of distance are marked with the most exquisite gradations of aerial perspective. Now it will be found universally, in the works of Claude, Gaspar, and Salvator, that the boughs do *not* get in the least complex or multiplied towards the extremities; that each large limb forks only into two or three smaller ones, each of which vanishes into the air without any cause or reason for such unaccountable conduct, unless that the mass of leaves transfixed upon it or tied to it, entirely dependent on its single strength, have been too much, as well they may be, for its powers of solitary endurance. This total ignorance of tree-structure is shown throughout their works. The Sinon before Priam is an instance of it in a really fine work of Claude's, but the most gross examples are in the works of Salvator. It appears that this latter artist was hardly in the habit of studying from nature at all, after his boyish ramble among the Calabrian hills; and I do not recollect any instance of a piece of his bough-drawing which is not palpably and demonstrably a made up phantasm of the studio, the proof derivable from this illegitimate tapering being one of the most convincing. The painter is always visibly embarrassed to reduce the thick boughs to spray, and *feeling* (for Salvator naturally had acute feeling for truth) that the bough was wrong when it tapered suddenly, he accomplishes its diminution by an impossible protraction; throwing out shoot after shoot until his branches straggle all across the picture, and at last disappear unwillingly where there is no room for them to stretch any farther. The consequence is, that whatever leaves are put upon such boughs have evidently no adequate support, their power of leverage is enough to uproot the tree; or, if the boughs are left bare, they have the look of the long tentacula of some complicated marine monster, or of the waving endless threads of bunchy sea-weed, instead of the firm, upholding, braced, and bending grace of natural boughs. I grant that this is in a measure done by Salvator from a love of ghastliness, and is in scenes of this sort in a measure allowable: but it is in a far greater degree done from pure ignorance of tree-structure [. . .] I have seen more spectral character in the real limbs of a blasted oak, than ever in Salvator's best monstrosities; more horror is to be obtained by right combination of inventive line, than by drawing tree branches as if they were wing-bones of a pterodactyle. All departure from natural forms to give fearfulness is mere Germanism; it is the work of fancy, not of imagination, and instantly degrades whatever it affects to a third-rate level. [. . .]

§ 13. Impossibility of the angles of boughs being taken out of them by wind

Now *the fiercest wind that ever blew upon the earth could not take the angles out of the bough of a tree an inch thick.* The whole bough bends together, retaining its elbows, and angles, and natural form, but affected throughout with curvature in each of its parts and joints. That part of it which was before perpendicular being bent aside, and that which was before sloping being bent into still greater inclination, the angle at which the two parts meet remains the same; or, if the strain be put in the opposite direction, the bough will break long before it loses its angle. You will find it difficult to bend the angles out of the youngest sapling, if they be marked; and absolutely impossible, with a strong bough. You may break it, but you will not destroy its angles. And if you watch a tree in the wildest storm, you will find that though all its boughs are bending, none lose their character, but the utmost shoots and sapling spray. Hence Gaspar Poussin, by his bad drawing, does not make his storm strong, but his tree weak; he does not make his gust violent, but his boughs of India-rubber.

§ 14. Bough drawing of Titian

These laws respecting vegetation are so far more imperative than those which were stated respecting water, that the greatest artist cannot violate them without danger, because they are laws resulting from organic structure which it is always painful to see interrupted; on the other hand, they have this in common with all laws, that they may be observed with mathematical precision, yet with no right result; the disciplined eye and the life in the woods are worth more than all botanical knowledge. For there is that about the growing of the tree trunk, and that grace in its upper ramification, which cannot be taught, and which cannot even be seen but by eager watchfulness. There is not an exhibition passes, but there appear in it hundreds of elaborate paintings of trees, many of them executed from nature. For three hundred years back, trees have been drawn with affection by all the civilized nations of Europe, and yet I repeat boldly, what I before asserted, that no men but Titian and Turner ever drew the stem of a tree. Generally, I think the perception of the muscular qualities of the tree trunk incomplete, except in men who have studied the human figure; and in loose expression of those characters, the painter who can draw the living muscle seldom fails; but the thoroughly peculiar lines belonging to woody fibre can only be learned by patient forest study. And hence in all the trees of the merely historical painters, there is fault of some kind or another; commonly exaggeration of the muscular swellings, or insipidity and want of spring in curvature, or fantasticism and unnaturalness of arrangement, and especially a want of the peculiar characters of bark which express the growth and age of the tree; for bark is no mere excrescence, lifeless and external, it is a skin of especial significance in its indications of the organic form beneath; in places under the arms of the tree it wrinkles up and forms fine lines *round* the trunk, inestimable in their indication of the direction of its surface; in others, it bursts or peels longitudinally, and the rending and bursting of it are influenced in direction and degree by the undergrowth and swelling of the woody fibre, and are not a mere roughness and granulated pattern of the hide. Where there are so many points to be observed, some are almost always exaggerated, and others missed, according to the predilections of the painter. Albert Dürer has given some splendid examples of woody structure, but misses the grace of the great lines. Titian took a larger view, yet (as before noticed), from the habit of drawing the figure, he admits too much flaccidity and bend, and sometimes makes his tree trunks look flexible like sea-weed. There is a peculiar stiffness about the curves of the wood, which separates them completely from animal curves, and which especially defies recollection or invention; it is so subtle that it escapes but too often, even in the most patient study from nature; it lies within the thickness of a pencil line. Farther, the modes of ramification of the upper branches are so varied, inventive, and graceful, that the least alteration of them, even the measure of a hair's-breadth, spoils them; and though it is sometimes possible to get rid of a troublesome bough, accidentally awkward, or in some minor respects to assist the arrangement, yet so far as the real branches are copied, the hand libels their lovely curvatures even in its best attempts to follow them.

[. . .]

§ 16. Leafage. Its variety and symmetry

LET us, however, pass to the leafage of the elder landscape-painters, and see if it atones for the deficiencies of the stems. One of the most remarkable characters of natural leafage is the constancy with which, while the leaves are arranged on the spray with exquisite regularity, that regularity is modified in their actual effect. For as in every group of leaves some are seen sideways, forming merely long lines, some foreshortened, some crossing each other, every one differently turned and placed from all the others, the forms of the leaves, though in themselves similar, give rise to a thousand strange and differing forms in the group; and the shadows of some, passing over the others, still farther disguise and confuse the mass, until the eye can distinguish nothing but a graceful and flexible disorder of innumerable forms, with here and there a perfect leaf on the extremity, or a symmetrical association of one or two, just enough to mark the specific character and to give unity and grace, but never enough to repeat in one group what was done in another, never enough to prevent the eye from feeling that, however, regular and mathematical may be the structure of parts, what is composed out of them is as various and infinite as any other part of nature. Nor does this take place in general effect only. Break off an elm bough three feet long, in full leaf, and lay it on the table before you, and try to draw it, leaf for leaf. It is ten to one if in the whole bough (provided you do not twist it about as you work) you find one form of a leaf exactly like another; perhaps you will not even have *one* complete. Every leaf will be oblique, or foreshortened, or curled, or crossed by another, or shaded by another, or have something or other the matter with it; and though the whole bough will look graceful and symmetrical, you will scarcely be able to tell how or why it does so, since there is not one line of it like another.

§17. Perfect regularity of Poussin

Now go to Gaspar Poussin and take one of his sprays where they come against the sky; you may count it all round: one, two, three, four, one bunch; five, six, seven, eight, two bunches; nine, ten, eleven, twelve, three bunches; with four leaves each; and such leaves! every one precisely the same as its neighbour, blunt and round at the end (where every forest leaf is sharp, except that of the fig-tree), tied together by the stalks, and so fastened on to the demoniacal claws above described, one bunch to each claw.

§ 18. Exceeding intricacy of nature's foliage

But if nature is so various when you have a bough on the table before you, what must she be when she retires from you, and gives you her whole mass and multitude? The leaves then at the extremities become as fine as dust, a mere confusion of points and lines between you and the sky, a confusion which, you might as well hope to draw sea-sand particle by particle, as to imitate leaf for leaf. This, as it comes down into the body of the tree, gets closer, but never opaque; it is always transparent with crumbling lights in it letting you through to the sky: then out of this, come, heavier and heavier, the masses of illumined foliage, all dazzling and inextricable, save here and there a single leaf on the extremities: then, under these, you get deep passages of broken irregular gloom, passing into transparent, green-lighted, misty hollows; the twisted stems glancing through them in their pale and entangled infinity, and the shafted sunbeams, rained from above, running along the lustrous leaves for an instant; then lost, then caught again on some emerald bank or knotted root, to be sent up again with a faint reflex on the white undersides of dim groups of drooping foliage, the shadows of the upper boughs running in grey network down the glossy stems, and resting in quiet chequers upon the glittering earth; but all penetrable and transparent, and, in proportion, inextricable and incomprehensible, except where across the labyrinth and the mystery of the dazzling light and dream-like shadow, falls, close to us, some solitary spray, some wreath of two or three motionless large leaves, the type and embodying of all that in the rest we feel and imagine, but can never see.

§ 19. How contradicted by the tree-patterns of G. Poussin

Now, with this much of nature in your mind, go to Gaspar Poussin's view near Albano, in the National Gallery. It is the very subject to unite all these effects, a sloping bank shaded with intertwined forest. And what has Gaspar given us? A mass of smooth, opaque, varnished brown, without one interstice, one change of hue, or any vestige of leafy structure, in its interior, or in those parts of it, I should say, which are intended to represent interior; but out of it, over it rather, at regular intervals, we have circular groups of greenish touches, always the same in size, shape, and distance from each other, containing so exactly the same number of touches each, that you cannot tell one from another. There are eight or nine and thirty of them, laid over each other like fish-scales; the shade being most carefully made darker and darker as it recedes from each until it comes to the edge of the next, against which it cuts in the same sharp circular line, and then begins to decline again, until the canvas is covered, with about as much intelligence or feeling of art as a house-painter has in marbling a wainscot, or a weaver in repeating an ornamental pattern. [. . .]

§ 21. Perfect unity in nature's foliage

But nature observes another principle in her foliage more important even than its intricacy. She always secures an exceeding harmony and repose. She is *so* intricate that her minuteness of parts becomes to the eye, at a little distance, one united veil or cloud of leaves, to destroy the evenness of which is perhaps a greater fault than to destroy its transparency. [. . .]

§ 22. Total want of it in Both and Hobbima

It is here that Hobbima and Both fail. They can paint oak leafage faithfully, but do not know where to stop, and by doing too much, lose the truth of all, lose the very truth of detail at which they aim, for all their minute work only gives two leaves to nature's twenty. They are evidently incapable of even thinking of a tree, much more of drawing it, except leaf by leaf; they have no notion nor sense of simplicity, mass, or obscurity, and when they come to distance, where it is totally impossible that leaves should be separately seen, being incapable of conceiving or rendering the grand and quiet forms of truth, they are reduced to paint their bushes with dots and touches expressive of leaves three feet broad each. Nevertheless there is a genuine aim in their works, and their failure is rather to be attributed to ignorance of art, than to such want of sense for nature as we find in Claude or Poussin: and when they come close to home, we sometimes receive from them fine passages of mechanical truth.

§ 23. How rendered by Turner

BUT let us oppose to their works the group of trees on the left in Turner's Marly. We have there perfect and ceaseless intricacy to oppose to Poussin, perfect and unbroken repose to oppose to Hobbima; and in the unity of these the perfection of truth. This group may be taken as a fair standard of Turner's tree-painting. We have in it the admirably drawn stems, instead of the claws or the serpents; full, transparent, boundless intricacy, instead of the shell pattern; and misty depth of intermingled light and leafage, instead of perpetual repetition of one mechanical touch.

[. . .]

§ 25. Universal termination of trees in symmetrical curves

THE last and most important truth to be observed respecting trees is, that their boughs always, in finely grown individuals, bear among themselves such a ratio of length as to describe with their extremities a symmetrical curve, constant for each species; and within this curve all the irregularities, segments, and divisions of the tree are included, each bough reaching the limit with its extremity, but not passing it. When a tree is perfectly grown, each bough starts from the trunk with just so much wood as, allowing for constant ramification, will enable it to reach the terminal line; or if, by mistake, it start with too little, it will proceed without ramifying till within a distance where it may safely divide; if on the contrary it start with too much, it will ramify quickly and constantly; or, to express the real operation more accurately, each bough growing on so as to keep even with its neighbours, takes so much wood from the trunk as is sufficient to enable it to do so, more or less in proportion as it ramifies fast or slowly. In badly grown trees the boughs are apt to fall short of the curve, or at least there are so many jags and openings that its symmetry is interrupted; and in young trees, the impatience of the upper shoots frequently breaks the line: but, in perfect and mature trees, every bough does its duty completely, and the line of curve is quite filled up, and the mass within it unbroken, so that the tree assumes the shape of a dome as in the oak, or, in tall trees, of a pear with the stalk downmost.

§ 26. Altogether unobserved by the old masters. Always given by Turner

THE old masters paid no attention whatsoever to this great principle. They swing their boughs about, anywhere and everywhere; each stops or goes on just as it likes; nor will it be possible, in any of their works, to find a single example in which any symmetrical curve is indicated by the extremities. But I need scarcely tell any one in the slightest degree acquainted with the works of Turner, how rigidly and constantly he adheres to this principle of nature; taking in his highest compositions the perfect ideal form, every spray being graceful and varied in itself, but inevitably terminating at the assigned limit, and filling up the curve without break or gap; in his lower works, taking less perfect form but invariably hinting the constant tendency in all; and thus, in spite of his abundant complexity, he arranges his trees under simpler and grander forms than any other artist, even among the moderns.

Note

1 Ruskin's footnote: It sometimes happens that a morbid direction of growth will cause an exception here and there to this rule, the bough swelling beyond its legitimate size: knots and excrescences, of course, sometimes interfere with the effect of diminution. I believe that in the laurel, when it grows large and old, singular instances may be found of thick upper boughs and over-quantity of wood at the extremities. All these accidents or exceptions are felt as such by the eye. They may occasionally be used by the painter in savage or grotesque scenery, or as points of contrast, but are no excuse for his ever losing sight of the general law.

53

JOSEPH DALTON HOOKER, *FLORA ANTARCTICA: THE BOTANY OF THE ANTARCTIC VOYAGE OF H.M. DISCOVERY SHIPS EREBUS AND TERROR IN THE YEARS 1839–1843 UNDER THE COMMAND OF CAPTAIN SIR JAMES CLARK ROSS*, 3 VOLS, VOL. 1. BOTANY OF LORD AUCKLAND'S GROUP AND CAMPBELL'S ISLAND

(London: Reeve Brothers, 1844)

Summary of the Voyage

In the beginning of the year 1839, the British Government having determined on fitting out an Expedition, for the purpose of investigating the phenomena of Terrestrial Magnetism in various remote countries, and for prosecuting Maritime Geographical Discovery in the high southern latitudes, H.M. Ships Erebus and Terror, commissioned by Captain Sir James Clark Ross, sailed from Chatham on the 29th of September 1839. In addition to carrying out the above-mentioned leading views, it was enjoined to the officers, that they should use every exertion to collect the various objects of Natural History which the many heretofore unexplored countries about to be visited would afford.

On the outward voyage we touched at most of the Atlantic Islands, making a longer stay at some of them than is usual, on account of the nature of the observations that were instituted. At Madeira, which was the first visited, we called in the middle of October, and remained eleven days; and then made Teneriffe and the Cape de Verds, whence we sailed for and landed upon St. Paul's Rocks, under the

DOI: 10.4324/9780429355653-59

Line, in long. 29° W. St. Helena was the next destination, and the course which it was found necessary to follow took us to the Island of Trinidad off the Brazilian coast, lat. 20° S.

After spending a week at St. Helena, the vessels sailed for the Cape of Good Hope, arriving there on the 4th of April 1840. The Cape may be regarded as the starting-point, whence the real object of the voyage, namely that which included South Polar Discovery, would commence.

[. . .]

Part II, Botany of Fuegia, the Falklands, Kerguelen's Land, etc.

MOST of the materials in this Part were amassed by myself, with the kind aid of Captain Sir James Ross, Lieutenant Smith, Mr. Davies, and particularly of Dr Lyall, to whose exertions I feel constrained throughout to acknowledge my obligations. For many important additions to the plants of Fuegia and particularly of the west coast of Patagonia, I am indebted to Captain King and Mr. Darwin, both of whom most generously confided their collections to me for the purpose of examination and description. Captain King's is certainly the most complete flora ever formed in those countries, whether in number of species or specimens of the flowering plants. To Dr Lemann I owe the use of another set of the same plants, gathered by Mr Anderson, the gardener who accompanied Captain King, and to Commodore Sulivan, a collection formed by his son, Captain Sulivan, during Captain Fitzroy's voyage. With all these advantages the materials for a Fuegian flora would still be incomplete, without the plants discovered by Menzies dining Vancouver's expedition; and still more valuable is the access afforded by the kindness of Mr. Brown and Mr. Benettt, to the specimens, drawings, and manuscripts of Banks and Solander, who preceded all other botanists, except Commerson, in the investigation of Natural History in the high southern regions.

The collections of Banks and Solander, wherever formed and under whatever difficulties, are lasting proofs not only of the extraordinary zeal and ability of those distinguished individuals, but of the spirit which pervaded every member of the gallant band that Cook led in his path of discovery. Our knowledge of the Botany of New Zealand is still mainly due to the labours of the companions of Cook's first voyage, for no subsequent travellers or even residents in that country have made equally extensive collections; and that their researches in Tierra del Fuego were no less eminently successful, the constant mention of their names in this volume will abundantly prove. Valuable as the dried plants are, their utility is doubly increased by the excellent descriptions and by the beautiful coloured drawings executed on the spot, which accompany them, and were made at Sir Joseph Banks' own expense. There are daily occurring instances, to the honour of the British nation be it mentioned, of individuals who undertake and conduct scientific expeditions on their own resources, and who return richly laden to reap the honours that await themselves as the projectors and commanders of their several efforts; but how few examples have we of men of birth and fortune, who like Banks will peril a life and spend a fortune as the zealous cooperator in an expedition not his own, and the main glory of which justly belongs to another. In scientific as in all other pursuits there are ever many to lead, but few who will stoop to be followers. This just tribute to the memory of Banks is peculiarly due from me, who owe so much to his labours in the Southern Ocean.

[. . .]

II. Magnoliaceæ, *DC*

1. Drimys, Ford

1. *Drimys Winteri*, Forst. *Gen.* p. 84. t. 42. *Linn. Fil. Suppl.* p. 269. *Lamarck, Dict.* vol. ii. p. 331. DC. *Syst. Veg.* vol. i. p. 443. *ProDr* vol. i. p. 78. D. punctata, *Lam. Dict.* vol. ii. p. 330. *Illust.* t. 494. f. 1. *Winterana aromatica, Soland. Med. Obs.* vol. v. p. 46. t. 1. *Wintera aromatica, Murray, Syst.* 507. *App. Med.* vol. iv. p. 557. *Humb. et Bonpl.* vol. i. p. 209.

HAB. Strait of Magalhaens and Fuegia; first noticed by John Winter who accompanied Drake's voyage in 1577, and since by all voyagers and collectors.

A very abundant tree throughout the western and southern parts of Fuegia, even in Hermite Island ascending to 1000 feet. The natives use the stems of the young trees, rudely fashioning them into handles sometimes ten feet long, for their harpoons; but the wood is too soft and supple. The bark has proved a most useful stomachic and antiscorbutic to various voyagers, and especially to a portion of the crew of the 'Beagle' during Capt. King's arduous surveying voyage [. . .]

After a careful examination of a very extensive suite of examples, I have come to the conclusion that there is but one South American species of this genus. There is a dissimilarity in the form of the foliage, even between the North and South Fuegian states, the former having longer and more membranous leaves, differing in no respect from specimens gathered near Valparaiso [. . .] which generally pass under the name of *D. Chilensis*, DC. From Juan Fernandez again, the plants collected by the two last-mentioned travellers belong to the same species: though the leaves are generally more linear, they are not so much so as in some of the continental states. In Brazil, the variety, called *D. Granatensis*, L. fil., is found over the whole of that vast empire, and equally occurs in New Grenada and the province of Santa Fe in Colombia. Mr. Gardner's number 5675 precisely accords with the Juan Fernandez plant. St. Hilaire and Cambessèdes describe four and give figures of three varieties; herein they differ from Martius, who considers it the same as *D. Winteri*, but these authors do not state their reasons ('Plantes Usuelles de Bresil', Tab. 26–28), and neither in the plates or descriptions do any characters appear which are not common to some of the Chilian and Fuegian specimens: their *var. sylvatica* coincides with Juan Fernandez specimens; the *var. montana* has smaller leaves than any found on the west coast of the continent. [. . .] The effuse panicle and larger flowers are more characteristic of the northern states of the tree, but these peculiarities afford no specific distinction. A singular state, with small narrow leaves, remarkably revolute at the margins, has been gathered in Brazil by Claussen. The variations in the foliage are too gradual to admit even of the forms being grouped into varieties indicative of countries or of other peculiarities, and the glaucous hue of the under surface of the leaves is equally apt to mislead.

I feel little doubt that this plant extends over no less than 86° of latitude, forming at the southern limit of its growth one of the three trees that advance the nearest to the Antarctic circle, and reaching as high a latitude as any flowering plant, save the solitary grass of the South Shetland Islands. No vegetable production of its size affords a parallel case to this, either in America or any other country. Such an extraordinarily extended range is in part obviously due to some peculiarities in the form and surface of South America, where under every degree of latitude there are large areas either at the level of the ocean or at an elevation where such a tree can enjoy a climate that is equable. To the influence of the like causes I should attribute the specific identity between some high northern and southern species, which like the *Gentiana prostrata*, *Trisetum subspicatum*, and other plants mentioned in the former part of this work, pass along the Andes from the northern temperate or frigid point to the southern extreme of America.

The *Drimys Winteri* is one of those plants which is represented by two closely allied species in other quarters of the globe, one in Tasmania, the *Tasmania aromatica*, and the *Drimys axillaris* in New Zealand. There are many instances of genera having representatives in those three botanical regions, the species being in general mutually more related than to any others, such are afforded by the genera *Fagus*, *Astelia*, *Abrotanella*, by shrubby *Veronicas* and many others. This similarity in some of the botanical productions of countries, otherwise unlike in vegetation, is far more remarkable than a total dissimilarity between lands so far separated, or even than a positive specific identity would be at first sight; because it argues the operation of some agent far above our powers of comprehension, and far other from what we commonly observe to affect geographical distribution.

III. Berberideæ. *Vent.*

1. Berberis, Linn.

1. Berberis *ilicifolia*, Forst.; erecta, spinis tripartitis, foliis obovatis grosse spinoso-dentatis, pedunculis folio brevioribus 4–6-floris, pedicellis elongatis subcorymbosis, floribus majusculis, baccis late ovatis lagenæforrmibus. B. ilicifolia, *Forst. Comm.* vol. ix. p. 28. *Linn. Fil. Suppl.* p. 210. *DC. Syst.* vol. li. p. 12. *ProDr* vol. i. p. 107. B. lagenaria, *Poir: Dict.* vol. viii. p. 619. (Tab. LXXXVI.)

HAB. Strait of Magalhaens on both sides and throughout Fuegia; *Commerson*, *Forster*, and all future collectors.

This is certainly the handsomest species of the genus, forming a straggling bush, eight feet high, with deep green shining leaves and very conspicuous golden yellow flowers. The wood is pale yellow, affording a gamboge coloured dye, the berries of a deep steel blue colour, and few in comparison to the size of the flower.

2. Berberis buxifolia, Lam.

[. . .]

HAB. Strait of Magalhaens and throughout Fuegia; *Commerson*, and all subsequent collectors.

This is a variable species, especially in the foliage, exhibiting a different aspect at different seasons of the year. In spring, when the flowering commences, fascicles of new leaves are produced, which are pale green, membranous, and entire; at this period the leaves of the former season begin falling while those of the present year gradually become larger, stiffer, coriaceous, and generally mucronate or pungent at the apex. They are not fully developed till autumn, when they are generally quite entire, attenuated at the base, and shortly petiolate, about half an inch long, rigid and coriaceous, reticulated on the upper surface; during the following spring these in their turn fall away. In seedling plants the leaves are larger than at any future time, on long petioles, broader, and here and there furnished with spinous teeth. The flowers are generally in threes, but sometimes solitary, pale yellow. The berries, about the size of a small pea, were much used for tarts by the officers of the 'Beagle' and found excellent. The B. *dulcis*, of Sweet, agrees with the common form of this plant, except that the flowers are larger in that author's figure and the pubescence of the pedicels not visible in the wild specimens. The B. *inermis* seems a variety, some of the specimens being quite unarmed; indeed the spines of this genus afford but an inconstant character.

[. . .]

3. Berberis empetrifolia, Lam.

[. . .]

HAB. Strait of Magalhaens; common in alpine woods; *Commerson.* Port Famine; *Capt. King.*

This species is more characteristic of a dry climate than of the moist wooded country of Fuegia and South-west Chili. The Strait seems to be its southern limit; it inhabits neither the east nor west coasts, but is confined to the Cordillera itself, from many elevated parts of which range we have received it [. . .]; it very probably therefore is a native of the whole length of that range, from lat. 34.° to lat. 54°, descending to the level of the sea at Port Famine, to which point the mountains are continued in one unbroken chain.

54

WILLIAM JACKSON HOOKER (ED.), *NIGER FLORA; OR, AN ENUMERATION OF THE PLANTS OF WESTERN TROPICAL AFRICA, COLLECTED BY THE LATE DR THEODORE VOGEL, BOTANIST TO THE VOYAGE OF THE EXPEDITION SENT BY HER BRITANNIC MAJESTY TO THE RIVER NIGER IN 1841*

(London: Hippolyte Bailliere, 1849)

TO CAPTAIN HENRY DUNDAS TROTTER
R.N. &c, &c., &c.

Commander of the Expedition sent by Her Britannic Majesty. Queen Victoria, to the River NIGER, with the view of obtaining information respecting the adjacent countries, and of forming treaties with the native Chiefs against the Slave Trade, as well as promoting Agriculture and Commerce; under whose auspices most of the Collections described in the following pages were formed;– this work is dedicated,

With sentiments of the highest regard and respect,

By his faithful friend and servant,

THE EDITOR.

ROYAL GARDENS, KEW,

Nov. 1, 1849

 DOI: 10.4324/9780429355653-60

Journal of the Voyage to the Niger

Tuesday, July 6th

Towards the evening of Tuesday, July 6th, we left Monrovia [. . .] We then proceeded, by ourselves, to Grand Bassa, where we anchored on Friday morning, for the purpose of taking in fuel. We stayed several days; not one of which passed without rain, sometimes most violent throughout the entire day. This, and other circumstances, limited my researches to the immediate vicinity of the shore; where, however, I found more plants than I was able to preserve. I made a collection of about a hundred specimens, at the risk of losing everything by the wet. Many plants, especially the *Monocotyledone*, were not yet in flower; and I regretted this most especially in the case of the numerous parasitical *Orchideæ*. The shore is flat and sandy; and the sand has drifted so far inland, that I never got beyond it. There were no forests, only bushes, intermingled with isolated high trees; which I could not determine, for they were all without blossom or fruit. The African *Bombax* appeared amongst them, and the same *Spondias* as at Sierra Leone, forming a considerable tree; respecting which I feel doubtful whether it be identical with *S. Myrobalanus*. The pride of this coast is the *Elais*, often growing in clumps of twelve or more, exhibiting under different circumstances a different habit, and giving a considerable variety of aspect to the country. This *Palm* is of generally moderate height, and constitutes with various *Fici*, the chief masses of wood. The underwood consists of close-growing shrubby *Rubiabeæ*, with shining leaves, intermingled with *Gloriosa superba*, *Cissi*, *Leguminosæ*, *Banisteriæ*, as creepers, leaving hardly room for *Melastoma* and other low plants that peep through with their fine blossoms. It is a very interesting sight, that of a few *Oil Palms* growing in a clump; the ribs of the lower leaves still adhering to the stems, which are clothed with a fresh verdure of parasitical *Ferns* and *Orchidaceæ*; whilst other parasites, such as *Ferns*, *Pothos*, *Anonæ*, *Commelinæ*, small *Rubiaceæ* and *Leguminosæ*, choose the airy shelter of the foliage for their habitation. Of single plants one might specify *Sarcocephalus*, which occurs frequently, the same *Phyllanthus* as in Liberia, *Schmidelia Africana*, a genus of *Apocyneæ*, apparently new and near *Tabernæmontana*, remarkable for its double fruit as large as a child's head, the seeds nestling in the almost woody pulp, wild *Sugar-cane*, not in blossom, *Conocarpus erectus*, var. *β*. a small shrub, a probably new *Cassytha*, *Scævola* (really different from *S. Lobelia*?), *Indigoferæ* sp. *Cannæ* sp. *Cassia occidentalis*, *Borreria Kohautiana*, &c. The *Stylosanthes* forms a close jungle, with its erect and much branched stem, about 1½ foot high, along the sandy shore. A few open spaces amongst the shrubby woods were covered, as if cultivated, with *Cyperaceæ*; amongst which a species of *Eriocaulon* is frequent. A few more watered spots showed *Grasses*, with a beautiful *Orchidea* 2 or 3 feet high. Near the village, I found *Eurphorbia drupifera*, Schum. An excursion to the river enabled me to examine the *Manghrove woods*, where a *Rhizophora* (different from R. *Mangle*?), but not yet in ripe fruit, formed the bulk of the wood: amongst it an *Avicennia*, judging by the leaves, different from that at Sierra Leone. Intermixed with these, *Drepanocarpus lunatus* rendered my progress very difficult. *Pandanus Candelabrum*, without leaves, occurs here, for the first time, in swamps. An *Anona* (a tree 10 to 12 feet high), in fruit, and apparently very similar to *chrysocarpa*, Lepr., if not the same, was not uncommon in these swamps. Leguminous trees seem rare, and do not attain a large size: there are no *Mimosae* or *Cæsalpiniæ*. Of cultivated plants, the *Sweet Cassava* is most valued and grown; also *Rice*, various sorts of *Capsicum*, *Papaw* and *plantains*, and *Holcus* here and there, with *Ananas* in large quantities amongst the shrubs [. . .]

Wednesday, July 14

WE left in the afternoon, and anchored on Friday, July 16th, about ten o'clock, A.M., off Cape Palmas, to take in a fresh supply of fuel. The Cape is formed by a narrow projection into the sea; on the foremost part of which, the houses of the American colony have been built. The dwellings of the fishermen are situated on the part nearest the main land. Their huts are very different from those of the Kroomen of Grand Bassa, being without raised floors, and having much more pointed roofs. The buildings of the American colony are straggling, and they extend, I was told, about four miles into the interior. There are none but people of colour at the Cape; the only whites, if I understand rightly, being a few missionaries, who devote all their attention to the natives. At this colony, the soil is very bad: the rock, frequently protruding through it, consists of hornblende (micaceous slate). The soil is a very hard iron-clay, in small clumps, originating, according to Rosher's statement, in the *debris* of decomposed granite veins transversing the rock; but to me it appears that the rock itself has much to do with the formation. Further up the stream, the land is said to be good. North of Cape Palmas, the river, according to the statement of the Governor, is navigable for seven miles with canoes, and empties itself into the sea, through several mouths. From a distance, the Cape has an agreeable aspect: the isthmus is well clothed with vegetation, and beyond it the beautiful forms of the *Oil* and *Fan-palms* are seen.

My excursions were limited to the isthmus and nearest parts. On the isthmus grows *Phoenix spinosa*, Th., a low shrub: beyond the river it is said to produce flowers and fruit. A few *Cocoas* had been planted, some years back, and were still small, as were the trees of *Anona muricata*. The plants chiefly cultivated seemed to be *Cassava*, *Sweet-potato*, *Bananas*, *Plantains*, *Indian Corn*, and *Rice*; whole *Cassia occidentalis* was seen in every cultivated spot: the same *Spondius* as before grows also here: *Coffee* had been introduced from Monrovia: here and there the indigenous species of *Cotton* had been raised: *Arachis hypogæe* (*Africana*?) I found planted in one place. In the native Flora, which, however, I have hardly seen, *Rubiaceæ*, *Convulvulaceæ*, *Leguminosæ* were chiefly conspicuous. The same *Anona* (near *chrysocarpa*) as in Grand Bassa grew here: *Jatropha Curcas* was frequently employed for fences. Amongst the underwood I found a small shrubby tree, related to *Belvisia* (*Napoleona*), and probably a distinct genus nearly approaching it, it bore blossom and fruit; the latter convinced me that I had seen the same, and a species but little differing from it, at Grand Bassa [. . .]

Thursday, August 5

WE left Accra after midnight, and cast anchor on Sunday, the 9th, at the mouth of a river, supposed to be the Nun. The weather was gloomy, and a dense rain falling all day, caused the wet to make its way through the shutters so that it was difficult to find a dry place, even for standing room. We stayed there the whole day, and sailed next morning for the mouth of the Nun, anchoring about nine miles off it [. . .]

Sunday, August 15

WE quitted our anchorage at half past eleven, A.M., and crossed without difficulty the bar; beyond which we cast anchor [. . .] at about a quarter to two, P.M. Here we stopped four days; during which I could only examine the right bank of the river, because I had no boat to get to the opposite side; where the greater extent of land and a village seemed to offer more interest. The river is here perhaps 10,000 yards wide; and the stream carries down a great deal of sand. The tide showed itself very distinctly, running perhaps three or four knots an hour, and the current seeming to set more on the left shore, which appears to be a mere sandbank, or sandy foreland, than on the right, which is covered with jungle, immediately beyond the sandy strand. The mouth of the Nun looks like a Delta, on a small scale; at least now, during the rainy season, being intersected by many shallow watercourses, forming, further on, low lands covered with *Mangroves*, similar to what I observed at Bassa Cove (Grand Bassa). The *Avicennia* appeared to prove, that the one hitherto seen, with quite naked leaves (*A. nitida*?) at Grand Bassa, is but a variety of that at Sierra Leone. In these Mangrove swamps, the *Oil palm* often grew, covered with parasitical Ferns (I found only two species of Ferns besides those, which are terrestrial), and on somewhat higher ground, *Drepanocarpus lunatus*, *Ormocarpus verrucosus*, a few shrubby *Rubiaceæ*, and a few *Mimoseæ*. Of the trees, intermixed with the Mangroves, little can be said: they were not many, and all covered, to the very top, with parasites. Some belonged to the genus *Bombax*. This land, if it can be so called, was but a few feet above high-water mark, and consisted of sea-sand and vegetable remains. The beach was quite flat, hardly higher than the sea, covered in many places with water, and formed of sand, mixed with mica, probably carried down by the Niger, and giving its shore a shining and peculiar appearance. In some places, the strand is clothed with jungle close to the sea, consisting of *Chrysobalanus icaco* and *Ecastophyllum brownei*; the fruits of the former, of a beautiful red, were very conspicuous. Intermingled with these grew *Melastomaceæ*, *Diodia maritima*, Th., some other small *Rubiaceæ*, and *Scoparia dulcis*; while the border, towards the higher woods, was frequently ornamented with the beautiful yellow flowers of *Hibiscus tiliaceus*. Amongst these shrubs, spots might be seen, here and there, covered with tall rough Grass and *Cyperaceæ*, to the height of a man, and higher, bound together by *Convulvuli*, *Cassytha*, and other *Lianes*, rendering them perfectly impenetrable. I found several places closely matted with *Stylosanthes guineensis*, forming carpets; upon which one might cross pools without observing them. The most barren and sandy places were much overgrown with a *Teleianthera*, R. Br., (*Illecebrum*, Schum.?) an *Eurphorbia* (*trineveria*, Schum.?) but especially with a yellow-flowered creeping *Dolichos* and *Convulvulus pes capræ*, (*rotundifolia*, Schum.), which latter is diffused over the whole coast from Monrovia. An *Umbellifera* (*Hydropcotyle interrupta*, *β*. *Platyph*. DC.), grew every where on the beach amongst the Mangroves, and seems to overspread the whole coast. A species of *Malaghetty Pepper*, differing from that in Grand Bassa by the long beak of fruit, was frequent.

On one spot, amongst the Mangroves, I noticed, on the decaying roots, a delicate white plant, having white scales instead of leaves, and three flowers: it was a parasite on the roots, but sent forth roots of its own. I have preserved a few specimens in spirits. Upon the whole, I have seen too little of the vegetation here, to compare it with that of any place hitherto visited on the coast. On the opposite shore, they cultivate *Cocoa Palms*, of which the natives brought us the nuts: on the right bank, where we did not now see any inhabitants, the *Cassava* showed traces of abandoned plantations. The scenery is not remarkable. At the entrance, the left side presented a pleasant prospect, from the familiar forms of the forest and brushwood, there appeared a sort of lagoon; while behind that the Mangroves rose into an erect and lofty-stemmed wood.

55

THOMAS EWBANK, *THE WORLD A WORKSHOP; OR, THE PHYSICAL RELATIONSHIP OF MAN TO THE EARTH*

(New York: D. Appleton and Company, 1855)

Chapter VI
The Three Storehouses of Matter

2. Vegetable Products

THE factory [Earth] would have been a failure if its operatives had had only minerals to work in. Properties not found in them were a necessity; and unless substances possessing them were to be had, the work would stop. The VEGETABLE department, by its contributions, prevented that. One prime desideratum was realized in the various kinds of wood – a species of matter eliminated by different processes than minerals, but one that added to their value by facilitating the means to procure and employ them.

Now, as wood could not be produced underground, and as, if it were developed like minerals in strata or wide layers, there was not room for it upon the ground, what was to be done? What would have been our suggestions on the dilemma? The needful quantity was so large that, if trees had grown horizontally, the earth had long ere now put on the appearance of a spherical raft of closely interlaced timber, floating in space as if for a market. There had been no room for man nor for his operations; and yet he requires almost the whole. Observe, then, with what singular wisdom and economy the exigence was met, and only a moderate, indeed a very small space, taken up. The new material was made to rise in vertical columns, whose lower ends only occupied the ground, all the rest being above, where there was nothing to interfere with or be incommoded by its extension; the intermediate spaces on the ground being, moreover, as usefully occupied as if the boles of trees had not interrupted them.

Another point of economy is exhibited in the general form of boles of trees. Their sections are circles; hence the greatest possible quantity of timber is compressed into the least possible space. Had they grown up in wide slabs, in square or angular masses, the same quantity of material would have taken up a great deal

DOI: 10.4324/9780429355653-61

more space, and would in other respects have been inconvenient. They would have been less able to resist storms of wind, and would have seriously interrupted the flight of animals through forests. Does the reader think there was abundance of room, and therefore no need of contrivances to make the most of it? Let us look at it: More than two thirds of the earth's surface are covered by oceans, seas, lakes, rivers, and other bodies of water-sources and distributing reservoirs of moisture and fertility, and at the same time theatres of aqueous and sub-aqueous life. Less than one third, then, only is left (and in it are included immense deserts of sand, wide regions of barren rocks, extensive moors, and arid wastes) for the swarming myriads of animals for which it has to provide room and food; for the moving millions of men, and their dwellings, farms, cities, factories, roads, and all other undertakings requiring room; for vegetable products required in a thousand manufactures, &c.; and the wonder will be how such masses of material are raised, and such inconceivable legions of living beings accommodated, on so small an area. Moreover, it is not merely what has been and is now required, but what will be when our species, instead of numbering one thousand, will amount to three, and perhaps to five thousand millions.

It would be superfluous to remark that woods most useful most abound, and that every climate has those most durable in it; nor need reference be made to the functions this glorious material fulfils; that to it we are indebted for social, civil, and manufacturing architecture; for navigation and its wondrous appurtenances; for implements and mechanisms without number; and for some of the most salutary and refining influences that pervade society.

As in the case of the metals, timber is provided in manageable masses. The size of trees is adapted for human not Cyclopean artisans. Had they generally approached in dimensions the great Californian cedar 325 feet high and 92 feet in circumference at the ground; 88 feet at four feet, and 66 feet at ten feet above the ground – what could have been done with them – with logs, one of which laid along the pavement of some streets, would fill them to the roofs of three-story houses! The difficulties of felling, transporting, handling, and slitting such into beams and boards, would have been seriously embarrassing.

[. . .]

The general truth remains that the largest trees are light and easily worked. The acacia is one of Asia's towering boles, but it weighs only 23 lbs to the cubic foot. The great firs of Europe only 17 lbs. But it may be said – How is it when from trees of the largest girth some of the hardest timber is derived? Why then the boles are decidedly shortened. Thus in the account of the woods of Africa in the report of the juries of the London exhibition, the ironwood tree, it is said, acquires a diameter of four feet, but the height of the trunk varies between 25 and 45 feet. The blaauw-bosch attains to eight or nine feet in diameter: as the wood is hard and heavy, masts of it from 200 to 300 feet long would be all but unmanageable; but what is their height? It varies between five and twelve feet! Another species has a diameter of seven feet, and the height of its bole confined to twelve, but more generally to ten feet? The boschquarry has a diameter of six feet ten inches, with a trunk six feet eight inches! Again, the seybast is seven feet nine inches in diameter, and in height seven feet ten inches.

The baobab, or monkey-bread tree, is among the most colossal of vegetable columns, but they are rather Doric than Corinthian in their proportions. In Senegambia trunks of great antiquity have been found from twenty to thirty feet in diameter, their heights not exceeding twice their thickness. One, thirty-two feet in diameter, is supposed to be from five to six thousand years old. Finally, yew trees are often of immense dimensions, but chiefly in their girth. The famous chestnut trees of Etna are so short that at a distance each has been taken for a group.

From what is at present known of the forests of the earth, this precious material is evidently prepared for us in masses perfectly within ordinary efforts and appliances to dispose of. Then, a perpetual supply is secured. Man cannot, if he would, waste his mineral stock; it costs too much to raise it to allow him to do that. But of timber and other vegetable products he might, through indolence and carelessness, bring about a scarcity; and so he would, were it not for the provision, that they have in themselves the elements of their preservation and multiplication. Acorns drop and take root without his care.

[. . .]

In the wonderful compounds of matter that make up the second great division of material, how much there is to elicit and exhaust admiration. A new world of thought and of art was opened in wood simply: so different from minerals, in its being developed before our eyes, in the system of perpetuating its varieties, in the diverse magnitudes of trees and their variegated crowns of foliage; in the mechanical properties of the ligneous fibre; in its diverse degrees of hardness, softness, flexibility, elasticity, and texture; every feature offering a class of advantages in the arts: in its ornamental attributes, too, as exhibited in colors – jet in ebony, black and dark brown in walnuts and oaks, purple and light greens in the munjaddy and myle-ellah of India, red in mahogany and cedar; yellow in box, satin-wood, and the maples; then there is the red ebony of Australia, the cream-tinted and snow-white tulip tree, and every shade and tint in others. [. . .]

Then woods, besides furnishing examples of painting in colors, provide us with material for giving to other substances colors which they do not always themselves possess. Each pigment, too, besides imparting its every tint, contributes to develop other and very different colors. Logwood yields blacks and purples; fustic, olive-browns and yellows; barwood, camwood, Brazil, and sappan woods impart reds, blacks, and browns; woad and indigo, blues and greens; madder, the brilliant scarlet or turkey red; turmeric, bright yellows; orchil, purples, reds, and blues; annatto, orange; safflower, crimson, scarlet, and pink. There is the green ebony, and a thousand more dyewoods, known and unknown.

From the same shelves in the vegetable storehouse we receive the great preservatives of colors, in the various lacs prepared from gums and resins.

Now, what is the language of these particulars, and of forest timber at large? That wood is specially designed for man – that no other occupant of the earth comprehends its worth, or can use it, or the lacs and dyes which it furnishes. Had it all been light and porous as the sycamore or cork tree, or, on the other hand, had it been heavy and dense as lignum-vitæ, it had been of comparative little value to man. But we are ordained to be elaborators in wood as in the metals; and hence the facilities for its acquisition, its varieties of masses, properties, and adaptations.

What proportion the amount of timber employed in the arts bears to that of mineral bodies has not, I suppose, been ascertained, perhaps not thought of. The beams, floors, partitions, stairs, window-frames, wainscotting, doors, roofing, and furniture, approach in bulk the material of the walls of most houses; but take the dwellings of man at large, and the greater part are built wholly of wood; add that which is consumed in them as fuel, and all the firewood used up in the earth's factories; then take into the account the timber worked up by ship-builders, carriage-makers, and other workmen in it: and the amount will equal, perhaps exceed, in cubic feet, all the materials drawn out of the earth. In any point of view the amount is extraordinary, and especially so, considering the very limited area of the earth's surface – a mere fraction – from which it is taken. But the secret is in its development: had it matured slowly, as minerals, not a ripe tree had been left standing, in the face of the enormous and incessant demand. How simply and beautifully are all difficulties arising from rapid demand met by rapid production. By this, supplies are secured, and will be secured to artisans, although civilization levels, and will continue to level, so many forests.

Of the annual accounts of ship-building in the United States, the tonnage for 1852 amounted to 35,149.41 tons. The lumber trade of a single town in the United States – Bangor, in Maine – amounted in 1853 to 182,942,284 cubic feet.

After timber was added to the artisan's stock, many chasms remained to be filled from the same department. One or two may be noticed. Wood and metals serve admirably to transmit force from one end of a bar to an object at the other, as in a crowbar or lever, the handle of a spade or a hammer; but in a wide class of cases it is necessary to send forces over distances, and in directions, where inflexible rods or shafts could not apply; as in hoisting a ship's sails or anchor, raising coal and ores from mines. A material, light, soft, pliable, tenacious, and easily handled, was wanted – *i.e.* a material for ropes; and how varied and inexhaustible the sources of supply are everyone knows. Into few things was man more early initiated than in the use of ropes. Long before a tree was cut down he employed them. In tropical forests especially, natural ropes abound, everywhere pendent from the highest trees, and running along the ground, thousands of feet in length, uniform in thickness, and varying in dimensions from cables to whip-cord. The prosperity and progress of the arts depended, and depend no little on cordage: not by any conceivable possibility could they have come up to what they are had ropes and pulleys never been known.

Another primitive and permanent application of vegetable matter constitutes the world's wicker, basket, chip and straw-plaited wares; embracing agricultural and mechanical implements – *e.g.* the bodies of carts, the unique tepiti or mandioca press of the Caribs and of South American Indians; and an infinity of personal, social, and domestic articles and utensils. In the manufacture of straw-plait alone, seventy thousand persons find employment on one of the earth's small islands. 1,577 cwts of plait were imported into England in 1852 for home consumption.

The vegetable world furnishes the most of our clothing. The annual produce of thread is, in its lineal extent, all but inconceivable. 1,481,000,000 – one billion four hundred and eighty-one millions of pounds of cotton were worked up into it in 1852. At the London Exhibition one manufacturer furnished samples of one pound of cotton spun into 900 hanks of 840 yards each, making nearly 430 miles. Another firm exhibited 4,200 hanks of the same numbers of yards each, making 2000 miles from a single pound of cotton! If we therefore multiply the above amount only by 430, the length of thread that a single crop of cotton could make, would be over six hundred billions of miles, or sufficient for a web of stout calico, a yard wide, and containing 85 threads to the inch, that would be more than enough to reach from us to the sun.

And yet all this is from cotton alone. Hemp and flax in some measure rival it: of them there were raised in the U. States in 1850, not less than 1,860,000,000 of pounds. In the rapidly increasing demand for material for woven fabrics and for machinery to manufacture it, but a few years would be required for our looms to fill an order for webs of double belting, sufficiently long to connect the Sun with each of the planets, in the way motion is communicated from the large drum of a factory to a number of smaller ones. We inclose our bodies in artificial cocoons:– In winter a lady is enwrapped in a hundred miles of thread; she throws over her shoulders from thirty to fifty in a shawl. A gentleman winds between three and four miles round his neck and uses four more in a pocket-handkerchief. At night he throws off his clothing and buries himself like a larva, in four or five hundred miles of convolved filaments.

Still, exceedingly few of the fibre-yielding plants have been taken up by manufacturers, and yet they abound everywhere – in weeds, sedges, coarse grasses, and in the leaves of some of the commonest shrubs and trees. The banana and its relatives have recently been named as examples, which besides fruit would yield from 9,000 to 12,000 lbs per acre, of fibre, fit for fabrics of every degree of fineness, from muslin to ropes. Countless millions of tons of this and kindred substances spontaneously shoot up every year and sink again into the ground, neglected by man.

For her factories, England imported in 1851, 1,301,488 cwts of hemp and 11,194,184 cwts of flax = nearly 700,000 tons of 2000 lbs each.

[. . .]

The very first of necessities is also supplied from this department. It is here that man's chief pantry as well as his wardrobe is placed. A description of aliments stored in it, their varieties, abundance, and means to improve them, cannot of course be attempted in these pages. Reference to some items will serve to show how liberally the Proprietor of the factory has victualled it. There were raised in 1850 in the U. States upwards of 592 million bushels of maize or Indian corn. Counting the bushel at 14 cubic feet, the grain would have filled a store-room, twenty feet wide, ten feet deep, and seven hundred miles in length. The yield of wheat in 1851 (125,607,000 bushels) would require an additional twenty miles to the structure; rye thirteen, buckwheat nine, barley four, between eight and nine for peas and beans, three or four for rice, and not less than five hundred for potatoes, beets, and other tubers. Partitions, miles apart, would be also required for apples, peaches, grapes, plums, cherries, and orchard produce; for sugar (over 200,000,000 lbs), nuts, strawberries, gooseberries, currants; for peppers, mustard, spices, and condiments, and all the produce of market gardens, over a thousand miles more would be taken up-to say nothing of tanks for molasses, wines (in 1853, two millions of gallons), ale, cider, and other drinks. But figures soon lose their force on the mind, and it is the same with magnitudes when repeated. Besides, to acquire definite ideas of the riches of vegetation, definite quantities should be ascertained and considered in reference to the areas whence they are taken. When this is done with respect to every item in our world's delicious and plenteous bill of fare, torpid must be the souls that peruse it without emotions of admiration and gratitude.

Of tea, England imported in 1853, 66,360,555 lbs; coffee, the world's product is between three and four hundred thousand tons. The world's crop of sugar from cane, beetroot, and maples, cannot be less than 900,000 tons, since the amount recognised in commerce is 840,365 tons. The demand is rapidly swelling, but however much it may increase, there are no limits to the means of supply.

The bread boundary of the earth is the widest of zones, extending from 45° north to 50° south of the equator while within it are others that foster maize, mandioca, yams, plantains, bread-trees, cocoas, sago, and others, so that man's larder is fully supplied with bread, and quite as generously with fruits, fish, and meats.

Sicily, Barbary, and Egypt were formerly the granaries of Europe, but are not now, because of the decay, not of the land, but of the people. The supplies now come from the south and south-east plains of the Baltic. Among the places of export is Dantzic, and it has sent out a million of tons of wheat and rye in a year. The average of Russian exports of the same cereals, from 1838 to 1840, was 4,500,000 tons a year.

When man enters this dépot for supplies, something more is required of him than when he applies at the storehouse of minerals. The latter are produced without his assistance; the long periods required for them to mature in, put it out of his power to affect them. He can neither change their quantities nor their qualities, while in vegetables he can do both. Had the nature of minerals been such as to admit of his labor in their preparation, it had certainly been required of him, since the purport of his existence was to be attained through his acquaintance with the compositions and evolutions of matter; but the processes of their formation are so slow, that had the tenure of his life extended into centuries, he could not have biased their development. Now the producing powers of vegetation are so active as to induce greater changes in a day than do those that form minerals in a thousand years, so that he has every opportunity to impress himself on them.

But is it his duty, and has he the power? Undoubtedly; although it may be there are those who think he cannot meddle with nature's works without marring them. A great mistake. In this department she produces nothing absolutely perfect without him, and she will not. Designed for a nursery, it requires nurserymen. Forests and prairies are at large, what neglected farms are in little. They cover the ground with things growing rank and wild, and choking each other; they are what he himself is before being drawn out of the jungles of ignorance and improved by cultivation. The principles at work and the soil they work on are at his service; but like tools in a machinist's shop, their profitable employment rests with himself. They will cover his fields with wheat and fill his gardens with fruit, if he so wills, by properly exciting them. If he fold his arms in indolence, they will expend themselves in weeds. In this department man was to acquire a very large portion of his knowledge of matter, and of his experience as a manipulator; hence, whatever he has the ability to do, is left for him to do. Perfectly developed organisms are not produced for him, but their germs are supplied, and agencies to unfold and ripen them. Spontaneous growth shows the workings of these agencies, but not their perfect working; that is left for him to bring out.

In some respects, planters and farmers, florists and fruiterers surpass other elaborators, inasmuch as they join nature and improve her products before leaving her hands. They cause qualities to appear where they were not, and in greater or less quantities where they were. They diversify dimensions, colors, and texture. The variations and multiplications of plants, extinction of old and introduction of new ones, are with them.

It is, then, by bringing in plants from the wilderness, domesticating and carefully cultivating them, that they are to be improved. It is thus that the sloe-bush has been changed into the plum tree, and grasses into corn bearing cereals. Vegetables are literal mechanisms for elaborating matter; and to improve their products, they themselves must be improved or changed, just as artificial contrivances are, when required to turn out better goods. Indeed, there have been made as great and beneficial changes in natural as in artificial mechanisms, and the future will no doubt record equally new achievements in both. The former can be multiplied indefinitely, and with infinitely greater facility than additions are made to looms and spindles. New varieties, moreover, produced by hybridity, and corresponding with new inventions in the arts, will never cease. Then as regards economy of space, a matter of the first importance in reference to a future densely populated world, it has been shown (in the Kew Gardens of London) that on an area where not over two hundred plants would grow in a wild state, twenty thousand have been made to flourish.

[. . .]

Some political economists have declaimed against foreign commerce that exchanges flour, corn, and other products for hardware, dry-goods, and fancy merchandise; the fertility of land, the essence of it, being bartered for things that return nothing to replenish it. The fruitfulness of an island or a continent, it is said, may thus be exhausted. Can this be? Is there no compensating principle at work in nature to prevent so serious an evil, and one that might derange the whole economy of the earth? Surely there is, and is it not in the atmosphere? All plants derive their food directly or indirectly from it; and do not trade-winds, aerial storms, ordinary gales, and ceaseless movements in it tend to maintain an equal distribution of the matter it holds in suspension – sufficient at least to counteract general and permanent if not local and temporary irregularities. Plants raised in glass vases from earth weighed and dried in an oven were found to take nothing or next to nothing from the earth. The rich palm-oil of Africa is from trees that luxuriate in hot dry white sand; so it is with many or most of the cacti. The olives of Sicily flourish on rocks. It is indeed an agricultural axiom that a numerous people can never be absolutely dependent on the soil of other countries for food. In Great Britain and Ireland, are unproductive lands sufficient to feed over eight millions of additional population; and of England alone, the most cultivated of the two islands, it has been stated by competent authorities that the produce might be doubled.

On the limited state of our knowledge of vegetable arithmetic, Humboldt observes, that if we had sufficient grounds for believing that one half of the phænogamous plants were known, and taking the known at 160,000 according to one estimate, or at 213,000 at another, we should have to add from 25,000 to 35,000 species of the grasses. And as these appear to form one twelfth of the Earth's plants, the united numbers would only amount to one eighth or one tenth of the species that now exist?

The assumption that we already know half the existing species of phænogamous plants is further opposed by the following considerations. Several thousand species of Monocotyledons and Dicotyledons, and among them tall trees, have been discovered in regions considerable portions of which had been previously examined by distinguished botanists. The portions of the great continents which have never even been trodden by botanical observers, considerably exceed in area those which have been traversed by such travellers, even in a superficial manner. The greatest variety of phænogamous vegetation, the greatest number of species on a given area, is found be tween the tropics, and in the sub-tropical zones. This last mentioned consideration renders it so much more important to remember how almost entirely unacquainted we are, on the New Continent north of the equator, with the floras of Oaxaca, Yucatan, Guatemala, Nicaragua, the Isthmus of Panama, Choco, Antioquia, and the Provincia de los Pastos; and south of the equator, with the floras of the vast forest region between the Ucayale, the Rio de la Madeira, and the Tocantin (three great tributaries of the Amazon), and with those of Paraguay and the Provincia de los Missiones. In Africa, except in respect to the coasts, we know nothing of the vegetation from 15° north to 20° south latitude; in Asia we are unacquainted with the floras of the south and south-east of Arabia, where the highlands rise to about 6,400 English feet above the level of the sea; of the countries between the Thianschan, the Kuenlien, and the Himalaya, all the west part of China, and the greater part of the countries beyond the Ganges. Still more unknown to the botanist are the interior of Borneo, New Guinea, and part of Australia.

As, therefore, by the progressive exploration of new countries, we gradually exhaust the remaining unknown species of any of the great families, the previously assigned lowest limit rises gradually higher and higher; and since the forms reciprocally limit each other, in conformity with still undiscovered laws of universal organization, we approach continually nearer to the solution of the great numerical problem of organic life.

Part 6

ZOOLOGY

Zoology

THIS section explores Victorian interest in the scientific study of animals, but this also overlaps with Parts 4 (Comparative Anatomy), 7 ('New World' Environments and Exploration), and 9 (Evolutionary Thought Before *Origin of Species*) of this volume and with many parts of Volume II. The extracts included here are testament to the same aims evident in Part 5 (Botany) in terms of enlarging knowledge of species, providing clearer classification, and examining the relationship between the form and function of organisms (and how this is affected by their relations within specific environments). Like the previous section, it also bears witness to increasing scientific professionalisation and specialisation (the emergence or consolidation of dedicated fields for the study of different animal groups), to the increasingly global reach of zoology, and to a desire to exploit animal life for economic ends. As well as reflecting attitudes of human sovereignty over animal life, however, some of the extracts speak to an emerging interest in, and wonder at, the diversity, variety, and selfhood of animals.

The first extract, from Gilbert White's *A Natural History of Selborne*, represents the status of late-eighteenth century zoology but reflects the author's pioneering attitudes. One of the excitements of reading White is the sense that he is involved in tentative, provisional production of knowledge. He makes postulations on the basis of his own close observations and from what he gathers from correspondents or reads in major works (he frequently consults Ray and Linnaeus), but he laments the lack of like-minded locals with whom he can discuss natural history. His writing is sometimes astonishingly ground-breaking – he offers the earliest description of the harvest mouse, for example – while elsewhere he appears, to modern readers, to err almost comically – as when speculating about toad venom or suggesting that snakes eat during only one season. Above all, though, there is a vivid impression of daily efforts to engage with his local environment and to understand the lived relationships of animals (and particularly his beloved birds) to their surroundings. There are running threads in the letters – discussions returning to the at-the-time unsettled questions of bird migrations, for example, or the identification of specimens that White has shot or gathered. There is little interest in the comparative anatomical work that would follow in subsequent decades: White is interested in whole organisms but is not, in general, a dissector. At the same time, his activities (including taking a harvest mouse nest with eight young inside) or shooting birds for specimens, speak of eighteenth-century attitudes as yet untouched by early movements for animal protection (a subject to which we will turn in Volume II). There is a melancholy sense for modern readers in hearing of the abundance of bird life in White's district – the 'vast flocks of the common *linnets*' that he records can, for example, no longer be witnessed there, and his work stands as a record of biodiversity loss that speaks mournfully of the impact of the nineteenth and twentieth centuries on our zoological neighbours.

A few explanatory notes are necessary. The harvest mouse (*Micromys minutus*) was first described here and in subsequent letters by White, but credit for its formal naming in a scientific journal goes to Peter Simon Pallas. The botanical calendar by a 'Swedish naturalist' to which White refers is A.M. Berger's *Calendarium Florae* (1756). In his discussion of birds of prey in Letter XII, he adds a footnote identifying 'this hawk' as '*falco peregrinus*' (peregrine falcon). Throughout the letters, White consistently misspells 'its' as 'it's', and this has been preserved in the extracts, along with his other variant spellings.

William Smellie's *The Philosophy of Natural History* (1791) provided an extract in Part 2 of this volume. We return to this important work now, for Smellie's remarks on the beaver in his chapter on 'Habitations of Animals', to find another example of sympathetic engagement with zoological life. Like White, Smellie is less interested in classification and anatomy than in habits and behaviours. Drawing on Buffon, he provides an intimate picture of the daily community lives of beaver colonies and speaks with warmth and admiration of their cooperativeness, dexterity, and ingenuity. Without ever lapsing into anthropomorphism, Smellie depicts beavers as capable of constructing habitations and colonies that are almost human in their level of sophistication. In this sense of wonder and admiration, and in the manner in which he seems to proclaim the species' right to selfhood and autonomy, Smellie's work anticipates much later manifestations of environmental engagement.

The third extract, from the first volume of John Curtis's *British Entomology* (1824), offers a glimpse of the pre-Victorian emergence of the scientific study of insects. The volume is devoted to the Coleoptera (beetles). The value of Curtis's work lay in helping to establish a scientific basis for insect classification and to resolve differences of opinion on the status of various species, many of which had by this point been given numerous names by different experts. The descriptions included, of the spotted elephant and spotted elephant hawk moth, are largely dry, laden with specialist anatomical terms – in this sense, a zoological counterpart to the standard botanical practices of John Lindley and W.J. Hooker (see Part 5 of this volume) and testament to the ways in which scientists were pursuing similar aims and methodologies across different fields. Disputes in natural sciences like zoology and botany largely arose over details (the names of species, their placement within particular genera or families, etc.) rather than more fundamental issues. These would arise in time – due to the seismic impact of evolutionary thought on all fields of natural history. Works like Curtis's exist in a period of relative calm before this storm, although nomenclatural disputes were often heated.

The fourth extract permits us to enter a field that was gaining greater prominence during the Victorian period, the study of microscopic animal life. Andrew Pritchard's *The Natural History of Animalcules* (1834) is by no means the first work of its kind, but Pritchard was a leading figure in the development of British microscopy and of its applications to natural history. The extract includes some of his Preface and introductory remarks in Book 1 and a representative example of his detailed description of one genus (*Volvox*). This example is also instructive: although Pritchard describes animalcules as 'living creatures inhabiting fluids' and as 'diminutive animals', he includes *Volvox*, in fact a species of algae (i.e., a plant). Such basic errors give a reasonable sense of the emerging status of the field at this period. Although investigations of 'animalcules' have a long history, dating back to the pioneering work of Antonie van Leeuwenhoeck in the 1600s, there is a strong sense in Pritchard's work of a field still being established, in which discoveries were being made and principles and systems worked out. The first decades of the nineteenth century were key in this because of advances in optical instruments. As Kate Flint (Further Reading) has shown, technologies like microscopy and telescopy were part of a much wider cultural re-invention of the notion of sight in the period.

Compared to many of the zoological and botanical works extracted in this volume, Pritchard's work is less interested in remorseless classification and enumeration of anatomical features and more involved in general descriptions. Directed at both scientific and general readers, it is at pains to underline Pritchard's own sense of excitement in a field of study that 'must carry with it an intensity of interest to the mind of every human being'. There is also a sense of Natural Theological intent in his opening remarks, speaking as they do of opportunities 'to recognize and adore the hand that guides her through all the vast variety of her stupendous operations'. This view was not universally shared, however: many expressed revulsion at the realisation that glasses of water, foods, the surfaces of human flesh, and the insides of human bodies swarmed with millions of creatures that could only be seen by the aid of the microscope and that appeared to many monstrous when viewed under these conditions. Much of this revulsion lay in microscopic investigations of water – that most needful of substances to humans – at a time when pollution of water supplies was a pressing concern: pioneers like Arthur Hill Hassall would soon cast a microscopic eye over the drinking water, dinner plates, and larders of his Victorian readers (see Volume II of this anthology). In an 1851 article, Charles Kingsley drew on his own recent experiences of microscopy as he represented London's notoriously filthy water supply as overburdened with monstrous bodies:

> You are literally filled with the fruit of your own devices, with rats and mice and such small deer, paramecia and entomostraceae, and kicking things with horrid names, which you see in microscopes at the Polytechnic, and rush home and call for brandy – without the water – with stone, and gravil [sic], and dyspepsia, and fragments of your own muscular tissue tinged with your own bile.
>
> (1851, 229)

Alert to the dangers of foul water, Kingsley also connects it to the complicit human body via a nauseatingly open system that recirculates human excrement through watercourses and the canal of the body. Although there was at the time of Pritchard's work no inkling of the waterborne nature of the transmission of diseases such as cholera and typhoid, the world of microscopic creatures was a source of dread to many – particularly as the revelation of this previously little-studied corner of the animal kingdom seemed to render God's work of creation mysterious in its intentions. At the same time, this registered a marked expansion in the scope of animal life that for others, including Pritchard, was a source of wonder.

Pritchard draws upon the prior work of Otto Friedrich Müller, Baron Gleichen, and Christian Gottfried Ehrenberg, who exemplify the dominance of German states in nineteenth-century microscopy, but Pritchard contributes new knowledge, clarifying many points in relation to the motion, digestion, and reproduction of various species. He also cites Cuvier and is clearly influenced by comparative anatomy in his examination of relationships between the forms and functions of animalcules. *The Natural History of Animalcules* is as much a practical as a scientific guide, however, showing readers how to obtain, prepare, and view specimens; obtain equipment; and undertake mathematical computations. Pritchard refers to his own earlier work, *The Microscopic Cabinet* (1832), which contains guidance on building microscopic equipment as well as detailed descriptions of species; to various new microscopic models and contrivances; and to 'the common Argand lamp', an advanced and bright oil lamp invented by François-Pierre-Amédée Argand.

The fifth extract introduces us to Charles Waterton, a significant figure in early nineteenth-century zoology famed for his explorations (in his 1825 work, *Wanderings in South America*, which includes what was for his readers a stirring description of wrestling a cayman); for his pioneering attempts to establish a walled bird preserve at his Walton Hall estate; and for his personal eccentricities, which supposedly included sleeping on the floor with his dogs.

Waterton's *Essays in Natural History* (1838) brings together various previously published pieces and new materials. He ranges widely across the globe and the animal kingdom, but, like Gilbert White, he is particularly fond of birds. The short extract chosen – 'On the Habits of the Rook' – exemplifies his enthusiasm. Waterton belongs firmly to the Gilbert White tradition of gentlemen amateurs, and his approach is by no means in tune with the professionalisation and specialisation of nineteenth-century natural science. In particular, there is a strikingly conversational tone and a scene-setting that belongs firmly within the English 'nature writing' tradition of White, Richard Jefferies, W.H. Hudson, Robert Macfarlane, et al. In setting the scene for his writing, Waterton makes poetical allusions. Speaking of 'November's dark and stormy nights [being] close at hand', he conjures 'such a night, probably, as that in which Tam O'Shanter unfortunately peeped into Kirk Alloway', referring here to Robert Burns's 1791 narrative poem, *Tam O'Shanter*. Waterton grounds his discussion in the local, making many references to places most familiar to him, including Walton Hall and Nostell Priory, West Yorkshire, but as an inveterate traveller and reader, he also invokes international locations in his essays.

While Waterton reaches back to an older tradition, his work is strikingly modern in its focus on bird behaviour. Waterton's astonishing patience and observational capacity provides a thorough account of the habits and behaviours of rooks in ways that foreshadow future directions. He has a conservationist eye, arguing against criticisms of rooks for allegedly feeding on harvest cereals by pointing out the beneficial effects of their predation on insects harmful to crops. His accompanying remarks on grackles (*Quiscalus*), which he calls grakles, are particularly fine examples of the attention that was beginning to be paid to the complex interactions of species within a particular habitat and the ways in which the decline of a population (in this case, through the extermination of grackles) leads to the proliferation of others (in this case, unwelcome crop-eating insects). As Richard H. Grove (1995, 61–72) points out, European interventions in the New World, and especially the occupation, deforestation, and cash-crop cultivation of islands in the Caribbean and Indian Ocean, had profound environmental impacts of these kinds. By intervening in such matters, Waterton also works in territory that would be occupied by Charles Darwin and other evolutionary scientists and by the ecologists who followed. The interventionist nature of the experiment he then describes, involving taking and replacing corvid eggs, is deeply distasteful, but it is also testament to Waterton's desire to learn from direct observation of living creatures and to advance ornithological knowledge.

The sixth extract returns us to Charles Darwin's *Journal of Researches into the Geology and Natural History of the Various Countries Visited by H.M.S. Beagle* (1839), a book to which we will return in subsequent sections. The extract here, from Darwin's travels in Argentina, is a global counterpart to Waterton: just as Waterton turns an observant eye on the life, habits, and behaviours of the rook, so Darwin does the same for the ostrich and the rhea. His account is extraordinarily rich and vivid, and it is little surprise that *Journal of Researches* quickly proved a bestseller: Darwin combines cutting-edge scientific discovery with compelling descriptions of what were still to his readers unfamiliar and 'exotic' landscapes. The depth of his engagement with South America outstrips that of Waterton a decade earlier, but although he is a more advanced scientist, the similarities between their two ornithological accounts are striking.

While the first half of the extract offers a richly detailed account of the ostrich, the second turns to Darwin's attempts to find and identify a species frequently discussed by his gaucho guides, who described it as a smaller species of ostrich. Ultimately able to locate them in Patagonia and to collect specimens, Darwin enabled the discovery of a new species – now known as Darwin's Rhea – an attribution confirmed by Darwin's London-based collaborator, the celebrated ornithologist John Gould, who in 1837 gave it the name *Rhea darwinii* in Darwin's honour (although Alcide D'Orbigny's ascription of the species that year as *Rhea pennata* ultimately prevailed).

Gould's significance to Darwin's ability to turn his *Beagle* discoveries into evolutionary science will be treated in Part 9 of this volume, but he also provides the seventh extract in Part 6, another example of ornithology. Gould produced several large works covering the birds of various countries or continents. *Birds of Australia* (1848) underlines the massive surge in zoological knowledge as a result of overseas voyages and the growth or establishment of European and American scientific institutions, museums, and private collections. It also shows Gould transforming from an arranger of Darwin's discoveries to a naturalist-explorer in his own right. The extract includes Gould's reasons for pursuing the opportunity to add to his celebrated achievement, *Birds of Europe* (1832–37), with a whole new volume of fascinating and novel birds. The extract includes descriptions of the quintessentially Australian parakeets and bower birds, examples of the kinds of striking exotic discoveries that were exciting both middle-class drawing rooms, specimen collectors, and scientists. Unfortunately, it is beyond the scope of this project to include the fabulously lavish illustrations that accompanied this and all of Gould's work. This volume is tinged with personal sadness, being the first produced after the death in 1840 of his wife, Elizabeth, his principal collaborator, who produced many illustrations for this and other volumes.

Gould's work is a good indicator of the prominence of science in nineteenth-century society: diverse figures, from Michael Faraday and Humphrey Davy early in the century to Gould, Richard Owen, Philip Gosse, Charles Darwin, and others, became household names, their works feted and widely read by a newly forged middle-class readership eager for instruction and entertainment and for books that could demonstrate their refinement and education. Gould's skilful combination of the personae of the scientist and the showman is evident in his preface, in which he declares that the study of Australian birds 'was comparatively a new one, and of no ordinary degree of interest' because 'its natural productions [are] as remarkable for the anomalous nature of their forms, as for their beauty, and the singularity of their habits'. Gould sells the opportunity to experience the strange animal lives of a faraway antipodean land from the comfort of one's own home and to gaze on forms deliciously 'exotic' to a British eye. He also burnishes his credentials as an adventurer, indicating that as he was not content to simply examine limited collections of Australian bird specimens available in Britain, he determined on a two-year exploration of a land mass still at this time largely unmapped. Gould's extensive namedropping of the illustrious colonial administrative and military figures who rendered him assistance in this task also gives a sense of the strengthening grasp of the U.K. government of a colony that was beginning to fulfil its economic potential but which also involved the task of repressing and marginalising its aboriginal populations.

Gould acknowledges using data gathered by Captain Ross's southern explorations on the *Erebus* and *Terror*, thus overlapping with J.D. Hooker's account of that voyage in *Flora Antarctica* (see Parts 5 and 8 of this volume). The power of these formal and informal scientific networks was remarkable, a sign of British and European power that was also manifest in its increasing control over lands, environments, and populations. After boasting of having penetrated 400 miles into the interior, Gould regrets the loss of his assistant, Mr Gilbert, in a subsequent journey, 'the party being treacherously attacked by the natives'. Gould contrasts the 'treachery' of 'savage' natives with the noble pursuit of knowledge undertaken by European naturalists. Tireless, Gould did not even rest during the journey to and from Australia, instead gathering available information on marine birds. His antipodean work doubled the number of known species of Australian birds and advanced knowledge of their status, habits, and behaviours – just as European occupation of the continent had begun the process that would ultimately compromise its biodiversity. There are allusions to the growing propensity for the collection of live as well as killed animal specimens in Gould's regretful remarks that a pair of satin birds destined for the Earl of Derby's 'magnificent aviary at Knowsley' did not survive the journey. As the century proceeded, the trade in live animals would become an enormous and sordid business, a zoological counterpart to the slave trade in its scale and remorseless exploitation.

The final extract sees us return to Thomas Ewbank's *The World a Workshop* (1855), already featured in two previous sections, moving from the earth's botanical 'storehouse of matter' to its zoological equivalent. While strictly speaking barely a work of zoology, it is included as a stark example of the manner in which fauna were routinely regarded as 'products' or, in Ewbank's phrase, as 'working stock' of 'elaborating machines', divinely designed for the sole benefit of humankind. The degree to which Ewbank is wedded to a model of human sovereignty based partly on Biblical precept and partly on arguments for human exceptionality is clear in his claim that 'though man cannot originate living organisms, he can control them so far as essentially to modify the products they yield him', a vision of the natural world as both blank canvas and machine-like resource.

In Ewbank's vision, humans are 'as tenant and manager of the factory' of organic and inorganic life that is his workshop world. That this is a workshop of slaughter and destruction is starkly revealed by his statistics on food production and resource exploitation. Ordained within this bloody priesthood, *Homo sapiens* is thus permitted not only to take but to modify animal life through breeding in order to increase its yields of 'matter': 'nothing', he proclaims, 'could more emphatically proclaim him a manufacturer than the power given him over the development of most of the substances' of animal and vegetable life. Ewbank's triumphal enumeration of animals kept and slaughtered in the United States, and of the economic benefits accruing from their exploitation, offers a grim picture of accelerating mechanisation that anticipates the direction that would lead to the hellish vision of the Chicago stockyards in Upton Sinclair's *The Jungle* (1906) (see Volume IV of this anthology). What is obviously missing from Ewbank's celebration of the productive capacities of the zoological storehouse, and of human ability to exploit it, is any sense of the costs of this approach to non-human others. Less than a decade after the publication of *The World a Workshop*, George Perkins Marsh's pioneering work of ecology and environmentalism, *Man and Nature* (1864), would take a very different approach, providing evidence of the ways in which human activities had already, since Classical times, caused desertification, declining fertility, loss of habitats, and species extinctions and declines. While Marsh's work (see Volume III of this anthology) would provide both an elegiac account of loss and a warning about the future, Ewbank can see nothing but progress and triumph, glibly proclaiming that in the United States, 'the buffaloes alone would form an unbroken phalanx round the earth, and the wild horses another'. The cost of their devastating exploitation would be a tale told in the second half of the century – but its lessons would be drawn from Marsh's ecological approach rather than Ewbank's economic gaze. The juxtaposition within a few years of mid-century of these two works tells us a great deal about changing attitudes to environment at this time. Ewbank's confident assertion that the thousands of tons of fish, whales, and other sea creatures gathered by human enterprise 'form but a small item in the annual yield of the deep sea, shore, lake, and river fisheries of the United States, and an insignificant one in that of the earth's fishery as a whole' has proved fatally flawed in our age of ocean crisis. His remarks on the widespread use of feathers 'employed . . . chiefly for various articles of female attire and accompaniments – tippets, boas, muffs, &c.' bring to mind the fact that later in the century, groups such as the Royal Society for the Protection of Birds were formed precisely to protect species, such as the great-crested grebe, from extinction threatened by the exploitation of their plumage for ornament.

Although this is one of the shorter sections of the anthology, its preoccupations are taken up in a number of the following sections, where the various ways of analysing, classifying, observing, and exploiting animals and their environments are approached from somewhat different angles. In short, though, the story of early nineteenth-century engagements with zoology constitutes a major discourse centred on a logic of human sovereignty and imperialism and a minor, emerging, and hesitant discourse involved in querying the certainties of this anthropocentric and Old World-centric perspective.

Further reading

Amato, Sarah, *Beastly Possessions: Animals in Victorian Consumer Culture* (Toronto: University of Toronto Press, 2015).

Armstrong, Patrick, *The English Parson-Naturalist: A Companionship Between Science and Religion* (Leominster: Gracewing, 2000).

Barber, Lynn, *The Heyday of Natural History 1820–1870* (London: Cape, 1980).

Clifford, David, Elizabeth Wadge, Alexandra Warwick and Martin Willis (eds.). *Repositioning Victorian Sciences: Shifting Centres in Nineteenth Century Scientific Thinking* (London: Anthem Press, 2006).

Colley, Ann C., *Wild Animal Skins in Victorian Britain: Zoos, Collections, Portraits, and Maps* (Farnham: Ashgate, 2014).

Flint, Kate, *The Victorians and the Visual Imagination* (Cambridge: Cambridge University Press, 2000).

Grove, Richard H., *Green Imperialism: Colonial Expansion, Tropical Islands, Edens and the Origins of Environmentalism, 1600–1860* (Cambridge: Cambridge University Press, 1995).

Hoage, R. J. and William A. Deiss (eds.), *New Worlds, New Animals: From Menagerie to Zoological Park in the Nineteenth Century* (Baltimore: Johns Hopkins University Press, 1996).

Huggan, G and H. Tiffin (eds.), *Postcolonial Ecocriticism: Literature, Animals, Environment* (Abingdon: Routledge, 2010).

Kingsley, Charles, 'The Water Supply of London', *North British Review* 25:29 (1851), 228–53.

Lightman, Bernard V., *Victorian Science in Context* (Chicago: Chicago University Press, 1997).

Mazzeno, Laurence and Ron Morrison (eds.). *Animals in Victorian Literature and Culture: Contexts for Criticism* (Basingstoke: Palgrave Macmillan, 2017).

Ritvo, Harriet, *The Animal Estate: the English and Other Creatures in the Victorian Age* (Cambridge, MA: Harvard University Press, 1987).

Rohman, Carrie. 'Animals', Sacha Bru, Ben De Bruyn, and Michel Delville (eds.), *Literature Now: Key Terms and Methods for Literary History*, (Edinburgh: Edinburgh University Press, 2016).

Sinclair, Upton, *The Jungle* (New York: Doubleday, Page & Co., 1906).

Smith, Jonathan, 'Gender, Royalty, and Sexuality in John Gould's *Birds of Australia*', *Victorian Literature and Culture* 35:2 (2007), 569–87.

Walls, Laura Dassow, 'Natural History in the Anthropocene', *A Global History of Literature and the Environment*, ed. John Parham and Louise Westling (Cambridge: Cambridge University Press, 2017), 187–200.

Worster, Donald, *Nature's Economy: A History of Ecological Ideas* (Cambridge: Cambridge University Press, 1994).

56

GILBERT WHITE, *THE NATURAL HISTORY AND ANTIQUITIES OF SELBORNE, IN THE COUNTY OF SOUTHAMPTON: WITH ENGRAVINGS AND AN APPENDIX*

(London: T. Bensley, 1789)

LETTERS TO THOMAS PENNANT
LETTER X

SIR, August 4, 1767

It has been my misfortune never to have had any neighbours whose studies have led them towards the pursuit of natural knowledge: so that, for want of a companion to quicken my industry and sharpen my attention, I have made but slender progress in a kind of information to which I have been attached from my childhood.

As to swallows (*hirundines rusticæ*) being found in a torpid state during the winter in the Isle of *Wight*, or any part of this country, I never heard any such account worth attending to. But a clergyman, of an inquisitive turn, assures me, that, when he was a great boy, some workmen, in pulling down the battlements of a church tower early in the spring, found two or three *swifts* (*hirundines apodes*) among the rubbish, which were, at first appearance, dead; but, on being carried toward the fire, revived. He told me that, out of his great care to preserve them, he put them in a paper-bag, and hung them by the kitchen fire, where they were suffocated.

Another intelligent person has informed me that, while he was a schoolboy at *Brighthelmstone*, in *Sussex*, a great fragment of the chalk-cliff fell down one stormy winter on the beach; and that many people found swallows among the rubbish: but, on my questioning him whether he saw any of those birds himself; to my no small disappointment, he answered me in the negative; but that others assured him they did.

Young broods of swallows began to appear this year on July the eleventh, and young *martins* (*hirundines urbicæ*) were then fledged in their nests. Both species

DOI: 10.4324/9780429355653-63

will breed again once. For I see by my fauna of last year, that young broods came forth so late as *September* the eighteenth. Are not these late hatchings more in favour of hiding than migration? Nay, some young martins remained in their nests last year so late as *September* the twenty-ninth; and yet they totally disappeared with us by the fifth of *October*. How strange is it that the *swift*, which seems to live exactly the same life with the *swallow* and *house-martin*, should leave us before the middle of *August* invariably! while the latter stay often till the middle of *October*; and once I saw numbers of house martins on the seventh of *November*. The *martins* and *red-wing fieldfares* were flying in fight together; an uncommon assemblage of summer and winter-birds!

[. . .]

Numbers of *snipes* breed every summer in some moory ground on the verge of this parish. It is very amusing to see the cock bird on wing at that time, and to hear his piping and humming notes. I have had no opportunity yet of procuring any of those *mice* which I mentioned to you in town. The person that brought me the last says they are plenty in harvest, at which time I will take care to get more; and will endeavour to put the matter out of doubt, whether it be a non-descript species or not.

[. . .]

LETTER XII
TO THE SAME

SIR, November 4, 1767.

It gave me no small satisfaction to hear that the *falcos* turned out an uncommon one. I must confess I should have been better pleased to have heard that I had sent you a bird that you had never seen before; but that, I find, would be a difficult task.

I have procured some of the *mice* mentioned in my former letters, a young one and a female with young, both of which I have preserved in brandy. From the colour, shape, size, and manner of nesting, I make no doubt but that the species is non descript. They are much smaller, and more slender, than the *mus domesticus medius* of *Ray*; and have more of the squirrel or dormouse colour: their belly is white; a straight line along their sides divides the shades of their back and belly. They never enter into houses; are carried into ricks and barns with the sheaves; abound in harvest; and build their nests amidst the straws of the corn above the ground, and sometimes in thistles. They breed as many as eight at a litter, in a little round nest composed of the blades of grass or wheat.

One of these nests I procured this autumn, most artificially platted, and composed of the blades of wheat; perfectly round, and about the size of a cricket-ball; with the aperture so ingeniously closed, that there was no discovering to what part it belonged. It was so compact and well filled, that it would roll across the table without being discomposed, though it contained eight little mice that were naked and blind. As this nest was perfectly full, how could the dam come at her litter respectively so as to administer a teat to each? perhaps she opens different places for that purpose, adjusting them again when the business is over: but she could not possibly be contained herself in the ball with her young, which moreover would be daily increasing in bulk. This wonderful procreant cradle, an elegant instance of the efforts of instinct, was found in a wheat-field suspended in the head of a thistle.

[. . .]

Some birds, haunting with the *missel-thrushes*, and feeding on the berries of the yew-tree, which answered to the description of the *merula torquata* or *ring-ouzel*, were lately seen in this neighbourhood. I employed some people to procure me a specimen, but without success [. . .]

Query – Might not *Canary* birds be naturalized to this climate, provided their eggs were put, in the spring, into the nests of some of their congeners, as *goldfinches*, *greenfinches*, &c.? Before winter perhaps they might be hardened, and able to shift for themselves.

About ten years ago I used to spend some weeks yearly at *Sunbury*, which is one of those pleasant villages lying on the *Thames*, near *Hampton-court*. In the autumn, I could not help being much amused with those myriads of the *swallow* kind which assemble in those parts. But what struck me most was, that, from the time they began to congregate, forsaking the chimnies and houses, they roosted every night in the osier-beds of the aits of that river. Now this resorting towards that element, at that season of the year, seems to give some countenance to the northern opinion (strange as it is) of their retiring under water. A *Swedish* naturalist is so much persuaded of that fact, that he talks, in his calendar of Flora, as familiarly of the swallows going under water in the beginning of September, as he would of his poultry going to roost a little before sunset.

An observing gentleman in *London* writes me word that he saw an *house-martin*, on the twenty-third of last *October*, flying in and out of it's nest in the *Borough*. And I myself, on the twenty ninth of last *October* (as I was travelling through *Oxford*), saw four or five swallows hovering round and settling on the roof of the county-hospital.

Now is it likely that these poor little birds (which perhaps had not been hatched but a few weeks) should, at that late season of the year, and from so midland a county, attempt a voyage to *Goree* or *Senegal*, almost as far as the *equator*?

I acquiesce entirely in your opinion – that, though most of the *swallow* kind may migrate, yet that some do stay behind and hide with us during the winter.

As to the short-winged soft-billed birds, which come trooping in such numbers in the spring, I am at a loss even what to suspect about them. I watched them narrowly this year, and saw them abound till about *Michaelmas*, when they appeared no longer. Subsist they cannot openly among us, and yet elude the eyes of the inquisitive: and, as to their hiding, no man pretends to have found any of them in a torpid state in the winter. But with regard to their migration, what difficulties attend that supposition! that such feeble bad fliers (who the summer long never flit but from hedge to hedge) should be able to traverse vast seas and continents in order to enjoy milder seasons amidst the regions of *Africa*!

LETTER XIII
TO THE SAME

SIR, SELBORNE, Jan, 22, 1768

[. . .]

For many years past I have observed that towards Christmas vast flocks of *chaffinches* have appeared in the fields; many more, I used to think, than could be hatched in any one neighbourhood. But, when I came to observe them more narrowly, I was amazed to find that they seemed to me to be almost all hens. I communicated my suspicions to some intelligent neighbours, who, after taking pains about the matter, declared that they also thought them all mostly females; at least fifty to one. This extraordinary occurrence brought to my mind the remark of *Linnæus*; that 'before winter all their hen chaffinches migrate through *Holland* into *Italy*'. Now I want to know, from some curious person in the north, whether there are any large flocks of these *finches* with them in the winter, and of which sex they mostly consist? For, from such intelligence, one might be able to judge whether our female flocks migrate from the other end of the island, or whether they come over to us from the continent.

We have, in the winter, vast flocks of the common *linnets*; more, I think, than can be bred in any one district. These, I observe, when the Spring advances, assemble on some tree in the sunshine, and join all in a gentle sort of chirping, as if they were about to break up their winter quarters and betake themselves to their proper summer homes. It is well known, at least, that the *swallows* and the *fieldfares* do congregate with a gentle twittering before they make their respective departure.

[. . .]

What you suggest, with regard to *Spain*, is highly probable. The winters of *Andalusia* are so mild, that, in all likelihood, the soft-billed birds that leave us at that season may find insects sufficient to support them there.

[. . .]

As to the small *mice*, I have farther to remark, that though they hang their nests for breeding up amidst the straws of the standing corn, above the ground; yet I find that, in the winter, they burrow deep in the earth, and make warm beds of grass: but their grand rendezvous seems to be in corn-ricks, into which they are carried at harvest. A neighbour housed an oat-rick lately, under the thatch of which were assembled near an hundred, most of which were taken; and some I saw. I measured them; and found that, from nose to tail, they were just two inches and a quarter, and their tails just two inches long. Two of them, in a scale, weighed down just one copper halfpenny, which is about the third of an ounce avoirdupois: so that I suppose they are the smallest quadrupeds in this island. A full-grown *mus medius domesticus* weighs, I find, one ounce lumping weight, which is more than six times as much as the mouse above; and measures from nose to rump four inches and a quarter, and the same in it's tail. [. . .]

LETTER XVI
TO THE SAME

DEAR SIR, SELBORNE, April 18, 1768

The history of the *stone curlew, charadrius oedicnemus*, is as follows. It lays it's eggs, usually two, never more than three, on the bare ground, without any nest, in the field; so that the countryman, in stirring his sallows, often destroys them. The young run immediately from the egg like *partridges*, &c. and are withdrawn to some flinty field by the dam, where they sculk among the stones, which are their best security; for their feathers are so exactly of the colour of our grey spotted flints, that the most exact observer, unless he catches the eye of the young bird, may be eluded. The eggs are short and round; of a dirty white, spotted with dark bloody blotches. Though I might not be able, just when I pleased, to procure you a bird, yet I could shew you them almost any day; and any evening you may hear them round the village, for they make a clamour which may be heard a mile. *Oedicnemus* is a most apt and expressive name for them, since their legs seem swoln like those of a gouty man. After harvest I have shot them before the pointers in turnip-fields.

I make no doubt but there are three species of the *willow-wrens*: two I know perfectly; but have not been able yet to procure the third. No two birds can differ more in their notes, and that constantly, than those two that I am acquainted with; for the one has a joyous, easy, laughing note; the other a harsh loud chirp. The former is every way larger, and three quarters of an inch longer, and weighs two drams and an half; while the latter weighs but two: so the songster is one fifth heavier than the chirper. The chirper (being the first summer-bird of passage that is heard, the *wryneck* sometimes excepted) begins his two notes in the middle of *March*, and continues them through the spring and summer till the end of *August*, as appears by my journals. The legs of the larger of these two are flesh-coloured; of the less, black.

[. . .]

LETTER XVII
TO THE SAME

DEAR SIR, SELBORNE, June 18, 1768

On *Wednesday* last arrived your agreeable letter of *June* the 10th. It gives me great satisfaction to find that you pursue these studies still with such vigour, and are in such forwardness with regard to *reptiles* and *fishes*.

The *reptiles*, few as they are, I am not acquainted with, so well as I could wish, with regard to their natural history. There is a degree of dubiousness and obscurity attending the propagation of this class of animals, something analagous to that of the *cryptogamia* in the sexual system of plants: and the case is the same with regard to some of the fishes; as the eel, &c.

The method in which toads procreate and bring forth seems to be very much in the dark. Some authors say that they are viviparous: and yet *Ray* classes them among his oviparous animals; and is silent with regard to the manner of their bringing forth. [. . .] The copulation of *frogs* (or at least the appearance of it; for *Swammerdam* proves that the male has no *penis intrans*) is notorious to every body: because we see them sticking upon each others backs for a month together in the spring: and yet I never saw, or read, of *toads* being observed in the same situation. It is strange that the matter with regard to the venom of *toads* has not been yet settled. That they are not noxious to some animals is plain: for *ducks*, *buzzards*, *owls*, *stone curlews*, and *snakes*, eat them, to my knowledge, with impunity. And I well remember the time, but was not eye-witness to the fact (though numbers of persons were) when a quack, at this village, ate a *toad* to make the country-people stare; afterwards he drank oil.

I have been informed also, from undoubted authority, that some ladies (ladies you will say of peculiar taste) took a fancy to a *toad*, which they nourished summer after summer, for many years, till grew to a monstrous size, with the maggots which turn to flesh flies. The *reptile* used to come forth every evening from an hole under the garden-steps; and was taken up, after supper, on the table to be fed. But at last a tame *raven*, kenning him as he put forth his head, gave him such a severe stroke with his horny beak as put out one eye. After this accident the creature languished for some time and died.

[. . .]

Providence has been so indulgent to us as to allow of but one venomous *reptile* of the serpent kind in these kingdoms, and that is the *viper*. As you propose the good of mankind to be an object of your publications, you will not omit to mention common sallad-oil as a sovereign remedy against the bite of the *viper*. As to the *blind worm* (*anguis fragilis*, so called because it snaps in sunder with a small blow), I have found, on examination, that it is perfectly innocuous. A neighbouring yeoman (to whom I am indebted for some good hints) killed and opened a female *viper* about the twenty-seventh of *May*: he found her filled with a chain of eleven eggs, about the size of those of a *blackbird*; but none of them were advanced so far towards a state of maturity as to contain any rudiments of young. Though they are oviparous, yet they are viviparous also, hatching their young within their bellies, and then bringing them forth. Whereas snakes lay chains of eggs every summer in my melon beds, in spite of all that my people can do to prevent them; which eggs do not hatch till the spring following, as I have often experienced. Several intelligent folks assure me that they have seen the *viper* open her mouth and admit her helpless young down her throat on sudden surprises, just as the female *opossum* does her brood into the pouch under her belly, upon the like emergencies; and yet the *London* viper-catchers insist on it, to Mr. *Barrington*, that no such thing ever happens. The serpent kind eat, I believe, but once in a year; or, rather, but only just at one season of the year. Country people talk much of a water-snake, but, I am pretty sure, without any reason; for the common snake (*coluber natrix*) delights much to sport in the water, perhaps with a view to procure frogs and other food.

I cannot well guess how you are to make out your twelve species of reptiles, unless it be by the various species, or rather varieties, of our lacerti, of which *Ray* enumerates five. I have not had opportunity of ascertaining these; but remember well to have seen, formerly, several beautiful green *lacerti* on the sunny sandbanks near *Farnham*, in *Surrey*; and *Ray* admits there are such in *Ireland*.

[. . .]

LETTER XXXVII
TO THE SAME

DEAR SIR, SELBORNE, 1771

On the twelfth of *July* I had a fair opportunity of contemplating the motions of the *caprimulgus*, or *fern-owl*, as it was playing round a large oak that swarmed with *scarabœi solstitiales*, or *fern-chafers*. The powers of it's wing were wonderful, exceeding, if possible, the various evolutions and quick turns of the swallow genus. But the circumstance that pleased me most was, that I saw it distinctly, more than once, put out it's short leg while on the wing, and, by a bend of the head, deliver somewhat into it's mouth. If it takes any part of it's prey with it's foot, as I have now the greatest reason to suppose it does these chafers, I no longer wonder at the use of it's middle toe, which is curiously furnished with a serrated claw.

Swallows and *martins*, the bulk of them I mean, have forsaken us sooner this year than usual; for, on *September* the twenty second, they rendezvoused in a neighbour's walnut-tree, where it seemed probable they had taken up their lodging for the night. At the dawn of the day, which was foggy, they arose all together in infinite numbers, occasioning such a rushing from the strokes of their wings against the hazy air, as might be heard to a considerable distance: since that no flock has appeared, only a few stragglers.

Some *swifts* staid late, till the twenty-second of August – a rare instance! for they usually withdraw within the first week.

On *September* the twenty-fourth three or four *ring-ousels* appeared in my fields for the first time this season: how punctual are these visitors in their autumnal and spring migrations!

LETTERS TO DAINES BARRINGTON
LETTER IV

DEAR SIR, SELBORNE, Feb. 19, 1770

Your observation that 'the *cuckoo* does not deposit it's egg indiscriminately in the nest of the first bird that comes in it's way', but probably looks out a nurse in some degree congenerous, with whom to intrust it's young', is perfectly new to me; and struck me so forcibly, that I naturally fell into a train of thought that led me to consider whether the fact was so, and what reason there was for it. When I came to recollect and inquire, I could not find that any cuckoo had ever been seen in these parts, except in the nest of the *wagtail*, the *hedge-Sparrow*, the *titlark*, the *white-throat*, and the *redbreast*, all soft-billed insectivorous birds. The excellent Mr. *Willughby* mentions the nest of the *palumbus* (ring-dove), and of the *fringilla* (chafinch), birds that subsist on acorns and grains, and such hard food: but then he does not mention them as of his own knowledge; but says afterwards that he saw himself a *wagtail* feeding a *cuckoo*. It appears hardly possible that a soft-billed bird should subsist on the same food with the hard-billed: for the former have thin membranaceous stomachs suited to their soft food; while the latter, the granivorous tribe, have strong muscular gizzards, which, like mills, grind, by the help of small gravels and pebbles, what is swallowed. This proceeding of the *cuckoo*, of dropping it's eggs as it were by chance, is such a monstrous outrage on maternal affection, one of the first great dictates of nature; and such a violence on instinct; that, had it only been related of a bird in the *Brasils*, or *Peru*, it would never have merited our belief. But yet, should it farther appear that this simple bird [. . .] may be still endued with a more enlarged faculty of discerning what species are suitable and congenerous nursing-mothers for it's disregarded eggs and young, and may deposit them only under their care, this would be adding wonder to wonder, and instancing, in a fresh manner, that the methods of Providence are not subjected to any mode or rule, but astonish us in new lights, and in various and changeable appearances.

[. . .]

Query.– Does each female cuckoo lay but one egg in a season, or does she drop in several different nests according as opportunity arises?

57

WILLIAM SMELLIE, *THE PHILOSOPHY OF NATURAL HISTORY*

(Edinburgh, 1791)

Chapter 8, 'Habitations of Animals'

[. . .] The Beaver is about three feet in length, and its tail, which is of an oval figure, and covered with scales, is eleven inches long. He uses his tail as a rudder to direct his course in the water. In places much frequented by man, the beavers neither associate nor build habitations. But in the northern regions of both continents, they assemble in the month of June or July, for the purpose of uniting into society, and of building a city. From all quarters they arrive in numbers, and soon form a troop of two or three hundred. The operations and architecture of the beavers are so well described by the Count de Buffon, that we shall lay it before our readers nearly in his own words. The place of rendezvous, he remarks, is generally the situation fixed upon for their establishment, and it is always on the banks of waters. If the waters be flat, and seldom rise above their ordinary level, as in lakes, the beavers make no bank or dam. But in rivers or brooks, where the water is subject to risings and fallings, they build a bank, which traverses the river from one side to the other, like a sluice, and is often from eighty to a hundred feet long, by ten or twelve broad at the base. This pile, for animals of so small a size (the largest beavers weighing only fifty or sixty pounds), appears to be enormous, and presupposes an incredible labor. But the solidity with which the work is constructed, is still more astonishing than its magnitude. The part of the river where they erect this bank is generally shallow. If they find on the margin a large tree, which can be made to fall into the river, they begin, by cutting it down, to form the principal basis of their work This tree is often thicker than a man's body. By gnawing it at the bottom with their four cutting teeth, they in a short time accomplish their purpose, and always make the tree fall across the river. They next cut the branches from the trunk to make it lie level. These operations are performed by the joint industry of the whole community. Some of them, at the same time, traverse the banks of the river, and cut down smaller trees, from the size of a man's leg to that of his thigh. These they cut to a certain length, dress them into stakes, and first drag them by land to the margin of the river, and then by water to the place where the building

DOI: 10.4324/9780429355653-64

is carrying on. These piles they sink down, and interweave the branches with the larger stakes. In performing this operation, many difficulties are to be surmounted. In order to dress these stakes, and to put them in a situation nearly perpendicular, some of the beavers must elevate, with their teeth, the thick ends against the margin of the river, or against the cross tree, while others plunge to the bottom, and dig holes with their fore feet to receive the points, that they may stand on end. When some are laboring in this manner, others bring earth in their mouths and with their fore feet, and transport it in such quantities, that they fill with it all the intervals between the piles. These piles consist of several rows of stakes of equal height, all placed opposite to each other, and extend from one bank of the river to the other. The stakes facing the lower part of the river are placed perpendicularly; but those which are opposed to the stream slope upward, to sustain the pressure of the water; so that the bank, which is ten or twelve feet wide at the base, is reduced to two or three at the top. Near the top, or thinnest part of the bank, the beavers make two or three sloping holes, to allow the surface water to escape. These they enlarge or contract in proportion as the river rises or falls; and when any breaches are made in the bank by sudden or violent inundations, they know how to repair them when the water subsides.

Hitherto all these operations were performed by the united force and dexterity of the whole community. They now separate into smaller societies, which build cabins or houses. These cabins are constructed upon piles near the margin of the river or pond, and have two openings, one for the animals going to the land, and the other for throwing themselves into the water. The form of these edifices is either round or oval, and they vary in size from four or five to eight or ten feet in diameter. Some of them consist of three or four stories. Their walls are about two feet thick, and are raised perpendicularly upon planks, or plain stakes, which serve both for foundations and floors to their houses. When they consist of but one story, they rise perpendicularly a few feet only, afterwards assume a curved form, and terminate in a dome or vault, which answers the purpose of a roof. They are built with amazing solidity, and neatly plastered with a kind of stucco both within and without. In the application of this mortar the tails of the beavers serve for trowels, and their feet for plashing. Their houses are impenetrable to rain, and resist the most impetuous winds. In their construction, they employ different materials, as wood, stone, and a kind of sandy earth, which is not liable to be dissolved in water. The wood they use is generally of the light and tender kinds, as alders, poplars, and willows, which commonly grow on the banks of rivers, and are more easily barked, cut, and transported, than the heavier and more solid species of timber. They always begin the operation of cutting trees at a foot or a foot and a half above the ground. They labour in a sitting posture; and, beside the convenience of this posture, they enjoy the pleasure of gnawing perpetually the bark and wood, which are their favourite food. Of these provisions they lay up ample stores in their cabins to support them during the winter. Each cabin has its own magazine, which is proportioned to the number of its inhabitants, who have all a common right to the store, and never pillage their neighbours. Some villages are composed of twenty or twenty-five cabins. But these large establishments are not frequent; and the common republics seldom exceed ten or twelve families, while each have their own quarter of the village, their own magazine, and their separate habitation. The smallest cabins contain two, four, or six, and the largest eighteen, twenty, and sometimes thirty beavers. As to males and females, they are almost always equally paired. Upon a moderate computation, therefore, the society is often composed of a hundred and fifty or two hundred, who all, at first, labour jointly in raising the great public building, and afterwards, in select tribes or companies, in making particular habitations. In this society, however numerous, an universal peace is maintained. Their union is cemented by common labours; and it is perpetuated by mutual conveniency, and the abundance of provisions which they amass and consume together. A simple taste, moderate appetites, and an aversion to blood and carnage, render them destitute of the ideas of rapine and war. Friends to each other, if they have any foreign enemies, they know how to avoid them. When danger approaches, they advertise one another, by striking their broad tail on the surface of the water, the noise of which is heard at a great distance, and resounds through all the vaults of their habitations. Each individual, upon these occasions, consults his own safety; some plunge into the water; others conceal themselves within their walls, which can be penetrated only by the fire of heaven, or the steel of man, and which no animal will attempt either to open or to overturn. These retreats are not only safe, but neat and commodious. The floors are spread over with verdure; the branches of the box and of the fir serve them for carpets, upon which they permit not the smallest dirtiness. The window that faces the water answers for a balcony to receive the fresh air, and for the purpose of bathing. During the greater part of the day, the beavers sit on end, with their head and the anterior parts of their body elevated, and their posterior parts sunk in the water. The aperture of this window is sufficiently raised to prevent its being stopped up with the ice, which, in the beaver climates, is often two or three feet thick. When this accident happens, they slope the sole of the window, cut obliquely the stakes which support it, and thus open a communication with the unfrozen water. They often swim a long way under the ice. In September, the beavers collect their provisions of bark and of wood. Till the end of winter, they remain in their cabins, enjoy the fruits of their labours, and taste the sweets of domestic happiness. This is their time of repose. In the spring they separate; the males retire into the country, to enjoy the pleasures and fruits of spring. They return occasionally, however, to their cabins; but dwell there no more. The females continue in the cabins, and are occupied in nursing, protecting, and rearing their young, which in a few weeks are in a condition to follow their dams. The beavers assemble not again till autumn, unless their banks or cabins be injured by inundations; for, when accidents of this kind happen, they suddenly collect their forces, and repair the breaches that have been made.

This account of the society and operations of beavers, however marvellous it may appear, has been established and confirmed by so many credible eye-witnesses, that it is impossible to doubt of its reality.

58

JOHN CURTIS, *BRITISH ENTOMOLOGY; BEING ILLUSTRATIONS AND DESCRIPTIONS OF THE GENERA OF INSECTS FOUND IN GREAT BRITAIN AND IRELAND: CONTAINING COLOURED FIGURES FROM NATURE OF THE MOST RARE AND BEAUTIFUL SPECIES, AND IN MANY INSTANCES OF THE PLANTS UPON WHICH THEY ARE FOUND, VOL. I. COLEOPTERA*

(London: Sherwood, Gilbert, and Piper, 1824)

3.
DEILEPHILA EUPHORBIÆ
Spotted Elephant
ORDER Lepidoptera Fam. Sphingidæ *Lat.*
Type of the Genus Sphinx Elpenor *Linn.*

DELLEPHILA *Ochsenheimer*. Sphinx *Linn.*

Antennæ composed of many joints, with the club prismatic, and appearing hooked, it being terminated by a long, subulated, naked joint: upper side thickly covered with scales: under side ciliated.
Labrum and mandibles attached to the head.
Mandibles parallel, curved inward, furnished internally with brushes of very strong hair.

 DOI: 10.4324/9780429355653-65

Maxillæ (forming the proboscis) very long and spiral [. . .]
Labial Palpi broad if seen in front, covered with short close scales, the first joint very much bent, second very large, somewhat oval, third tuberculiform, scarcely distinct [. . .]

Wings horizontal, or deflexed in repose; a hook or catch at the exterior edge of the lower wings to retain those above.
Caterpillars with 6 *anterior*, 8 *abdominal*, and 2 *anal* feet.

D. EUPHORBIÆ *Ochs.*
Sphinx Euphorbiæ *Linn. Syst. Nat.* 2.802.19. *Fab. Syst. Nat. tom.* 3. p. 367. 37. *Haw. Lep. Brit.* p. 61. 8. *Roesel. Ins. v.* 1. t. 3. p. 17.

Head and thorax white, the centre fuscous-green. Abdomen above fuscous-green, sides of the first 3 segments white, with pure black spots upon the first 2, the next 3 segments having narrow white spots on their sides. Upper wings fuscous-green, white at their base, with a black spot; the posterior margin white; a rosy fascia extending from the posterior margin to the apex, very deeply sinuated above and undulated beneath, and a darker rosy margin from the apex to the posterior angle; under wings black, whitish internally, with a deep rose-coloured fascia in the centre, and another along the external margin: the whole Insect beneath clouded rose colour, with 2 obscure black spots in the upper wings. Antennæ white above and fuscous beneath: legs white, first pair fuscous-green above. The male has much less black in the under wings, the antennæ are thicker, and the abdomen more dilated with hair at the apex, than in the female.

In the Cabinets of Mr. Raddon and the Author.

DEILEPHILA is derived from the Greek, and means Lovers of Evening. It was a genus proposed I believe by Hubner, and established by Ochsenheimer in his 'Die Schmetterling von Europa'. It contains the following British species: *D. Celerio, Elpenor, Porcellus, lineata, Galii, and Euphorbiæ*, which are all rare excepting the second. These insects, which have been called Hawk-moths, fly about sun-set, darting from flower to flower, and hovering over the most fragrant with their long proboscis extended to extract the honey deposited in the nectaries. *Deilephila Euphorbiæ* is eminently beautiful both in its larva and imago states; and although it has been met with by the earlier collectors, I am indebted to the assiduity and liberality of my friend Mr Raddon for being able to give its history, as well as figures of the larva, and the plant upon which it feeds.

During a long residence in Devonshire, that gentleman visited occasionally the extensive sand-hills at Appledore and Braunton Burrows near Barnstaple, where *Euphorbia Paralias* grows in great abundance; and from the size and beauty of the caterpillar it would be imagined that it might readily be found: but in the young state they are not easily discoverable; and when more advanced, they become so conspicuous that their numbers are reduced by marine birds which feed upon them:– sometimes they may be traced by their soil, at other times they may be seen far from the spot where they fed, at the extremity of a tall rush. They are full-grown about the middle of September, when they descend into the sand and change into chrysalides, forming a loose case of earth around them, from which they emerge the beginning of the following June. Sometimes, however, they remain in the pupa state two seasons, as many other Lepidoptera do;– a wise provision of Nature to prevent any accident from destroying the whole brood. The sand-hills where the larva is found are of great extent and magnitude, and must have been collected by the winds and storms to which they are constantly exposed: during the winter the whole soil is frequently removed, so as completely to alter the surface of the country; a great number of the pupa must consequently be destroyed or buried at a great depth below the surface, where probably they lie hid until they are brought to light and life by the influence of the elements.[1]

Dr. Schwægrichen of Leipsic informs me that in Germany *D. Euphorbiæ* feeds upon *Euphorbia esula* and *E. Cyparissias*, plants of the same division as *E. Paralias* (Sea Spurge).

DEILEPHILA EUPHORBIÆ
Spotted Elephant Hawk-Moth
Order Lepidoptera. Fam. Sphingidæ *Lat.*
Type of the *Genus* Sphinx Elpenor *Linn.*

[. . .]

Head and thorax olivaceous, margined with white. Abdomen rosy, the back olivaceous, sides of the first 3 segments white, with intense black spots upon the two first, and a white stripe on each side of the 3 following. Superior wings dull rosy, with a black and white spot at their base; the costa, a large spot attached near the base and another towards the disk olivaceous, and a stripe arising from the interior margin (which is white) and attenuated to the apex olivaceous also; posterior margin rosy inclining to olive colour. Inferior wings deep rose colour, white next the abdomen, black at their base, with a black sinuated narrow fascia parallel to the posterior margin: cilia white. Antenna white above, fuscous beneath. Legs white, anterior pair fuscous on the underside. Beneath rose-coloured, a little spotted with olive, with an obscure blackish spot in the centre of each of the upper wings. The males have less black in the under wings than the females, and a variety of the former sex has occurred, with the fascia on the inferior wings of a dark-rose colour instead of black.

Note

1 Curtis's footnote: 'I think it probable that the larva found in marshy ground at Barnscray near Crayford in Kent, and figured by Harris, as well as those recorded by De Geer as feeding upon a common Galium, were the caterpillars of *D. Galii*, especially as that species has been frequently confounded with *D. Euphorbiæ*'.

59

ANDREW PRITCHARD, *THE NATURAL HISTORY OF ANIMALCULES: CONTAINING DESCRIPTIONS OF ALL THE KNOWN SPECIES OF INFUSORIA; WITH INSTRUCTIONS FOR PROCURING AND VIEWING THEM, &C, &C., &C.*

(London: Whittaker and Co, 1834)

Preface

FEW branches of science hold out stronger inducements for their study than the Natural History of Animalcules, which, while it is pursued with great facility, affords, at the same time, by reason of the singular forms and diverting habits of these creatures, a degree of interest scarcely to be exceeded. For this reason, and to reply to the many inquiries addressed to me, such as – Where can you refer me to a description of animalcules? – What magnifying powers are the best to view them with? – What are their comparative sizes? – I have found a very curious creature of such and such a form; is it known? Where can I obtain drawings of such as are known? &c. &c. – I have ventured to take a general survey of the subject [. . .]

Book I

OF the multiplicity of objects, which the almost in credible powers of the Microscope have brought under our observation and scrutiny, perhaps that class of animated beings denominated Animalcules may be considered the most remarkable. The bare knowledge that there are myriads of atoms (and in the scale of living creatures we can call them nothing else) existing in a single drop of water, recreating and executing all their various functions and evolutions with as much rapidity and apparent facility as if the range afforded them were as boundless as the ocean, must carry with it an intensity of interest to the mind of every human being; of every one, at least, who is at all accustomed to meditate on the

DOI: 10.4324/9780429355653-66

perfections of Nature, and to recognize and adore the hand that guides her through all the vast variety of her stupendous operations. As our acquaintance with the minute portions of the creation is exclusively dependent upon the properties of the Microscope, every refinement of this interesting and valuable instrument must necessarily contribute something to our stock of knowledge on the subject; and indeed it is entirely owing to the very great perfection which it has now acquired, that a fresh spirit of research is widely extending itself,– research into those recondite truths which may lead not merely to the gratification of our curiosity but to some of the most important and scientific results.

For a lapse of years, after the publication of Müller's 'Animalia Infusoria', in 1786, this branch of Natural History remained stationary, if not utterly disregarded; nor, indeed, did it, until very recently, assume what may be termed a regular form, constructed of materials the most precious of all, viz. truths brought together by practical and diligent investigation. To Dr Ehrenberg's late observations, although they apply to such only as belong strictly to the Phytozoa, we are greatly indebted. Lamarck, too, in 1815, and Cuvier, in 1817, made considerable advances in classification: but then, as the systems of these two last-mentioned Naturalists were not founded upon a rigid inspection of the Animalcules themselves, I have deemed it advisable, in the general arrangement of this little Treatise, to abide by those of Müller and Ehrenberg.

The term Animalcule, which implies nothing more than the *diminutive* of animal, has been commonly used to denote those living creatures inhabiting fluids, which are too minute to be scanned, or even seen by the naked eye: such, for instance, as those produced in inconceivable numbers from infusions of animal and vegetable matter: it comprehends as well such as are found in, and are peculiar to, the bodies of larger animals: this latter class, however, does not fall within the province of this work.

In the variety of systems that have been put forth respecting these extraordinary creatures, the main characteristics of each have referred either to a difference in their size, or to the general appearance of their external forms: the present design, however, is not to investigate the value of these. Until the introduction of vegetable colouring matter into the fluid, which supplies them with food,– an experiment that has been attended with very successful results,– these creatures were commonly supposed to be entirely devoid of internal organization, and to be nourished by the simple process of cuticular absorption. By the application of coloured substances, which, moreover, have been found to invigorate rather than to depress the animalcule, and to maintain it in the full exercise of all its functions, this erroneous notion is set at rest, and an internal structure is discerned in some, equal, if not surpassing that of many of the larger invertebrated animals, and comprising a muscular, nervous, and, in all probability, vascular system; all wonderfully contrived for the performance of their respective offices.

The most obvious portion of their internal structure is undoubtedly that connected with the digestive functions; and hence it is that Ehrenberg has selected this as the leading feature of his arrangement, denominating his two grand divisions of the Phytozoa, Polygastrica and Rotatoria; the former of which includes such as are possessed of several distinct stomachs or digestive sacs; and the latter such as have true alimentary canals and rotatory organs provided with a number of cilia aptly disposed for promoting the objects of life: these two grand divisions of the Phytozoa are afterwards subdivided into families and other minor branches [. . .]. The cilia, in their different combinations, supply the means of locomotion, propelling the creature in many cases with great rapidity through the water: they are apparently stiff, like eye-lashes; and from Dr E.'s description of some of the larger ones, they issue from bulbous substances at their bases, and being acted upon by muscular fibres are capable of being moved to and fro in particular directions, so as to occasion a current of the fluid to flow towards the mouth of the animalcule, by which it is furnished with fresh water, or food. They are sometimes disposed, as before stated, round certain organs of a circular form, which, on account of their peculiar vibrations giving the appearance of a rotatory action, are termed rotatory organs [. . .]. A second curious feature in the construction of some of these minute creatures are the *setæ*, or bristles, attached to the surface of their bodies: these short moveable hairs in all probability act as fins, and contribute greatly to their means of motion. The third feature are the *uncini*, or hooks, setaceous appendages curved at their extremities, and serving the creature to attach itself to any object it chooses. A fourth are the styli, jointed at their bases, and differing from the cilia in respect of their being unable to effect a rotatory motion: these, however, are more flexible, and have more play than the *setæ*. Independently of these peculiarities, some animalcules possess the extraordinary faculty of thrusting out, or elongating, portions of their bodies at various points, which, assuming the appearance either of legs or fins, are termed variable *processes*, and enable the creature to walk or swim. [. . .] With one more observation respecting the caudal appendages of animalcules – viz. that in many cases they have important functions to fulfil – I shall close this cursory view of the external structure of these little beings, remarking, at the same time, that the power and goodness of the Almighty are as clearly evinced by the humble but efficient means afforded these living atoms of pervading the narrow limits of their sphere of action, to provide for their wants and pleasures, as by the more exalted gifts He has graciously bestowed upon the intellectual part of the creation, whose occupations are so manifold, and whose views are as boundless as their thoughts. [. . .]

It was a favourite hypothesis with Naturalists, some years ago, that the class of animalcules under consideration was entirely nourished by cutaneous absorption, and that no suitable organs for transmitting and digesting food were discoverable. Baron Gleichen was the first who brought the truth of this theory to the test; for having tinged some water containing animalcules with carmine, he found on the second day that only some distinct cavities in the interior of their bodies were filled with the colouring matter, evidently demonstrating the existence of an alimentary structure: here, however, he left the subject, and it is to Dr Ehrenberg's further investigation of it that we are indebted for an accurate description of their different forms [. . .] In more recent experiments it has been found advisable to employ vegetable colouring substances in their pure state; such, for instance, as sap green and indigo, which, together with the valuable acquisition of an excellent instrument, enabled the Doctor to contribute much to our previously imperfect knowledge of this branch of Natural History. Before I explain the manner of *Viewing these creatures under the Microscope*, I shall venture a few remarks on the *Method of Procuring them*. In the selection of vegetable substances for infusions, such as stalks, leaves, flowers, seeds of plants, &c. care must be taken that there be no admixture of quinine in them, or the intention will be frustrated. Immerse these, whatever they may be, for a few days, in some clear water, when, if the vessels which contain them be not agitated, a thin pellicle or film will be discerned on the surface, which, under the microscope, will be seen to be inhabited by several descriptions of animalcules: the first produce are commonly those of the simplest kind, such as the Monads. In a few days more, their numbers will increase to such an amazing extent, that it would be utterly impossible to compute those in a single drop of the fluid. After this, again, they will begin to diminish in numbers, and I have generally observed them supplanted by others of a larger species and more perfect organization; such as the Cyclidia, Paramesia, Kolpodæ, &c. It is worthy of remark here, however, that in their production they do not pursue any regular order, even in similar infusions. If the vessel be large, and the circumstances under which it is placed sufficiently favourable, a still higher description of animalcules will succeed, viz. the Vorticella, and, lastly, the Brachioni; and thus a single infusion will repay for the little trouble of making it, with a great variety of species. Water in which flour has been steeped will be found to abound also with animalcules: and it is remarked by G. Leach, Esq. that the leaden troughs, constantly appropriated for birds to drink out of, contain several descriptions of them, and more especially those of the wheel genus. In ponds, too, especially in the shallow parts, near their edges, and in the immediate vicinity of water-plants, prodigious quantities of all kinds may be easily procured; so that possessing as we do such myriads of them all around us, that they impregnate almost everything that we eat and drink, touch and breathe, an anxiety to know more about them, and the effects they produce, cannot but be regarded as rational and laudable.

[. . .] Some animalcules resemble spheres, others are egg-shaped; others again represent fruits of various kinds; eels, serpents, and many of the invertebrated animals; funnels, tops, cylinders, pitchers, wheels, flasks, &c. &c.; all of which are found to possess their own particular habits, and to pursue a course of life best adapted to their peculiar constructions: thus, for instance, while some move through the water with the greatest imaginable rapidity, darting, leaping, or swimming, others merely creep or glide along; and many are altogether so passive that it requires long and patient observation to discover any of their movements at all. One description are perceptibly soft, and yield easily to the touch; another are covered with a delicate shell or horn-like coat. Of the latter order there are different degrees of density, as in the Volvox, Gonium, &c. where the envelope is comparatively thick; and where, strange to say, the internal substance separates by the mode of propagation into several portions, forming so many distinct young ones, which at their birth burst the envelope, and the parent becomes entirely dissipated. In others of this order the shell is merely a plate covering the body, resembling that of the tortoise: sometimes it includes the body, so as to leave only two small apertures at the extremities, and at others it is bivalve, and encloses the creature like that of the oyster or muscle.

[. . .] All vertebrated animals are either oviparous or viviparous, which terms sufficiently designate their modes of production: but it is not so with animalcules; for, in addition to these two methods. Animalcules propagate by a spontaneous scissure, or division of their bodies into two or more portions, each one forming a new creature, which, on its arrival at maturity, pursues the same course. These divisions take place in some genera symmetrically, as in the Gonia, &c.; in others, by transverse, longitudinal, or diagonal sections. In these latter cases the produce have forms differently proportioned from those of the creatures from which they spring [. . .] This circumstance, we may observe, renders it sometimes difficult to determine the species. 2. They propagate, in the manner before mentioned of the Volvox, and some other genera, by a distribution of the internal substance of the parent into a proportionate number of young ones, all of which at their birth issue forth, and leave behind them nothing but the envelope, soon to be dissolved. 3. They are produced from germs, shooting forth from the parent's sides [. . .]. 4. From spawn, which, in the act of being shed, carries along with it a portion of the parent animalcule [. . .].

With respect to the mode of viewing animalcules under the Microscope; I would direct that they be placed in what are termed aquatic live-boxes (described in the 'Microscopic Cabinet'), or on a slip of glass, in which case they should be covered with a thin plate of mica, which will have the effect of preventing the small quantity of water put with them from evaporating, and of rendering the surface perfectly plane for the purpose of observation: an additional advantage, however, will be obtained by making use of my Aplanatic Engiscope, (or that of Dr Goring), which is easily arranged to shew these creatures and their domestic habits in the phials they are kept, adhering, as many of them will, to its sides. Although no English work on this subject has previously given any admeasurement of the sizes of the different species, yet as I conceive it to be a matter of some importance, I shall point out the method I have pursued in accomplishing it. Thus; a system of graduation is easily formed on the aquatic live-boxes, or slips of glass, by means of cross-lines intersecting each other at right angles; and under a series of these lines, say 1-500th of an inch apart, an animalcule may be so examined as to afford a very accurate estimate of its dimensions: for instance, suppose it to occupy one-half the interval between two of these lines, then it is clear that the creature measures 1-1000th of an inch: and if it occupy two divisions, compute it at 1-250th of an inch in length. In this manner, with a little practice, and with the aid of a few micrometers of different grades, the intention will be readily effected.

Having selected and placed the object for examination on the stage of your microscope, the next consideration will be how to regulate the illumination, and to select a suitable magnifying power. These points must be carefully attended to, for on them, even with the best instruments, much of the beauty and effect will depend. The most intense and best description of light is to be derived from either a sperm or wax candle, or from what is perhaps on the whole most convenient, the common Argand lamp. Concentrate this light on the object with a proper condensing lens, taking care at the same time to reduce the quantity, if necessary, by means of diaphragms or stops placed under the stage: these should be rendered capable of adjustment as to distance from the object, &c. so as to transmit only a cone of rays of the proper dimensions.

A magnifying power of about 100 to 500 will be found to be sufficient for most purposes; although in an inspection of the Monads, and some minute portions of other objects, a stronger one will doubtless be required. I do not think, however, that any advantage will be gained from powers exceeding 800, as it is of far more importance to obtain a deep penetration and perfect definition than an excess of amplification. Apply in the outset, therefore, a low power, say 100, and if on trial it prove insufficient, double it, and proceed onwards until you are satisfied as to the result; taking it as a general rule, never to increase the power beyond what is absolutely requisite. As the expression, magnifying power, has reference to some standard or other for sight, it is necessary that I should inform the reader of the one I have adopted here; it is the decimal one, presuming that an object is always viewed without a magnifier under the angle subtended by it at ten inches from the eye: thus a single lens, which requires the object to be distant from it one inch (ie. of one inch focus), will magnify ten times in length and breadth, and 100 times in surface: and with respect to a system of lenses, arranged as in the compound Microscope, it will be easy enough to estimate their combined power, by first of all referring it to the relative power of a single lens, and then expressing it according to the scale just mentioned. Thus, if an Engiscope were equivalent to a single lens of 1-10th of an inch focus, we should call the power 100: if 1-20th, 200: 1-30th, 300: &c. &c.[. . .]

The Achromatic Engiscope has indisputably prodigious advantages over any other description of Microscope, and particularly that of being equally applicable to a review of many other classes of objects; this is especially a property in the one I have recently constructed. It affords a luxuriant field of view, and is managed with great facility. If this, however, cannot be obtained, a good Doublet, well put together, possesses sufficient penetration, and a definition scarcely surpassed by any other instrument. The estimable qualities in these magnifiers are, that they produce a clear and well-defined outline distance between the object and the anterior lens, and an uniformly colourless, and not a dingy yellow, field of view. Again, if a good Doublet cannot be procured, single lenses, suitably mounted, as in the new Vertical Microscope, display animalcules exceedingly well; and, when skilfully managed, will afford the possessor many an interesting exhibition, will beguile many a tedious hour, and store his mind with some useful truths.

[. . .]

Genus III

Volvox

THE animalcules belonging to this genus are of a globular form, and revolve in the water. Some of the species are so large as to be discerned by unassisted vision, while others are very diminutive. Ehrenberg has not demonstrated their digestive organization; but in a note to his table, conceives they ought to follow the monads. In this genus is included that beautiful animalcule, called the *Volvox globator*, which forms so interesting a spectacle in the Solar and Gas Microscopes.

25. *VOLVOX punctum. The point Volvox*.– This volvox takes its name from the appearance of a bright point in the centre. It is spherical, with one part opaque and black; the other portion transparent and colourless. A violent internal motion is often observed in the dark part. It swims in a tremulous manner, and often passes across the drop of water, and occasionally turns upon its axis. They congregate together, moving as in a little whirlpool, and then separate. Found very abundant on the surface of foetid sea-water.

26. *VOLVOX calamus*, (new species, Mihi). *The pipe Volvox* – This pretty little animalcule, about the 1-1500th of an inch in diameter, is generally of a bright red colour, of a globular form, with a small diaphanous tube protruding from one side, and slightly enlarged at its extremity, like a trumpet: its length is equal to the diameter of the animalcule. This pipe, I am inclined to believe, is its sucker, or proboscis, by which it imbibes nourishment. In swimming, it oscillates in front of the animalcule. They are found, during the spring and autumn, along with the green Cercaria, and are the prey of the genus Bursaria.

[. . .]

37. *VOLVOX globator. The globe Volvox*.– This popular and diverting animalcule was discovered by Leeuwenhoeck, and has been described by all subsequent writers on microscopic objects. As its name imports, it is of a globular form; its colour is usually a light green, though I have met with some of an orange brown, which, however, are generally smaller than the green ones. The envelope is composed of a diaphanous membrane, beneath the surface of which is disposed, at equal distances, small spherical bodies of a green colour. These granular bodies have been supposed to reside on the exterior, and by some have been mistaken for hairs; but that they are actually within the envelope is evident when the circumference of the globe is accurately brought into focus. The proximity of these pustules is greater the younger the specimen, and as these pustules contain the colouring matter of the animalcule, the young always appear more coloured than the old ones, as the transparent spaces between the pustules is augmented in the latter, and spread over a greater surface. Within the parent is often seen a number of (from six to forty) smaller ones, and even within these, when about to be excluded, another generation may be observed. The young within the parent, which, by the way, forms the most striking characteristic of the species, may be observed at first attached to the inside of the membranous covering, but long before their

birth revolving freely in the parent, and others again within them. In parturition, a portion of the parent globe is broken, and the young are gradually and slowly evolved; when this is completed, like the fabular Phoenix, the parent dies, and its body separates into numberless parts.

This singular animalcule, to use the words of Baker, 'moves in all directions; forwards, backwards, up and down, rolling over and over like a bowl, spinning horizontally like a top, or gliding along smoothly without turning itself: sometimes its motions are slow, at others rapid'.

The diameter of this animalcule, when full grown, is about 1-30th of an inch, and is therefore easily perceived by unassisted vision: a magnifying power of 100 times is sufficient.

It is found most abundant, during spring and summer, in ponds and stagnant water; and often in the same water with young lizards and frogs. Infusions of hemp-seed and tremella are said to abound with them.

60

CHARLES WATERTON, *ESSAYS IN NATURAL HISTORY, CHIEFLY ORNITHOLOGY, WITH AN AUTOBIOGRAPHY OF THE AUTHOR AND A VIEW OF WALTON HALL*

(London: Longman, Orme, Brown, Green, & Longman, 1838)

On the Habits of the Rook

LAST year I partly promised that, on some dismal winter's evening, I would sit me down, and write the history of the rook. The period has now arrived. Nothing can be more gloomy and tempestuous than the present aspect of the heavens. The wind is roaring through the naked branches of the sycamores, the rain beats fiercely on the eastern windows, and the dashing of the waves against the walls of the island, warns us that one of November's dark and stormy nights is close at hand; such a night, probably, as that in which Tam O'Shanter unfortunately peeped into Kirk Alloway. Foreigners tell us that on these nights Englishmen are prone to use the knife, or a piece of twisted hemp, to calm their agitated spirits. For my own part, I must say that I have an insuperable repugnance to such anodynes; and, were a host of blue devils, conjured up by November's fogs, just now to assail me, I would prefer combating the phantoms with the weapons of ornithology, rather than run any risk of disturbing the economy of my jugular vein, by a process productive of very unpleasant sensations, before it lulls one to rest.

According to my promise, I will now pen down a few remarks on the habits of the rook, which bird, in good old sensible times, was styled *frugilegus*. It is now pronounced to be *prædatorius*. Who knows but that our Great Ones in Ornithology, may ultimately determine to call it up to the house of hawks?

[. . .]

There is no wild bird in England so completely gregarious as the rook; or so regular in its daily movements. The ringdoves will assemble in countless multitudes, the finches will unite in vast assemblies, and waterfowl will flock in thousands to the protected lake, during the dreary months of winter: but, when the returning sun

DOI: 10.4324/9780429355653-67

spreads joy and consolation over the face of nature, their congregated numbers are dissolved, and the individuals retire in pairs to propagate their respective species. The rook, however, remains in society the year throughout, In flocks it builds its nest, in flocks it seeks for food, and in flocks it retires to roost.

About two miles to the eastward of this place are the woods of Nostell Priory, where, from time immemorial, the rooks have retired to pass the night. I suspect, by the observations which I have been able to make on the morning and evening transit of these birds, that there is not another roosting-place for, at least, thirty miles to the westward of Nostell Priory. Every morning, from within a few days of the autumnal, to about a week before the vernal equinox, the rooks, in congregated thousands upon thousands, fly over this valley in a westerly direction, and return, in undiminished numbers, to the east, an hour or so before the night sets in. In their morning passage, some stop here; others, in other favourite places, farther and farther on; now repairing to the trees for pastime, now resorting to the fields for food, till the declining sun warns those which have gone farthest to the westward that it is time they should return. They rise in a mass, receiving additions to their numbers from every intervening place, till they reach this neighbourhood in an amazing flock. Sometimes they pass on without stopping, and are joined by those which have spent the day here. At other times they make my park their place of rendezvous, and cover the ground in vast profusion, or perch upon the surrounding trees. After tarrying here for a certain time, every rook takes wing. They linger in the air for a while, in slow revolving circles, and then they all proceed to Nostell Priory, which is their last resting-place for the night. In their morning and evening passage, the loftiness or lowliness of their flight, seems to be regulated by the state of the weather. When it blows a hard gale of wind, they descend the valley with astonishing rapidity, and just skim over the tops of the intervening hills, a few feet above the trees: but, when the sky is calm and clear, they pass through the heavens at a great height, in regular and easy flight.

Sometimes these birds perform an evolution, which is, in this part of the country, usually called the shooting of the rooks. Farmers tell you, that this shooting portends a coming wind. He who pays attention to the flight of birds has, no doubt, observed this downward movement. When rooks have risen to an immense height in the air, so that, in appearance, they are scarcely larger than the lark, they suddenly descend to the ground, or to the tops of trees exactly under them. To effect this, they come headlong down, on pinion a little raised, but not expanded, in a zig-zag direction (presenting, alternately their back and breast to you), through the resisting air, which causes a noise similar to that of a rushing wind. This is a magnificent and beautiful sight to the eye of an ornithologist. It is idle to suppose for a moment that it portends wind. It is merely the ordinary descent of the birds to an inviting spot beneath them, where, in general, some of their associates are already assembled, or where there is food to be procured. When we consider the prodigious height of the rooks at the time they begin to descend, we conclude that they cannot effect their arrival at a spot perpendicular under them, by any other process so short and rapid.

Rooks remain with us the year throughout. If there were a deficiency of food, this would not be the case; for, when birds can no longer support themselves in the place which they have chosen for their residence, they leave it, and go in quest of nutriment elsewhere. Thus, for want of food, myriads of wild fowl leave the frozen north, and repair to milder climates; and in this immediate district, when there is but a scanty sprinkling of seeds on the whitethorn bush, our flocks of fieldfares and of redwings bear no proportion to those in times of a plentiful supply of their favourite food, But the number of rooks never visibly diminishes; and on this account we may safely conclude that, one way or other, they always find a sufficiency of food. Now, if we bring, as a charge against them, their feeding upon the industry of man, as, for example, during the time of a hard frost, or at seedtime, or at harvest, at which periods they will commit depredations, if not narrowly watched, we ought, in justice, to put down in their favour the rest of the year, when they feed entirely upon insects. Should we wish to know the amount of noxious insects destroyed by rooks, we have only to refer to a most valuable and interesting Paper on the Services of the Rook, signed T. G. Clitheroe, Lancashire, which is given in the Mag. of Nat. Hist., vol. vi. p. 142. I wish every farmer in England would read it: they would then be convinced how much the rook befriends them.

Some author (I think, Goldsmith) informs us, that the North American colonists got the notion into their heads that the purple grakle was a great consumer of their maize; and these wise men of the west actually offered a reward of threepence for the killed dozen of the plunderers. This tempting boon soon caused the country to be thinned of grakles, and then myriads of insects appeared, to put the good people in mind of the former plagues of Egypt. They damaged the grass to such a fearful extent, that, in 1749, the rash colonists were obliged to procure hay from Pennsylvania, and even from England. Buffon mentions, that grakles were brought from India to Bourbon, in order to exterminate the grasshoppers. The colonists, seeing these birds busy in the new-sown fields, fancied that they were searching for grain, and instantly gave the alarm. The poor grakles were proscribed by Government, and in two hours after the sentence was passed, not a grakle remained in the island. The grasshoppers again got the ascendency, and then the deluded islanders began to mourn for the loss of their grakles. The governor procured four of these birds from India, about eight years after their proscription, and the state took charge of their preservation. Laws were immediately framed for their protection; and, lest the people should have a hankering for grakle pie, the physicians were instructed to proclaim the flesh of the grakle very unwholesome food. When ever I see a flock of rooks at work in a turnip-field, which, in dry weather, is often the case, I know that they have not assembled there to eat either the turnips or the tops, but that they are employed in picking out a grub, which has already made a lodgment in the turnip.

Last spring, I paid a visit, once a day, to a carrion crow's nest on the top of a fir tree. In the course of the morning in which she had laid her fifth egg, I took all the eggs out of the nest, and in their place I put two rooks' eggs, which were within six days of being hatched. The carrion crow attended on the stranger eggs, just as though they had been her own, and she raised the young of them with parental care. When they had become sufficiently large, I took them out of the nest, and carried them home. One of them was sent up to the gamekeeper's house, with proper instructions; the other remained with me. Just at this time an old woman. had made me a present of a barn-door hen. 'Take it, sir', said she, 'and welcome; for, if it stays here any longer, we shall be obliged to kill it. When we get up to wash in the morning, it crows like a cock. All its feathers are getting like those of a cock; it is high time that it was put out of the way, for when hens turn cocks people say that they are known to be very unlucky; and, if this thing is allowed to live, we don't know what may happen. It has great spurs on its legs, and last summer it laid four eggs. If I had had my own way, it would have been killed when it first began to crow'. I received the hen with abundant thanks; and, in return, I sent the old woman a full-bred Malay fowl. On examining the hen, I found her comb very large; the feathers on the neck and rump much elongated; the spurs curved, and about 14 inch long; the two largest feathers in her tail arched, and four or five smaller arched ones, of a beautiful and glossy colour, hanging down on each side of the tail. In a word, this hen had so masculine an appearance, that, when strangers looked at her, they all took her to be a cock, and it was with difficulty I persuaded them that she was a hen. We allowed her the range of a sheltered grass-plot, flanked on one side by holly trees, and open to the lake on the other. Here, also, was placed, in a cage, the young rook which I had taken from the nest of the carrion crow. The hen showed such an antipathy to it, that, whenever I held it to her, she would immediately fly at it. When visitors came to inspect her, I had only to take the rook out of the cage, and pit it against her, when she would stand upright, raise the long feathers on her neck, and begin to cackle, cluck, and crow. One morning the rook had managed to push aside a bar in front of its cage. A servant, in passing by, looked into it, and missed the bird. The hen had also disappeared. On search being made, they were both found floating side by side, dead, in the lake below. We conjectured that the hen had pursued the rook after its escape from the cage, and that the wind, which blew very strong that morning, had forced them both into a watery grave. I had still one rook left at the gamekeeper's. It was kept in a cage, which was placed on a little stand in his garden; and I had given orders that upon no account was it to be allowed to go at large. The feathers remained firm at the base of the bill till the 15th of August; on which day the keeper perceived that a few feathers had dropped from the lower mandible, and were lying at the bottom of the cage. In a couple of weeks more, the lower mandible had begun to put on a white scurfy appearance, while here and there a few feathers had fallen from the upper one. This is the purport of the keeper's information to me, on my return home from Bavaria. On the 31st of the same month, a terrible storm set in. By what the keeper told me, the night must have been as dark and dismal as that in which poor King Lear stood in lamentation, and exposed his hoary locks to the four rude winds of heaven. A standard white-heart cherry tree, perhaps the finest in Yorkshire, and which, for many generations, had been the pride and ornament of this place, lost two large branches during the gale; and in the morning, when the keeper rose, he found the cage shattered and upset, and driven to the farthest corner of his garden. The rook was quite dead. It had lost its life, either through the inclemency of that stormy night, or through bruises received in the fall of the cage. Thus both the rooks were unlucky. The old woman, no doubt, could clearly trace their misfortunes to her crowing hen. However, the experiment with the two young rooks, though not perfect, has nevertheless been of some use. It has shown us that the carrion crow makes no distinction betwixt its own eggs and those of the rook; that it can know nothing of the actual time required to sit upon eggs in order to produce the young; that the young of the rook will thrive under the care of the carrion crow, just as well as under that of its own parents; and finally, that the feathers fall off from the root of the rook's bill, by the order of nature, as was surmised by the intelligent Bewick, and not by the process of the bird's thrusting its bill into the earth, in search of food, as is the opinion of some naturalists.

The rook advances through the heavens with a very regular and a somewhat tardy beat of wing; but it is capable of proceeding with great velocity when it chooses: witness its pursuit and attack on the sparrowhawk and kestrel. It is apt to injure, in the course of time, the elm trees on which it builds its nest, by nipping off the uppermost twigs. But this, after all, is mere conjecture. The damage may be caused by an accumulation of nests, or by the constant resort of such a number of birds to one tree. Certain, however, it is, that, when rooks have taken possession of an elm tree for the purpose of incubation, the uppermost branches of that tree are often subject to premature decay.

Though the flocks of rooks appear to have no objection to keep company, from time to time, with the carrion crows, in a winter's evening, before they retire to roost, still I can never see a carrion crow build its nest in a rookery. There was always a carrion crow's nest here, in a clump of high Scotch pines, near the stables, till the rooks got possession of the trees; the carrion couple then forsook the place: the rooks were dislodged from this clump of trees; and then a pair of carrion crows (the same, for aught I know to the contrary) came and built their nest in it.

The rook lays from three to five eggs, varying much, like those of the carrion crow, in colour, shape, and size. After the rooks have built, and even lined their nests, they leave them, on the approach of night, to repair to the general rendezvous at Nostell Priory; but, as soon as they begin to lay, they then no longer quit the trees at night, until they have reared their young. When this has been effected, we see large flocks of them resorting to the different woods of the neighbourhood to pass the night. This they continue to do, till a few days before the autumnal equinox, when, for reasons which baffle all conjecture, they begin to pass over this valley every morning in a westerly direction, and return in the evening to their eastern roosting place in the woods of Nostell Priory.

Rooks are observed to keep up a very close and friendly intercourse with starlings and jackdaws; but, on looking at them in the fields, the observer will perceive, that, while the jackdaws mix promiscuously with the rooks, both in their flight and in searching for food, the starlings always keep in their own flock. This circumstance has long engaged my attention; but I am no farther advanced in the investigation than I was on the first day on which I set out. It is one of the many secrets in the habits of birds, which will, perhaps, be for ever concealed from our view.

61

CHARLES DARWIN, *JOURNAL OF RESEARCHES INTO THE GEOLOGY AND NATURAL HISTORY OF THE VARIOUS COUNTRIES VISITED BY H.M.S. BEAGLE UNDER THE COMMAND OF CAPTAIN FITZROY, R.N. FROM 1832 TO 1836*

(London: Henry Colburn, 1839)

Chapter 5, Bahia Blanca

THE Beagle arrived on the 24th of August, and a week afterwards sailed for the Plata. With Captain Fitzroy's consent I was left behind, to travel by land to Buenos Ayres. I will here add some observations, which were made during this visit, and on a previous occasion, when the Beagle was employed in surveying the harbour [. . .]

I will now give an account of the habits of some of the more interesting birds, which are common on these wild plains; and first of the *Struthio Rhea*, or South American ostrich. This bird is well known to abound over the plains of Northern Patagonia, and the united provinces of La Plata. It has not crossed the Cordillera; but I have seen it within the first range of mountains on the Uspallata plain, elevated between six and seven thousand feet. The ordinary habits of the ostrich are familiar to every one. They feed on vegetable matter; such as roots and grass; but at Bahia Blanca, I have repeatedly seen three or four come down at low water to the extensive mud-banks which are then dry, for the sake, as the Gauchos say, of catching small fish. Although the ostrich in its habits is so shy, wary, and solitary, and although so fleet in its pace, it falls a prey, without much difficulty, to the Indian or Gaucho armed with the bolas. When several horsemen appear in a semicircle, it becomes confounded, and does not know which way to escape. They generally prefer running against the wind; yet at the first start they expand their wings, and like a vessel make all sail. On one fine hot day I saw several ostriches enter a bed of tall rushes, where they squatted concealed, till quite closely approached. It

DOI: 10.4324/9780429355653-68

is not generally known that ostriches readily take to the water [. . .] When swimming, very little of their bodies appear above water, and their necks are extended a little forward: their progress is slow. On two occasions, I saw some ostriches swimming across the Santa Cruz river, where its course was about four hundred yards wide [. . .]

The inhabitants who live in the country readily distinguish, even at a distance, the cock bird from the hen. The former is larger and darker-coloured, and has a bigger head. The ostrich, I believe the cock, emits a singular deep-toned, hissing note. When first I heard it, standing in the midst of some sand-hillocks, I thought it was made by some wild beast, for it is a sound that one cannot tell whence it comes or from how far distance. When we were at Bahia Blanca in the months of September and October, the eggs, in extraordinary numbers, were found all over the country. They either lie scattered single, in which case they are never hatched, and are called by the Spaniards, huachos; or they are collected together into a shallow excavation, which forms the nest. Out of the four nests which I saw, three contained twenty-two eggs each, and the fourth twenty-seven. In one day's hunting on horseback sixty-four eggs were found; forty-four of these were in two nests, and the remaining twenty scattered huachos. The Gauchos unanimously affirm, and there is no reason to doubt their statement, that the male bird alone hatches the eggs, and for some time afterwards accompanies the young. The cock when on the nest lies very close; I have myself almost ridden over one. It is asserted that at such times they are occasionally fierce, and even dangerous, and that they have been known to attack a man on horseback, trying to kick and leap on him. My informer pointed out to me an old man, whom he had seen much terrified by one chasing him [. . .]

The Gauchos unanimously affirm that several females lay in one nest. I have been positively told, that four or five hen birds have been seen to go, in the middle of the day, one after the other, to the same nest. I may add, also, that it is believed in Africa, that two females lay in one nest. Although this habit at first appears very strange, I think the cause may be explained in a simple manner. The number of eggs in the nest varies from twenty to forty, and even to fifty [. . .] Now although it is most probable, from the number of eggs found in one district being so extraordinarily great, in proportion to that of the parent birds, and likewise from the state of the ovarium of the hen, that she may in the course of the season, lay a large number, yet the time required must be very long [. . .] If the hen were obliged to hatch her own eggs, before the last was laid the first probably would be addled; but if each laid a few eggs at successive periods, in different nests, and several hens, as is stated to be the case, combined together, then the eggs in one collection would be nearly of the same age. If the number of eggs in one of these nests is, as I believe, not greater on an average than the number laid by one female in the season, then there must be as many nests as females, and each cock bird will have its fair share of the labour of incubation; and that during a period when the females could not sit, on account of not having finished laying. I have before mentioned the great numbers of huachos, or scattered eggs; so that in one day's hunting the third part were found in this state. It appears odd that so many should be wasted. Does it not arise from the difficulty of several females associating together, and persuading an old cock to undertake the office of incubation? It is evident that there must at first be some degree of association, between at least two females; otherwise the eggs would remain scattered over the wide plains [. . .] Some have believed that the scattered eggs were deposited for the young birds to feed on. This can hardly be the case in America, because the huachos, although oftentimes found addled and putrid, are generally whole.

When at the Rio Negro, in Northern Patagonia, I repeatedly heard the Gauchos talking of a very rare bird which they called Avestruz Petise. They described it as being less than the common ostrich (which is there abundant), but with a very close resemblance. They said its colour was dark and mottled, and that its legs were shorter, and feathered lower down than those of the common ostrich. It is more easily caught by the bolas than the other species. The few inhabitants who had seen both kinds, affirmed they could distinguish them apart from a long distance. The eggs of the small species appeared, however, more generally known; and it was remarked, with surprise, that they were very little less than those of the Rhea, but of a slightly different form, and with a tinge of pale blue [. . .] Mr Martens shot an ostrich; and I looked at it, forgetting at the moment, in the most unaccountable manner, the whole subject of Petises, and thought it was a two-third grown one of the common sort. The bird was cooked and eaten before my memory returned. Fortunately the head, neck, legs, wings, many of the larger feathers, and a large part of the skin, had been preserved. From these a very nearly perfect specimen has been put together, and is now exhibited in the museum of the Zoological Society. Mr Gould, who in describing this new species, did me the honour of calling it after my name, states, that besides the smaller size and different colour of the plumage, the beak is of considerably less proportional dimensions than in the common Rhea; that the tarsi are covered with differently-shaped scales, and that they are feathered six inches beneath the knee. In this latter respect, and in the broader feathers of the wing, this bird perhaps shows more affinity to the gallinaceous family than any other of the Struthonidæ.

[. . .] At Santa Cruz we saw several of these birds. They were excessively wary: I think they could see a person approaching when he was so far off as not to distinguish the ostrich. In ascending the river few were seen; but in our quiet and rapid descent, many, in pairs and by fours or fives, were observed. It was remarked, and I think with truth, that this bird did not expand its wings, when first starting at full speed, after the manner of the northern kind.

62

JOHN GOULD, *AN INTRODUCTION TO THE BIRDS OF AUSTRALIA*

(London: Richard and John E. Taylor, 1848)

Preface

Having in the summer of 1837 brought my work on the 'Birds of Europe' to a successful termination, I was naturally desirous of turning my attention to the Ornithology of some other region; and a variety of opportune and concurring circumstances induced me to select that of Australia, the birds of which, although invested with the highest degree of interest, had been almost entirely neglected. Dr Shaw, in his 'Zoology of New Holland', had devoted a few plates to the subject, from specimens collected by Sir Joseph Banks during the first voyage of Captain Cook; the 'Birds of New Holland' by Lewin comprised not more than twenty-six plates; and figures and descriptions of a few species were given in the earlier voyages of Phillip, White and Collins, and the more recent one of King. At a subsequent period the late Mr Vigors and Dr Horsfield commenced an elaborate memoir on the Collection of Australian Birds in the possession of the Linnean Society; but unfortunately, they did not proceed farther than the *Meliphagidæ* and the non-completion of their labours is the more to be regretted, inasmuch as the Linnean Society's collection of Australian birds, at that time the finest extant, comprised many species collected by Mr Brown during his voyage with the celebrated navigator Flinders, and was moreover enriched with some interesting notes by the late Mr George Caley, by whom the collection was chiefly formed. Descriptions of many Australian birds were also included in the works of Latham, Shaw, Cuvier and Vieillot, as well as in several of the recent French voyages of discovery; still no general work on the subject had been undertaken, and nearly all that had been recorded by the various writers above enumerated, had reference almost exclusively to the productions of New South Wales and Van Diemen's Land, these being almost the only explored portions of that great country. In the absence, then, of any general work on the Birds of Australia, the field was comparatively a new one, and of no ordinary degree of interest, from the circumstance of its being one of the finest possessions of the British Crown, and from its natural productions being as remarkable for the anomalous nature of their forms, as for their beauty,

DOI: 10.4324/9780429355653-69

and the singularity of their habits. In the attempt to supply this desideratum I commenced publishing from the materials then accessible, but soon found, from the paucity of information extant upon the subject, that it could not be executed in a manner that would be satisfactory to my own mind, or commensurate with the exigencies of science; I therefore determined to proceed to Australia and personally investigate (so far as a stay of two years would allow) the habits and manners of its birds in a state of nature. I accordingly left England in May 1838, provided, by the liberality of Government, with letters from Lord Glenelg, at that time Secretary of State for the Colonies; Sir George Grey, Bart., and Gordon Gairdner, Esq., of the Colonial Office, recommending me to the countenance and protection of the various Governors, and requesting them to afford me such aid and assistance in furtherance of my objects as they might have it in their power to render; similar favours were also granted me by the authorities of the Admiralty, who, through their Secretary, Sir John Barrow, directed the captains and commanders of Her Majesty's ships and vessels employed on the coasts of Australia to further my views, by giving myself and my assistant a passage to such part of the coasts as either of us might be desirous of visiting, only stipulating that the ships under their command should not be detained on any parts of the coasts they were not ordered to visit. His late Royal Highness the Duke of Sussex, in his capacity of President of the Royal Society, was pleased to favour me with a letter addressed to the authorities, civil and military, of Her Majesty's Colonies, recommending me to their kind offices and protection, as he felt assured that my exertions would materially promote the interests of Natural History. I was also under considerable obligations to the kindness of Captain Washington, R.N., at that time Secretary of the Royal Geographical Society, who furnished me with introductions to Captains Sir John Franklin and Sir Gordon Bremer, R.N., and other influential persons. Having thus acknowledged the facilities afforded me by the home authorities, it becomes my pleasing duty to state that their recommendations and wishes were responded to in the warmest manner by Captain Sir John Franklin, R.N., Governor of Van Diemen's Land; Sir George Gipps, Governor-General of New South Wales; Lieut-Colonel Gawler, Governor of South Australia; John Hutt, Esq., Governor of Western Australia; and Captain M'Arthur, Commandant at Port Essington; all of whom rendered me every assistance compatible with the instructions under which they were acting. I should be wanting, however, both in courtesy and gratitude, did I not especially acknowledge the warm friendship, and the many acts of genuine kindness which I received at the hands of my valued friends Sir John and Lady Franklin, who, besides facilitating my views in every way, both publicly and privately, with the most generous hospitality received myself and family into their house, where Mrs. Gould and my eldest son, who had accompanied us, remained for nearly ten months, while I pursued my researches in various parts of Van Diemen's Land and the continent of Australia; and it is only by those who, like myself, have had the advantage of residing with that amiable family, that the kindness of their nature and the goodness of their hearts can be duly appreciated, and which can never be erased from my memory. [. . .] After exploring Van Diemen's Land, the islands in Bass's Straits, South Australia, and New South Wales, into the interior of which country I penetrated to the distance of nearly four hundred miles from the coast-line, I despatched my able assistant, Mr Gilbert, to explore the western and northern portions of the country, and returned to England in August 1840; I immediately commenced the work *de novo*, and the result of my labours is now before the public. It fortunately happened that at the commencement and during the progress of the work, Her Majesty's ships the Beagle, under the command of Captains Wickham and Stokes; the Britomart, under Captain Stanley; the Fly, under Captain Blackwood, and the Pelorus, under Captain Chambers, were employed in surveying the northern and north-western coasts of Australia; and the Erebus and Terror under Captain Sir James C. Ross, in a voyage of discovery towards the south pole. While engaged in the performance of their arduous duties the officers of those vessels succeeded in procuring many interesting novelties, which were with the greatest liberality communicated to me for the present work, whereby its value has been much enhanced. [. . .]

After spending two years in Western and Northern Australia, Mr Gilbert returned to England in September 1841, bringing with him the result of his labours, which proved of sufficient value and importance to induce me to believe that much yet remained to be discovered in those countries, and to direct him to return thither, which he accordingly did in the ensuing spring; and after again visiting Swan River, and sedulously exploring the interior so far as practicable, he proceeded to Sydney, and, unfortunately for himself, allowed his love of science, in the advancement of which no one was more ardent, to induce him to join Dr Leichardt in his overland journey from Moreton Bay to Port Essington. On this expedition he, as usual, displayed his wonted zeal and activity until the 28th of June, when, the party being treacherously attacked by the natives, his valuable life was sacrificed, I lost a most able coadjutor, and science has to deplore one of her most devoted servants; fortunately, however, in despite of the many difficulties and dangers which beset the party during the remainder of their journey, his journals and notes, together with the specimens he had been able to procure, were preserved and transmitted to me by Dr Leichardt, and proved of valuable assistance in determining the range of many of the species.

My own researches commenced immediately after passing the Equator, from whence, throughout the entire route to Australia, I omitted no opportunity of studying the habits, and collecting the different species of the oceanic birds that came under my notice: these observations were again resumed on my return from thence to England; and as the outward passage was by the Cape of Good Hope, and the homeward one by Cape Horn, they extended round the globe, and, as will be seen in the course of the work, have led to some important results [. . .]

At the commencement of the work it was not expected that it would prove so extensive as it has become, since not more than about 300 species were then known, which number has now been increased, by the united efforts of myself and those who have so kindly aided my views, to upwards of 600 species, among which are comprised many forms remarkable for their novelty, the anomalous character of their structure, and the singularity of their habits, such as the Bower Birds (*Ptilonorhynchi* and *Chlamyderæ*) and the Mound-raising Birds (*Talegalla*, *Leipoa* and *Megapodius*). The singular runs or bowers of the *Chlamyderæ* were considered by some explorers to be the cradles of the infants of the aborigines, and the mounds of the *Megapodius* to be tumuli, errors which have been rectified in the present work.

It is not to be supposed that an undertaking of such magnitude as the present could have been brought to a successful termination by the unaided efforts of a single individual, and I have, therefore, very great pleasure in stating that my views were most ably seconded by everyone with whom the nature of my investigations brought me in contact; but by none more than by the Rev Thomas James Ewing, who, besides manifesting the warmest friendship, has ever taken especial interest in promoting the success of the present work [. . .] Much valuable information has been communicated to me by George Grey, Esq. (now Governor of New Zealand), whose exertions during his expedition along the north western coasts of Australia were characterized by a degree of energy of character and perseverance but rarely equalled; whose ornithological collection made during this arduous enterprise, although small, was by no means destitute of interest; and who, upon succeeding Colonel Gawler in the Governorship of South Australia, found time amidst his multifarious occupations to devote considerable attention to Natural History, and to send me some interesting drawings and other details respecting the mounds raised by the *Leipoa*, &c. In South Australia I received many acts of kind attention and assistance from my friend Captain Sturt, whom I accompanied on one of his expeditions into the interior [. . .]

[. . .]

At the conclusion of my 'Birds of Europe', I had the pleasing duty of stating that nearly the whole of the plates had been lithographed by my amiable wife. Would that I had the happiness of recording a similar statement with regard to the present work; but such, alas! is not the case, it having pleased the All-wise Disposer of Events to remove her from this sublunary world within one short year after our return from Australia, during her sojourn in which country an immense mass of drawings, both ornithological and botanical, were made by her inimitable hand and pencil, and which has enabled Mr H.C. Richter, to whom, after her lamented death, the execution of the plates was entrusted, to perform his task in a manner highly satisfactory to myself, and I trust equally so to the Subscribers.

[. . .]

JOHN GOULD
June 12, 1848

Family PARADISEIDÆ

I certainly consider the accounts I have given of the extraordinary habits of the *Chlamyderæ* and *Ptilonorhynchi* as some of the most valuable and interesting portions of my work, and however incredible they may appear I am happy to say they have been fully confirmed by other observers.

Genus CHLAMYDERA
Generic characters

Bill moderate, *culmen* elevated, and arched to the tip which is emarginated, compressed on the sides; *gonys* slightly advancing upwards; *nostrils* basal, lateral, exposed, rounded, and pierced in a membrane; *wings* long and pointed, first primary short, second primary shorter than the third and fourth, which are equal, and the longest; *tail* long and slightly rounded; *tarsi* robust, defended anteriorly with broad scuta; *toes* long and strong; outer toe longer than the inner, hind-toe long and robust; *claws* long, curved, and acute.

258. *Chlamydera maculata,* Gould. Vol. IV. Pl. 8

INHABITS South Australia, New South Wales, and according to Mr. Gilbert's Journal of his overland journey to Port Essington, the intertropical regions of the east coast.

In one of Mr. Gilbert's many interesting letters received since the account above referred to was printed, he says, 'the question as to the nidification of *Chlamydera* is now settled by Mr C. Coxen having found a nest in December with three young birds; in form it was very similar to that of the common Thrush of Europe, being of a cup shape, constructed of dried sticks with a slight lining of feathers, and fine grass, and was placed among the smaller branches of an Acacia overhanging a pool of water'.

259. *Chlamydera nuchalis* Vol. IV. Pl. 9

'I found matter for conjecture', says Captain Stokes, 'in noticing a number of twigs with their ends stuck in the ground, which was strewed over with shells, and their tops brought together so as to form a small bower; this was 2½ feet long, 1½ foot wide at either end. It was not until my next visit to Port Essington that I thought this anything but some Australian mother's toy to amuse her child; there I was asked, one day, to go and see the birds' playhouse', when I immediately recognised the same kind of construction I had seen at the Victoria River; the bird (*Chlamydera nuchalis* of Mr. Gould's work) was amusing itself by flying backwards and forwards, taking a shell alternately from each side, and carrying it through the archway in its mouth'. – *Discoveries in Australia,* vol. ii. p. 97.

Genus Ptilonorhynchus

260. *Ptilonorhynchus holosericeus, Kuhl.* Vol. IV. Pl. 10

THAT this bird continues its singular habits under the disadvantages of captivity, I learn from the following passage in a letter lately received from Mr. Strange of Sydney.

'My aviary is now tenanted by a pair of Satin Birds, which I had hoped would have bred, as for the last two months they have been constantly engaged in constructing bowers, which I find are built for the express purpose of courting the female in. Both sexes assist in their erection, but the male is the principal workman. At times the male will chase the female all over the aviary, then go to the bower, pick up a gay feather or a large leaf, utter a curious kind of noise, set all his feathers erect, and run round the bower, into which at length the female proceeds, when he becomes so excited that his eyes appear ready to start from his head, and he continues opening first one wing and then the other, uttering a low whistling note, and like the common Cock, seems to be picking up something from the ground, until at last the female goes gently towards him, when, after two turns round her, he suddenly makes a dash and the scene ends'. This pair of birds was sent to England by Mr. Strange for the Earl of Derby, and had they not unfortunately died from cold when rounding Cape Horn, they would doubtless have continued their singular habits in his lordship's magnificent aviary at Knowsley.

The habitat of this species appears to be confined to the south eastern part of New South Wales, for it has not as yet been found in any other portion of the country.

[. . .]

Family Psittacidæ

No one group of birds gives to Australia so tropical and foreign an air as the numerous species of this great family, by which it is tenanted, each and all of which are individually very abundant. Immense flocks of white Cockatoos are sometimes seen perched among the green foliage of the *Eucalypti*; the brilliant scarlet breasts of the Rose-hills blaze forth from the yellow flowering *Acaciæ*; the *Trichoglossi* or Honey-eating Parrakeets enliven the flowering branches of the larger *Eucalypti* with their beauty and their lively actions; the little grass Parrakeets rise from the plains of the interior and render these solitary spots a world of animation; nay the very towns, particularly Hobart Town and Adelaide, are constantly visited by flights of this beautiful tribe of birds, which traverse the streets with arrow-like swiftness, and chase each other precisely after the manner the Swifts are seen to do in our own islands. In the public roads of Van Diemen's Land the beautiful *Platycerci* may be constantly seen in small companies, performing precisely the same offices as the Sparrow in England. I have also seen flocks of from fifty to a hundred, like tame pigeons at the barn-doors in the farm yards of the settlers, to which they descend for the refuse grain thrown out with the straw by the threshers. As might naturally be expected, the agriculturist is often sadly annoyed by the destruction certain species effect among his newly-sown and ripening corn, particularly where the land has been recently cleared and is adjacent to the brushes. Fifty-five well-defined species of this great family are figured and described in the present work. They appear to constitute four great groups, each comprising several genera, nearly the whole of which are strictly and peculiarly

Australian; for instance, neither *Calyplorhynchus*, *Platycercus*, *Euphema*, *Psephotus*, *Melopsittacus*, or *Nymphicus* have been found in any other country; and whether we consider the elegance of their forms or the beauty of their plumage, they may vie with the members of this extensive family from any part of the world [. . .]

Genus *Calyptorhynchus*

THE members of this genus are strictly arboreal, and are evidently formed to live upon the seeds of the *Banksiæ*, *Eucalypti*, and other trees peculiar to the country they inhabit; they diversify their food by occasionally devouring large caterpillars; they can scarcely be considered gregarious, but move about in small companies. Their flight is rather powerful, but at the same time laboured and heavy; and their voice is a low crying call, totally different from the harsh screaming notes of the *Cacatua*. Each division of the country, from the north coast of the continent to Van Diemen's Land, is inhabited by its own peculiar species.

I have never seen a member of this genus from any other country than Australia, but I have heard that an extraordinary Parrot, said to be larger than any at present in our collections, inhabits New Guinea, and which, from the description given of it, will probably be of this form. The *Calyptorhynchi* lay from two to four eggs in the holes of trees.

368. *Calyptorhynchus Banksii,* Vol. V. Pl. 7

369. *Calyptorhynchus macrorhynchus, Gould* Vol. V. Pl. 8

INHABITS the north coast, where it represents the *C. Banksii* of the eastern and the *C. naso* of the western coasts.

370. *Calyptorhynchus naso, Gould* Vol. V. Pl. 9

THIS species, which is confined to Western Australia, is rendered conspicuous by the small size of its crest, and by its bill being nearly as large as that of *C. macrorhynchus*, while its wings are much shorter than those of that species.

371. *Calyptorhynchus Leachii* Vol. V. Pl. 10

Banksianus Australis, Less. Traité d'Orn. p. 180, Atlas, pl. 18.

Inhabits the south-eastern parts of the continent, and differs from all the others in its smaller size, the gibbose form of its bill, and in the paucity of its crest.

372. *Calyptorhynchus funereus,* Vol. V. Pl. 11

CONFINED, I believe, to New South Wales, and South Australia?

373. Calyptorhynchus xanthonotus, Gould Vol. V. Pl. 12

THE true habitat of this species is Van Diemen's Land, but I have lately received a specimen from Port Lincoln, which proves that its range extends to South Australia. It is distinguished from *C. funereus* by its much smaller size, and by the uniformity of the yellow colouring of the tail.

374. Calyptorhynchus Baudinii, Vig. Vol. V. Pl. 13

INHABITS Western Australia, and is distinguished by its small size and by the white marks on the tail.

63

THOMAS EWBANK, *THE WORLD A WORKSHOP; OR, THE PHYSICAL RELATIONSHIP OF MAN TO THE EARTH*

(New York: D. Appleton and Company, 1855)

Chapter VII
The Three Storehouses of Matter

3. Animal Products

In this third department matter appears in types and forms widely unlike mineral and vegetable bodies. In one of those it is inert; in the other it has motion, but is tethered to the ground; in this it is locomotive. In the first it crystallizes, in the second germinates, in the third it lives, being endued with sensitive organs, and impelled by instinctive impulses. The change of an impalpable air – of matter so attenuated that leagues of it offer no obstacle to vision passing through it – to a gross liquid or solid, is marvellous; but not more so than the metamorphosis of common earth into fruit trees and flowers, into insects, reptiles, fishes, birds, quadrupeds, and men.

There is no medium adapted to sustain life but what is pervaded with it. While the torrid and temperate zones teem with living forms, the polar ice resounds with the cries or songs of birds, and the hum of insects. The atmospheric ocean, from its bottom on which we move to the elevations in which the condor soars, is redolent of life. But utterly innumerable as visible living forms are, their numbers are insignificant compared to the legions that the microscope reveals. Nor is it known where life is most abundant – on land, or in the unfathomed depths of the oceans.

As with plants, so it is with animals: they are natural apparatus for supplying man with materials for his fabrics, such as he could not elsewhere obtain, as wool, hair, feathers, down, silk, leather, glue, horn, ivory, wax, oils, furs, coloring matters, bone, pearl, tortoise and other shells, sperm, whalebone, isinglass, &c., &c., – their numbers filling important pages in the catalogue of his working stock.

Now, though man cannot originate living organisms, he can control them so far as essentially to modify the products they yield him: a circumstance wonderful in itself, although being common no one wonders at it, and yet nothing less than a verbal or written declaration from above could more emphatically proclaim him

 DOI: 10.4324/9780429355653-70

a manufacturer than the power given him over the development of most of the substances just enumerated, and over others in the vegetable world. By them he learns that the two active departments of Nature unite with the passive (mineral) one in preparing materials for him. It seems strange that he should be trusted with the awful power to contract and expand the area of existence, and, consequently, its enjoyments; but as tenant and manager of the factory, and alone responsible for the manner of conducting it, he was left to determine what living aids he should reject or employ. As in the vegetable department, so in this, he was to have the means within himself of increasing and diminishing the materials he wanted.

Elaborators who provide materials for clothing and food operate chiefly on the soil and its products. While some devote themselves to fibrous substances, to cereals, sugar, fruits, and roots, others convert grass, corn, and potatoes into beef, mutton, and pork, into wool, hair, horns, and hides – operations that are as truly arts and manufactures as the casting of types or building of ships. Indeed, those, whose producing apparatus are plants and cattle are, in some respects, superior to other elaborators, since they deal with matter in its highest forms of development. They diversify products, and evolve them with equal certainty and uniformity as operatives on inert matter. They improve them too, as do engineers and manufacturers, who, to obtain better results, alter or exchange their machinery. Precisely on the same ground do planters and herdsmen introduce new seeds and breeds.

A few items of the statistics of animals and of animal products will lighten the subject by breaking the monotony of its disquisitions.

The milch cows of the U. States in 1850 numbered 6,385,094, working oxen and other black cattle 11,393,813, sheep 21,723,220, swine 30,354,213, laboring horses, mules, and jacks, 4,335,669. There were, exclusive of the above, animals slaughtered for food, whose value was nearly 112,000,000 dollars, and whose numbers could not therefore have been under 20,000,000. The numbers for 1853, according to the Patent Office Report, were five millions of horses and mules, twenty millions of horned cattle, thirty-two millions of sheep, and twenty three millions of swine. Arrange these flocks and herds after the manner of Jacob, allow ten feet for the larger and eight for the smaller, to prevent their treading on each other – mount five millions of drovers, being one to every sixteen animals and the line would extend several times round the globe. But what are they compared to the wild animals that range the forests and prairies of the U. States! The buffaloes alone would form an unbroken phalanx round the earth, and the wild horses another. Still, to the quadrupeds of the world they are little more than specimens in a menagerie.

Of animal products leather is an ordinary one. Of the amount made in the U. States in 1850 we have no account, but the value of that worked into the single article of shoes in a single state (Massachusetts) is set down at twelve millions of dollars. The leather manufactures of England stand third or fourth on the list, being inferior only in point of value and extent to those of cotton, wool, and iron. She imported in 1851, and tanned 2,330,901 hides. She uses up yearly 60,000,000 lbs of leather, and the value of the manufactured article is 70,000,000 dollars.

In 1800, England had 26,147,743 sheep and lambs. In 1854, the number in the three kingdoms was 32,000,000. She now imports (chiefly from Australia) 70,000,000 lbs of wool, and clips from her own sheep 120,000,000 lbs, making one hundred and ninety millions of pounds of the fibre spun and woven by one people in each year. The clip of wool in the U. States in 1850 was 52,516,959 lbs France ten years ago (1843) worked up 45,000,000 lbs of wool annually.

Such are at best but a few solitary specimens of products furnished by domestic land animals. More might have been added, as hair, horn, and tallow – of the last, Russia, after supplying herself, sent 137,160,000 lbs to other peoples.

Animals are not confined to land, water is an immense theatre of vitality. Why were two thirds or three fourths of the earth permanently flooded with this liquid? Was so large an evaporating surface required to supply rain to fertilize the soil and aid in breaking down mountains? Was so wide a receptacle necessary to receive the debris washed into it through one geological period, in order to digest and prepare undisturbed the sediments for reappearance in stratified layers in another? Had it special reference to aqueous life, or was it determined by these and by other requirements? Whatever it was that led to the present proportions betwixt land and water, they are doubtless beneficial in the highest degree, and perfectly fulfil every organic and inorganic condition.

Water is a distinct arena for man's enterprise, and as such it has incalculably extended the range of his thoughts and of his acquisitions. It teems with substances precious to him, and he can travel over it easier than on land. It presents no impenetrable thickets nor inaccessible mountains; and although as yet the basins of oceans and seas are imperfectly known, the same remark applies, though in a less degree, to many of the great slopes of continental valleys. In full one half of the earth there remains as much to be explored as in some of the seas. Subaqueous researches are not to be expected of man in the morning of his career; still he has turned and is turning his attention to them. Explorations of wrecks and other matters at the bottom of the sea are becoming as common as descents into mines.

Between the ocean and the land are other analogies; the face of the former varies in color from the white surf and pale green along shores, to olive green and deep blue further out. Sometimes it is colorless and transparent, at others dull and opake. The Greenland sea changes from ultramarine blue to green and grey; at one time pellucid, and muddy at another. The Mediterranean puts on a purple hue, the Gulf of Guinea appears white, and the waters about the Maldives look black. As in landscapes, the colors are often due to the soil and to surface vegetation. Then the surface, as on shore, is broken by undulations, and at night lit up by phosphorescent animalcula – analogues of fire-flies. Like continents, seas and oceans are fertile fields of labor, and in some respects the most profitable, since their crops are raised without man's care, and he has only to reap them, as in whale, seal, cod, mackerel, herring, shad, oyster, salmon, coral, pearl, and other fisheries. The fecundity of the ocean equals, perhaps surpasses, that of the land. Its contributions to the arts are numerous and unique, while the streams of food it turns into man's garners never cease to flow. The fishermen of one nation have taken in one season, from the banks of Newfoundland, a million and a half of quintals of one kind of fish: in 1853 United States whalers brought in 363,191 barrels of whale and sperm oil, besides five and a half millions of pounds of whalebone. In 1850 the English home fisheries yielded 340,256 barrels of herrings. In 1852 the Scotch fishermen took 112,000 cwts of cod and ling (mostly dried). The Dutch fisheries probably were more productive. The Scotch fishermen sent 3,192,672 lbs of salmon to London in 1841. Yet all these put together form but a small item in the annual yield of the deep sea, shore, lake, and river fisheries of the United States, and an insignificant one in that of the earth's fishery as a whole.

Birds:– The three general forms of matter are the media of life on our planet, and on and in them respectively flourish the three great cohorts of living beings – of creatures that walk, swim, and fly. Air and water, essential to the concrete portion of the earth, presented opportunities for diversifying the forms and functions of life, and of immensely swelling the amount of sensuous enjoyment; hence, without affecting the applications of air and water to other purposes, they are made to sustain the countless hosts that live in them. In thus doubling, or perhaps quadrupling the numbers of sensitive creatures, the beneficence of the Creator is as conspicuous as his wisdom and power.

[. . .] Their subdivisions are such that they draw pleasure from every part of vegetation, and by feeding on its enemies they are its great conservators. But for them the earth had been barren as granite or a desert of sand; nor had there been a man living to till it. Scratchers seek their food about the roots, climbers hunt insects in the boles and stems, perchers feed and warble on the foliage, waders stalk among aquatic plants, while swimmers forage in deeper water. The swallow tribes course insects through the atmosphere; the accipitrines soar over all, prey upon all, and act as scavengers for all.

Birds do not elaborate as much matter for manufacturers as quadrupeds, but as protectors of vegetation the value of their labors is incalculable. Besides the flesh and eggs of those used as food, feathers are their chief offering. With those of water-fowl, pillows and beds are stuffed; and from the soft delicacy and non-conducting properties of down, it also is employed for the same purpose, but chiefly for various articles of female attire and accompaniments – tippets, boas, muffs, &c. Plumes of the ostrich and of other birds have always been worn as ornaments and insignia. Fans and screens are made of feathers, and entire dresses of them are common among Indians. From the variety and richness of colors in which birds are draped, artificial flowers have, for ages, been made of feathers.

Quills have been split and made into cheap and durable brushes, and also into hats and other parts of dress. But the pen is the most memorable application of the quill. In it birds have contributed more than all other classes to what has been deemed the highest of arts, the recording of thought. By the instruments of their flight man has soared into higher regions than the material atmosphere. In a mechanical and engineering view, the structure and material of feathers and of pinion quills are of surpassing interest. Of a singular composition of matter and moulded into peculiar forms, they, the latter particularly, combine unequalled strength and elasticity with the least weight of material; the qualities required in propelling organs. Every feather is a study in itself, and in the mechanism and movements of wings lessons of the highest practical value are to be learned.

The most singular, unexpected, and the largest contribution of birds to the arts is guano: a substance so rich in fertilizing power as to have become a staple item in commerce, and to afford employment for ships of all nations. In it the economy that pervades this heritage of ours is very obvious. Volcanic and other rocks appear here and there in the midst of oceans. Without vegetation no land animals can live on them, but they are the very places for oceanic birds to sleep and breed on, and for amphibia that seek dry land on which to expire. Hence guano has been accumulating on islet rocks for unknown periods of time, and is now dug out of beds varying from 50 to 100 feet in depth. It is found in African and Australian islands as well as in the Pacific. Samples from the Cape of Good Hope and Van Dieman's Land were at the London Exhibition. Thus have oceanic birds, in remote ages, whom men never saw, been laying up material for us.

The varieties of birds now on the planet are unknown; one naturalist has recorded 3,800 species.

Insects are interesting as birds. The most unattractive, if fully known, would appear beautiful as golden pheasants. A bee, a minnow, a beetle, and a common house-fly, are as great miracles of chemical and mechanical actions and motions as the universe itself is. Without irreverence, we may believe that in no world can there be more wonderful illustrations of what mechanism can do. Statuaries go into raptures over the Elgin marbles, and old men and youths gaze and copy their moving outlines of horses and men, while inventors turn not so much as an eye aside to study infinitely higher lessons that concern and are daily pressed on them. The time will come when their successors will be sensible of the advantages of contemplating the chefs-d'oeuvre of God. Had the earth produced but single specimens of reptiles, insects, fishes, and birds, philosophers would have gone to the antipodes to witness their movements when living, and to dissect them when dead. One might almost be tempted to think it an error in the owner of the earth in giving us such numbers of his devices, since the more he sends the less they are regarded. But the days of sciolism are passing, and men will seek for higher lessons in manufacturing mechanisms than they learn from inorganic matter.

Animals are elaborating machines, and by carefully studying their construction and their actions, we may as certain how each does its work, and imitate it. Their organs of elaboration, their processes, and the materials they use are as purely mechanical, chemical, and common, as those employed in artificial works; and so it is, that every class, order, genus, and species invites us into fields of knowledge that will never become barren of mechanical and chemical novelties; no, not if harvests be reaped every hour.

As yet man has pressed but few insects into his service, because his limited wants and researches have not made him acquainted with the nature and uses of ten thousand substances which these minute and most dexterous and delicate of elaborators produce. It is not, however, improbable that in time they will induce as great changes in the artificial as they have induced in the natural world. No works on earth approach in magnitude and durability some of theirs; and none tend more fully to illustrate our globe as one vast factory.

Silk-worms man has long employed. They are among the most industrious and profitable of his assistants. The fine, soft, and hitherto unparalleled thread they yield him is extraordinary; it were futile to attempt its measurement by yards, or even leagues. In the United States, in 1850, the census gives only 10,843 lbs, while in 1844 the amount was 396,790 lbs. We know not what quantities are raised in Canada, Mexico, South America, and the rest of this western hemisphere. In Great Britain, about 5,000,000 lbs are annually worked up; in France, much more; then large quantities are raised and woven in Austria, Prussia, Spain, Portugal, Greece, Switzerland, Italy, Holland, Russia, Turkey, Persia, &c. Yet the aggregate product of Europe and America must be very small compared to that spun by the worms of India, China, and the rest of Asia, by those of Africa and Oceanica.

Every thing about these creatures is surprising. How singular the contrast presented by their delicate natures and ephemeral duration with the strength and durability of their products! No sooner are they grown caterpillars, than they begin their filamental bobbins, and they live but a few days after finishing them – hastening to be dissolved in air to make room for fresh legions. But the thread they leave is lasting. Ladies' silk-dresses, after being worn through life, often descend as heir-looms through several generations. Then what is more remarkable than their numbers! There may be throughout the world fifty millions of persons engaged more or less directly in the manufacture of fibrous materials. Suppose one tenth of our living species, or a hundred millions, thus engaged. That number is truly a great one, but square it – multiply it by itself – and the quotient would not equal the hosts which the Proprietor of this factory has sent to make even one kind of thread for us. And moreover, swarming as they do, it rests with us to multiply them, and that indefinitely. To other tribes of insect spinners little attention has yet been given.

As elaborators of rich chemical compounds, bees have from early times been kept at work by man. The silk was exported from the country some years ago. The amount is now probably much greater. The amounts of wax and honey they yield is prodigious. Those given in the U. States seventh census are alone at hand, and certainly do not represent a tithe of what might be procured. Nearly fifteen millions of pounds of wax and honey are given as the product for the year 1850. It would be interesting in these and all other articles, whether of food or clothing, to know how much is produced and how much might be produced, that by dividing the amounts by the number of our species we might ascertain how much each individual receives or might receive.

Dyes are mostly from other departments, but not altogether so. The famous Tyrian purple of old was obtained from a sea shell. The modern color is derived from an insect. Besides the cochineal the grana kermes furnishes another red dye. The lac dye is produced on trees by an order of insects. The amount of coloring matter these minute creatures furnish is also remarkable. Nearly two millions and a half of pounds' weight of cochineal were imported into England in 1850, and over two millions of lac dye. Thus some of our richest colors are from insects, and in the vegetable world from lichens, masses of weeds from things insignificant and worthless in appearance.

It has been well remarked that numbers of new coloring materials have been in late times discovered and made available; so that the dyer of the present times employs substances of the existence of which his practical predecessors were wholly ignorant: and just such remarks will be made hereafter with regard not to our dyers only, but to almost every other profession.

Is it asked of what use are invisible animated molecules ever likely to be to man? What effect can they produce on matter serviceable to him? Why, there are examples of the accumulation of matter by them that are perfectly startling – that put at utter defiance the efforts of large animals and of man himself. Ehrenberg has shown that not only on several microscopic infusoria do others live as parasites, but such is the prodigious power of development or capability of division of the gallionellæ that in four days an animalcule, invisible to the naked eye, can form two cubic feet of the Bilin polishing slate, or tripoli!

[. . .]

Some ask, what elaborators can learn from the organic world? Almost everything. We know nothing, or next to nothing, of the principles by which matter is elaborated in wool, hair, feathers, scales, horn, ivory, &c., &c., nor how colors are evolved, defined, limited, and mingled in the bodies of animals, birds, and flowers: how the metallic lustre in the peacock and other creatures is produced, how perfumes are drawn out of common earth in a word, how every object in nature is produced. This knowledge is to be acquired. Man is not for ever to be empiric.

Part 7

'NEW WORLD' ENVIRONMENTS AND SCIENTIFIC EXPLORATION

'New World' Environments and Scientific Exploration

THE previous sections on geology, botany, and zoology have in places drawn upon tales of global voyages of exploration. This section turns to this subject in more detail, underlining the international reach of European sciences and offering examples of engagements with the so-called 'new world', ranging across virtually every continent and taking us from the 1790s to the 1850s. It is worth underlining that just as scientific works found wide readerships, so travel literature expanded massively in response to a rapidly growing middle-class readership. For the Victorians, explorers of every stamp had the status of movie stars, whose works were bestsellers and whose public engagements were sold out. The late eighteenth-century voyages of James Cook and Louis-Antoine de Bougainville made them celebrities, and many others followed in their wake, including Mungo Park, Meriwether Lewis and William Clark, and Charles Sturt. This section turns to a sub-group of nineteenth-century voyagers, the natural historians, zoologists, botanists, and geographers who joined expeditionary ventures.

While often more liberal in their attitudes than many other explorers, these exclusively male figures participated in a cultural phenomenon based largely on the expropriation of indigenous objects, knowledge, and organic specimens, situating the so-called 'new world' as a site in which to mine materials for the burgeoning museum collections of Europe and the United States. The doleful effects of European travel on indigenous peoples and landscapes is neatly summarised by Charles Darwin in *Voyage of the Beagle*, where he argues that 'wherever the European had trod, death seems to pursue the aboriginal'. Whether in 'the Americas, Polynesia, the Cape of Good Hope, and Australia', he notes, 'we find the same result'. It is worth noting that European global expansionism began as early as the fifteenth century and that what we see in our period is an intensification of a well-established pattern. As Alfred Crosby (2004, 5) points out, more than 50 million Europeans migrated to the 'new world' between 1820 and 1930. According to Crosby (2004, 7), 'the success of European imperialism' had 'a biological, an ecological component'. It is beyond the purview of this volume to fully consider the impacts of colonialism and imperialism on indigenous peoples, but these are touched upon in places and should be situated as part of a broader exploitative urge that turned its gaze on indigenous environments. As Laura Dassow Walls (2011, xv) suggests, nineteenth-century 'industrial, consumer capitalism was reshaping the globe through expanding circuits of capital and commerce, and reshaping the planet too, felling forests, draining marshes, and levelling mountains'. This section reflects these contexts and vindicates John Miller's argument (2012, 96) that this 'most concentrated period of British expansionist energy' marks out the Victorian period as 'a pivotal stage in the nexus of ecological and political violence'. Attitudes to environments certainly vary in the extracts that follow – from wonder to wilderness-making and from sympathy and concern for flora and fauna to an unthinking tendency to shoot 'specimens' and consume wildlife.

Although the first extract falls just before the period covered in this volume, it is included because of its seminal role in intensifying interest in scientific voyages to parts of the globe hitherto unexplored or little explored by Europeans. Sir Joseph Banks's journal of 1768–71 covers his time as one of two botanists on Captain Cook's *HMS Endeavour*, calling at Brazil, Cape of Good Hope, Tahiti, New Zealand, Australia, and Indonesia. An 1896 edition, edited by Joseph Dalton Hooker, is used, and the extract is from Chapter 8, covering Banks's impressions of New Zealand on what was the first European contact with ancient Māori cultures since Abel Tasman's 1642 voyage. The extract records a series of cultural encounters which were largely peaceful, perhaps as a result of the mediation of the *Endeavour*'s remarkable Polynesian navigator, Tupaia (or Tupia). Cook's time in New Zealand was generally peaceful, and he made efforts to behave compassionately and considerately to the peoples he encountered, learning from other examples of conflict during the voyage. The journal also records Banks and fellow botanist Daniel Carlsson Solander collecting zoological and botanical specimens new to European science.

Banks returned to Britain to immediate fame, and with a substantial collection of 30,000 plants, including 1,400 new discoveries. His influence in scientific circles, and his role as a patron of the natural sciences over the coming decades, was unmatched during his lifetime, but he also inspired a growing public interest in global travel and natural history. Accounts of voyages such as this whetted an appetite for what was constructed as the 'exotic' and underlined the degree to which Georgian Britain was increasingly bound up – for good and ill – in international colonial, economic, and scientific networks.

Solander's position alongside Banks is one indication of the importance of Sweden to the development of European natural history since the influential role of Linnaeus in the eighteenth century. Both Solander and the author of the following extract, Carl Peter Thunberg, were amongst the influential disciples of Linnaeus who continued his practices in natural history. Thunberg's first volume of *Travels in Europe, Africa, and Asia* (1795) aims for a work of natural history travel both accessible and scientifically rigorous. His account of his time at Cape Colony is of interest for the light it sheds on the practicalities and difficulties of travel but also of the emergence of European networks of trade and influence in the Southern Hemisphere. Thunberg's work follows on from *A Voyage to the Cape of Good Hope* (1789) by fellow Linnaean Anders Sparrman, who is regularly cited by Thunberg. The Dutch East India Company established a colony at Cape of Good Hope in 1652 as a vital mid-station for their trading routes to the far east. The often-painful history of the colony is analysed in Chapter 3 of Grove's *Green Imperialism*, and he points out that while Dutch environmental policies were far more enlightened than those of the British on St Helena and Barbados, deforestation and exploitation of land were key features. By the 1690s, the Dutch were already embarking on extensive inland expeditions (1995, 140). Thunberg, then, was often following paths already well trodden, both by natives and Europeans. The date of his *Travels*, it should also be noted, coincided with British conquest of the Cape, a reflection that this part of Southern Africa has a long and complex imperial history.

The first part of the extract deals with the organisation of European households at the Cape, including their reliance both on slave labour (East African and Asian) and on the importation of European crops and seeds, the latter shedding light on just how quickly Europeans began the often-disastrous process of importing non-native species into their colonies. While by no means virulently racist, Thunberg's Eurocentric descriptions of engagements with 'new world' environments and peoples see both figured as spaces destined for improvement and exploitation by more enlightened races. That conflicts resulted from European activities is evident in the second part of the extract, an account of the capture and punishment of 'Hottentots' (an increasingly pejorative term used to denote the native Khokoi tribes). While Thunberg describes their violent resistance as 'crimes', he also notes their claim that 'they acted so in their own defence, the Europeans making every year fresh encroachments upon their lands and possessions, and forcing them continually further up into the country'. While being thus given voice, they are nonetheless rendered savage and exotic in Thunberg's subsequent description of their appearance, dress, and habits.

The final, and longest, part of the extract is taken from Thunberg's journey into the interior, where his interests are mainly botanical and zoological. The account gives a strong impression of the challenges of such travels, including the reliance on hired 'Hottentots' to commandeer the oxen used to transport large volumes of food, clothing, and equipment inland. The land had already 'been much inhabited and cultivated by European colonists', primarily Dutch cattle farmers, early 'pioneers' of the exploitation of Africa that would follow later in the century. Thunberg describes numerous birds, including flamingos, and (presumably erroneously) records witnessing a dead tiger, as well as providing an account of seal-hunting. His botanical notes are particularly focused on medicinal properties, and his approach is rooted in eighteenth-century models rather than the later, cutting-edge science of explorers like Hooker and Darwin (see previous sections of this volume) and far more clearly rooted in an exploitative model of engagement.

First published just over a decade after Thunberg's work, *Personal Narrative of Travels to the Equinoctial Regions of the New Continent During the Years 1799–1804* (1814–29) by Alexander von Humboldt and Aimé Bonpland, feels strikingly modern by comparison, both in social attitudes and scientific insights. This is little surprise given von Humboldt's stellar place in nineteenth-century science, in particular in the establishment of an approach to natural history that comprised geology, geography, botany, zoology, and meteorology. His importance to biogeography will be further explored in extracts in Part 8 of this volume, while his achievement of providing the first statement of anthropogenic climate change will be a focus in Volume II. Humboldt's collaboration with Bonpland arose because of the latter's role as botanist on a Spanish Bourbon Empire-approved expedition to South America. Their friendship and shared love of science proved supportive of their expeditions and subsequent writing, although Humboldt was always the more energetic partner. His fame, and a wide readership, had already been guaranteed by the publication of *Views of Nature* (1807) and *Political Essay on the Kingdom of New Spain* (1810). *Personal Narrative* was therefore eagerly anticipated and warmly received, but the journeys it described provided the data and experience that established Humboldt as a scientific pioneer. A lengthy chapter in Volume 1 (unexcerpted) describes the enormous quantities of scientific equipment and materials with which the scientists travelled. The success of their voyage owed much to the highly systematic manner in which they recorded and measured phenomena using advanced instrumentation and the meticulous methods used for gathering and preserving specimens.

The extract, from Chapter 6 of the first volume, covers initial impressions and analyses of the elevated woody countryside around Cumanacoa, inland from Venezuela's Caribbean coast, which had been exploited for its salt pans by the Spanish and Dutch since the seventeenth century. Although their travels were expedited by the relatively progressive attitudes of the new Bourbon government, Humboldt and Bonpland are not reticent about criticising the doleful effects of Spanish settlement on South America since the Conquistadores. In particular, they are highly critical of 'the trade in the copper-coloured Indians [that] was accompanied by the same acts of inhumanity as that in the African negroes' and the manner in which Spanish missionaries had diminished amongst the native Chayma Indian populations 'that vigour of character, and that natural vivacity, which in every state of society are the noble fruits of independence'. Driven by Enlightenment and Abolitionist sentiments, Humboldt and Bonpland criticise the manner in which the Indians 'are kept in a state remote from improvement' while also largely avoiding the common Romantic trap of idealising a constructed 'noble savage'. The remainder of the expedition offers a fine example of the naturalist mode of the *Personal Narrative* via descriptions of travel, flora, fauna, and geography.

The next two extracts, both from works by English naturalist Charles Waterton, continue the focus on South American travel. That Waterton's *Wanderings in South America* (1825) was, like Humboldt and Bonpland's *Personal Narrative*, a bestseller is testament to the appetite of the European middle classes for travel literature: Waterton offered a riveting combination of science, spectacle, and adventure. The extract is taken from his description of his first journey, in Dutch Guiana (Surinam). Waterton's work is also included in Volume II, as his work is both scientific and popular. The part selected here involves Waterton's address to the reader – including his already old-fashioned use of 'thee' and 'thou'. Waterton was a gifted stylist, able to change tone, and here he conjures his own sense of wonder and reverence within an ancient tropical forest environment. Invoking the name of Mungo Park, the Scottish explorer of Africa who did much to spur interest in 'New World' explorations, Waterton also seeks inspiration in 'the genius which presides over these wilds'. In anticipation of his later conservation work in Yorkshire, while he accepts the need to kill 'a pair of doves in order to enable thee to give mankind a true and proper description of them', he deplores indiscriminate shooting of wildlife 'through wantonness, or to show what a good marksman thou art'.

Waterton also largely resists the European tradition of constructing 'New World' landscapes as wildernesses empty of human inhabitants: as well as celebrating the company of the animals of the forest, Waterton is at pains to show the lives of native Indians, doing so in ways that are just as sympathetic as Humboldt and Bonpland. Although speaking of 'seeing man in his rudest state', the subsequent description emphasises the cleanliness, social order, and customs of the Indians, as well as the effectiveness of their engagements with forest environments.

The second Waterton extract, from *Essays in Natural History* (1838), continues the coverage of Guiana but this time focusing not on the transcendent pleasures of journeys into its forests but on what he describes as the strangely pleasurable torments of one of its native insects, known as a chegoe, chigoe flea, or jigger: *Tunga penetrans*, a parasitical flea that burrows under the surface of the skin of its animal hosts, feeds on their blood while maintaining a skin lesion through which it can breathe, defecate, mate, and disseminate its eggs. A parasitical infestation (tungiasis) causes mild to serious pain, lesions, secondary infections, toe deformation, and loss of toenails. During its infestation, the insect swells from 1 mm to the size of a pea. Waterton's absorbing account of 'this insidious miner' describes its activities in gruesome detail and its impacts on himself, on imported African slaves, and on dogs, although *Tunga* affects a considerable range of South American mammals. The chapter offers vividly powerful insights into the challenges and dangers facing European explorers in South America and into Waterton's characteristically phlegmatic responses to them.

Although Charles Darwin's *Journal of Researches* (*Voyage of the Beagle*) is famous for its engagements with South America, the following extract from this work turns to his time on the Falkland Islands. It is chosen because it is typical of the author's approach to travel writing in combining scenes of excitement with discussions of natural history. The former is provided by accounts of the dexterity of Darwin's gaucho guides in using lassoes and bolas to capture wild cattle, while the latter is more wide ranging, including an account of the island's bleak topography and limited fauna. Darwin was writing only twelve years after Britain had reasserted its claims to sovereignty of the islands, previously variously held or claimed by Britain, France, Spain, and Argentina, and at this period, they had a reputation for lawlessness associated with their recent use as a penal colony.

Following von Humboldt, whom he cites regularly in the *Journal*, Darwin takes a biogeographical approach to the Falklands, just as Joseph Dalton Hooker would do in *Flora Antarctica* (see Parts 5 and 7 of this volume) in classifying the flora of the Falklands as belonging to the Antarctic Zone. Darwin's account of the first two days of his island trips also contains allusions to the various ways in which the islands were being commercially exploited – by whalers, sealers, and gauchos. The degree to which scientific travellers were either following or further opening up networks of international trade and commerce is a key question, an indication of the proliferation of capitalist economics and the modes of social and environmental exploitation on which they relied. An example of the costs of European expansion is evident in Darwin's mournful and sadly accurate prediction that the native Falklands Islands wolf or warrah (*Dusicyon australis*) would go the same way as the dodo of Mauritius, which became extinct within a century of Dutch settlement in 1598, largely because of the introduction of European animal species. There is also, perhaps, a glimmer of Darwin's later theories in his other prediction, that 'at some future period the southern hemisphere probably will have its breed of Falkland ponies, as the northern has that of Shetland'.

George Gardner's *Travels in the Interior of Brazil* (1846) forms the next extract, offering insights into the degree to which much of South America was by this time heavily marked by a long period of Spanish and Portuguese settlement and already a major site of economic activity and commercial exchange. While the journey Gardner describes into the Organ Mountains (Serra dos Órgãos), north of Rio de Janeiro, is certainly arduous, particularly for the mules used for much of the journey, it is also clear that these lands were already gathering travel and leisure infrastructure. Gardner first passes through a health resort and cottages much used by English residents of Rio as well as an English 'farm for the breeding of horses and mules, and a large garden from which the Rio market is regularly supplied with European vegetables'. Comparisons can be made with Thunberg's earlier account of the Cape of Good Hope. Both texts reveal the reliance of colonial economies on slave labour and expropriation of lands, a process of Europeanisation that sought to efface aboriginal cultures. Gardner reports the construction of a new road across the Organ mountains to provide better connection to the mining districts beyond – reminding one that the original attraction of South America to the Conquistadores lay as much in its precious metals as in any desire to disseminate Christianity.

The extract is also of interest as botanical travel literature. Gardner brings his expertise to bear on a range of native flora, from mangrove swamps and jasmines to *Furcraea* and cacti, but particularly to the forests through which he passes on his ascent to the Organ range. To modern readers, there is melancholy in Gardner's delighted references to 'one dense forest, the magnificence of which cannot be imagined by those who have never seen it, or penetrated its recesses', which he compares favourably to 'those remnants of the virgin forest which still stand in the vicinity of the capital' which 'become insignificant when compared with the mass of giant vegetation which clothes the sides of the Organ Mountains'. That the Serra dos Órgãos National Park exists today, and manages to preserve much of its lowland and highland forests, is a small cause for celebration in a country so marked by agriculture-led campaigns of Amazonian deforestation. Gardner's account of the complexities of the forest's flora is proto-ecological in its sensitive attention to organic relationships.

While wealthier and more intrepid Victorian readers might realistically aspire to follow in Gardner's footsteps, the same could not be said for readers of *Niger Flora* (1849), a work already extracted in Part 5. There, we turned to the journals of the expedition's botanist, Theodore Vogel. Here, I include Hooker's prefatory explanation of the voyage's purposes and activities under the auspices of the African Civilization Society and Hooker's memoir of Vogel's life. Together these offer valuable insights into the motivations and practicalities of such voyages, to their travails and dangers, and of the Eurocentric attitudes revealed even in a self-proclaimed philanthropic exercise such as this. The expedition's aim was to support Africa at a time when transatlantic slavery still continued in the United States by fostering viable opportunities for trade and commerce and by supporting efforts to direct African agriculture to these ends. The stated aims of the voyage were 'to penetrate by this vast navigable river [the Niger], into the interior of this little-known country, to make treaties with the inhabitants, and to establish an emporium at some suitable place'.

Although well-meaning, the notion that Africa required civilisation – and that it had therefore never experienced civilisation before – is writ large in the establishment of an African Civilization Society. The voyage's aims thereby blended condescending attitudes to the inhabitants and history of the 'dark continent' with a series of scientific objectives. In Hooker's memoir of Vogel, we glimpse his time in a colony whose aim was 'to teach the Africans active habits and to christianize them'. The sense that Europeans were sacrificing themselves for noble causes is clear in Hooker's lament over the deaths of so many of the voyagers that 'the European constitution is incapable of withstanding the effect of that deadly atmosphere'. Here, writ large, is a prevailing narrative of European engagements with African environments, figured as they were as forbidding and dangerous, excessive in terms of climate, the accelerated growth rates of tropical plants, and the large size of many African animals. In so many colonial engagements with the 'New World', including the Caribbean, Africa, the Indian subcontinent, and South-East Asia, the essentially orientalist assumption was that European temperate climates led to temperate personalities and societies, while tropical environments led to excessive, intemperate peoples, societies, and behaviours (a theory propounded directly by Henry Buckle in Part 8). The project to bring the 'benefits' of European governance, social institutions, customs, economic models, and religions to places like Africa rested on an assumption of the right of the more 'civilised' to exercise sovereignty over the peoples and environments of the earth's 'uncivilised' realms. It is in this context that we must read Hooker's reasonably progressive claim that 'the only hope of enlightening the sons of Africa is by *native* agency', for this native agency is, according to Hooker, in need of the support of superior British science and ingenuity. By the standards of his day, it must be said, Hooker is relatively liberal in claiming of the 'two native missionaries' whom he had subsequently charged with transporting 'a considerable collection of useful Tropical plants for introduction into Africa' that 'he knows not any well-educated Europeans more competent to estimate the value of such importations, or likely to feel more interest in their successful cultivation and use'. At the same time, the benefits of the mutual association of British science and African native agency include, as Hooker notes, the expansion of his own herbarium collection at Kew and the future promise of a thorough study of the botany of Africa built upon the foundations of *Niger Flora*.

The memoir of Vogel draws extensively upon his journals and gives a clear, often grim depiction of the difficulties of his journeys. In a fine example of Victorian travel writing, Vogel records the landscapes and peoples he encounters, as well as the difficulties that tropical weather threw in the way of his attempts to assemble and maintain a botanical collection and sometimes simply to get about: 'the African brooks, when they are swollen with rain, assume the privilege of making their way down the footpaths; and I was therefore obliged for hours to wade up to the knees in water' so that 'I was indeed, in general, whether at sea or on land, as wet as it was possible to be'. He also records encounters with various 'natives' including the 'Kroo', who are 'tolerably docile, and are therefore hired by the coasters to perform such hard labours as are considered prejudicial to Europeans'. After being carried fourteen miles into the interior by hired 'natives', Vogel expresses his delight in reaching a Danish coffee-plantation with 'a house arranged with European accommodations, where we were surrounded with all the luxury of the civilized world, and had for dinner French asparagus'. Underlining the delights of this Europeanised idyll, Vogel proclaimed it 'lovely, pleasantly varied with hill and dale', adding that while the Danes still kept slaves, these 'seem well off, and were merry and cheerful beings'. Speaking of free tribes near Iddah, where the expedition purchased land for a colony, Vogel notes their accomplishments in spinning, weaving, iron work, and agriculture but regrets that they do not 'progress' in these arts because 'they lack the spiritual energy which renders every acquisition a step to further advancement'.

This section closes with an example of Asian travel, from Thomas Thomson's *Western Himalaya and Tibet; A Narrative of a Journey Through the Mountains of Northern India, During the Years 1847–8* (1852), a work exemplifying growing fascination with a region that was spurred by Britain's colonisation of India, which would soon become formally absorbed within the British Empire as 'the Raj' following the brutal suppression of the Indian Rebellion (Indian Mutiny) of 1857. Thomson's journey was enabled by his role as an assistant surgeon of the British Bengal Army, during which time he saw service in the Afghan campaigns (1839–42), the Sutlej campaign (1845–6), and the second Sikh War (1848–9). Later promoted to surgeon major, he subsequently oversaw the East India Company's Botanic Gardens, Calcutta, and transitioned, on his return to Britain, to a botanical career.

Thomson's travel narrative involves the years he spent in between the Sutlej and Sikh campaigns, during which time he travelled extensively following his appointment to what he describes as 'the Tibet Mission', not to be confused with the British Expedition to Tibet (1903–4), which was effectively a temporary invasion. Thomson notes that the 1847 mission, sponsored by the governor-general of India, set out to 'despatch across the Himalaya Mountains into Tibet', but he gives no further indication of the remit of the expedition. It was certainly the case that in this period British imperial anxieties were beginning to grow because of the Russian Empire's eastward expansion and the possibility of invasion from the north, but this was also a time in which early European Catholic missions to Tibet were being established. Thomson's scientific accounts of Tibet and the Western Himalayas are highly empirical, a far cry from the later 'Shangri-La' mysticism that would encapsulate western responses to the region following James Hilton's much later novel, *Lost Horizons* (1933). Despite travelling through extremely dramatic landscapes, Thomson eschews dramatic prose and lacks the writing skills of a Waterton or Darwin. Nonetheless, his attention to detail and focus on geology, geography, botany, zoology, and meteorology make his work a substantial contribution to European literature on the region.

The extract is taken from Chapter 14, in which Thomson gives an account of a wild 'journey of twenty days through uninhabited regions' towards the high Karakoram Pass, a 18,176-ft (5,540-m) pass between India and China, deep in the Karakoram Range, which includes eighteen summits higher than 24,600 ft (7,500 m), including K2, the world's second-highest mountain. While Thomson was by no means the first European to travel through the pass, this part of the world was still largely inaccessible to outsiders. Describing the first days of the journey, Thomson shows little interest in either local inhabitants or his own guides but offers a clear picture of the difficulties of passing through such trying terrain and notes the considerable number of equine skeletons lining the road, the horses having perished during their journeys. The second part of the extract resumes Thomson's journey as he approaches and reaches the Karakoram Pass, many days after first setting off and having passed through many miles of arduous country to reach altitudes that he estimates as in excess of 17,000 ft. Thomson notes the deleterious effects of the rarefied air and offers a typically precise description of the Shayuk (Shyok) river (which translates as 'the river of death'), a tributary of the Indus.

Like all of the extracts included in this section, Thomson's work is a testament to the ingenuity and energy of European explorers in this age of expansion. Like all of the other extracts, it invites questions about the attitudes of such explorers (and of the wider cultural drive in which they were operating) to the lands and peoples they encountered – questions about their right to 'penetrate' territories, to describe them as 'virgin' or 'wilderness', and to impose European attitudes on all they surveyed. At the same time, many of the explorers featured in this section articulate relatively liberal attitudes and evince a genuine desire to expand the frontiers of knowledge. This Enlightenment urge, founded as it is in a belief in the inexorable onward march of human progress, led by western models of thought and enquiry, is writ large across all of these pages – with all of the good and ill involved. In terms of what these extracts tell us of attitudes to environment, there is no simple generalisation, for they are in part bound up in attitudes of human environmental sovereignty and in part reach, however tentatively and imperfectly, for alternatives to that vision. The work of these scientific travellers revealed the range and considerable complexity of the globe's environments, and this offered multiple opportunities – for the expansion of knowledge, the interchange of cultures, and the expansion of economic networks. The principal result of this inexorable outward movement, however, is ecocrisis.

Further reading

Banks, Sir Joseph, *The Letters of Sir Joseph Banks: A Selection, 1768–1820* (London: Imperial College Press; River Edge, NJ, 2000).

Barber, Lynn, *The Heyday of Natural History 1820–1870* (London: Cape, 1980).

Blackburn, Julia, *Charles Waterton, 1782–1865: Traveller and Conservationist* (London: Vintage, 1997).

Crosby, Alfred W., *Ecological Imperialism: The Biological Expansion of Europe, 900–1900*, 2nd ed. (Cambridge: Cambridge University Press, 2004).

Drayton, Richard Harry, *Nature's Government: Science, Imperial Britain, and the 'Improvement' of the World* (New Haven: Yale University Press, 2000).

Endersby, Jim, *Imperial Nature: Joseph Hooker and the Practices of Victorian Science* (Chicago: University of Chicago Press, 2008).

Fara, Patricia, *Sex, Botany and Empire: The Story of Carl Linnaeus and Joseph Banks* (Cambridge: Icon Books, 2003).

Fulton, Richard, Peter Hoffenberg, Stephen Hancock, and Allison Poynter (eds.). *South Seas Encounters: Nineteenth-Century Oceania, Britain, and America* (London: Routledge, 2018).

Grove, Richard H., *Green Imperialism: Colonial Expansion, Tropical Islands, Edens and the Origins of Environmentalism, 1600–1860* (Cambridge: Cambridge University Press, 1995).

Hoage, R. J. and William A. Deiss (eds.), *New Worlds, New Animals: From Menagerie to Zoological Park in the Nineteenth Century* (Baltimore: Johns Hopkins University Press, 1996).

Huggan, G and H. Tiffin (eds.), *Postcolonial Ecocriticism: Literature, Animals, Environment* (Abingdon: Routledge, 2010).

Ingleby, Matthew and Kerr, Matthew P.M. (eds.). *Coastal Cultures of the Long Nineteenth Century* (Edinburgh: Edinburgh University Press, 2018).

Keynes, Richard, *Charles Darwin's Beagle Diary* (Cambridge: Cambridge University Press, 2001).

McKay, Alex, *Their Footprints Remain: Biomedical Beginnings Across the Indo-Tibetan Frontier* (Amsterdam: Amsterdam University Press, 2007).

Miller, John, 'Postcolonial Ecocriticism and Victorian Studies', *Literature Compass* 9:7 (2012), 476–88.

Nixon, Rob, *Slow Violence and the Environmentalism of the Poor* (Cambridge: Harvard University Press, 2011).

O'Brian, Patrick, *Joseph Banks: A Life* (London: David R. Godine, 1993).

Oelschaeger, M. *The Idea of Wilderness: From Prehistory to the Age of Ecology* (New Haven and London: Yale University Press, 1991).

Rietbergen, P., 'Becoming Famous in the Eighteenth Century: Carl Peter Thunberg Between Sweden, the Netherlands and Japan', *De Achttiende Eeuw* 36:1 (2004), 50–61.

Walls, Laura Dassow. 'Foreword', *Victorian Ecocriticism: The Politics of Place and Early Environmental Justice*, ed. Dewey W. Hall (Lanham, MD: Lexington Books, 2011), xiii–xvii.

Wisnicki, Adrian S., *Fieldwork of Empire, 1840–1900: Intercultural Dynamics in the Production of British Expeditionary Literature* (New York and London: Routledge, 2019).

Wulf, Andrea, *The Invention of Nature: Alexander von Humboldt's New World* (London: Vintage, 2016).

64

JOSEPH DALTON HOOKER (ED.), *JOURNAL OF THE RIGHT HONOURABLE SIR JOSEPH BANKS*

(1896) (covering the voyages of *HMS Endeavour*, 1768–71)

In sailing along shore, we could clearly see several cultivated spots of land, some freshly turned up, and lying in furrows, as if ploughed; others with plants growing upon them, some younger and some older. We also saw in two places high rails upon the ridges of hills, but could only guess that they are a part of some superstition, as they were in lines not inclosing anything.

15*th*. Snow was still to be seen upon the mountains inland. In the morning we were abreast of the southernmost cape of a large bay, the northernmost of which was named Portland Isle. The bay itself was called Hawke's Bay. The southern point was called Cape Kidnappers, on account of an attempt made by the natives to steal Tayeto, Tupia's boy. He was employed in handing up the articles which the natives were selling, when one of the men in a canoe seized him and pushed off. A shot was fired into the canoe, whereupon they loosed the boy, who immediately leaped into the water and swam to the ship. When he had a little recovered from his fright, Tayeto brought a fish to Tupia, and told him that he intended it as an offering to his *eatua,* in gratitude for his escape. Tupia approved it, and ordered him to throw it in the water, which he did.

16*th*. Mountains covered with snow were in sight again this morning, so that a chain of them probably runs within the country. Vast shoals of fish were about the ship, pursued by large flocks of brownish birds a little bigger than a pigeon (*Nectris munda*). Their method of fishing was amusing enough: a whole flock of birds would follow the fish, which swam fast; they continually plunged under water, and soon after rose again in another place, so that the whole flock sometimes vanished altogether, and rose again, often where you did not expect them; in less than a minute's time they were down again, and so alternately as long as we saw them. Before dinner we were abreast of another cape, which made in a bluff rock, the upper part of a reddish-coloured stone or clay, the lower white.

DOI: 10.4324/9780429355653-72

Beyond this the country appeared pleasant, with low smooth hills like downs. The captain thought it not necessary to proceed any farther on this side of the coast, so the ship's head was turned to the northward, and the cape thence called Cape Turnagain. At night we were off Hawke's Bay and saw two monstrous fires inland on the hills. We are now inclined to think that these, and most if not all the great fires that we have seen, are made for the convenience of clearing the land for tillage, but for whatever purpose they are a certain indication that where they are the country is inhabited.

20th. Several canoes followed us, and seemed very peaceably inclined, inviting us to go into a bay they pointed out, where they said was plenty of fresh water. We followed them in, and by eleven came to an anchor. We then invited two, who seemed by their dress to be chiefs, to come on board; they immediately accepted our invitation. In the meantime those who remained in the canoes traded with our people very fairly for whatever they had in their boats. The chiefs, who were two old men, the one dressed in a jacket ornamented after their fashion with dog skin, the other in one covered almost entirely with some tufts of red feathers, received our presents, and stayed with us till we had dined.

21st. At daybreak the waterers went ashore, and soon after Dr. Solander and myself did the same. There was a good deal of surf upon the beach, but we landed without much difficulty. The natives sat by our people, but did not intermix with them. They traded, however, for cloth chiefly, giving whatever they had, though they seemed pleased with observing our people, as well as with the gain they got by trading with them; yet they did not neglect their ordinary occupations. In the morning several of their boats went out fishing, and at dinner-time all went to their respective homes, returning after a certain time. Such fair appearances made Dr. Solander and myself almost trust them; we ranged all about the bay and were well repaid by finding many plants, and shooting some most beautiful birds. In doing this we visited several houses, and saw a little of their customs, for they were not at all shy of showing us anything we desired to see, nor did they on our account interrupt their meals, the only employment we saw them engaged in.

Their food at this time of the year consisted of fish, with which, instead of bread, they eat the roots of a kind of fern, *Pteris crenulata,*[1] very like that which grows upon our commons in England. These were slightly roasted on the fire and then beaten with a stick, which took off the bark and dry outside; what remained had a sweetish, clammy, but not disagreeable taste. It might be esteemed a tolerable food, were it not for the quantity of strings and fibres in it, which in quantity three or four times exceed the soft part. These were swallowed by some, but the greater number spit them out, for which purpose they had a basket standing under them to receive their chewed morsels, in shape and colour not unlike chaws of tobacco. Though at this time of the year this most homely fare was their principal diet, yet in the proper seasons they certainly have plenty of excellent vegetables. We have seen no sign of tame animals among them, except very small and ugly dogs. Their plantations were now hardly finished, but so well was the ground tilled that I have seldom seen land better broken up. In them were planted sweet potatoes, cocos, and a plant of the cucumber kind, as we judged from the seed leaves which just appeared above ground.

The first of these were planted in small hills, some in rows, others in quincunx, all laid most regularly in line. The cocos were planted on flat land, and had not yet appeared above ground. The cucumbers were set in small hollows or ditches, much as in England. These plantations varied in size from 1 to 10 acres each. In the bay there might be 150 or 200 acres in cultivation, though we did not see 100 people in all. Each distinct patch was fenced in, generally with reeds placed close one by another, so that a mouse could scarcely creep through.

When we went to their houses, men, women and children received us; no one showed the least signs of fear. The women were plain, and made themselves more so by painting their faces with red ochre and oil, which was generally fresh and wet upon their cheeks and foreheads, easily transferable to the noses of any one who should attempt to kiss them, not that they seemed to have any objection to such familiarities, as the noses of several of our people evidently showed. But they were as coquettish as any Europeans could be, and the young ones as skittish as unbroken fillies. One part of their dress I cannot omit to mention: besides their cloth, each one wore round the waist a string made of the leaves of a highly-perfumed grass, to which was fastened a small bunch of the leaves of some fragrant plant. Though the men did not so frequently paint their faces, yet they often did so; one especially I observed, whose whole body and garments were rubbed over with dry ochre; of this he constantly kept a piece in his hand, and generally rubbed it on some part or other.

In the evening, all the boats being employed in carrying on board water, we were likely to be left ashore till after dark. We did not like to lose so much of our time for sorting our specimens and putting them in order, so we applied to our friends the Indians for a passage in one of their canoes. They readily launched one for us; but we, in number eight, not being used to so ticklish a conveyance, overset her in the surf, and were very well soused. Four of us were obliged to remain, and Dr. Solander, Tupia, Tayeto and myself embarked again, and came without accident to the ship, well pleased with the behaviour of our Indian friends, who would a second time undertake to carry off such clumsy fellows.

24*th*. Dr. Solander and I went ashore botanising, and found many new plants. The people behaved perfectly well, not mixing with or at all interrupting our people in what they were about, but on the contrary selling them whatever they had for Otahite cloth and glass bottles, of which they were uncommonly fond.

In our walks we met with many houses in the valleys that seemed to be quite deserted. The people lived on the ridges of hills in very slightly-built houses, or rather sheds. For what reason they have left the valleys we can only guess, maybe for air, but if so they purchase that convenience at a dear rate, as all their fishing tackle and lobster pots, of which they have many, must be brought up with no small labour.

We saw also an extraordinary natural curiosity. In pursuing a valley bounded on each side by steep hills, we suddenly saw a most noble arch or cavern through the face of a rock leading directly to the sea, so that through it we had not only a view of the bay and hills on the other side, but an opportunity of imagining a ship or any other grand object opposite to it. It was certainly the most magnificent surprise I have ever met with; so much is pure nature superior to art in these cases. I have seen such places made by art, where from an inland view you were led through an arch 6 feet wide, and 7 feet high, to a prospect of the sea; but here was an arch 25 yards in length, 9 in breadth, and at least 15 in height.

In the evening we returned to the watering-place, in order to go on board with our treasure of plants, birds, etc., but were prevented by an old man who detained us some time in showing us their exercises with arms, lances, and *patoo patoos*. The lance is made of a hard wood, from 10 to 14 feet long, and very sharp at the ends. A stick was set up as an enemy; to this he advanced with a most furious aspect, brandishing his lance, which he held with great firmness; after some time he ran at the stick, and, supposing it a man run through the body, immediately fell upon the upper end of it, dealing it most merciless blows with his *patoo patoo,* any one of which would have probably split most skulls. From this I should conclude that they give no quarter.

25*th*. Went ashore this morning and renewed our search for plants, etc., with great success. In the meantime Tupia, who stayed with the waterers, had much conversation with one of their priests; they seemed to agree very well in their notions of religion, only Tupia was much more learned than the other, and all his discourse was received with much attention. He asked them in the course of his conversation many questions, among the rest whether or no they really ate men, which he was very loth to believe; they answered in the affirmative, saying that they ate the bodies only of those of their enemies who were killed in war.

Among other knicknacks, Dr. Solander bought a boy's top, which resembled those our boys play with in England, and which they made signs was to be whipped in the same manner.

28*th*. On an island called Jubolai we saw the largest canoe which we had met with; her length was 68½ feet, her breadth 5 feet, and her height 3 feet 6 inches. She was built with a sharp bottom, made in three pieces of trunks of trees hollowed out, the middlemost of which was much longer than either of the other two; their gunnel planks were in one piece 62 feet 2 inches in length, carved prettily enough in bas-relief; the head also was richly carved in their fashion. We saw also a house larger than any we had seen, though not more than 30 feet long; it seemed as if it had never been finished, being full of chips; the woodwork was squared so evenly and smoothly that we could not doubt of their having very sharp tools. All the side-posts were carved in a masterly style of their whimsical taste, which seems confined to making spirals and distorted human faces; all these had clearly been moved from some other place, so that such work probably bears a value among them.

While Mr. Sporing was drawing on the island he saw a most strange bird fly over his head. He described it as being about as large as a kite, and brown like one; his tail, however, was of so enormous a length that he at first took it for a flock of small birds flying after him: he who is a grave thinking man, and is not at all given to telling wonderful stories, says he judged it to be yards in length.

Note

1 The same plant as the British bracken, *Pteris aquilina*.

65

CARL PETER THUNBERG, 'THE CAPE', *TRAVELS IN EUROPE, AFRICA, AND ASIA*

(London: F. and C. Rivington, 1795)

THE houses are all of brick, white-washed, and one, seldom two, but very rarely three stories high, and covered in for the most part with flat roofs of brick-work, or a kind of grass indigenous to this country (*restio tectorum*) laid upon very low frame work. On account of the violence of the winds that prevail here, the roofs cannot be tiled over, nor raised higher. The house of the lieutenant-governor, and the company's warehouse, were the only houses that were three stories high.

The domestics here do not consist of Europeans, but of black or tawny slaves from *Malabar, Madagscar,* or other parts of *India*. These, in general, speak either broken Portuguese, or else the Malabar, seldom the Dutch language, and learn various trades, by which they bring their masters considerable profit, especially such as are taylors, carpenters, bricklayers, or cooks. The slaves are let out by the month, week, or day, during which term they are to earn for their masters a certain fixed sum *per diem*. The male slaves wear their own hair, upon which they set a great value, wrapped up in a twisted handkerchief like a turban, and the females wreath up their hair and fix it on their heads with a large pin. Trowsers constitute the other part of their dress; and as a token of their servile condition, they always go barefoot, and without a hat.

Previous to the company's sitting down to meals, either dinner or supper, a female slave brings a wash-hand bason and towel, to wash their hands, which is also done on the company's rising from table. In the houses of the wealthy, every one of the company has a slave behind his chair to wait on him. The slave has frequently a large palm leaf in his hand, by way of a fan, to drive away the flies, which are as troublesome here as they are in Sweden.

As well within as without the town, neat and excellent gardens are laid out, both for fruit and culinary vegetables, being watered by the streams that run down from the mountains. Among these that extensive and beautiful garden belonging to the company distinguishes itself, like an old oak amidst a thicket of bushes. It is from these gardens that the stranger, on his arrival, meets with his first refreshments; and from their superfluous stock the Dutch and other ships are supplied with stores for their voyage. The garden-seeds must be brought every year fresh from

DOI: 10.4324/9780429355653-73

Holland, as they otherwise, almost all, degenerate in time, excepting the seeds of cauliflowers, which are brought to great perfection here, and on that account exported from hence to Holland, where they gradually degenerate.

Apples, pears, and other European fruits, are mellower and riper, but have not that flavour which they have in Europe, neither will they keep long. Nor are the peaches produced here equal in goodness to those of the south of Europe. They are sometimes dried like pears, with or without their stones.

The trees imported from Europe, such as oak (*quercus robur*), the white poplar (*populus alba*), and others, shed their leaves in the winter, as they do in their native places, whereas the African trees do not part with theirs. It is not long, however, before they recover their leaves again. This circumstance is singular enough; first, because the cold here in winter is not more severe than it is in Sweden in the autumn; and in the second place, because they shed their leaves to the southward of the equator at the very time that they put them forth to the northward of it. The lime-trees (*tilia Europœa*) do not thrive well, on account of the violent winds that rage here; and the same may be said of the hazel (*corylus avellana*), cherry-tree (*prunus cerasus*), gooseberry-tree (*ribes grossularia & ribes crispa*), currant-tree (*ribes rubrum & nigrum*), all of which degenerate, and seldom yield any fruit.

The myrtle (*Myrtus communis*) grows to the height of a tree, though its stem is neither thick nor stiff, nor does it throw out many branches. For this reason it seems to be proper, and indeed, is frequently used, for forming high hedges, in a country subject to violent winds, as its supple stem bends to the storm.

The foot of the mountain, or the hills round the town, consisted of a red flame-coloured clay, which proceeds from the water's running down the cracks, and tinging the earth with its acid, charged with ferruginous particles. Higher up on the hills, lie scattered without order, stones of all sizes, that have been rolled down from the mountains.

[. . .]

On my return to the Cape, I saw, towards the end of June, a body of Hottentots, men, women, and children, to the number of nine and fifty, brought up about one hundred and fifty miles from the interior part of the country, where they had committed various acts of violence against the colonists. They had been taken by a Hottentot captain, of the name of Kees, in the cleft of a mountain, where they had concealed and fortified themselves against a party of peasants and soldiers ordered out against them, and had for a long time defended themselves, by rolling large stones down upon their enemies. In two villages they had carried off the cattle, killed the inhabitants, plundered the houses, and taken possession of several fire-arms. They did not deny their crimes, but asserted that they acted so in their own defence, the Europeans making every year fresh encroachments upon their lands and possessions, and forcing them continually farther up into the country, whence they were driven back again by the other Hottentots, or else killed. These Hottentots were *Boshiesmen,* of a dark brown complexion, some of whom were naked, wearing only a band round their waists, which covered the pudenda before. Others wore, hanging loose over their shoulders, a sheep's skin, the ends of which scarcely met before, the upper part going, like a calash, over the head. The women had their little ones hanging behind on their shoulders; and girls eleven or twelve years of age had already children. The women were adorned with ear-rings, and broad rings of metal round their wrists. Their mouths and cheek-bones were very prominent, so that they bore the strongest resemblance imaginable to apes. After these Hottentots had been confined for some time at the Cape, they lost their colour, and became almost white.

In the *month of August* the winter drew near to its end, and the fields began to be decorated with flowers; it therefore now became necessary for me to think of such preparations as would be useful and requisite for me in my approaching long journey into the interior part of the country, a journey, relative to which a promise had been given me, that I should make it, in a great measure, at the company's expence.

I therefore provided myself with necessary clothes, as well as with boxes and bags, for collecting roots and seeds, with boxes and pins for insects, a keg of arrack for preserving serpents and amphibious animals, cotton and boxes for stuffing and keeping birds in, cartridge-paper for the drying of plants, tea and biscuits for my own use, and tobacco to distribute among the Hottentots, together with fire arms, and a large quantity of powder, ball, and shot of various kinds. Shoes for the space of four months were no inconsiderable article in this account, as the leather prepared in the Indies, is by no means strong besides, that it is quite cut to pieces, or soon worn out, by the sharp stones that occur every where in the mountains.

My equipage consisted of a saddle-horse, a cart covered with sail-cloth, like an ammunition-waggon, and three yoke of oxen, by which it was to be drawn through the whole of the journey. My travelling companions were AUGE, the gardener, who had before made eighteen journies of different lengths into the country, and was now to be my sure and faithful guide; M. IMMELMAN, a youth, the son of a lieutenant in the army, together with LEONHARDI, a serjeant, who undertook this tedious journey for the sake of shooting the larger animals and birds; and lastly, two domesticated Hottentots, one of whom was to drive, and the other to lead our oxen.

Every one that travels in this country, performs his journey pretty nearly in the following way. A large waggon, worth from one hundred and twenty to two hundred dollars, and covered with a large tilt of sail-cloth, is commonly drawn by five or six yoke of oxen, which are driven by a man with a long whip, but led through the rivulets and by the farms, The horses are but weak in this part of the world, and find no where in Africa either pasture or water; consequently they cannot be used for long journies. Nor are horses employed for carrying any wares on their backs up to town from the farms that lie near it, but there are some few wealthy persons that now and then put two or three pair of them into a waggon for short journies. Still, however, they are made use of all over the country for riding on. When farmers, that live far down in the country, go up to the Cape, they generally take five or six spare oxen with them, for the purpose of changing cattle in a journey which lasts several weeks. The whip is an instrument which might seem to command respect not only from the oxen, for whose service it is principally destined, but from every one else.

Thus equipped, I set out with my company from the Cape on the *7th of September* for *Jan Besis Kraal,* a small grazing farm belonging to the company, and situated by the sea-side, where we arrived at eleven o'clock.

All over the sandy fields the *protea hypophylla* was seen creeping and procumbent, with its leaves standing up erect on each side of it. Near *Eland Fontain,* (or Elk's Fountain) a plant of this species was seen standing upright like a bush, much resembling the former, but with broader leaves.

We proceeded on our journey till twelve o'clock, when we came to another farm belonging to the company, called *Riet Valley:* afterwards to MOSTERT's Farm, and lastly, passing by *Brack Fontain* we came to *Groene Kloof* (the Green Valley), a considerable grazing farm belonging to the company, at the distance of eight hours journey from the Cape. In this pleasant place we remained a whole week, as well because we found a great deal to collect here, as because that, in consequence of the refraction of the sun-beams from the burning sand, I was unfortunately attacked with a very violent inflammation in my eyes, which I did not easily get rid of.

The country has indeed been much inhabited and cultivated by the European colonists, but as yet no mile-stones have been set up, nor have the farms and rivers every where received suitable names. The farms are frequently called after their owners, and the distances between places are measured by the time required to travel over them in a waggon drawn by oxen, which answers pretty exactly to a sea-league per hour. All this occasions travellers a great deal of trouble, and is the cause that I am obliged to call the places, which I passed in my travels, by the Dutch names, by which they are known on the spot.

The sandy and low plains, which we traversed, abounded at this time in bulbous plants, besides others which were now sprung up in consequence of the heavy rains that had fallen during the winter, and which with their infinitely varied flowers decorated these otherwise naked heaths.

The roots (*bulbi*) of the *iris edulis,* when boiled and served up at table, tasted much like potatoes.

The African flowers vary greatly as to colour, especially on the upper part, and are more constant on the under part.

Flamingoes (*phœnicopterus ruber*) were seen in abundance, wading every where in the ponds and puddles, in which were found also ducks and snipes (*scolopax capensis*). In the plains were heard among the bushes the korrhaan (*otis*). The *baantje* (a small bird), and deer of various kinds were seen running about, such as harte-beests (*capra dorcas*), steen-boks (*capra grimmia*), divers (*capra* —), as well as the stately ostrich, distinguished by its black feathers from its grey females.

A clay, impregnated with sulphur, was shown me, which is to be found near a fountain hard by *Paard Mountain.*

The seed-vessels of a species of *Euphorbia,* pulverized, were used for poisoning wolves.

Here I saw, for the first time, the *oleum Ricini,* or castor oil. The seeds were said to be boiled in water, and the oil is skimmed off as it rises, which is taken, in as large a dose as that of a tea-cup full, for a gentle purge. The leaves of the shrub dried, and applied round the head, were affirmed to be serviceable in the headach.

On the 14*th,* we passed *Oranie Fontein,* or *Orange Fountain,* and *Uyle Kraal,* or the *Owl's Kraal,* in our way to *Thé Fontein,* or *Tea Fountain,* a journey of six hours; and afterwards passing by *Elk's Fountain* (Eland's Fontein), got to *Saldabna Bay* the next day.

The farmers on this side of the Cape have neither vineyards nor much arable land, but instead of these plenty of cattle. Butter is made here every day, in a churn like a pump; and the butter-milk, excellent as it is, is thrown out to the calves and dogs. Indeed, they scarcely, allow their milk to cream beforehand. As to household furniture, they were in great want of it.

We left our saddle-horses at a farmer's house; after which we crossed the harbour in a vessel to the Company's Post, where we staid several days.

Here was plenty of game, consisting of antelopes, ducks, and other animals.

The expressed juice of the sow-thistle (*sonchus oteraceus*) was used for cleansing and healing ulcers.

The black juice of the cuttle-fish (*sepia*) mixed up with vinegar, was used for making ink. This animal has real eyes, consisting of a cornea, choroidea, and a crystalline lens, with all the humours usually found in the eye.

Among the servants I found Elisaeus Hyphoff, who was in the capacity of a cook here, and was the son of M. Hyphoff, director of the bank in Stockholm.

The *albuca major* grew in this neighbourhood tall, straight, and elegant. Its succulent stalk, which is rather mucilaginous, is chewed by the Hottentots and other travellers, by way of quenching their thirst.

There were a great many sand-banks in the harbour, which were seen at low water.

Grass grew on the islands in abundance; but there were neither sheep nor oxen in them.

While I was botanizing, I found a dead tiger near the shore. He had probably been eating some poisonous plants, and afterwards went in quest of water, before he fell down.

On the islands without and round about *Saldahna Bay,* seals (*phoca*) were caught in abundance, from the blubber of which a good and useful oil was prepared. The skins of the smaller sort of these animals are used only for shooting-bags and tobacco-pouches. The large seals, I was told, would weigh fourteen or fifteen hundredweight. With respect to these creatures, a disagreeable accident had happened here lately: a soldier was sent out to shoot them, and having wounded one of them, which lay as though it were dead, he went to open a vein in order to draw off its blood, as the oil is supposed to be the better for this operation, when on a sudden the seal caught hold of his hand, which the soldier pulling back in haste, his thumb was bitten off, and the tendon drawn out to a great length.

From *Saldahna Bay* we returned to *Thé Fontein,* and at a farm there had an opportunity of seeing with what dexterity the peasants perform the castration of their oxen, fifty of which, two years old, and one at three years, went through this operation in one evening. The cord of a whip was fastened round the horns, and a rope round one of the hind-legs. The animal being by this means thrown down on one side, its four legs were tied together. They then cut with a knife on the exterior side through all the integuments quite to the testicle; after this they laid hold of the testicle and scraped the funiculus, continually twisting it at the same time, till the testicle came away.

Great complaints were made of the seed-vessels of the *rumex spinosus* (*dubelties*), which grew very common here, as the sharp prickles of them cut the feet of the slaves and others, who walked bare-footed.

In wet years, the *pharnaceum mollugo* (*muggekruyd*) grows copiously here, and is said to make the cattle, that feed on it, very fat.

Difficult as it is to come within reach of it, we at last shot a *korhaan,* a bird which in its flight cries *kok-karri, kok-carri.*

The *secretary bird* (*falco secretarius*), made its appearance frequently, with its beautiful head and long legs; it runs very fast, and lives on the serpents it catches. I was told, that its young are not reared without difficulty, as they are very apt to break their legs. Yet I saw at *Constantia* an old bird that was tame. They lay two or three eggs, and are said to build their nests with twigs upon bushes. They are almost always found solitary, and in no great abundance.

66

ALEXANDER VON HUMBOLDT AND AIMÉ BONPLAND, *PERSONAL NARRATIVE OF TRAVELS TO THE EQUINOCTIAL REGIONS OF THE NEW CONTINENT DURING THE YEARS 1799–1804, VOL. 3*

(London: Longman, Hurst, Rees, Orme, and Brown, 1818) [1814])

Chapter VI

Mountains of New Andalusia. – Valley of Cumanacoa. – Summit of the Cocollar. – Missions of the Chayma Indians.

Our first visit to the peninsula of Araya was soon succeeded by a longer and more instructive excursion to the interior of the mountains of the missions of the Chayma Indians, where a variety of interesting objects claimed our attention. We entered on a country studded with forests; and visited a convent surrounded by palm trees and arborescent fern, situate in a narrow valley, where we felt the enjoyment of a cool and delicious climate, in the centre of the torrid zone. The surrounding mountains contain caverns haunted by thousands of nocturnal birds; and, what affects the imagination more than all the wonders of the physical world, we find beyond these mountains a people so lately nomade, and still nearly in a state of nature, savage without being barbarous, and stupid rather from ignorance than long rudeness. This interesting meditation was blended involuntarily with historical remembrances. It was in the promontory of Paria that Columbus first recognized the continent: there terminate these valleys, alternately devastated by the warlike anthropophagical Carib, and by the commercial and polished nations of Europe. At the beginning of the sixteenth century, the unhappy Indians of the coasts of Carupano, of Macarapan, and of Caraccas, were treated in the same

DOI: 10.4324/9780429355653-74

manner as the inhabitants of the coast of Guinea in our days. The soil of the islands was cultivated, the vegetables of the ancient continent were transplanted thither; but the regular system of colonization remained long unknown on the continent. If the Spaniards visited its shores, it was only to procure, either by violence or exchange, slaves, pearls, grains of gold, and dye-woods. The motives of this insatiable avarice seemed to be ennobled by the pretence of an enthusiastic zeal for religion; for every age has its peculiar tint, and a character appropriate to itself.

The trade in the copper-coloured Indians was accompanied by the same acts of inhumanity as that in the African negroes; and had also the same result, in rendering both the conquerors and the conquered more ferocious. Thence wars became more frequent among the natives; prisoners were dragged from the inland countries to the coast, in order to be sold to the whites, who loaded them with chains in their ships. Yet the Spaniards were at this epocha, and long after, one of the most polished nations of Europe. The resplendent light, which arts and literature then shed over Italy, has been reflected on every country, of which the language emanated from the same source as that of Dante and Petrarch. It might have been thought, that a general melioration of manners would be the natural consequence of this noble awakening of the mind, this sublime soaring of the imagination. But in distant climates, wherever the thirst of wealth has introduced the abuse of power, the nations of Europe, at every period of their history, have displayed the same character. The illustrious era of Leo X was signalized in the new world by acts of cruelty, that seemed to belong to the most barbarous ages. We are less surprised, however, at the horrible picture with which the conquest of America presents us, when we recollect what still takes place on the western coasts of Africa, notwithstanding the benefits of a more humane legislation.

The principles adopted by Charles V had long abolished the slave-trade on the continent. But the *Conquistadores,* by the continuation of their incursions, prolonged this system of ravaging, which has diminished the American population, perpetuated national animosities, and during a long period crushed the seeds of rising civilization. At length the missionaries, under the protection of the secular arm, spake words of peace. It was the privilege of religion, to console humanity for a part of the evils committed in it's name; to plead the cause of the natives before kings, to resist the violence of the commendataries, and to assemble wandering tribes into small communities, which are called *Missions;* and the existence of which favours the improvement of agriculture. Thus were insensibly founded, though by a uniform and premeditated progress, those vast monastic establishments, that singular system, which continually tends to insulate itself, and places countries four or five times more extensive than France under the control of religious orders.

Institutions, thus useful in stopping the effusion of blood, and in laying the first basis of society, have become in their result hostile to its progress. The effects of this insulated system have been such, that the Indians have remained in a state little different from that in which they existed, when their scattered dwellings were not yet collected round the habitation of a missionary. Their number has considerably augmented, but the sphere of their ideas is not enlarged. They have progressively lost that vigour of character, and that natural vivacity, which in every state of society are the noble fruits of independence. By subjecting to invariable rules even the slightest actions of their domestic life, they have been rendered stupid, by the effort to render them obedient. Their subsistence is in general more certain, and their habits more pacific; but subject to the constraint and the dull monotony of the government of the missions, they discover by their gloomy and reserved looks, that they have not sacrificed their liberty to their repose without regret. The monastic system confined to the cloister, while it deprives the state of useful citizens, may however sometimes contribute to calm the passions, to sooth incurable sorrows, and fit the mind for meditation; but transplanted into the forests of the new world, applied to the numerous relations of civil society, it has consequences so much the more fatal, as its duration is prolonged; it enchains from generation to generation the intellectual faculties, interrupts the intercourse of nations, and is hostile to whatever elevates the mind, or enlarges its conceptions. From these united causes, the natives who inhabit the Missions are kept in a state remote from all improvement; and which we should call stationary, if societies did not follow the course of the human mind, and must therefore be said to retrograde, whenever they cease to go forward.

On the 4th of September, at five in the morning, we began our journey to the Missions of the Chayma Indians, and the group of lofty mountains which traverse New Andalusia. We had been advised, on account of the extreme difficulties of the road, to reduce our baggage to a small bulk. Two beasts of burden were indeed sufficient to carry our provision, our instruments, and the paper necessary to dry our plants. One chest contained a sextant, a dipping-needle, an apparatus to determine the magnetic variation, thermometers, and Saussure's hygrometer. We always selected these instruments in excursions of short duration. The barometer requires more attention even than the time-keeper; and it may be well to add, that this instrument embarrasses travellers more than any other. We confided it during five years to a guide, who followed us on foot; and this precaution, which was expensive, did not always secure it from accidents. Having determined with precision the period of the atmospheric tides, that is, the hours at which the mercury rises and falls regularly every day under the tropics, we ascertained the possibility of taking the level of the country by means of the barometer, without employing correspondent observations at Cumana. The greatest changes in the pressure of the air in these climates, on the coasts, arise only to 1–1.3 of a line; and if at any given hour, or place, the height of the mercury be once marked, we may with some probability determine the variations, which this height experiences throughout the whole year, at every hour of the day or night. Hence it results, that, under the torrid zone, the want of correspondent observations can scarcely produce an error exceeding 12 or 15 toises, which is of small importance relative to geological levelling, or the influence of height on the climate, and the distribution of plants.

The morning was deliciously cool. The road, or rather path, that leads to Cumanacoa, follows the right bank of the Manzanares, passing by the hospital of the Capuchins, situate in a small wood of lignum vitæ and arborescent capparis. On leaving Cumana we enjoyed during the short duration of the twilight, from the top of the hill of San Francisco, an extensive view over the sea, the plain covered with *bera* and its golden flowers, and the Mountains of the Brigantine. We were struck at the great proximity in which the Cordillera presented itself, before the disk of the rising sun had reached the horizon. The tint of the summits is of a deeper blue, their outline is more strongly marked, and their masses are more detached, as long as the transparency of the air remains undisturbed by the vapours, which, accumulated during the night in the vallies, rise in proportion as the atmosphere acquires warmth.

At the hospital of the *Divina Pastora,* the path turns to the north-east, and stretches for two leagues over a soil without trees, and formerly levelled by the waters. We there found not only cacti, tufts of cistus-leaved tribulus, and the beautiful purple euphorbia, cultivated in the gardens of the Havanna under the singular name of *dictamno real,* but also the avicennia, the allionia, the sesuvium, the thalinum, and most of the portulaceous plants, that grow on the banks of the gulf of Cariaco. This geographical distribution of plants appears to designate the limits of the ancient coast; and to prove, as we have already observed, that the hills, of which we went along the southern side, formed heretofore a small island, separated from the continent by an arm of the sea.

After walking two hours, we arrived at the foot of the high chain of the interior mountains, which stretches from the east to the west; from the Brigantine to the Cerro de San Lorenzo. There, new rocks appear, and with them another aspect of vegetation. Every object assumes a more majestic and picturesque character; the soil, watered by springs, is furrowed in every direction; trees of gigantic loftiness, and covered with lianas, rise from the ravins; their bark, black and burnt by the double action of the light and the oxygen of the atmosphere, forms a contrast with the fresh verdure of the pothos and dracontium, the tough and shining leaves of which are sometimes several feet long. The parasite monocotyledones take between the tropics the place of the moss and lichens of our northern zone. As we advanced, the forms and grouping of the rocks reminded us of the scenes of Switzerland and the Tyrol. In these American Alps, the heliconia, costus, maranta, and other plants of the family of the balisiers *(canna indica),* which near the coasts vegetate only in damp and low places, flourish here at considerable height. Thus in the torrid zone, by a singular similitude, under the influence of an atmosphere continually loaded with vapours, as in the north of Europe on a soil moistened by melting snows, the vegetation of the mountains offers the same aspect, as characterises the vegetation of marshes.

67

CHARLES WATERTON, 'FIRST JOURNEY', *WANDERINGS IN SOUTH AMERICA*

(London: B. Fellowes, 1828) [1825])

Courteous reader, here thou hast the outlines of an amazing landscape given thee; thou wilt see that the principal parts of it are but faintly traced, some of them scarcely visible at all, and that the shades are wholly wanting. If thy soul partakes of the ardent flame which the persevering Mungo Park's did, these outlines will be enough for thee; they will give thee some idea of what a noble country this is; and if thou hast but courage to set about giving the world a finished picture of it, neither materials to work on, nor colours to paint it in its true shades, will be wanting to thee. It may appear a difficult task at a distance; but look close at it, and it is nothing at all; provided thou hast but a quiet mind, little more is necessary, and the genius which presides over these wilds will kindly help thee through the rest. She will allow thee to slay the fawn, and to cut down the mountain-cabbage for thy support, and to select from every part of her domain whatever may be necessary for the work thou art about; but having killed a pair of doves in order to enable thee to give mankind a true and proper description of them, thou must not destroy a third through wantonness, or to show what a good marksman thou art; that would only blot the picture thou art finishing, not colour it.

Though retired from the haunts of men, and even without a friend with thee, thou wouldst not find it solitary. The crowing of the hannaquoi will sound in thine ears like the daybreak town clock; and the wren and the thrush will join with thee in thy matin hymn to thy Creator, to thank him for thy night's rest.

At noon the Genius will lead thee to the troely, one leaf of which will defend thee from both sun and rain. And if, in the cool of the evening, thou hast been tempted to stray too far from thy place of abode, and art deprived of light to write down the information thou hast collected, the fire-fly, which thou wilt see in almost every bush around thee, will be thy candle. Hold it over thy pocket-book, in any position which thou knowest will not hurt it, and it will afford thee ample light. And when thou hast done with it, put it kindly back again on the next branch to thee. It will want no other reward for its services.

When in thy hammock, should the thought of thy little crosses and disappointments, in thy ups and downs through life, break in upon thee, and throw

DOI: 10.4324/9780429355653-75

thee into a pensive mood, the owl will bear thee company. She will tell thee that hard has been her fate too; and at intervals, 'Whip-poor-Will', and 'Willy come go', will take up the tale of sorrow. Ovid has told thee how the owl once boasted the human form, and lost it for a very small offence; and were the poet alive now, he would inform thee, that 'Whip-poor-Will', and 'Willy come go', are the shades of those poor African and Indian slaves, who died worn out and brokenhearted. They wail and cry, 'Whip-poor-will', 'Willy come go', all night long; and often, when the moon shines, you see them sitting on the green turf, near the houses of those whose ancestors tore them from the bosom of their helpless families, which all probably perished through grief and want, after their support was gone.

About an hour above the rock of Saba, stands the habitation of an Indian, called Simon, on the top of a hill. The side next the river is almost perpendicular, and you may easily throw a stone over to the opposite bank. Here there was an opportunity of seeing man in his rudest state. The Indians who frequented this habitation, though living in the midst of woods, bore evident marks of attention to their persons. Their hair was neatly collected, and tied up in a knot; their bodies fancifully painted red, and the paint was scented with hayawa. This gave them a gay and animated appearance. Some of them had on necklaces, composed of the teeth of wild boars slain in the chase; many wore rings, and others had an ornament on the left arm, midway betwixt the shoulder and the elbow. At the close of day, they regularly bathed in the river below; and the next morning seemed busy in renewing the faded colours of their faces.

One day there came into the hut a form which literally might be called the wild man of the woods. On entering, he laid down a ball of wax, which he had collected in the forest. His hammock was all ragged and torn; and his bow, though of good wood, was without any ornament or polish; 'erubuit domino, cultior esse suo'. His face was meagre, his looks forbidding, and his whole appearance neglected. His long black hair hung from his head in matted confusion; nor had his body, to all appearance, ever been painted. They gave him some cassava bread and boiled fish, which he ate voraciously, and soon after left the hut. As he went out, you could observe no traces in his countenance or demeanour, which indicated that he was in the least mindful of having been benefited by the society he was just leaving.

The Indians said that he had neither wife, nor child, nor friend. They had often tried to persuade him to come and live amongst them; but all was of no avail. He went roving on, plundering the wild bees of their honey, and picking up the fallen nuts and fruits of the forest. When he fell in with game, he procured fire from two sticks, and cooked it on the spot. When a hut happened to be in his way, he stepped in, and asked for something to eat, and then months elapsed ere they saw him again. They did not know what had caused him to be thus unsettled; he had been so for years; nor did they believe that even old age itself would change the habits of this poor, harmless, solitary wanderer.

From Simon's, the traveller may reach the large fall, with ease, in four days.

The first falls that he meets are merely rapids, scarce a stone appearing above the water in the rainy season; and those in the bed of the river, barely high enough to arrest the water's course, and by causing a bubbling, show that they are there.

With this small change of appearance in the stream, the stranger observes nothing new till he comes within eight or ten miles of the great fall. Each side of the river presents an uninterrupted range of wood, just as it did below. All the productions found betwixt the plantations and the rock Saba, are to be met with here.

From Simon's to the great fall, there are five habitations of the Indians. Two of them close to the river's side; the other three a little way in the forest. These habitations consist of from four to eight huts, situated on about an acre of ground, which they have cleared from the surrounding woods. A few pappaw, cotton, and mountain cabbage-trees, are scattered round them.

At one of these habitations, a small quantity of the wourali poison was procured. It was in a little gourd. The Indian who had it, said that he had killed a number of wild hogs with it, and two tapirs. Appearances seemed to confirm what he said; for on one side it had been nearly taken out to the bottom, at different times, which probably would not have been the case had the first or second trial failed.

68

CHARLES WATERTON, 'NOTES ON THE HABITS OF THE CHEGOE OF GUIANA, BETTER KNOWN BY THE NAME OF JIGGER, AND INSTANCES OF ITS EFFECTS ON MAN AND DOGS', *ESSAYS IN NATURAL HISTORY*

(London: Longman, Orme, Brown, Green, & Longman, 1838)

'Priore relictâ
Sede, novis domibus habitant, vivuntque receptæ'.

Leaving their former haunts, beneath the skin
They form new settlements, and thrive within.

THIS apparently insignificant insect far outdoes the bug in the exercise of its noxious qualities. The bug attacks you in an open manner, makes a hearty meal, and then retires to enjoy it: but the chegoe commences its operations upon you so gently, that they are scarcely felt; and it terminates them in a way that calls for your most serious attention. In a word, it approaches you with such insinuating address, that you absolutely feel a kind of gratification at the very time that it is adopting measures which will infallibly end in your certain torment.

Soon after the chegoe has entered your skin, you experience a pleasant itching kind of sensation, by which you begin to suspect that all is not right; and, on taking a nearer view of the part, you perceive that the skin is somewhat discoloured.

DOI: 10.4324/9780429355653-76

I know it is supposed by some people, that the accounts concerning the chegoe have been much exaggerated. I am not of this way of thinking, for I myself have smarted under its attacks; and I have minutely inspected the foot of a negro, which was a mass of ulcers, formed entirely by the neglected ravages of the chegoe.

Guiana is the native country of this insect. In that hot and humid region, which is replete with every thing that can please our imagination, or administer to our wants, we must not be surprised to find here and there some little drawback; some few obstructions in our way; some thorny plants to impede our journey as we wander on.

The chegoe resembles a flea: and, had you just come out of a dovecot, on seeing it upon your skin, you might easily mistake it for a small pigeon flea; although, upon a closer inspection, you would surmise that it is not capable of taking those amazingly elastic bounds, so notorious in the flea of Europe.

Not content with merely paying you a visit, and then taking itself off again, as is the custom of most insects, this insidious miner contrives to work its way quite under your skin, and there remains to rear a numerous progeny. I once had the curiosity to watch the movements of a chegoe on the back of my hand, a part not usually selected by it to form a settlement. It worked its way pretty rapidly for so small an insect. In half an hour it had bored quite through the skin, and was completely out of sight. Not wishful to encourage its intended colony, 'Avast, there! my good little fellow', said I; 'we must part company without loss of time. I cannot afford to keep you, and a numerous family, for nothing: you would soon eat me out of house and home'. On saying this, I applied the point of my penknife to the place where the chegoe had entered, and turned it loose upon the world again.

In the plantations of Guiana, there is generally an old negress, known by the name of Granny, a kind of 'Junonis anus', who loiters about the negro yard, and is supposed to take charge of the little negroes who are too young to work. Towards the close of day, you will sometimes hear the most dismal cries of woe coming from that quarter. Old Granny is then at work, grubbing the chegoe nests out of the feet of the sable urchins, and filling the holes with lime juice and Cayenne pepper. This searching compound has two duties to perform: first, it causes death to any remaining chegoe in the hole; and, secondly, it acts as a kind of birch-rod to the unruly brats, by which they are warned, to their cost, not to conceal their chegoes in future: for, afraid of encountering old Granny's tomahawk, many of them prefer to let the chegoes riot in their flesh, rather than come under her dissecting hand.

A knowing eye may always perceive when the feet of negroes are the abode of the chegoe. They dare not place their feet firmly on the ground, on account of the pain which such a position would give them; but they hobble along with their toes turned up: and by this you know that they are not suffering from tubboes (a remnant of the yaws), but from the actual depredations of the chegoes, which have penetrated under the nails of the toes, and there formed sores, which, if not attended to, would, ere long, become foul and corroding ulcers. As I seldom had a shoe or stocking on my foot from the time that I finally left the sea coast in 1812, the chegoe was a source of perpetual disquietude to me. I found it necessary to examine my feet every evening, in order to counteract the career of this extraordinary insect. Occasionally, at one overhauling, I have broken up no less than four of its establishments under the toe nails.

In 1825, a day or two before I left Guiana, wishful to try how this puny creature and myself would agree during a sea voyage, I purposely went to a place where it abounded, not doubting but that some needy individual of its tribe would attempt to better its condition. Ere long, a pleasant and agreeable kind of itching under the bend of the great toe informed me that a chegoe had bored for a settlement. In about three days after we had sailed, a change of colour took place in the skin, just at the spot where the chegoe had entered, appearing somewhat like a blue pea. By the time we were in the latitude of Antigua, my guest had become insupportable; and I saw there was an immediate necessity for his discharge. Wherefore, I turned him and his numerous family adrift, and poured spirits of turpentine into the cavity which they had occupied, in order to prevent the remotest chance of a regeneration.

The Indian and negro wenches perform the operation of extracting chegoes with surprising skill. They take a pin, and, by a very slow process, they lay the part bare, and contrive to work quite round the bag which contains the chegoe and its offspring. As soon as this has been effected, they turn the bag out, whole and uninjured; by which means none are left in the hole to form a new colony.

For my own part, I never troubled these gentle operators; although I have looked on many a time, and admired their exquisite skill, whilst they were fingering the toes of my acquaintance. Once, however, I had it not in my power to be my own surgeon, and, on that occasion, a faithful old negro performed the friendly office.

I was descending the Demerara, with an inveterate tertian ague; and I was so much exhausted by sitting upright in the canoe, that I no sooner got ashore at an Indian's hut, than I lay down on the ground at full length. Sickness had pressed so heavily on me, that I was callous to the well-known feeling which the chegoe causes. I was quite unconscious that there were nine thriving nests of chegoes in my back, until one was accidentally observed by the old negro; and this led to the discovery of the rest. I handed him my penknife, and told him to start the intruders. Sick as I was, I wished an artist were present at the operation. The Indian's hut, with its scanty furniture, and bows and arrows hanging round; the deep verdure of the adjoining forest; the river flowing rapidly by; myself wasted to a shadow; and the negro grinning with exultation, as he showed me the chegoes' nests which he had grubbed out; would have formed a scene of no ordinary variety.

Dogs are often sorely tormented by the chegoe; and they get rid of them by an extremely painful operation. They gradually gnaw into their own toes, whining piteously as they do it, until they get at the chegoe's nest. Were it not for this singular mode of freeing themselves from the latent enemy, dogs would absolutely be cripples in Guiana.

But it is time to stop. I have penned down enough to give the reader a tolerably correct idea of one of the smallest, and, at the same time, one of the most annoying, insects, which attack both man and beast in the interminable region of Guiana.

69

CHARLES DARWIN, *JOURNAL OF RESEARCHES INTO THE GEOLOGY AND NATURAL HISTORY OF THE VARIOUS COUNTRIES VISITED BY H.M.S. BEAGLE*

(London: Henry Colburn, 1839)

Falkland Islands

MARCH 16TH, 1834. – The Beagle anchored in Berkeley Sound, in East Falkland Island. This archipelago is situated in nearly the same latitude as the mouth of the Strait of Magellan. It covers a space of about 120 by 60 geographical miles, and is a little more than half the size of Ireland. After the possession of these miserable islands had been contested by France, Spain, and England, they were left uninhabited. The government of Buenos Ayres then sold them to a private individual, but likewise used them, as old Spain had done before, for a penal settlement. England claimed her right and seized them. The Englishman who was left in charge of the flag was consequently murdered. A British officer was next sent, unsupported by any power: and when we arrived, we found him in charge of a population, of which rather more than half were runaway rebels and murderers.

The theatre is worthy of the scenes acted on it. An undulating land, with a desolate and wretched aspect, is every where covered by a peaty soil and wiry grass, of one monotonous brown colour. Here and there a peak or ridge of gray quartz rock, breaks through the smooth surface. Every one has heard of the climate of these regions; it may be compared to that which is experienced at the height of between one and two thousand feet, on the mountains of North Wales; having however less sunshine and less frost, but more wind and rain.

MARCH 16TH. – I will now describe a short excursion which I made round a part of this island. In the morning I started with six horses and two Gauchos: the latter were capital men for the purpose, and well accustomed to living on their own resources. The weather was very boisterous and cold, with heavy

DOI: 10.4324/9780429355653-77

hailstorms. We got on, however, pretty well; but excepting in the geology, nothing could be less interesting than our day's ride. The country is uniformly the same undulating moorland; the surface being covered by light brown withered grass and a few very small shrubs, all springing out of an elastic peaty soil. In the valleys here and there might be seen a small flock of wild geese, and every where the ground was so soft, that the snipe was able to feed. Besides these two kinds of birds, there were few others. There is one main range of hills, nearly two thousand feet in height, and composed of quartz rock, the rugged and barren crests of which gave us some trouble to cross. On the south side we came to the best country for wild cattle; we met however no great number, for they had lately been much harassed.

In the evening we came across a small herd. One of my companions, St. Jago by name, soon separated a fat cow; he threw the bolas, and it struck her legs, but failed in becoming entangled. Then dropping his hat to mark the spot where the balls were left, while at full gallop he uncoiled his lazo, and after a most severe chase, again came up to the cow, and caught her round the horns. The other Gaucho had gone on ahead with the horses, so that St. Jago had some difficulty in killing the furious beast. He managed to get her on a level piece of ground, by taking advantage of her as often as she rushed at him; and when she would not move, my horse, from having been trained, would canter up, and with his chest give her a violent push. But when on level ground it does not appear an easy job for one man to kill a beast mad with terror. Nor would it be so, if the horse, when left to itself without its rider, did not soon learn, for its own safety, to keep the lazo tight; so that, if the animal moves forward, the horse moves just as quickly so much away; otherwise, it stands motionless leaning on one side. This horse, however, was a young one, and would not stand still, but gave in to the cow as she struggled. It was admirable to see with what dexterity St. Jago dodged behind the beast, till at last he contrived to give the fatal touch to the main tendon of the hind leg; after which, driving his knife into the head of the spinal marrow, the cow dropped as if struck by lightning. He cut off pieces of flesh with the skin to it, but without any bones, sufficient for our expedition. We then rode on to our sleeping-place, and had for supper 'carne con cuero', or meat roasted with the skin on it. This is as superior to common beef, as venison is to mutton. A large circular piece taken from the back, is roasted on the embers with the hide downwards and in the form of a saucer, so that none of the gravy is lost. If any worthy alderman had supped with us that evening, 'carne con cuero', without doubt, would soon have been celebrated in London.

During the night it rained, and the next day (17th) was very stormy, with much hail and snow. We rode across the island to the neck of land which joins the Rincon del Toro (the great peninsula at the S.W. extremity) to the rest of the island. From the greater number of cows which have been killed, there is a large proportion of bulls. These wander about by twos and threes, or by themselves, and are very savage. I never saw such magnificent beasts; they truly resembled the ancient sculptures, in which the size of the neck and head is but seldom equalled among tame animals. The young bulls ran away for a short distance, but the old ones did not stir a step, except to rush at man and horse; and many of the latter have thus been killed. One old bull crossed a boggy stream, and took up his stand on the opposite side to us. We in vain tried to drive him away, and failing, were obliged to make a large circuit. The Gauchos in revenge determined to render him for the future innocuous. It was very interesting to see how art completely mastered force. One lazo was thrown over his horns, as he rushed at the horse, and another round his hind legs: in a minute the monster was stretched harmless on the ground. After the lazo has once been tightly drawn round the horns of a furious animal, it does not at first appear an easy thing to disengage it again; nor, I apprehend, would it be so, if the man was by himself, and he did not wish to kill the beast. By the aid, however, of a second person throwing his lazo, so as to catch both hind legs, it is quickly managed: for the animal, as long as its hind legs are kept outstretched, is quite powerless, and the first man can with his hands loosen his lazo, and then quietly mount his horse; but the moment the second man, by backing ever so little, relaxes the strain, the lazo slips off the legs of the struggling beast, which thus rises free, shakes himself, and vainly rushes after his antagonist.

During our whole ride we only saw one troop of wild horses. These animals, as well as the cattle, were introduced by the French in 1764, since which time they have greatly increased. It is a curious fact, that the horses have never left the eastern end of the island, although there is no natural boundary to prevent them from roaming, and that part of the island is not more tempting than the rest. The Gauchos, though asserting this to be the case, are unable to account for the circumstance. The horses appear to thrive well, yet they are small sized, and have lost so much strength, that they are unfit to be used in taking wild cattle with the lazo. In consequence, it is necessary to go to the great expense of importing fresh horses from the Plata. At some future period the southern hemisphere probably will have its breed of Falkland ponies, as the northern has that of Shetland.

The rabbit is another animal which has been introduced, and has succeeded very well; so that they abound over large parts of the island. Yet, like the horses, they are confined within certain limits; for they have not crossed the central chain of hills; nor would they have extended even so far as the base, if, as the Gauchos informed me, small colonies had not been carried there. I should not have supposed that these animals, natives of northern Africa, could have existed in a climate so extremely humid as this, and which enjoys so little sunshine that even wheat ripens only occasionally. It is asserted that in Sweden, which any one would have thought a more favourable climate, the rabbit cannot live out of doors. The first few pair moreover had here to contend against pre-existing enemies, in the fox, and some large hawks. The French naturalists have considered the black variety a distinct species, and called it *Lepus Magellanicus*.[1] They imagined that Magellan, when talking of an animal under the name of 'conejos', in the Strait of Magellan, referred to this species; but he was alluding to a small cavy, which to this day is thus called. The Gauchos laughed at the idea of the black kind being different from the gray, and they said that at all events it had not extended its range any further than the other; that the two were never found separate; and that they readily bred together, and produced piebald offspring. Of the latter I now possess a specimen, and it is marked about the head, differently from the French specific description. This circumstance shows how cautious naturalists should be in making species; for even Cuvier, on looking at the skull of one of these rabbits, thought it was probably distinct.

The only quadruped native to the island, is a large wolf-like fox,[2] which is common to both East and West Falkland. I have no doubt it is a peculiar species, and confined to this archipelago; because many sealers, Gauchos, and Indians, who have visited these islands, all maintain that no such animal is found in any part of South America. Molina, from a similarity in habits, thought this was the same with his 'culpeu';[3] but I have seen both, and they are quite distinct. These wolves are well known, from Byron's account of their tameness and curiosity; which the sailors, who ran into the water to avoid them, mistook for fierceness. To this day their manners remain the same. They have been observed to enter a tent, and actually pull some meat from beneath the head of a sleeping seaman. The Gauchos, also, have frequently killed them in the evening, by holding out a piece of meat in one hand, and in the other a knife ready to stick them. As far as I am aware, there is no other instance in any part of the world, of so small a mass of broken land, distant from a continent, possessing so large a quadruped peculiar to itself. Their numbers have rapidly decreased; they are already banished from that half of the island which lies to the eastward of the neck of land between St. Salvador Bay and Berkeley Sound. Within a very few years after these islands shall have become regularly settled, in all probability this fox will be classed with the dodo, as an animal which has perished from the face of the earth. Mr. Lowe, an intelligent person who has long been acquainted with these islands, assured me, that all the foxes from the western island were smaller and of a redder colour than those from the eastern. In the four specimens which were brought to England in the Beagle[4] there was some variation, but the difference with respect to the islands could not be perceived. At the same time the fact is far from improbable.

At night (17th) we slept on the neck of land which forms the south-west peninsula. The valley was pretty well sheltered from the cold wind; but there was very little brushwood for fuel. The Gauchos, however, soon found what, to my great surprise, made nearly as hot a fire as coals; this was the skeleton of a bullock lately killed, from which the flesh had been picked by the Caracaras. They told me that in winter they had often killed a beast, cleaned the flesh from the bones with their knives, and then with these same bones roasted the meat for their suppers.

Notes

1 Lesson's Zoology of the Voyage of the Coquille, vol. i., p. 168. All the early voyagers, and especially Bougainville, distinctly state that the wolf-like fox was the only native animal on the island. The distinction of this rabbit as a species, is taken from peculiarities in the fur, from the shape of the head, and from the shortness of the ears. I may here

observe that the difference between the Irish and English hare, rests upon nearly similar characters, only more strongly marked.

2 I have reason to believe there is likewise a field-mouse. The common European rat and mouse have roamed from the habitations, and have settled themselves at various points. The common hog has also run wild.

3 The 'culpeu' is the *Vulpes Magellanicus* brought home by Captain King from the Strait of Magellan. It is common in Chile.

4 Captain FitzRoy has presented two of these foxes to the British Museum, where Mr. Gray had the kindness to compare them in my presence.

70

GEORGE GARDNER, 'JOURNEY TO AND RESIDENCE IN THE ORGAN MOUNTAINS', *TRAVELS IN THE INTERIOR OF BRAZIL*

(London: Reeve, Benham, and Reeve, 1849)

Preface

THE present volume is not given to the public, because the Author supposes it presents a better account of certain parts of the immense Empire of Brazil, than is to be found in the works of other travellers, but because it contains a description of a large portion of that interesting country, of which no account has yet been presented to the world. It has been his object to give as faithful a picture as possible of the physical aspect and natural productions of the country, together with cursory remarks on the character, habits, and condition of the different races, whether indigenous or otherwise, of which the population of those parts he visited is now composed. It is seldom that he has trusted to information received from others on those points; and he hopes that this fact will be considered a sufficient reason for his not entering into desultory details more frequently than he has done.

Ample opportunities were offered for studying the objects he had in view, of which he never ceased to avail himself. Besides visiting many places along the coast his journeys in the interior were numerous; and, although he never ventured, like Waterton – whose veracity is not to be doubted – to ride on the bare back of an alligator, or engage in single combat with a boa constrictor, yet he had his full share of adventure, particularly during his last journey, which extended, north to south, from near the equator to the twenty-third degree of south latitude; and east to west, from the coast to the tributaries of the Amazon. The privations which the traveller experiences in these uninhabited, and often desert countries, can scarcely be appreciated by those who have never ventured into them, where he is exposed at times to a burning sun, at others to torrents of rain, such as are only to be witnessed within the tropics, separated for years from all civilized society, sleeping for months together in the open air, in all seasons, surrounded by beasts of prey and hordes of more savage Indians, often obliged to carry a supply of water on horseback over the desert tracks, and not unfrequently passing two or three days

DOI: 10.4324/9780429355653-78

without tasting solid food, not even a monkey coming in the way to satisfy the cravings of hunger. Notwithstanding these, however, and one serious attack of illness, his enthusiasm carried him through all difficulties, and they have in some measure been repaid by the pleasure which such wanderings always afford to the lover of nature, and by the number of new species which he has been enabled to add to the already long list of organized beings.

The Author has only further to add, that the notes from which the Narrative has been drawn up, were, for the most part, written during those hours, which, under other circumstances, should have been devoted to sleep; and that the Narrative itself was principally compiled from them, during a voyage from England to the Island of Ceylon.

Kandy, Ceylon, January 1st, 1846.

Chapter II

Journey to and Residence in the Organ Mountains

THE collections which had accumulated during the period of my residence in the vicinity of Rio de Janeiro, having been put into a proper state and sent to England, I made arrangements for visiting the Organ Mountains. The peaks which receive this appellation form part of a mountain range, situated about sixty miles to the north of Rio, which, branching out in various directions, stretches from about Bahia, in lat. 12° S., to S. Catharina, in lat. 29° S. The name (Serra dos Orgãos) bestowed on them by the Portuguese, originated in a fancied resemblance which the peaks, rising gradually one above the other, bear to the pipes of an organ. About ten years before my visit a Sanatorium, or health station, had been established on this range, at about 3,000 feet above the level of the sea, in a beautiful valley behind the higher peaks. A large tract of country there belongs to Mr. March, an English gentleman, on which he has a farm for the breeding of horses and mules, and a large garden, from which the Rio market is regularly supplied with European vegetables. On this property a number of cottages have been erected, which are resorted to by the families of the English residents at Rio during the hot months. He also receives boarders into his own house, and it rarely happens that the place is without visitors. About one-third of the journey has to be performed by water, the other is accomplished on mules, which are sent down from Mr. March's farm (Fazenda).

As Mr. March happened to be in Rio at the time I purposed visiting the mountains, we started together, on the 24th of Dec., along with two or three English merchants, who were going up to spend the Christmas holidays with their families. It was mid-day before we could leave the city, and, under the influence of a strong sea-breeze, we reached Piedade, the landing-place, at half-past three o'clock, the distance being about twenty miles. The boat in which we embarked belongs to a class which is very common in the harbour, and much employed in conveying goods to the head of the bay, and produce from the interior, from thence to Rio. They are also much made use of by pleasure parties frequenting the islands and opposite shores of the bay. They are called Faluas, and are manned by six rowers, and a steersman who is called the Patrão. The latter is very frequently the owner, and most of them are natives of Portugal. They have two masts, each of which carries a large sail; the stern part is covered over and enclosed with curtains. The negroes who man these boats are generally strong muscular men. Seated on one thwart, they place their feet against another, and rise up at each stroke of the oar, keeping time to a melancholy chant all the while they are pulling. These boats can be hired for an entire day at about eighteen shillings.

The day was a most delightful one, the sun shining out brightly from a clear sky, and the air cooled by the fresh sea-breeze. We passed close to the Ilha do Governador, which is the largest island in the bay. It is about eight miles in length, but narrow in proportion, and thinly inhabited. Shortly before my arrival in the country, an Englishman commenced a soap and candle manufactory on it, both of which articles bring the same price in Rio as those imported from Europe. The muddy shores of this island, as well as those of the whole bay, abound with crabs of all sizes, and every variety of colour, from nearly black to a bright scarlet. On one occasion when I visited the island, I observed within a very short space about eight species. They are gregarious, and each kind inhabits a distinct colony; they burrow in the mud, under the shade and among the roots of the mangrove and other shore-loving trees. It was here that I first saw the apparent anomaly of trees bearing crops of oysters. These animals, when young, attach themselves to the lower part of the trunks, and long pendulous roots, of the mangrove and other trees, which grow in the sea even to low-water mark. The oysters are small and not well-flavoured. Others are found in the bay of enormous size, some of their shells, which I collected as specimens, measuring upwards of a foot in length. Near the head of the bay there are many little islands, some of which are inhabited, and present the agreeable prospect of cultivation, while others are little more than masses of rock, among the clefts of which grow a few stunted shrubs, and grotesque prickly pears.

At Piedade, mules from Mr. March's Fazenda were waiting for us and our luggage, and, after a short stay for the arrangement of the latter, we began the land part of our journey. At Piedade, which only consists of a few scattered houses, a large hotel was being erected by Col. Leite, a Brazilian gentleman, who, at his own expense, was then making a new road across the Organ Mountains, to join the one which leads to the mining districts from Porto de Estrella, another landing place at the head of the bay. The latter place has hitherto been the common harbour between Rio and the interior. The Colonel, however, expects that his new road will ultimately be preferred, as it is much shorter. Four years after when I again visited this part of the country, I found that this road was still in an unfinished state. To save the expense of an engineer he had traced the road himself, and the consequence was, that it afterwards required many alterations. The road from Piedade to Magé, a small town about four miles distant, leads through a flat, sandy, and, in several places, marshy plain, abounding with low trees and beautiful flowering shrubs. The hedges were covered with numerous climbers, one of them a small sweet-flowered kind of Jasmine, the only one which has hitherto been found in a wild state on the continent of America. In moist places, *Dichorizandra thyrsiflora,* with its spikes of azure blossoms, was not uncommon, while the sandy fields were covered with a large kind of *Cactus,* among which many plants of the aloe-like *Fourcroya gigantea* were to be seen throwing up their flowering stems to a height of thirty and forty feet.

The town of Magé is rather prettily situated on the banks of the Magé-assú, one of the many small rivers which take their rise in the Organ Mountains, and fall into the head of the bay. It contains a neat church, and a number of well-furnished shops. The river is navigable, for craft of a small size, about eight miles from its mouth. A considerable quantity of Farinha de Mandiocca (Cassava) is exported from this place to Rio. Its low situation, and the surrounding swamps, render it unhealthy at particular seasons; intermittent fevers are here common, and they frequently terminate in others of a more malignant nature. From Magé to Frechal, the place where we slept for the night, the distance is about fourteen miles. The road still continued flat, but wound round many low hills, the sides of which are covered with plantations of Mandiocca. We met several troops of mules coming down from the interior, loaded with produce. Unaccustomed to such a mode of transport, the European looks with astonishment at the great number of animals which are here required to carry what, in his own country, would scarcely form a load for one. Loaded mules start daily from Rio, Piedade, and Porto d'Estrella, to make journeys into the interior of from five hundred to two thousand miles and upwards. They seldom travel above twelve or sixteen miles a day, and the load allowed to each varies from six to eight arrobas of thirty-two pounds each. The loads are protected from the weather by dried ox-hides, which are strapped lightly over them. Frechal is a small village, consisting of a few scattered houses, and situated about two miles from the foot of the mountains. The place at which we put up for the night is a large kind of public house (Venda), where there is an open room for the accommodation of travellers; around this room a number of beds are arranged, which gives it very much the appearance of a hospital ward. Here, unlike most other places of the same kind between Rio and the mining districts, a very comfortable meal may always be obtained.

Next morning by break of day we again continued our journey. At about two miles from Frechal the ascent of the mountains begins. From thence to Mr. March's Fazenda, which stands at an elevation of upwards of 3,000 feet above the level of the sea, is twelve miles. During the whole way the road is very bad, and in many places so steep, that it is with considerable difficulty the mules make their way up it. Indeed, to one unused to travel on such paths, which have more the appearance of the bed of a mountain torrent than a road for beasts of burden, many parts of it appear impassable; but he is soon undeceived by the slow yet sure manner in which the mules pass along the worst portion of it, especially if left entirely to themselves. The whole length of the road is through one dense forest, the magnificence of which cannot be imagined by those who have never seen it, or penetrated into its recesses. Those remnants of the virgin forest which still stand in the vicinity of the capital, although they appear grand to the eye of a newly-arrived European, become insignificant when compared with the mass of giant vegetation which clothes the sides of the Organ Mountains. So far as I have been able to determine, the large forest trees consist of various species of *Palms, Laurus, Ficus, Cassia, Bignonia, Solanum, Myrtaceæ,* and *Melastomaceæ*. In temperate climates natural forests are mostly composed of trees which grow gregariously. In those of tropical countries it is seldom that two trees of a kind are to be seen growing together, the variety of different species is so great. Many of the trees are of immense size, and have their trunks and branches covered with myriads of those plants which are usually called parasites, but are not so in reality, consisting of *Orchideæ, Bromeliaceæ, Ferns, Peperomiæ,* &c., which derive their nourishment from the moisture of their bark, and the earthy matter which has been formed from the decay of mosses, &c. Many of the trees have their trunks encircled by twiners, the stems of which are often thicker than those they surround. This is particularly the case with a kind of wild fig, called by the Brazilians, Cipo Matador. It runs up the tree to which it has attached itself, and at the distance of about every ten feet throws out from each side a thick clasper, which curves round, and closely entwines the other stem. As both the trees increase in size, the pressure ultimately becomes so great, that the supporting one dies from the embrace of the parasite.

There is another kind of wild fig-tree, with an enormous height and thickness of stem, to which the English residents give the name of Buttress-tree, from several large thin plates which stand out from the bottom of the trunk. They begin to jut out from the stem at the height of ten or twelve feet from the bottom, and gradually increase in breadth till they reach the ground, where they are connected with the large roots of the tree. At the surface of the ground these plates are often five feet broad, and throughout not more than a few inches thick. The various species of *Laurus* form fine trees; they flower in the months of April and May, at which season the atmosphere is loaded with the rich perfume of their small white blossoms. When their fruit is ripe, it forms the principal food of the Jacutinga (*Penelope Jacutinga,* Spix), a fine large game bird. The large *Cassiæ* have a striking appearance when in flower; and, as an almost equal number of large trees of *Lasiandra Fontanesiana,* and others of the *Melastoma* tribe, are in bloom at the same time, the forests are then almost one mass of yellow and purple from the abundance of these flowers. Rising amid these, the pink-coloured flowers of the *Chorisia speciosa* – a kind of silk cotton-tree – can be easily distinguished. It is also a large tree, with a stem, covered with strong prickles, from five to eight feet in circumference unbranched to the height of thirty or forty feet. The branches then form a nearly hemispherical top, which, when covered with its thousands of beautiful large rose-coloured blossoms, has a striking effect when contrasted with the masses of green, yellow, and purple of the surrounding trees.

Many of these large trunks afford support to various species of climbing and twining shrubs, belonging to the natural orders *Bignoniaceæ, Compositæ, Apocyneæ,* and *Leguminosæ,* the stems of which frequently assume a very remarkable appearance. Several of them are often twisted together and dangle from the branches of the trees, like large ropes, while others are flat and compressed, like belts: of the latter description I have met with some six inches broad, and not more than an inch thick. Two of the finest climbers are the beautiful large trumpet-flowered *Solandra grandiflora,* which, diffusing itself among the largest trees of the forest, gives them a magnificence not their own; and a showy species of Fuchsia (*F. integrifolia,* Cambess.), which is very common, attaching itself to all kinds of trees, often reaching to the height of from sixty to one hundred feet, and then falling down in the most beautiful festoons.

71

SIR WILLIAM JACKSON HOOKER (ED.), *NIGER FLORA; OR, AN ENUMERATION OF THE PLANTS OF WESTERN TROPICAL AFRICA, COLLECTED BY THE LATE DR THEODORE VOGEL, BOTANIST TO THE VOYAGE OF THE EXPEDITION SENT BY HER BRITANNIC MAJESTY TO THE RIVER NIGER IN 1841*

(London: Hippolyte Bailliere, 1849)

Preface

THE majority of the Plants described in the following pages were entrusted to the Editor, for the purpose of publication, by the African Civilization Society, which, as well known, was formed in London in 1839, through the instrumentality of the late Sir Thomas Powell Buxton. That enlightened and philanthropic statesman, deeply impressed by the aggravated horrors of the Slave Trade, was extremely anxious to try, what appeared to him to be the only remedy, to put down that iniquitous traffic by the encouragement of lawful trade and the advancement of Africa itself to a condition in which she would no longer find it her interest to furnish the slavers with supplies for their market.

Many persons of influence and sound judgment, uniting with Sir Powell in his views, and Government having taken up the subject cordially, the 'NIGER EXPEDITION' was dispatched, under the command of Capt. H.D. Trotter, in 1841.

Dr Theodore Vogel, a German gentleman of high scientific attainments, was selected as chief botanist to the Expedition, and with him was associated Mr Ansell, strongly recommended by the Horticultural Society of London.

DOI: 10.4324/9780429355653-79

Very great and unusual pains were taken to render the service less dangerous to the health of those engaged in it, than had been the case with former attempts to explore intratropical Africa. Indeed, every precaution that could be thought of – every guard against the climate – were, as was believed, employed;– yet it cannot be denied, there was a failure, and it would truly appear from this, and from former voyages of a similar character, that the European constitution is incapable of withstanding the effect of that deadly atmosphere.

But while we deplore the loss of so many brave officers and men, engaged, voluntarily, in this most sacred cause, it would be unjust to shut our eyes to much good that has hereby been accomplished. It has proved to the natives the real intentions of the English, and convinced them of our sincerity in establishing mutual and beneficial, and a wholesome commerce, and that we have no sinister ends of our own to answer. Of this, too, they were the more convinced, when they saw their friends, who had been rescued from captivity, returning with the Expedition. It further showed, that the only hope of enlightening the sons of Africa is by *native* agency: and it is with no small pride that the Editor of this Work, in the capacity of Director of the Royal Gardens of Kew, is at this moment giving in charge a considerable collection of useful Tropical plants for introduction into Africa, to two native Missionaries (recently ordained by the Bishop of London),– than whom he knows not any well educated Europeans more competent to estimate the value of such importations, or likely to feel more interest in their successful cultivation and use.

Among those who fell victims to the climate of Niger, was Dr Vogel. Happily for science, he was not among the most early to be attacked by fever. He formed his collections with uncommon energy, while even a slight portion of health and strength remained to him; and the number of species amassed by him, in a short space of time, and under the most disadvantageous circumstances, reflects great credit upon his memory [. . .] Mr Ansell, though he fortunately survived the effects of the climate, was yet too ill, from a very early period of the voyage, to make any extensive or well-preserved collections. These facts must plead the apology for the imperfect nature of many of the descriptions. The work, however, the Editor is sure, will be hailed by every friend of Botany, and by everyone interested in the vegetable productions of Western Tropical Africa as a *Prodromus* of a Flora of that region; something upon which a more perfect superstructure will be hereafter built [. . .]

Memoir of the Life of Dr J.R.T. Vogel

Amongst the numerous sacrifices consequent on the unfortunate expedition to the Niger, science is not without her peculiar loss. Whatever reliance may be placed in wealth and a careful choice of means, it must be admitted that little has been accomplished by the numerous and deeply calculated plans for obtaining a more perfect knowledge of the interior of Africa. Amongst many other individuals, one of the naturalists of the expedition, to whose memory the following pages are dedicated, has succumbed to the destructive influence of the climate. If, however, according to the expression of a philosopher, it be the province of eloquence to commemorate illustrious minds, whose labours, owing to an unfortunate concurrence of circumstances, have not been productive of commensurate effects, and so, to compensate for the want of incident, a more skilful pen than mine is requisite. I must be contented to show what the world and what science have lost, by the simple relation of a few circumstances, and by extracts from the last official records of the deceased.

[. . .]

The British Government fitted out three steamers, destined to run into the Niger, or Quorra, at its entrance into the Bight of Benin, on the western coast of Central Africa, to penetrate by this vast navigable river, into the interior of this little-known country, to make treaties with the inhabitants, and to establish an emporium at some suitable place. A Botanist was needed to ascertain the vegetable productions of the country and the capabilities of the soil; and Dr Vogel was found willing to hold this office, hoping by these means to satisfy his eager to explore a rich and almost unknown vegetation. He [. . .] left Bonn on the 2nd of December, 1840, to enter upon his journey, having obtained from the proper authorities a two years' leave of absence.

The departure of the expedition, which, according to the first plan, was to be in the end of January 1841, was deferred from various circumstances and impediments to the third week in May [. . .] From Sierra Leone he wrote on the 30th of June, as follows:

'We sailed from Madeira by Teneriffe to St Vincent, one of the Cape de Verd Islands, and from thence came here. At Teneriffe we remained a day; but I was able to take only a cursory glance, since I was unwell on the passage from Madeira thither, and did not venture to leave the ship. We remained a fortnight off St Vincent: the island is small, but has an excellent harbour, and was therefore the rendezvous of the ships belonging to the expedition. Anything more comfortless than the view of this island, I never beheld: one might believe that after the formation of the world, a quantity of useless surplus stones was cast into the sea; and that thus the Island of St Vincent arose. There is nothing but hills and mountains (some of them 2500 feet high); with small valleys, which in the broader parts are very sandy, without a plant deserving the name of a tree: while the vallies themselves produce scarcely a species; for in my first excursion, I found in four hours only two species, of which one, a *lavandar*, was completely dried up. What had been wanting here, namely moisture, was in a few days too abundant. On the part of the coast where we are at present, the rainy season has begun; that is, the first portion of it, which announces itself by single thunder-storms with violent winds (tornados). Sometimes on the passage my cabin got very wet, and what was worse, my plants. Since we have been at Sierra Leone, the weather is generally clear by day; but towards evening there comes heavy rain or a thunder-storm, and last night we had one, such as I never witnessed before.

On entering the river at Free Town, the shore, on which the town stands, is bordered at a short distance by a range of hills, exhibiting a very pretty appearance with their gentle swelling summits and isolated lofty trees. A rich vegetation stretches from the shore upwards, which captivates the eye by its soft bright green, such as is only seen in the tropics, and gives the whole an incomparably charming character. I rushed into these woods, and much regret that the short time of our stay did not permit me to do more; for we were obliged to proceed. The object of the colony here is to teach the Africans active habits and to christianize them: there are, I think, about 40,000 in the colony, and many of their villages are built close to the town; so that, for miles, there is no cultivation. Since we left St Vincent, the temperature has been nearly the same. The thermometer there was generally 81° Fahr. in my cabin: here it is about 84°, and sometimes in the middle of the day reaches 86°. This heat is not greater than with us in summer; but the power of the sun, makes it seem often more intense than it is. An awning is spread over the deck, under which, when there is a breeze, it is always cool. I am very comfortable on board, except when my collections are lying about. When I return laden with plants, I have nowhere to prepare them; and when they are dry, the damp insinuates itself to such a degree, that I am compelled to redry them. This is very troublesome; and on board a ship, especially a man of war, there is no especial place for preparing or preserving plants. I am quite a nuisance to my messmates when I unpack; and so is the servant who announces breakfast, lunch, &c.; for the table must be cleared, and I must be off. Then I try to work on deck; but there the wind and rain attack me; so that I have to contend with all the elements. I am here quite amongst the negroes, for there are few white persons in the town; and during my excursions I frequently do not see one during the whole day. I cannot, however, say that this seems altogether strange to me: on our voyage outward, we had many black sailors in our ship; and their number has gradually increased in the course of our progress'.

From Cape-Coast Castle roads, where the ships belonging to the expedition arrived on the 24th of July, Vogel writes as follows:

'Our passage from Sierra Leone hither has been rather tedious. We set out from that port with but little fuel, and were therefore necessitated twice after we left Monrovia (Liberia), viz. at Grand Bassa and Cape Palmas, to cause wood to be felled, to enable us to proceed. Our voyage has been constantly along the coast; so that we have had ample opportunity for observing the remarkable nation of the Kroo: a people who dwell scattered along the coast, and often undertake long coasting voyages in small canoes. [. . .] The natives sit in them generally naked; they use broad oars and a very small rudder; and do not trouble themselves when the craft upsets; for they have commonly nothing to lose, and if they carry garments with them, they are soon dried. They have mostly a piece of cloth, bound around the head, which, when they come on board, they place round the loins, and think themselves fully dressed with great ivory rings round the ankles, and belts or chains round the foot or arm. We had many of their young people on board, for they are tolerably docile, and are therefore hired by the coasters to perform such hard labours as are considered prejudicial to Europeans. When they have earned so much money by their voyage, as will enable them to buy one or more wives, they return home, establish the women, and leave them for a new expedition, until they get eight or ten more wives, who must support them; for all field-labour, &c., is performed by females. Including these Kroos and other negroes, who are employed in various ways about the ship, we are now considerably more than one hundred men strong: frequently, therefore, when I have been for a time at that part of the vessel which they occupy and where alone smoking is allowed, and return to the quarter-deck where only the officers are, I feel quite relieved from the bustle. It is now the rainy season and we have had in Monrovia and Grand Bassa a week of continued rain; during which the sky has been for many successive days as dark as it can be with us in autumn only. Besides, the African brooks, when they are swollen with rain, assume the privilege of making their way down the footpaths; and I was therefore obliged for hours to wade up to the knees in water. I was indeed, in general, whether at sea or on land, as wet as it was possible to be. [. . .] At Cape Palmas we arrived at a spot where an intermission of the rainy season takes place, and from thence to this place, we have enjoyed delightful weather. The passage, however, was longer than we expected; so that water ran very short; and one day we were absolutely placed on half-allowance: otherwise we should scarcely guess that we were in a foreign zone. As regards meat and drink, we have several times a week salted beef or pork, and in general, other kinds of meat preserved in hermetically sealed cases. Hares, poultry, &c., prepared in this way, often appear at table. These ship-stores are preferred to the fresh provisions which are presented to us on landing [. . .] My health has been very good; and although there cannot but be some irksome hours to men shut up in a ship, I have yet, on the whole, felt happy and contented, and only look forward with impatience to the time when my own peculiar service will begin'.

The next letter from Vogel was written from Accra, on the 4th of August.

'We remain here but a few days, so that I can acquire only a very superficial view of the vegetation of the coast. Real forests lie at some distance in the interior, that is, about thirty English miles:– too long an excursion, even were it not desired that nobody should sleep on shore, for fear of fever. Yet I have been twelve or fourteen miles into the interior, in the district of Aquafim, to inspect a Danish settlement. There was a geologist with me, and we were received by the Danish Governor with the greatest civility. Such a journey on foot being considered too difficult for an European, large flat baskets, used here instead of sedan-chairs, were placed at our disposal, and four negroes to carry each basket. There were, besides, a number of negroes to take charge of our luggage; so that our caravan amounted to seventeen persons, besides ourselves. At the coffee-plantation there is a house arranged with European accommodations, where we were surrounded with all the luxury of the civilized world, and had for dinner French asparagus. The spot was lovely, pleasantly varied with hill and dale, mostly covered with savannas; where the grass is taller and stronger than in our own meadows, and between the tufts grew little buses, instead of flowers. [. . .] The negroes who accompanied us on this excursion were slaves; for the Danes still have slaves, but they seem well off, and were merry and cheerful beings. On the whole, I found in the short period of my acquaintance with them, no difference in their behaviour or dealing from the free negroes at Cape Coast Castle; except that the latter are shameless in demanding money for drink. At Cape Coast, it is absolutely necessary to keep an immoderate number of servants; and on an excursion from thence, our train of attendants consisted of thirty-six persons. There is no difficulty in this, for the blacks go as servants merely for food and clothing, which in this climate costs little; or they are sent when boys by their fathers to an European, that they may in this way learn something. The houses of Europeans here are very large, roomy, and well built, raised high above the ground to make them airy, and furnished with open verandahs for the same purpose. Europeans, however, do not in general remain long, since the climate on the coast is not suitable to their constitution. The few who are here seem to lead a miserable life: the society is very limited and monotonous, and their wishes are confined principally to making money; in which many fail. At Cape Coast, the small white shells which we use for ornamenting horses' bridles are given in exchange as coin; they are called cowries: a thousand of them are worth about a guelder, in the interior they are worth more: we have with us whole sacks of them. Gold-dust also appears at first a very curious medium of exchange; it is used especially in Cape Coast and Accra, where it is washed from the sand of the river banks which flows through the town. Every one of the market people carries a small pair of gold-scales: with which he weighs out for a silver-groschen, or perhaps for a sechser, its worth of gold-dust: they then take these very small grains with them, wrapped up in a piece of rag. All these market people are natives, and sell palm-oil, cocoa-nuts, different kinds of fruit, fish, home-woven cotton &c. The clothing of the men consists simply of a napkin round the loins; or, in addition, a long piece of cloth passed under one arm and over the other. They remove it from the shoulder when they meet a white man, and lay bare the heart by way of salutation. The women have these garments, and others in addition. The cloth round their loins is larger, and furnished behind with a monstrous bustle: the bigger this is, the more respectable is the woman, and the larger her family: in many it projects like a saddle. Little children are perfectly naked. So soon, however, as a young girl assumes a piece of cloth by way of clothing, it is furnished with a bustle, which with time is made gradually larger'.

[. . .]

'At Iddáh, the country which was before low and flat, begins to be elevated and rises in mountains 2000 feet high, which, with occasional interruptions, extend to this place, where they are confined to the right bank of the river. Here and there, spots occur, which remind one of the Rhine: the bed of the river is, however, too broad (generally above half a mile) to be picturesque, and is often broken and enlarged by various islands. The mountains are bare, without any signs of human industry: once only I saw a village on the top of a hill, which appeared very pretty. Mount Patteh, in whose neighbourhood we lie, is a quadrangular mountain on the right bank, rising precipitously on all sides about 1200 feet high, with many patches of forest, and thickly clothed everywhere with plants. At its foot grow many slender *Oil-palms*; so that the whole picture, painted with the fresh green which the rainy season has produced, is very lovely. As I sit under the awning on the quarter-deck, and look towards that spot, I cannot help being pleased with the view, beholding in the solitary *Baobabs*, and the *Oil-palms*, though familiar to me now for weeks, forms which still interest me from their novelty.

We have bought a piece of land on the right bank, extending from Mount Patteh to Beaufort Island, and at this moment are preparing a habitation for the person who is to have the charge of the station at the foot of the mountain. The land is decidedly of bad quality, and a better situation will be sought for; the other bank is far more suitable, but it has been rejected as too low; indeed, it is now under water. It is impossible for me, at present, to say anything of the nature of the vegetation. We certainly have not here the usual exuberance of the tropics; perhaps, since I have been on the river, I have collected three hundred species. No single family gives a peculiar character to the vegetation, but this depends on a mixture of many families. [. . .]

The natives, who come to us from far and near, behave extremely well; they have never shewn the slightest sign of enmity; on the contrary, they are rather too confiding. They are not of that deep black hue which is observable in other Africans, and in this neighbourhood they have often very good features. They understand spinning and making cloth: they know how to work in iron, to manufacture knives, sabres, nails, &c.: they cultivate also the fields with some degree of skill. It is sad, however, to think, that they have possessed the same aptness for these arts, probably from an almost inconceivable time, without making any improvement: they lack the spiritual energy which renders every acquisition a step to further advancement. We have a daily market on the shore; whither the inhabitants of a neighbouring village resort in great numbers, to sell or barter what they possess. Small looking-glasses, framed in paper, meet with very ready purchasers; and I shall never forget the joy which beamed in the eyes of many, when they first beheld their own faces in a mirror. The women, especially, cannot be satisfied with gazing on themselves, smeared with the powder of a red wood and their short hair standing upright in little tufts, so that they appear more like horned devils than human beings. In general, however, they prefer what is useful to trifles, provided the latter be not too dazzling and enticing; as, for instance, a bright red cap edged with gold.

72

THOMAS THOMSON, *WESTERN HIMALAYA AND TIBET*

(London: Reeve and Co., 1852)

Chapter XIV

HAVING at last completed the preparations necessary for a journey of twenty days through uninhabited regions, I started on the 9th of August from the village of Taksha. My first day's journey lay up the Nubra valley, which continued wide, though the alluvial platforms were destitute of cultivation, and quite barren. In several places (always opposite to ravines) they were covered with enormous boulders, which had all the appearance of having been brought to the position they occupied by glaciers. Two small villages were seen, both on the west bank of the river. Four miles from Taksha I crossed, by a good wooden bridge, a large stream which descended from the mountains on my right hand through an exceedingly rocky gorge. After seven miles and a half, I found that I had reached the point at which the road followed by the merchants in travelling from Le to Yarkand leaves the valley of Nubra. It was too late in the day to attempt the ascent of the ridge to the right; I therefore encamped in a grove of willows, which formed a belt along the margin of a stream whose bed was now quite dry, its scanty supply of water having been diverted into an artificial channel for the irrigation of a couple of fields of indifferent barley not far off.

In the valley of Nubra, beyond this encamping ground, which is known by the name of Changlung, there are, I believe, only three small villages, the most distant of which appeared to be not more than five or six miles off. In the direction of the valley, which was still north-north-west, very lofty mountains were visible at no great distance, all with snowy tops, and generally with heavy snow-beds and glaciers in their hollows; and according to the statement of my guides, the river at the distance of less than two days' journey issues from beneath a glacier, by which all passage is stopped.

On the 10th of August I started at daybreak, immediately commencing the ascent of the mountain range which enclosed the valley on the east. The mountain was exceedingly steep, indeed almost precipitous, and the road proceeded in a zigzag direction over bare granite rock, with scarce a vestige of vegetation. During the ascent I had a good view of the valley, and of the mountain range which bounded it on the south-west; large patches of snow lay on its peaks, and here and

 DOI: 10.4324/9780429355653-80

there I saw a small glacier in its ravines. The upper part of the valleys by which these mountains were furrowed had a very moderate slope, but from about 14,000 feet down to the bottom they were extremely abrupt.

After about 3500 feet of extremely laborious climbing, I arrived at a small level plain, perhaps two hundred yards long and forty or fifty wide, evidently much frequented as a resting-place by travellers, a small pool of water being the inducement. I here met a party of merchants on their way from Yarkand to Le. Their goods were conveyed by ponies, apparently much exhausted by their long journey through desert country. I had noticed, on the way up the mountain, that the road was lined by numerous skeletons and scattered bones of horses; I had also seen one or two of the same animals recently dead, and the appearance of these loaded ponies enabled me to understand the cause of the great mortality. Many of the unfortunate animals appeared scarcely to have strength to accomplish the few miles of descent which still intervened between them and plenty of food. The main reliance of the merchants for the support of their horses is on corn carried with them, to which there must be a limit, otherwise they would carry nothing but their own food.

[. . .]

The course of the Shayuk was visible for several miles, running nearly due west. Beyond that distance, it disappeared among rocky hills. Fording the river, I ascended a steep bank, to get upon a stony platform, over which I proceeded in a northerly direction, gradually approaching a small stream which came from the north to join the Shayuk. Passing a low rounded hill to the right, I descended after about two miles into the ravine excavated by this little stream, and, crossing it, encamped under low limestone rocks on its right bank after a march of twelve miles. I did not ascertain the elevation of this halting-ground, but, from the result of an experiment made at a place which appeared nearly midway (in point of elevation) between it and the bed of the Shayuk, where I got a boiling-point, indicating an elevation of 17,000 feet, I estimate the bed of the river at 16,800 feet, and my encamping-ground of the 18th at 17,200 feet. The plain all round seemed destitute of vegetation, so that, as on the two last days, there was a great scarcity of fuel, which had to be collected from a distance of many miles; and consisted only of the roots of a small bushy *Artemisia* or *Tanacetum,* which rose three or four inches above the ground. During these three days, I suffered very considerably from the effects of the rarefaction of the air, being never free from a dull headache, which was increased on the slightest exertion.

On the 19th of August, leaving my tent standing, I started to visit the Karakoram pass, the limit of my journey to the northward. The country round my halting-place was open, except to the north, where a stream descended through a narrow valley from a range of hills, the highest part of which was apparently about 3000 feet above me. All the rivers had formed for themselves depressions in the platform of gravel which was spread over the plain. At first I kept on the south bank of the river close to which I had halted, but about a mile from camp I crossed a large tributary which descended from the south-west, and soon after, turning round the rocky termination of a low range of hills, entered a narrow valley which came from a little west of north-west. At the foot of the rocky point of the range were three very small huts, built against the rock as a place of shelter for travellers, in case of stormy or snowy weather; and bones of horses were here scattered about the plain in greater profusion than usual.

I ascended this valley for about six miles: its width varied from 200 yards to about half a mile, gradually widening as I ascended. The slope was throughout gentle. An accumulation of alluvium frequently formed broad and gently sloping banks, which were cut into cliffs by the river. Now and then large tracts covered with glacial boulders were passed over; and several small streams were crossed, descending from the northern mountains through narrow ravines. About eight miles from my starting-point the road left the bank of the stream, and began to ascend obliquely and gradually on the sides of the hills. The course of the valley beyond where I left it continued unaltered, sloping gently up to a large snow-bed, which covered the side of a long sloping ridge four or five miles off. After a mile, I turned suddenly to the right, and, ascending very steeply over fragments of rock for four or five hundred yards, I found myself on the top of the Karakoram pass – a rounded ridge connecting two hills which rose somewhat abruptly to the height of perhaps 1000 feet above me. The height of the pass was 18,200 feet, the boiling-point of water being 180·8°, and the temperature of the air about 50°. Towards the north, much to my disappointment, there was no distant view. On that side the descent was steep for about 500 yards, beyond which distance a small streamlet occupied the middle of a very gently sloping valley, which curved gradually to the left, and disappeared behind a stony ridge at the distance of half a mile. The hills opposite to me were very abrupt, and rose a little higher than the pass; they were quite without snow, nor was there any on the pass itself, though large patches lay on the shoulder of the hill to the right. To the south, on the opposite side of the valley which I had ascended, the mountains, which were sufficiently high to exclude entirely all view of the lofty snowy mountain seen the day before, were round-topped and covered with snow. Vegetation was entirely wanting on the top of the pass, but the loose shingle with which it was covered was unfavourable to the growth of plants, otherwise, no doubt, lichens at least would have been seen. Large ravens were circling about overhead, apparently quite unaffected by the rarity of the atmosphere, as they seemed to fly with just as much ease as at the level of the sea.

Part 8

DEMOGRAPHICS, GEOGRAPHY, AND BIOGEOGRAPHY

Demographics, Geography, and Biogeography

This section begins with Gilbert White's *The Natural History and Antiquities of Selborne* (1788–9), a work that helpfully marks the beginning of many developments in the various environmental engagements occurring during the long nineteenth century. If geography is the study of physical spaces and of human interactions with environments, White was, amongst so many other things, a quintessential geographer before the professionalisation of the discipline had come to pass. The term geography (meaning 'earth description') comes to us from the Greeks, and its Classical roots go beyond the Greek world. Early pioneers like Anaximander and Hipparchus drew extensively on Babylonian cartography and mathematics, while from the third century CE onwards, the Romans and the Chinese took the lead ahead of a period of dominance of Arabic science in the Middle Ages.

The impetus for geographical studies in Europe owed much to a desire for reliable cartography to enable world exploration, colonisation, and exploitation; for knowledge of the earth to support agriculture; and more broadly to an Enlightenment desire to know and explain the world. It should not be forgotten that the initial impetus for that grand geographical project, the British Ordnance Survey (established 1791), was to provide the military with reliable maps and that the O.S. relied much on collaboration with the Royal Engineers.

Geography gradually became a separate discipline in the eighteenth and nineteenth centuries, leading to the establishment of bodies such as the French Société de Géographie and the British Royal Geographical Society in the early decades of the 1800s. Two Germans, Carl Ritter and Alexander von Humboldt, are rightly regarded as founders of modern geography, but many of the precursor figures we examined in Part 1 of this volume – Linnaeus, Buffon, and Smellie especially – had already contributed energy to the geographical idea that relationships between humanity and environment were worth studying. In Britain, the study of geography gained much from the founders of the Royal Geographical Society, one of whom, Alexander Machonocie, became the first professor of geography at a British institution (University College London) in 1833. The study of geography was commended and supported by many, but the role of Jeremy Bentham was particularly important in making it a university discipline. This is perhaps fitting given the strongly utilitarian aspects of the discipline. In addition, the philosopher Immanuel Kant produced a series of lectures on *Physical Geography*, excerpts of which were published in 1802 (but which are not available in a copyright-free English edition). The period would witness the increasing professionalisation of what had begun as a somewhat vague branch or element of the study of natural history. The formal division between physical geography and human geography would come later in the nineteenth century, and we will turn to some of the leading human geographers, like Élisée Reclus and Pyotr Kropotkin, in Volume III of this anthology. Many of the extracts included here show that early geographical practitioners worked across what became discrete aspects of the discipline.

In the study of geography, and particularly in its division into physical and human branches, lies an implicit articulation of a series of unfortunate assumptions that have driven and dominated human societies at least since the agricultural revolution and the establishment of 'civilisation': that the non-human world is separate from the human and can be studied separately; that 'natural history' is distinct from 'human history'; that the inorganic world is distinct from the organic. Another assumption arises from such binary constructions: the privileging of humankind within these constructions invites, permits, and justifies exploitation of environments. It is also true, however, that gradual challenges to all of these assumptions owe much to geography and its eagerness to intersect with the biological sciences. Evolutionary theory and ecology were as much reliant on geography as on geology and biology for the establishment of key ideas and principles, and geography remains at the forefront of ongoing academic and political efforts to disentangle the dangerous and stubborn divisions that remain between those realms we label 'culture' and 'nature', 'human' and 'non-human'. Geography, then, is in much the same position as many of the disciplines examined so far in this volume: at once complicit in environmental violence (maps and geographical knowledge permit colonial expansion and economic exploitation of 'new' and 'old' worlds, the enslavement of peoples, and the despoliation of their lands) *and* involved in attempts to better understand the earth and its inherent value as a life-supporting globe. In various ways, the different extracts in this section offer elements of this complex story while also showing the ways that geographers engaged with different parts of the globe, expanded knowledge, and constructed new conceptualisations of the earth.

This section also examines the development of two closely related fields – demographics and biogeography. The former becomes a major area of study following the work of Thomas Malthus and a site of considerable anxiety about the relationship between human populations and the environments, centred on issues of food supplies and population sustainability. Biogeography, a branch of its parent science, is essentially the study of species distributions in relation to different environments and was established gradually during the period under examination here. Although there were earlier pioneers, including Jean-Louis Soulavie, Humboldt is its most significant founding figure. Both fields intersect with the broader concerns of geography and are testament to the broadening of its vision and capabilities, and both fields, in different but sometimes related ways, later contributed to the development of evolutionary theories and ecology around mid-century.

The 'Advertisement' to White's work, the first part of the first extract, offers a vision of a 'parochial history' on a local scale that is implicitly geographical in its desire that gentlemen like himself should 'pay some attention to the districts on which they reside, and would publish their thoughts respecting the objects that surround them'. The environmentalist impulse is also religious: White hopes that his 'enlargement of the boundaries of historical and topographical knowledge' will induce his readers 'to pay a more ready attention to the wonders of the Creation, too frequently overlooked as common occurrences', an implied resistance to assumptions that the cultural is inherently more interesting than the natural. What follows, in Letter I to Thomas Pennant, is an opening survey of the geography of Selborne parish, a combination of topography and geology that also displays tentative biogeographical impulses: White anticipates the importance of studying associations between flora and habitats both here and in Letter V to Pennant, the final excerpt here. Its study of the results of human lanes on botanical habitats is conjoined to meteorological data, as well as some brief allusions to the problems of land management.

The second extract is also a foundational work. No survey of demographic discussion in the period would be viable without Malthus's enormously influential and controversial *Essay on the Principle of Population* (1798), a work that challenged Enlightenment-era confidence in the triumphant progression of humankind, that augured a dark picture of resource scarcity and overpopulation that has shaped demographic debate ever since, and that led to government action, not least in the creation of the British census in 1801 and in shaping debates about the Poor Laws. Malthus's work has been widely criticised – for its unsympathetic view of the working classes and its advocacy of late marriage as a way to curb birth rates amongst the poor – and for what many see as the unfounded nature of his central claim about the inability of humans to increase food supply in line with population. Malthusian ideas were ultimately co-opted to controversial developments in Social Darwinism, eugenics, and, via these, the race policies of German fascism. That criticisms of Malthus are sometimes excessive is borne out by closer inspection of his work, but it remains truly difficult to stomach the recommendations of self-restraint of a comfortably-off cleric. Recent work by Dean, Bashford and Chaplin, and Huzel (Further Reading) provides helpful contextualisations of Malthus and his impacts.

The extract comprises Malthus's Preface and Chapter 1, an overview of his theory which seeks to demonstrate that while food production can only at best increase arithmetically (i.e., by doubling), human population will grow exponentially (in Malthus's terms, geometrically). More specifically, Malthus argues that whenever a nation's food production increases, this in turn improves the well-being of the population, which in turn leads to population growth, ultimately forsaking gains in production levels. If population were controlled, a higher overall standard of living might be achieved and the 'Malthusian trap' of lost production gains averted. Without such wise responses, unrestrained action would ultimately mean that population would outstrip food supplies. Malthus's theory falters by not taking into account the advantages attendant on the Industrial Revolution in terms of agricultural production, transport, distribution, and trade but also in not foreseeing the enormous impact of the use of synthetic fertilisers and other advances in plant and animal breeding that lay ahead. It is instructive to read Malthus's pessimistic views alongside Thomas Ewbank's boundless optimism about the productive capacity of the earth in *The World a Workshop* (see extracts in Parts 5 and 6 of this volume): neither are realistic, but both rest upon a view that environment is ultimately a resource solely designed for the benefit of humankind. Whether the vision is bleakly Malthusian or confidently Ewbankian, what is occluded is recognition that *Homo sapiens* is only one of many species and that all rely on the earth's environments.

Given his singular importance to both geography and biogeography, Humboldt is excerpted twice in this section, having already appeared in the previous section, and he features again in Volume II. *Views of Nature* (1808) is the first published result of his extensive travels in South America, and in the extract, he turns to the subject of 'Steppes and Deserts' from a geographical and biogeographical perspective. Rather than simply providing a description of the arid lands of South America and Africa, Humboldt compares them, identifying reasons for similarities and differences. These are rooted in the effects of topography, geology, and meteorology on dry regions and their effects on the flora and fauna able to consist in these conditions. In the 1860s, as we shall see in Volume III, the American ecological pioneer and environmentalist George Perkins Marsh went further than Humboldt in tracing human causes for the desertification of the desert regions of north Africa and the middle east, but his work is clearly indebted to Humboldt's biogeographical insights.

The extract begins with a general description of the 'vast and boundless plain' of dry lands in the northerly and westerly portions of South America, immediately drawing attention to the contrast between the relative organic paucity of this region of 'more than three thousand English square miles' and the superabundance of the Amazon rainforest. His geographical impulses are also combined with Romanticist sympathies as he characterises the sublimity of a region that 'fills the mind with a sense of the infinite', so that 'the soul, freed from the sensuous impressions of space, expands with spiritual emotions of a higher order'. In this Burkean moment, Humboldt transports European readers to dry lands that are 'stretched before us, cold and monotonous, like the naked stony crust of some desolate planet'. Such passages indicate why Humboldt was able to captivate readers, but he skilfully subjoins sparkling prose to scientific intent, arguing that while 'in all latitudes nature presents the phenomenon of these vast plains [. . .] each has some peculiar character of physiognomy, determined by diversity of soil and climate, and by elevation above the level of the sea'. In the pages that follow, not excerpted here, Humboldt conducts readers on a bravura tour of the arid regions of the world, brilliantly demonstrating the particularities of the flora and fauna inhabiting different regions because of the conditions of each place. In returning to the South American steppe, where our extract resumes, his detailed survey attempts to augment the 'inaccurate geographical data' hitherto available. Only Humboldt's extensive travels, brilliance, and exemplary scientific techniques permit him to present such an authoritative overview. He then offers a detailed comparison of the steppe with the Sahara, arguing that the relative organic abundance of the former has its roots in different climatic, topographical, and geological conditions. In pursuing this 'somewhat difficult branch of physical geography', he eschews 'geological myths' in favour of empirical data and careful investigation.

The following two extracts, taking us forward twenty-five years, demonstrate the degree to which geography was becoming established. Both are from an 1833 issue of *The Asiatic Journal and Monthly Register for British and Foreign India, China, and Australasia*, a long-running imperial periodical that disseminated social, political, cultural, and scientific information from Britain's burgeoning Asian colonies. Amongst many semi-scientific pieces included in the journal, 'The Altai Mountains and Sources of the Ob' and 'Travels in Daghestan' indicate British appetite for Asian knowledge and a desire to vicariously explore areas little known or understood by Europeans. The short article on the Altai and the Ob, excerpted here in full, invokes geography in its first line and exemplifies Humboldtian methods. In a manner typical of such periodicals, the article is largely based on borrowings from other sources, in this case a letter from Dr Alexander von Bunge of the St Petersburg Academy of Sciences, who led several expeditions to Siberia and eastern Asia. von Bunge's description of travels in the Altai Mountains (then part of the Russian Empire but now in Mongolia) treats of its topography, geology, hostile climate, and animal populations. The article demonstrates that this was a period of ongoing discoveries for geographers – tasks such as tracing the source of rivers, fixing the altitudes of mountains and the depths of lakes, and providing cartographic guides continued to occupy geographers throughout the century and beyond, often for imperialist purposes.

'Travels in Daghestan', the second *Asiatic Journal* article, again draws upon the expertise of the St Petersburg Academy of Science, via the travel journals of Dr Aemilius Lenz, the geographer and physicist from an 1830 expedition to the Caucasus and the Caspian Sea. Travelling via Mt Elbruz, Grozny, and Dagestan *en route* to Baku, Lenz's account is a reminder that the Russian Empire was aggressively expanding into often hostile territories and that dangerous scientific journeys such as this required military assistance. Although primarily a physicist, Lenz's account is broadly geographical, noting the geology, climate, and botany of Dagestan and commenting on the declining water level of the Caspian Sea (a trend that has alarmingly accelerated in recent decades). Interestingly, this is also an early example of what would come to be known as human geography: its account of the Islamic inhabitants of Dagestan is detailed, anthropological, and orientalised.

Already excerpted in Part 5 of this volume, Joseph Dalton Hooker's *Flora Antarctica* (1844) is included here as a fine example of his biogeographical achievements. Reflecting his expeditionary travels throughout the southernmost portions of the globe in a voyage of extraordinary length and significance, it showcases Hooker as a multidisciplinary scientist, immersed in Lyell's *Principles of Geology*, Darwin's *Journal of Researches*, Humboldt's works, and an impressively wide range of other texts. His writings, particularly the lengthy footnotes, provide an exhilarating picture of knowledge provisionally but empirically in process of formation on the page. His main purpose in the extract is to compare the floras of Tierra del Fuego, Antarctica, and the South Atlantic islands; to trace affinities between floral populations; and to provide reasons for similarities and differences – to decide, for example, which plants belong in a distinctive 'Fuegian' flora, how far their range extends, and how its distribution progressed over time, including any barriers or enablers to their extension. This biogeographical endeavour is deeply rooted in previous southern hemisphere expeditions – from Cook, Bougainville, and Vancouver in the previous century to the *Beagle* in the 1830s. Hooker's commitment 'to follow the law of botanical affinity in preference to that of geographical position' is a clear statement of biogeographical endeavour, as is his interest in tracing the effects of specific details of climate, seasonality, geology, soils, topography, latitude, and zoology on the formation of groups.

Although initially overshadowed by his illustrious father, Sir W.J. Hooker, the son's work was at the cutting edges of Victorian science. His musings on the long geological history of the Kerguelen Islands provide support for Lyell's still-controversial theory of uniformitarianism (see Part 3 of this volume), while, as Darwin's closest friend, Hooker was part of the inner circle of confidantes who supported him in the years prior to the publication of *Origin of Species*. Contact with Darwin meant a shared commitment to conceptualising environments as dynamic systems and paying attention to their complex interactions. Part of an immense nineteenth-century project to gather environmental data, Hooker's findings in botanical biogeography provided supporting evidence for Darwin's 'development hypothesis'. Indeed, Hooker comes tantalisingly close to a statement of evolutionary principles himself as he speaks of island groups that were vital to the development of Darwinism:

> It might seem natural to suppose that a varied surface would have the effect of obliterating specific distinction, especially in small areas, as the Pacific Islands, the Galapagos, St. Helena, and the like, whose present contour is not the result of recent geological changes, and where time, the required element for developing such species as are the offspring of variation, has been granted.

As if this reference to relationships between time, development, and the 'offspring of variation' were not enticing enough, there are also occasional ecological and environmentalist impulses at work: observing the general tendency of islands to have limited numbers of species, Hooker correctly notes that the extremely limited flora and fauna of St Helena were the result of human interventions, including deforestation, agriculture, and introduced species. While this had been a feature of British control of the island since the 1600s, Hooker speaks of apparent extinctions or diminutions of certain species between his visits to the island. This proto-ecological interest in tracing dynamic environmental relationships over time suggests that the dividing line between biogeography and ecology is hardly substantial.

Hooker's approach and tone is different from that of works like *Niger Flora* (see Parts 5 and 7 of this volume). Overwhelmingly focused on plants, landscapes, and scientific investigation, his works are almost absent of aboriginal peoples or settlers. This may not necessarily indicate liberal attitudes, however. For one thing, depicting new world landscapes as empty of human life is part of a deeply problematic wilderness tradition in western representations, to which we will turn our attention more fully in the final two volumes of this edition. For another, we get scattered glimpses of colonial thinking, for example in Hooker's footnote about the disappearance of native St Helena flora and their replacement by introduced species, occurring

> exactly as is the case with various savage races of mankind, which do not suit themselves to the condition of the soil when altered by the European settler, but diminish in number and dwindle away even when violent measures have not been used for their extirpation.

Many of the works in this section owe a debt to Humboldt's pioneering work. His later book, *Cosmos: Sketch of a Physical Description of the Universe* (1845), attempts to encapsulate everything he had learnt during an illustrious career. The next extract is from the final section of this book and begins with Humboldt's overview of the preceding sections before announcing his intentions for the remainder. These opening lines encapsulate physical geography. What he then adds is the biogeographical element – the study of the interaction of flora and fauna with their environments, and vice versa. Biogeography is a modern perspective on environment because it implies or invites an argument that we cannot separate out the various elements of the globe, that the organic and inorganic interact in complex ways, so that plants and animals rely on the land, the soil, the climate, and the atmosphere while also affecting them in return. As Humboldt argues, 'the idea of life is so intimately connected with the moving, combining, forming, and decomposing forces which are incessantly in action in the globe itself' that it is impossible to consider them separately. While there are specifically biogeographical elements, particularly in tracing 'the distribution of organic beings over the surface of the globe', Humboldt's new science is a gateway to ecology, with which it essentially merges. There is in his approach a sense of wonder at the power and immensity of nature that is part of a broader ecologising of nineteenth-century consciousness. His remarks, for example, on the incessant powers of the vegetable kingdom are profoundly reverent but at the same time represent a re-ordering of human understanding of the earth and our relative position within it. Descriptions of the enormous volume and fecundity of vegetable life – from single-celled organisms to mighty trees – underline the long-established fact that 'the mass of vegetation on the earth very far exceeds that of the animal creation' but in ways that do not lead to Ewbankian transports of delight at the economic benefits to be gained. Instead, they suggest that we should take vegetation more seriously and understand the vital role it plays in the healthy functioning of environments.

The next extract, from Wilhelm Wittich, *Curiosities of Physical Geography* (1845), indicates the extent to which geography was beginning to infiltrate academic institutions. Although a linguist rather than a geographer, Wittich's role as professor of German at University College London meant that he rubbed shoulders with Alexander Maconachie, its first professor of geography, and it is clear that Wittich saw geography as vital to general education. His work surveys 'the phenomena which [. . .] may be considered as portions of Physical Geography', at this stage comprising both the study of planetary systems and of the Earth. *Curiosities* is primarily an overview of landscape types, including mountains, dry lands, and forests, and the work is a testament to the expansion of knowledge of the globe that has been evidenced throughout the volume so far. The extract exemplifies another area of the book's purview, the study of winds and currents, a cutting-edge field of enquiry. In the eighteenth century, Edmond Halley and George Hadley began to investigate trade winds: understanding these permanent east-west winds predominating in equatorial regions was vital to the ongoing expansion of global shipping. The Gulf Stream, the warm, rapid ocean east-west current of the North Atlantic that forms the subject of Wittich's chapter, was first recognised by Spanish sailors in the 1500s but imperfectly understood until the eighteenth and nineteenth centuries. As well as being of importance to navigators, ocean currents were of interest to geographers and meteorologists, given their profound impacts on weather and the biogeographical distribution of species. The British Meteorological Society, it should be noted, was only formed in 1850, while the forerunner of the Meteorological Office and modern weather forecasting emerged around the same time, much enabled by the expansion of the telegraph system which permitted the rapid dissemination of meteorological data to London.

Wittich's focus in studying the Gulf Stream is more broadly geographical, illustrating his desire to promote understanding of the profound influence of the ocean on the conditions of the land. The chapter evinces knowledge of the role of salt and fresh water in creating and sustaining ocean currents but also points out that 'it is not more than about seventy years that these currents have attracted the attention of navigators': Wittich reflects the latest science, particularly in his account of the river-like narrowing of the current in the Gulf of Mexico. Early studies of ocean currents such as these are of contemporary interest, particularly at a time when global heating and the melting of polar ice are predicted to lead to a permanent weakening of such currents, with disastrous consequences for life on earth.

Already excerpted in the Zoology section of this volume, John Gould's *An Introduction to the Birds of Australia* (1848) is included here as an example of the manner in which the influence of geography on flora and fauna was being more widely explored. Indicative of Gould's scientific brilliance and range, it touches upon geology, geography, and meteorology in offering a biogeographical account of Australian birds. Seeking to explain, a decade before *Origin of Species* elucidated such issues, the marked differences between the fauna of Australia and New Zealand, Gould argued that the different climates and topographies of the two islands were key. He also offers an overview of the existing physical geography of Australia, commenting on the crucial role of north-south mountain ranges in forming different biogeographical zones and considering the impact of an arid climate and intermittent flash flooding on the types and distribution of species. All in all, Gould's introduction, which (beyond our extract) goes on to survey the ornithology of Australia, is a bravura example of joined-up scientific thinking in ways that indicate that the gradual compartmentalisation of sciences from later in the century was not yet in place.

Richard Francis Burton's career reveals the degree to which the discipline of geography was still very much in formation during this period. A prolific writer from the 1850s to the 1880s, Burton's work is always a lively, conversational mixture of travel literature, anthropology, and geography. This is particularly true of his earliest work, *Goa and the Blue Mountains* (1851) – a chatty, informal collection of Burton's reflections on the period he spent on the island and the mainland at Kerala during a period of sick leave from his post as an officer of the East India Company army. This trip, it is worth noting, spurred his lifelong desire to travel. Not long after, he would make a memorable pilgrimage to Mecca and begin his long-term association with the Royal Geographical Society. Further extracts from his later works will be included in Volume III of this anthology.

For much of *Goa*, Burton is principally interested in urban environments and their inhabitants, and it is only here and there that more clearly geographical material is sustained. This is the case for the extract from Chapter 11, where he treats the topography, climate, soils, and agricultural economy of the Malabar coast of southwestern India in ways that demonstrate keen observation and an early elaboration of techniques that would become staples of human geography in the second half of the century. After providing mythological, etymological, and historical accounts of the region, he undertakes a geographical survey and an account of the administrative organisation of the English government. As well as speaking approvingly of the agriculture of this fertile region, he also offers recommendations for improvements to irrigation (another British reflection on Indian agriculture can be found in Part 10 of this volume). Although often critical of aspects of colonialism, Burton was a product of his nation, class, and status as a British military officer.

Published in the same year as *Goa*, the next extract is from a curious work, the full title of which – *System of Universal Geography; Founded on the Works of Malte-Brun and Balbi; Embracing the History of Geographical Discovery, the Principles of Mathematical and Physical Geography, and a Complete Description, From the Most Recent Sources, of all the Countries of the World* – indicates its peculiar diversity. The preface, not included here, states that the work began in attempts to translate influential works by the Danish-French geographer Conrad Malte-Brun and Adrian Balbi, Italian physicist and geographer. Gradually, the preface suggests, the project became a wider survey of geographical knowledge under the supervision of James Laurie, a rather mysterious figure described only as 'a gentleman well qualified to undertake the general superintendence', and including contributions from a range of authors. The resulting work includes an introductory section comprising a 'Historical Sketch of the Progress of Geographical Discovery' and sections on mathematical geography, inorganic physical geography, biological geography, and political geography, followed by what is in effect a gazetteer of 'descriptive geographies' of the various regions of the earth. Taken together, these indicate the enormous expansion of geographical knowledge by mid-century but also its depth and range.

The extract from *System* is written by the Scottish botanist John Hutton Balfour. It is taken from the section on biological geography, specifically on global flora, and is indebted to Humboldt, who is cited along with De Candolle and others. While this is plainly biogeographical work, as the chapter title's reference to 'the geographical distribution of vegetables, of animals, and of the human race' suggests, it is also rooted in the idea that geography had a role in supporting economic development. Balfour emphasises the global range and fecundity of vegetation in its marine and terrestrial forms, and the extract is remarkable in the precision of its categorisation of plant zones that occupy all parts of the planet, as well as the attentiveness to the role of climate, topography, and weather. Balfour focuses on natural and human causes of plant distribution; describes the relative numbers of flowering and non-flowering plants; and provides a list of floral zones created by Joakim Frederik Schouw, a pioneer of phytogeography, the sub-division of biogeography concerned with the geographical distribution of plants and the manner in which they modify environments. The related field of geobotany, also in formation in this period, focuses on the influence of environments on plants. When united, an ecological vision results.

These burgeoning areas of geographical enquiry are also the subject of the following extract, from Arthur Henfrey's now-obscure *The Vegetation of Europe, Its Conditions and Causes* (1852). As a popular work from an established botanist with scientific credentials, this book will also appear in Volume II. Henfrey's preface (not included here) makes clear his debt to the pioneering work of Humboldt, Robert Brown, and Schouw, and positions his own work as a decidedly biogeographical enterprise – an attempt to understand Europe's regional distributions of flora and to demarcate particular zones. The extract, taken from his chapter on Italy, exemplifies biogeographical methods in practice. Like other chapters, the first section (not included here) offers a detailed survey of the peninsula's geography, topography and climate, in this case emphasising the predominance of mountains, the relative aridity of the country, and the influence of the Mediterranean. Our extract begins with Henfrey's study of the resulting flora in various regions, noting a prevailing division between Alpine and Mediterranean floral zones but tracing exceptions and their causes.

The final extract in this section is from Henry Buckle's influential and problematic work, *A History of Civilization in England* (1857). Despite the title, this is as much a work of geography as of history, attempting as it does to trace the environmental causes of the development or lack of development of 'civilisation' in different parts of the world. The *History* was unfinished at the time of Buckle's death, and the published work was conceived merely as an introduction to his attempt to provide a scientific basis to historical studies. In a manner somewhat reminiscent of Lyell's uniformitarian claims, Buckle argued that human history was governed by fixed and regular laws and that these arose from our situatedness within environments, while his approach is also clearly influenced by a Malthusian focus on demographics and food supplies. Specifically, Buckle claimed that climate, soil, food production, and environment were the principal regulators of social and intellectual development. Most controversially, he then argued that European 'advancement' was due to its citizens being stronger than nature and able to subdue it to their purposes. Where humans were less powerful than their environments, they were less likely to advance and more likely to succumb to inefficient, corrupt, and authoritarian social and political systems. In classic imperial fashion, then, Buckle conflates European temperate climates with temperate moderation and tropical excesses of climate and environment with excessive and unruly societies. While not, strictly speaking, a work of geography, Buckle's *History* is included as an example of the political uses to which the discipline could be adapted.

The extract is from Buckle's chapter on physical causes affecting the 'progress' of civilisation. His combined focus on 'Climate, Food, Soil, and the General Aspect of Nature' is, in a scientific sense, progressive – part of a movement we have seen already to treat environments holistically – and he specifically invokes 'Physical Geography' as his guide. After forwarding his argument about the role of food surplus in creating wealth, and thus enabling the first steps of advancement, he turns to Europe, arguing that the heights of civilisation are to be found only in the middle latitudes of the continent, because the cold of Scandinavia and the heat of southern Europe both inhibit industrious behaviour. He is more damning still when turning to the history of Asia, where 'rude and wandering tribes', 'hordes', and 'roving savages' experience poverty as a result of existing in barren environments. Distastefully Eurocentric as this is, Buckle's approach is at least not essentialist: there is no inherent and enduring weakness in particular societies, he suggests, for when the Mongols, Tartars, and others were able to conquer more fertile territories, 'these barbarous tribes acquired for the first time some degree of refinement, produced a national literature, and organized a national polity'.

Buckle's work exemplifies a development that would gather pace in the second half of the century, one that we will trace further in Volumes III–IV: the various ways in which sciences and social sciences would render service to imperial and colonial agendas and to the broader project, of which this is a significant part, to subjugate, exploit, and exert sovereignty over the earth. Taken as a whole, the extracts in this section, all written by European authors, demonstrate the degree to which disciplines like geography were part of the confident self-definition of 'the west' as the leading agent of the world in the period. We have also seen, however, that geography in the period was not simply imperial and not simply exploitative, because it also contributed to changing conceptualisations of environment, particularly those that would result in ecological and holistic perspectives.

Further reading

Barber, Lynn, *The Heyday of Natural History 1820–1870* (London: Cape, 1980).

Bashford, Alison and Joyce E. Chaplin. *The New Worlds of Thomas Robert Malthus: Rereading the Principle of Population* (Princeton: Princeton University Press, 2016).

Brodie, Fawn McKay, *The Devil Drives: A Life of Sir Richard Burton* (London: Eland, 2002).

Clewis, Robert R., 'Kant's Physical Geography and the Critical Philosophy', *Epoche: A Journal for the History of Philosophy* 22:2 (2018), 411–27.

Clifford, David, Elizabeth Wadge, Alexandra Warwick and Martin Willis (eds.), *Repositioning Victorian Sciences: Shifting Centres in Nineteenth Century Scientific Thinking* (London: Anthem Press, 2006).

Clout, H. 'Geography at University College London: A Brief History' (London: UCL, 2003).

Dean, Mitchell. 'The Malthus Effect: Population and the Liberal Government of Life', *Economy and Society* 44 (2015), 18–39.

Dickinson, Robert E., *The Makers of Modern Geography* (Abingdon: Routledge, 2015) (first published 1969).

Endersby, Jim, 'Joseph Hooker: A Philosophical Botanist', *Journal of Biosciences* 33:2 (June 2008), 163–9.

Freeman, T.W., 'The Royal Geographical Society and the Development of Geography', *Geography: Yesterday and Tomorrow*, ed. E.H. Brown (Oxford: Oxford University Press, 1980).

Garske, Joseph P., 'Climate Change, Malthusian Catastrophe, and a Global Rule of Law', *Journal of Globalization Studies* 11:1 (May 2020), 3–15.

Huzel, James P., *The Popularization of Malthus in Early Nineteenth Century England: Martineau, Cobbett and the Pauper Press* (Aldershot: Ashgate, 2006).

Karlsohn, Thomas, Peter Josephson, and Johan Őstling (eds.), *The Humboldtian Tradition: Origins and Legacies* (Leiden: Brill, 2014).

Martin, Alison E. *Nature Translated: Alexander von Humboldt's Works in Nineteenth-Century Britain* (Edinburgh: Edinburgh University Press, 2018).

Parham, John (ed.), *A Global History of Literature and the Environment* (Cambridge: Cambridge University Press, 2016).

Winch, Donald, *Malthus: A Very Short Introduction* (Oxford: Oxford University Press, 2013).

Wulf, Andrea, *The Invention of Nature: Alexander von Humboldt's New World* (London: Vintage, 2016).

Young, S. *Richard Francis Burton: Explorer, Scholar, Spy* (New York: Marshall Cavendish, 2006).

73

GILBERT WHITE, *THE NATURAL HISTORY AND ANTIQUITIES OF SELBORNE*

(London: T. Bensley, 1789)

Advertisement

THE Author of the following Letters takes the liberty, with all proper deference, of laying before the public his idea of *parochial history,* which, he thinks, ought to consist of natural productions and occurrences as well as antiquities. He is also of opinion that if stationary men would pay some attention to the districts on which they reside, and would publish their thoughts respecting the objects that surround them, from such materials might be drawn the most complete county-histories, which are still wanting in several parts of this kingdom, and in particular in the county of *Southampton.*

[. . .]

If the writer should at all appear to have induced any of his readers to pay a more ready attention to the wonders of the Creation, too frequently overlooked as common occurrences; or if he should by any means, through his researches, have lent an helping hand towards the enlargement of the boundaries of historical and topographical knowledge; or if he should have thrown some small light upon ancient customs and manners, and especially on those that were monastic; his purpose will be fully answered. But if he should not have been successful in any of these his intentions, yet there remains this consolation behind – that these his pursuits, by keeping the body and mind employed, have, under Providence, contributed to much health and cheerfulness of spirits, even to old age: and, what still adds to his happiness, have led him to the knowledge of a circle of gentlemen whose intelligent communications, as they have afforded him much pleasing information, so, could he flatter himself with a continuation of them, would they ever be deemed a matter of singular satisfaction and improvement.

Letter I

To Thomas Pennant, Esquire

THE parish of SELBORNE lies in the extreme eastern corner of the county of *Hampshire,* bordering on the county of *Sussex,* and not far from the county of *Surrey*; is

DOI: 10.4324/9780429355653-82

about fifty miles south-west of *London,* in latitude 51, and near midway between the towns of *Alton* and *Petersfield.* Being very large and extensive it abuts on twelve parishes, two of which are in *Sussex,* viz. *Trotton* and *Rogate.* If you begin from the south and proceed westward the adjacent parishes are *Emshot, Newton Valence, Faringdon, Harteley Mauduit, Great Ward le ham, Kingsley, Hedleigh, Bramshot, Trotton, Rogate, Lysse,* and *Greatham.* The soils of this district are almost as various and diversified as the views and aspects. The high part to the south-west consists of a vast hill of chalk, rising three hundred feet above the village; and is divided into a sheep down, the high wood, and a long hanging wood called *The Hanger.* The covert of this eminence is altogether beech, the most lovely of all forest trees, whether we consider it's smooth rind or bark, it's glossy foliage, or graceful pendulous boughs. The down, or sheep-walk, is a pleasing park-like spot, of about one mile by half that space, jutting out on the verge of the hill-country, where it begins to break down into the plains, and commanding a very engaging view, being an assemblage of hill, dale, wood-lands, heath, and water. The prospect is bounded to the south-east and east by the vast range of mountains called *The Sussex Downs,* by *Guild-down* near *Guildford,* and by the *Downs* round *Dorking,* and *Ryegate* in *Surrey,* to the north-east, which altogether, with the country beyond *Alton* and *Farnham,* form a noble and extensive outline.

At the foot of this hill, one stage or step from the uplands, lies the village, which consists of one single straggling street, three quarters of a mile in length, in a sheltered vale, and running parallel with *The Hanger.* The houses are divided from the hill by a vein of stiff clay (good wheat-land), yet stand on a rock of white stone, little in appearance removed from chalk; but seems so far from being calcarious, that it endures extreme heat. Yet that the freestone still preserves somewhat that is analogous to chalk, is plain from the beeches which descend as low as those rocks extend, and no farther, and thrive as well on them, where the ground is steep, as on the chalks.

The cart-way of the village divides, in a remarkable manner, two very incongruous soils. To the south-west is a rank clay, that requires the labour of years to render it mellow; while the gardens to the north-east, and small enclosures behind, consist of a warm, forward, crumbling mould, called *black malm,* which seems highly saturated with vegetable and animal manure; and these may perhaps have been the original site of the town; while the woods and coverts might extend down to the opposite bank.

At each end of the village, which runs from south-east to north-west, arises a small rivulet: that at the north-west end frequently fails; but the other is a fine perennial spring, little influenced by drought or wet seasons, called *Well-head.* This breaks out of some high grounds joining to *Nore Hill,* a noble chalk promontory, remarkable for sending forth two streams into two different seas. The one to the south becomes a branch of the *Arun,* running to *Arundel,* and so falling into the *British* channel: the other to the north. The *Selborne* stream makes one branch of the *Wey*; and, meeting the *Black-down* stream at *Hedleigh,* and the *Alton* and *Farnham* stream at *Tilford-bridge,* swells into a considerable river, navigable at

Godalming; from whence it passes to *Guildford,* and so into the *Thames* at *Weybridge*; and thus at the *Nore* into the *German* ocean.

Our wells, at an average, run to about sixty-three feet, and when sunk to that depth seldom fail; but produce a fine limpid water, soft to the taste, and much commended by those who drink the pure element, but which does not lather well with soap.

To the north-west, north and east of the village, is a range of fair enclosures, consisting of what is called a *white malm,* a sort of rotten or rubble stone, which, when turned up to the frost and rain, moulders to pieces, and becomes manure to itself.

Still on to the north-east, and a step lower, is a kind of white land, neither chalk nor clay, neither fit for pasture nor for the plough, yet kindly for hops, which root deep into the freestone, and have their poles and wood for charcoal growing just at hand. This white soil produces the brightest hops.

As the parish still inclines down towards *Wolmer-forest,* at the juncture of the clays and sand the soil becomes a wet, sandy loam, remarkable for timber, and infamous for roads. The oaks of *Temple* and *Blackmoor* stand high in the estimation of purveyors, and have furnished much naval timber; while the trees on the freestone grow large, but are what workmen call *shakey,* and so brittle as often to fall to pieces in sawing. Beyond the sandy loam the soil becomes an hungry lean sand, till it mingles with the forest; and will produce little without the assistance of lime and turnips.

Letter V

To the same

Among the singularities of this place the two rocky hollow lanes, the one to *Alton,* and the other to the forest, deserve our attention. These roads, running through the malm lands, are, by the traffick of ages, and the fretting of water, worn down through the first stratum of our freestone, and partly through the second; so that they look more like water-courses than roads; and are bedded with naked *rag* for furlongs together. In many places they are reduced sixteen or eighteen feet beneath the level of the fields; and after floods, and in frosts, exhibit very grotesque and wild appearances, from the tangled roots that are twisted among the strata, and from the torrents rushing down their broken sides; and especially when those cascades are frozen into icicles, hanging in all the fanciful shapes of frost-work. These rugged gloomy scences affright the ladies when they peep down into them from the paths above, and make timid horsemen shudder while they ride along them; but delight the naturalist with their various botany, and particularly with their curious *filices* with which they abound.

The manor of *Selborne,* was it strictly looked after, with all it's kindly aspects, and all it's sloping coverts, would swarm with game; even now hares, partridges, and pheasants abound; and in old days woodcocks were as plentiful. There are few quails, because they more affect open fields than enclosures; after harvest some few land-rails are seen.

The parish of *Selborne,* by taking in so much of the forest, is a vast district. Those who tread the bounds are employed part of three days in the business, and are of opinion that the outline, in all it's curves and indentings, does not comprise less than thirty miles.

The village stands in a sheltered spot, secured by *The Hanger* from the strong westerly winds. The air is soft, but rather moist from the effluvia of so many trees; yet perfectly healthy and free from agues.

The quantity of rain that falls on it is very considerable, as may be supposed in so woody and mountainous a district. As my experience in measuring the water is but of short date, I am not qualified to give the mean quantity. I only know that

	Inch.	*Hund.*
From May 1, 1779, to the end of the year there fell	28	37!
From Jan. 1, 1780, to Jan. 1, 1781	27	32
From Jan. 1, 1781, to Jan. 1, 1782	30	71
From Jan. 1, 1782, to Jan. 1, 1783	50	26!
From Jan. 1, 1783, to Jan. 1, 1784	33	71
From Jan. 1, 1784, to Jan. 1, 1785	33	80
From Jan. 1, 1785, to Jan. 1, 1786	31	55
From Jan. 1, 1786, to Jan. 1, 1787	39	57

The village of *Selborne,* and large hamlet of *Oakhanger,* with the single farms, and many scattered houses along the verge of the forest, contain upwards of six hundred and seventy inhabitants.

74

THOMAS MALTHUS, *ESSAY ON THE PRINCIPLE OF POPULATION*

(London: J. Johnson, 1798)

Preface

It is an obvious truth, which has been taken notice of by many writers, that population must always be kept down to the level of the means of subsistence; but no writer that the Author recollects has inquired particularly into the means by which this level is effected: and it is a view of these means which forms, to his mind, the strongest obstacle in the way to any very great future improvement of society. He hopes it will appear that, in the discussion of this interesting subject, he is actuated solely by a love of truth, and not by any prejudices against any particular set of men, or of opinions. He professes to have read some of the speculations on the future improvement of society in a temper very different from a wish to find them visionary, but he has not acquired that command over his understanding which would enable him to believe what he wishes, without evidence, or to refuse his assent to what might be unpleasing, when accompanied with evidence.

The view which he has given of human life has a melancholy hue, but he feels conscious that he has drawn these dark tints from a conviction that they are really in the picture, and not from a jaundiced eye or an inherent spleen of disposition. The theory of mind which he has sketched in the two last chapters accounts to his own understanding in a satisfactory manner for the existence of most of the evils of life, but whether it will have the same effect upon others must be left to the judgment of his readers.

If he should succeed in drawing the attention of more able men to what he conceives to be the principal difficulty in the way to the improvement of society and should, in consequence, see this difficulty removed, even in theory, he will gladly retract his present opinions and rejoice in a conviction of his error.

June 7, 1798

Chapter 1

I think I may fairly make two postulata.

First, That food is necessary to the existence of man.

DOI: 10.4324/9780429355653-83

Secondly, That the passion between the sexes is necessary and will remain nearly in its present state.

These two laws, ever since we have had any knowledge of mankind, appear to have been fixed laws of our nature, and, as we have not hitherto seen any alteration in them, we have no right to conclude that they will ever cease to be what they now are, without an immediate act of power in that Being who first arranged the system of the universe, and for the advantage of his creatures, still executes, according to fixed laws, all its various operations.

I do not know that any writer has supposed that on this earth man will ultimately be able to live without food. But Mr Godwin has conjectured that the passion between the sexes may in time be extinguished. As, however, he calls this part of his work a deviation into the land of conjecture, I will not dwell longer upon it at present than to say that the best arguments for the perfectibility of man are drawn from a contemplation of the great progress that he has already made from the savage state and the difficulty of saying where he is to stop. But towards the extinction of the passion between the sexes, no progress whatever has hitherto been made. It appears to exist in as much force at present as it did two thousand or four thousand years ago. There are individual exceptions now as there always have been. But, as these exceptions do not appear to increase in number it would surely be a very unphilosophical mode of arguing, to infer merely from the existence of an exception, that the exception would, in time, become the rule, and the rule the exception.

Assuming then, my postulata as granted, I say that the power of population is indefinitely greater than the power in the earth to produce subsistence for man.

Population, when unchecked, increases in a geometrical ratio. Subsistence increases only in an arithmetical ratio. A slight acquaintance with numbers will shew the immensity of the first power in comparison of the second.

By that law of our nature which makes food necessary to the life of man, the effects of these two unequal powers must be kept equal.

This implies a strong and constantly operating check on population from the difficulty of subsistence. This difficulty must fall somewhere and must necessarily be severely felt by a large portion of mankind.

Through the animal and vegetable kingdoms, nature has scattered the seeds of life abroad with the most profuse and liberal hand. She has been comparatively sparing in the room and the nourishment necessary to rear them. The germs of existence contained in this spot of earth, with ample food and ample room to expand in, would fill millions of worlds in the course of a few thousand years. Necessity, that imperious all pervading law of nature, restrains them within the prescribed bounds. The race of plants and the race of animals shrink under this great restrictive law. And the race of man cannot, by any efforts of reason, escape from it. Among plants and animals its effects are waste of seed, sickness, and premature death. Among mankind, misery and vice. The former, misery, is an absolutely necessary consequence of it. Vice is a highly probable consequence, and we therefore see it abundantly prevail, but it ought not, perhaps, to be called an absolutely necessary consequence. The ordeal of virtue is to resist all temptation to evil.

This natural inequality of the two powers of population and of production in the earth and that great law of our nature which must constantly keep their effects equal form the great difficulty that to me appears insurmountable in the way to the perfectibility of society. All other arguments are of slight and subordinate consideration in comparison of this. I see no way by which man can escape from the weight of this law which pervades all animated nature. No fancied equality, no agrarian regulations in their utmost extent, could remove the pressure of it even for a single century. And it appears, therefore, to be decisive against the possible existence of a society, all the members of which should live in ease, happiness, and comparative leisure, and feel no anxiety about providing the means of subsistence for themselves and families.

Consequently, if the premises are just, the argument is conclusive against the perfectibility of the mass of mankind.

75

ALEXANDER VON HUMBOLDT, 'ON STEPPES AND DESERTS', *VIEWS OF NATURE*

(London: Henry G. Bohn, 1850 [1808])

On Steppes and Deserts

AT the foot of the lofty granitic range which, in the early age of our planet, resisted the irruption of the waters on the formation of the Caribbean Gulf, extends a vast and boundless plain. When the traveller turns from the Alpine valleys of Caracas, and the island-studded lake of Tacarigua, whose waters reflect the forms of the neighbouring bananas, – when he leaves the fields verdant with the light and tender green of the Tahitian sugar-cane, or the sombre shade of the cacoa groves, – his eye rests in the south on Steppes, whose seeming elevations disappear in the distant horizon.

From the rich luxuriance of organic life the astonished traveller suddenly finds himself on the dreary margin of a treeless waste. Nor hill, nor cliff rears its head, like an island in the ocean, above the boundless plain: only here and there broken strata of floetz, extending over a surface of two hundred square miles, (more than three thousand English square miles,) appear sensibly higher than the surrounding district. The natives term them *banks*, as if the spirit of language would convey some record of that ancient condition of the world, when these elevations formed the shoals, and the Steppes themselves the bottom, of some vast inland sea.

Even now, illusion often recalls, in the obscurity of night, these images of a former age. For when the guiding constellations illumine the margin of the plain with their rapidly rising and setting beams, or when their flickering forms are reflected in the lower stratum of undulating vapour, a shoreless ocean seems spread before us. Like a limitless expanse of waters, the Steppe fills the mind with a sense of the infinite, and the soul, freed from the sensuous impressions of space, expands with spiritual emotions of a higher order. But the aspect of the ocean, its bright surface diversified with rippling or gently swelling waves, is productive of pleasurable sensations, – while the Steppe lies stretched before us, cold and monotonous, like the naked stony crust of some desolate planet.

In all latitudes nature presents the phenomenon of these vast plains, and each has some peculiar character or physiognomy, determined by diversity of soil and climate, and by elevation above the level of the sea.

[. . .]

DOI: 10.4324/9780429355653-84

The Steppe extends from the littoral chain of Caracas to the forests of Guiana, and from the snow-covered mountains of Merida, on whose declivity lies the Natron lake of Urao, – the object of the religious superstition of the natives, – to the vast delta formed by the mouth of the Orinoco. To the southwest it stretches like an arm of the sea, beyond the banks of the Meta and of the Vichada, to the unexplored sources of the Guaviare, and to the solitary mountain group to which the vivid imagination of the Spanish warriors gave the name of *Paramo de la Suma Paz,* as though it were the beautiful seat of eternal repose.

This Steppe incloses an area of 256,000 square miles. Owing to inaccurate geographical data, it has often been described as extending in equal breadth to the Straits of Magellan, unmindful that it is intersected by the wooded plain of the Amazon, which is bounded to the north by the grassy Steppes of the Apure, and to the south by those of the Rio de la Plata. The Andes of Cochabamba and the Brazilian mountains approximate each other by means of separate transverse spurs, projecting between the province of Chiquitos and the isthmus of Villabella. A narrow plain unites the *Hylæa* of the Amazon with the Pampas of Buenos Ayres. The area of the latter is three times larger than that of the Llanos of Venezuela; indeed so vast in extent, that it is bounded on the north by palms, while its southern extremity is almost covered with perpetual ice. The Tuyu, which resembles the Cassowary, (Struthio Rhea,) is peculiar to these Pampas, as are also those herds of wild dogs, which dwell in social community in subterranean caverns, and often ferociously attack man, for whose defence their progenitors fought.

Like the greater part of the desert of Sahara, the Llanos, the most northern plains of South America, lie within the torrid zone. Twice in every year they change their whole aspect, during one half of it appearing waste and barren like the Lybian desert; during the other, covered with verdure, like many of the elevated Steppes of Central Asia.

The attempt to compare the natural characteristics of remote regions, and to pourtray the results of this comparison in brief outline, though a gratifying, is a somewhat difficult branch of physical geography.

A number of causes, many of them still but little understood, diminish the dryness and heat of the New World. Among these are: the narrowness of this extensively indented continent in the northern part of the tropics, where the fluid basis on which the atmosphere rests, occasions the ascent of a less warm current of air; its wide extension towards both the icy poles; a broad ocean swept by cool tropical winds; the flatness of the eastern shores; currents of cold sea-water from the antarctic region, which, at first following a direction from south-west to north-east, strike the coast of Chili below the parallel of 35° south lat., and advance as far north on the coasts of Peru as Cape Pariña, where they suddenly diverge towards the west; the numerous mountains abounding in springs, whose snow-crowned summits soar above the strata of clouds, and cause the descent of currents of air down their declivities; the abundance of rivers of enormous breadth, which after many windings invariably seek the most distant coast; Steppes, devoid of sand, and therefore less readily acquiring heat; impenetrable forests, which, protecting the earth from the sun's rays, or radiating heat from the surface of their leaves, cover the richly-watered plains of the Equator, and exhale into the interior of the country, most remote from mountains and the Ocean, prodigious quantities of moisture, partly absorbed and partly generated – all these causes produce in the flat portions of America a climate which presents a most striking contrast in point of humidity and coolness with that of Africa. On these alone depend the luxuriant and exuberant vegetation and that richness of foliage which are so peculiarly characteristic of the New Continent.

If, therefore, the atmosphere on one side of our planet be more humid than on the other, a consideration of the actual condition of things will be sufficient to solve the problem of this inequality. The natural philosopher need not shroud the explanation of such phenomena in the garb of geological myths. It is not necessary to assume that the destructive conflict of the elements raged at different epochs in the eastern and western hemispheres, during the early condition of our planet; or that America emerged subsequently to the other quarters of the world from the chaotic covering of waters, as a swampy island, the abode of crocodiles and serpents.

South America presents indeed a remarkable similarity to the south-western peninsula of the old continent, in the form of its outlines and the direction of its coast-line. But the internal structure of the soil, and its relative position with respect to the contiguous masses of land, occasion in Africa that remarkable aridity which over a vast area checks the development of organic life. Four-fifths of South America lie beyond the Equator, and therefore in a region which, on account of its abundant waters, as well as from many other causes, is cooler and moister than our northern hemisphere. To this, nevertheless, the most considerable portion of Africa belongs.

The extent from east to west of the South American Steppes or Llanos, is only one third that of the African Desert. The former are refreshed by the tropical sea wind, while the latter, situated in the same parallel of latitude as Arabia and Southern Persia, are visited by currents of air which have passed over heat-radiating continents. The venerable father of history, Herodotus, so long insufficiently appreciated, has in the true spirit of a comprehensive observer of nature, described all the deserts of Northern Africa, Yemen, Kerman, and Mekran (the Gedrosia of the Greeks), as far even as Mooltan in Western India, as one sole connected sea of sand.

To the action of hot land winds, may be associated in Africa, as far as we know, a deficiency of large rivers, of forests that generate cold by exhaling aqueous vapour, and of lofty mountains. The only spot covered with perpetual snow is the western portion of Mount Atlas, whose narrow ridge, seen laterally, appeared to the ancient navigators when coasting the shore, as one solitary and aërial pillar of heaven. This mountain range extends eastward to Dakul, where the famed Carthage, once mistress of the seas, lies in crumbling ruins. This range forms a far extended coast-line or Gætulian rampart, which repels the cool north winds and with them the vapours rising from the Mediterranean.

76

'THE ALTAI MOUNTAINS AND SOURCES OF THE OB', *ASIATIC JOURNAL* IX

(1833)

The Altai Mountains and Sources of the Ob

Geography teaches us that the true source of a river is that which is the farthest from its mouth. Yet there are many cases in which this rule, which ought to be universal, is departed from. Thus, it is said that the Ob, which is usually but erroneously called *Oby,* is formed by the union of the rivers Biya and Katoonya, which effect their junction below Biysk, in Siberia. The Katoonya (properly *Katoon gol,* 'river of the Princess') is formed in the lofty Alps of the government of Tomsk, by the union of the Koksoon (properly *Kookee oosoon,* 'blue-water'), which flows from west to east; and the Chooya (properly *Chooï*), which runs from east to west. The Biya issues from Lake Teletskoi of the Russians, the true name of which is *Altan kol* or *Al-aï ko,* 'gold lake'. This lake is situated to the north of the high chain of the Altaï or Altan tau; it receives a vast number of rivers of different magnitudes, amongst which the Bashkoosh and the Cholosba, vulgarly called Choolyshman, are the most considerable. The sources of the two latter and of the Chooi, are not far from each other, in the highest part of the snowy chain of the Altaï, which separates the government of Tomsk from the Chinese empire. Dr. Bunge, member of the academy of St. Petersburgh, who had previously made several journies into this part of Siberia, was despatched thither again by the Academy in 1832. In July last, he visited the sources of the Chooi, the Bashkoosh and the Cholosba, which are the affluents of the Ob the most remote from its embouchure. He has given some particulars of this expedition in a letter to the academy, an extract of which we have an opportunity of communicating to our readers: a few rectifications are added at the end: –

> 'I quitted my camp, on the banks of the little river Kara dyrghoon, and having turned a group of hills, situated on the left bank of the Chooya, I pursued my way over the plain, which rises gradually, and which is bounded on the right by the Chooya, and on the left by an abrupt chain of snowy mountains. After a journey of thirteen or fourteen miles across

DOI: 10.4324/9780429355653-85

this steppe, which is dry and scantily covered with vegetation, we passed the Chooya on horseback. This river is here not deep but very rapid: we ascended one of its principal affluents, the Boilooghem. Late in the evening, we reached the felt tents of Demeshi Chokhon, the independent chief of a camp of Kalmucks, established near a spring in a hollow of the Alps. I had hoped to find, amongst these good-natured inhabitants of the frontier, the assistance requisite for the execution of my plan, of penetrating as far as possible into the lofty mountains. They, indeed, furnished me with the necessary instructions respecting the localities, and for a trifling remuneration, they provided me with horses and guides for some days. I disembarrassed myself of everything not absolutely indispensable for the journey, and we quitted the banks of the Boilooghem next day. We directed our course across the steep mountains towards the Chooya, and reached its sources early in the afternoon. A little more than a mile to the left of these sources, there is on the crest of a high chain of snowy mountains an *obo,* or column marking the frontier between the Russian territories and the Chinese empire. This column bears an inscription in the Mongol language, importing that at this spot, where, on either side, the springs flow in opposite directions, the habitations of the Kalmucks, who pay contributions to the two empires, end, and those of the Mongol tribe of Soyon begin.

'After halting for some time on this elevated spot, we descended into a deep valley of the Alps, which extended very far. It forms an immense marshy plain, covered with innumerable lakes of different dimensions. Several clusters of snowy peaks rise in the midst of this plain, and the water which descends from them feeds the lakes, which are the origin of some considerable rivers. The extent of this valley is upwards of thirty-five miles from east to west; it is inhabited by a vast number of deer, reindeer, argalis, ibexes, wolves, foxes, &c. These animals prefer this spot, because it is rarely visited, except by some Kalmuck hunters, who, however, remain there but a short time, because they are strictly prohibited from taking up their residence there, although they are always anxious to do so, notwithstanding the severe cold, on account of the fine pasturage in the valley.

'We made a further march, the same day, of about twenty miles. We passed the sources of the Bashkan, an affluent of the Abakan, and we halted for the night near a small lake. Although it was the middle of July, and we were warmly clad, we endured much from the cold, for we could no where find firewood of any kind. Next day, a dense fog enveloped the lofty mountains in our front, which are those of the Altan tau, or the true grand Altaï, the Kim shan of the Chinese, which was the ultimate object of my journey. Happily, the fog dispersed when the sun rose, and we commenced our march for the pass of Mount Chapchal, the only practicable route to the Altan tau. After about ten miles, we arrived at the borders of a beautiful lake, named Yeiln kol ('lake of the evil genius'), which stretches at the foot of the Altan tau for nine or ten miles, its breadth being from three to three and a-half. From its north-west extremity issues the river Choolyshman, beyond which, the gigantic mass of the Altan tau rears itself, and like a sharp wall extends from N.N.W. to S.S.E., enclosing the valley in which the river flows. With great difficulty we succeeded in scaling the advanced mountains of the chain. We passed immense fields of snow, and at length found ourselves in a narrow path, which is the real defile of the Chapchal, which gives its name to the whole mountain. This terrific path is, as it were, suspended on the southern slope, looking over a precipice of massive rocks. It runs over detached and not very firm fragments of *glimmerschiefer* (a species of talc), of which the whole mountain consists. The golden lustre which proceeds from this rock, when the sun shines upon it, has caused the name of *Altan tau,* or 'golden mountain', to be given to this vast chain. In ascending this path, we perceived in the ravine below piles of bones of men and horses, who had been precipitated down, which admonished us of the temerity of our undertaking. We at length gained the summit, whence I enjoyed a most extensive view. To the west we perceived the wide valley which we had lately traversed, full of lakes, amongst which we distinguished the Yeiln kol, close at our feet, and to the left, a larger one, named Kyndiktoo kol, still partly covered with ice. To the north-east, appeared a deep ravine, bounded by abrupt peaks covered with snow, whence originates the Tsooi, which falls into the Kem choog, one of the constituents of the Yeniseï. We were told that, about forty-six miles from the place where we then were, there was a Mongol-Soyon town, situated on its banks, also called Kemchoog.

'In respect to botany, these mountains present little novelty. Its vegetation is the same as that on the banks of the Kooraï. The crests are entirely destitute of vegetation and covered with perpetual snow'.

77

'TRAVELS IN DAGHESTAN', *ASIATIC JOURNAL* IX

(1833)

Travels in Daghestan

MR. LENZ, of the Imperial Academy of Sciences of St. Petersburgh, who accompanied Mr. Kupffer, a few years back, in his journey to Mount Elbrouz, the loftiest peak of the Caucasus, was despatched by the Academy, in the early part of 1830, to explore the provinces situated on the W. shore of the Caspian Sea, known under the names of Daghestan and Shirwan. On his return, he drew up a detailed report of this expedition to the Academy, from which we extract the following passages:

> 'On the 18th January 1830, I once more approached Elbrouz, by Georgievsk, whose lofty crest I had had the good fortune to scale, and determined its height by the barometer, six months before. A thick fog, which enveloped me during my fifteen days' journey along the new military line of the Caucasus, did not allow me, for a single moment, to enjoy the sublime spectacle which these Caucasian Alps present to the traveller. At Grosnaya, a fort situated on the extreme frontier of the Chechentses, I remained for eight days, with General Engelhardt, who provided excellent arrangements to make my journey less perilous. The bold and rapacious mountaineers had then recommenced their forays across the icy Terek, and fell suddenly, in bands of 300, upon travellers or the inhabitants, carrying the former to their inaccessible mountains as slaves, and plundering the latter of their horses and cattle. I was consequently obliged to be attended through this dangerous country by an escort of fifty foot-soldiers, with a piece of cannon. These escorts have frequent skirmishes with the resolute mountaineers, which contests, whilst it makes the service of these frontier-garrisons very painful, causes the troops to be the best and bravest of the Russian army.
>
> 'From the line of the Caucasus there are two routes to Bakoo; one traverses the chain near Mount Mquinvari, improperly termed by the Russians *Kazbek,* one of the loftiest summits, and runs in the direction of Tiflis, – which is called the military road of Georgia; – the other passes through Daghestan, and along the W. coast of the Caspian Sea. On the

 DOI: 10.4324/9780429355653-86

right of this route, in a S.W. direction, is the extreme branch of the principal chain of the Caucasus, which extends as far as Bakoo and Salian, and, gradually lowering, there terminates in hills of slight elevation. For several reasons, I determined to choose the less-frequented route of Daghestan.

'Journeying through this country is very troublesome, especially for those who carry much baggage, and as I had my instruments with me, I could not travel on horseback. Daghestan is not wide; it is bounded to the E. by the Caspian Sea, and to the W. by the branch of the Caucasus just mentioned. A number of small streams, the most considerable of which is the Samoor, traverse it from W. to E. Although their depth is trifling, they acquire such rapidity by falling from heights, that their beds are everywhere obstructed by rocky fragments, several feet in diameter, and at certain seasons, as in May and June, when the snow melts in the high mountains, all communications are intercepted or rendered very dangerous.

'The principal rock of the Daghestan Mountains is a shelly limestone, which, as is observed near Terek, rests upon a grey and friable freestone. The strata of these two species of rock are alike inclined in a S.E. direction, at an angle of about 30°, but which I remarked varied from 15° to 45°. Vegetation is sickly, and during the months of July and August, the heat is so intense, that the very shoots of the plants are burnt up. The mountains of Daghestan approach the coast at three points, dividing the country by two arched curves into two natural portions, one of which is Northern the other Southern Daghestan, which are separated by the narrow defile of Derbend. The three points are near Tarkoo, Derbend, and Mount Besh-barmak.

'Northern Daghestan, which extends likewise beyond Tarkoo, to the N., comprehends the possessions of the Shamkhal and the territory of Derbeud, formerly dependent upon the Ootmie. The city of Tarkoo is the residence of the Shamkhal. Like almost all the small towns of Daghestan, it is built upon the slope of a steep hill; the summit is occupied by the Russian fort of Boornaya, which denotes that the prince of Tarkoo is subject to Russia. Although this city holds in Daghestan a rank immediately after Derbend, I was nevertheless unable to discover any of those beautiful edifices which M. Gamba pretends to have found there. Even the palace of the Shamkhal differs from the other houses only by its extent: its architecture and the arrangement of its interior are not superior to those of the rest. In winter, owing to the clayey soil being thoroughly saturated with frequent rains, it is difficult to pass along the crooked lanes which wind about the mountains in all directions, and what renders them less practicable is a rivulet, which falls from the top of the mountain, and runs through the streets, in order to supply the different quarters of the city with water. The inhabitants have not quite so wild an exterior as those of the neighbouring towns; for the Russian fortress, which commands the town, keeps them in tolerable order. Pillage and murder, however, are heard of even here; but it is principally in the towns of Booinak and Kaya-kand that travellers run most risk of being plundered. On leaving Tarkoo, I halted for the night in their vicinity: the former of these towns belongs to the Shamkhal, the other is a dependency of Derbend. Although I was attended by natives of the country, to whom the commandant of Boornaya had given the strictest orders to provide me with whatever I wanted, I was, nevertheless, heartily rejoiced to escape safe and sound from the clutches of these brigands, with only a few threats of being poignarded, because I was unable to give them any brandy.

'In general, the inhabitants of Daghestan, in those parts where their manners have not been ameliorated by a residence in towns, and by commerce and industry, still wear the stamp of their primitive ferocity. They are mostly of the middle height, with broad shoulders, a deep complexion, and wild exterior. They are always armed with a long two-edged dagger, and when they leave their villages, they are further armed with a musket, a pistol, and a long sabre, slightly curved. They are generally on horseback. Their warlike character does not prevent them from receiving a stranger, when necessary, with an affability exceeding, perhaps, that of a civilized European. They are vindictive, and do not fail to wash out a wrong done them in the blood of the offender. At Velikend, a town ten miles from Derbend, I was told an instance of this thirst of vengeance. During the period of the Russian conquest of this country, an individual, named Nawrooz-bek, was at war with a certain chief, who rendered himself formidable by plunder and audacity. Nawrooz-bek, determined to destroy him, proceeded with one of his sons into his enemy's camp, and reaching his dwelling, laid his musket at his feet in token of peace; but, at the very moment when he was swearing to bury the wrong he had received in oblivion, his son deposited beneath the dwelling a barrel of gunpowder, and stuck a lighted match in it. The father and son retired, and waited at a little distance the effect of their stratagem. The house was blown up into the air, carrying with its fragments the mutilated bodies of its tenants. This action drew upon Nawrooz-bek the hatred of the other mountaineers, and it was with difficulty, and only through his own courage and the devotion of his five sons, who guarded him whilst he slept, that he escaped the efforts made to take his life. I paid a visit to this celebrated personage. I beheld an old man of very agreeable exterior, with manners exhibiting all the polish of the best bred European. He treated me with the utmost kindness, offered to accompany me with his sons as far as the frontier of Daghestan, employing the most select and well-turned expressions. The figurative and poetic style, which characterises

Orientals, surprised me in all my intercourse with the inhabitants of these parts. Even in business, the style they use is singular. Thus a moollah, in my presence, having received an order from the commandant to cite a native to trial, who was accused of murder, issued the following summons to him: 'Come to me, and should you be about lifting a glass of water to your lips, leave it and come; I have something to communicate to you'.

'The narrow defile of Derbend, which separates the two Daghestans, is entirely formed by the city itself and by its two walls to the N. and S. These walls, with their gates of iron, begin at the crest of the mountain, which inclines towards the sea, and descend by their other extremities to the water. Tradition attributes their construction to Alexander the Great; and, although it can be proved that this conqueror never penetrated so far, the manner in which they are constructed is not unworthy of him. They consist of enormous masses of stone, which, as far as can be perceived, are not connected together by any cement. Formerly, they extended so far into the sea, that a chain thrown between them closed the entrance of the port. But the general and well-ascertained subsidence of the waters of the Caspian, deposits, and the fall of a portion of the walls into the sea, now enable a person to walk dryshod round the enclosure. Derbend, by its site and walls, heretofore regarded as impregnable, once served as a powerful barrier to the flourishing provinces of Persia against the incursions of the northern tribes, separating, as it were, the civilized south from the yet barbarous north. This circumstance, and the celebrity of the port in ancient times, rendered this city famous; but at present, since neither of these causes any longer exists, the port being ruined, Derbend is an insignificant place.

'Kooba, the capital of Southern Daghestan, is still less important than Derbend and Tarkoo; but being nearer the mountains, and on more elevated ground, it enjoys a healthier climate. The route from Derbend to Kooba is one of the most difficult in all Daghestan, owing to the number of streams which intersect it, amongst which is the rapid Samoor, with its three arms. This route traverses forests, which cannot, however, be compared with the woods of lofty forest-trees in our country; on the contrary, the trees are thinly scattered, and of stunted form and growth. Such is also the route between Kooba and Bakoo, which tends towards the coast, which it reaches near the Besh-barmak, or "Five-fingered Mountain", also called Shatagan. The Besh-barmak has received this name from its singular aspect, which, however, can only be assimilated to a hand by the imagination of a Tatar. This mountain is composed of a hard calcareous rock, much resembling a small-grained sandstone; its height is only 3,000 feet; at its foot is a caravanserai, occupied by a post of Cossacs. The gates of the caravanserai are covered with inscriptions engraved by travellers who have passed here. I sought in vain for the name of the learned Kæmpfer, which Gmelin discovered here: it has probably been covered with plaster when the gate was last repaired.

'From Besh-barmak, the road runs at first along the shore, then, intersecting the peninsula of Absheron, it runs over hills of sandy clay as far as Bakoo, the end of my journey. I performed this route through Daghestan in twelve days'.

78

JOSEPH DALTON HOOKER, *FLORA ANTARCTICA: THE BOTANY OF THE ANTARCTIC VOYAGE OF H.M. DISCOVERY SHIPS EREBUS AND TERROR IN THE YEARS 1839–1843 UNDER THE COMMAND OF CAPTAIN SIR JAMES CLARK ROSS*, 3 VOLS, VOL. 1. BOTANY OF LORD AUCKLAND'S GROUP AND CAMPBELL'S ISLAND

(London: Reeve Brothers, 1844)

Part II, Botany of Fuegia, the Falklands, Kerguelen's Land, Etc

THE pages of the present portion of the work are destined to contain descriptions of all the plants ascertained to exist in what we may term the Antarctic regions, (Lord Auckland's and Campbell's Islands excepted), viz. Fuegia and some part of the south-west coast of Patagonia, the Falkland Islands, Palmer's Land, and the adjoining groups, as the South Shetlands, South Georgia, &c, and (proceeding eastward) Tristan d'Acunha and Kerguelen's Land. I shall preface the Flora of these widely severed, and in some cases very isolated spots, with a few remarks upon each, and on the general character of the whole as forming one great botanical region.

It may appear paradoxical, at first sight, to associate the plants of Kerguelen's Land with those of Fuegia, separated by 140 degrees of longitude, rather than with those of Lord Auckland's group, which is nearer by about 50 degrees. But the features of the Flora of Kerguelen's Land are similar to, and many of the species identical with, those of the American continent, constraining me to follow the law of

DOI: 10.4324/9780429355653-87

botanical affinity in preference to that of geographical position. Two alternatives presented themselves, each possessing some advantages over the course which is now adopted, of dividing the Antarctic Botany into two large sections: one, to consider each little island or group as a separate flora; but this would lead to much repetition, and is not warranted by the amount of novelty exhibited in any of the groups: the other, to unite all under one head; a plan certainly accompanied with many advantages, but counterbalanced by the consequent delay of the work, for it would have obliged the author to study the plants of two very different botanical regions at the same time. The remarkable beauty and novelty of the vegetation in Lord Auckland's and the neighbouring Islands also merited particular consideration. As it is, some plants described in Part I. will re-appear in the present; very few, however; so few as to excite surprise, when it is remembered that lands, far more remote from Tierra del Fuego than those to the south of New Zealand, possess the characteristics of the Fuegian Flora.

A certain affinity in botanical productions has often been traced in widely severed countries, and Professor E. Forbes[1] has lately brought geological causes to bear immediately upon this subject. In reference to this curious topic I would adduce, as corroborative perhaps of his speculations, the general geographical arrangement of those islands, whose botany I am about to describe as that of one country. They stretch from Fuegia on the west, to Kerguelen's Land on the east, between the parallels 45° and 04° of south latitude. Throughout this portion of the world the land exhibits a manifest tendency eastward, from the extreme south of the American continent; for there are no fewer than five detached groups of islands between Fuegia and Kerguelen's Land, but none between the latter island and the longitude of Lord Auckland's group, nor between this last again and the western shores of Fuegia and Patagonia.

Tierra del Fuego and the neighbouring southern extremity of the American continent appear to be the region of whose botanical peculiarities all the other Antarctic Islands, except those in the vicinity of New Zealand, more or less evidently partake. It presents a Flora, characterizing isolated groups of islands extending for 5000 miles to the eastward of its own position; some of these detached spots are much closer to the African and Australian continents, whose vegetation they do not assume, than to the American; and they are all situated in latitudes and under circumstances eminently unfavourable to the migration of species, save that their position relatively to Fuegia is in the same direction as that of the violent and prevailing westerly winds.

Tierra del Fuego itself is a crowded archipelago, forming the southern extremity of America; it is of an irregularly four-sided figure, bounded on the north by the strait of Magalhaens, and on the east and west respectively by the South Atlantic and South Pacific Oceans, whilst its southern shores are washed by the Antarctic Sea; the main body of land lies between the 53rd and 56th parallels of latitude and the 64th and 70th degrees of west longitude, and its greatest extension is from east to west, indicated by a diagonal of 500 miles. The general appearance of the whole has been aptly compared, by Mr Darwin, to what would be presented by a partially submerged chain of mountains. These islands are, in fact, formed by the southern termination of the great Cordillera that traverses both Americas, which here trends to the eastward, and whose further extension is probably indicated by South Georgia in the same latitude; and possibly also by Prince Edward's Island, the Crozets, and Kerguelen's Land still more to the east, situated though these be in another ocean. The natural features of Fuegia have been admirably described by various voyagers, and more particularly by Cook, King, Fitzroy, and Darwin, to whose writings I would refer for more particular information. The exposed mountain-tops rise to a height of 7000 feet above the level of the sea, and the lower limit of perpetual snow is reckoned at 3500–4000 feet.

The botanical features exhibited by this country are not circumscribed by its geographical limits; along the north-east shores the very distinct Flora of East Patagonia accompanies the geological formation prolonged there from the Patagonian plains. On the south-west and south sides again, the vegetation is a continuation of that of West Patagonia, and is characteristic of the western flank of the Cordillera, from South Chili to Cape Horn. Thus it is that we find the Andes dividing two botanical regions from the North Polar almost to the Antarctic circle. The greater part of Fuegia is formed by the Andes alone; but the plants of the north-east portion, where the granitic formation of Patagonia introduces a change in the vegetation foreign to that of Tierra del Fuego, will be necessarily included in the present Flora.

The Deciduous Beech (*Fagus antarctica*), is the most distinguishing botanical production of this country. In company with the Evergreen Beech (*F. Forsteri*), it covers the land, especially on the west coasts, as far north as the Chonos Archipelago, in latitude 45° south. It is hardly seen in the north-east portions of Fuegia proper, northward of Staten Land, and though abundant on the west flanks of the Andes, through fourteen degrees of latitude, is unknown on the Atlantic side of Patagonia.[2] I have assumed therefore the shores of the strait of Magalhaens to be the northern limit of the Fuegian Flora eastward of Port Famine, and have included in, or rather added to that Flora, all the known plants of the Pacific side of the Andes, reaching north to the Chonos Archipelago. The latter position is peculiar, in the Beech being there replaced, at the level of the sea, with other trees; by the sudden change in the aspect of the coast vegetation that the flora of Chiloe, immediately to the northward, presents; and by its being only a few miles beyond the 'glacier-bound Gulf of Pehas', where perennial ice descends to the level of the ocean in a latitude nearly midway between the Equator and the Antarctic Pole.

The successive labours of Commerson, Banks and Solander, and of Menzies, early called the attention of Botanists to the singular aspect of the Fuegian Flora, apparently incompatible in its luxuriance with the rigour of the climate. The subsequent exertions of Captain King and Mr Anderson, and of Darwin, during the voyages of Captain Fitzroy, of D'Urville, and the officers of our own late Antarctic Expedition, have nearly exhausted the Phaenogamic productions. Much remains, however, to be done amongst the lower Orders, for the last-named expedition procured from a small island in the immediate vicinity of Cape Horn, more than twice as many Cryptogamic species as had been previously detected in the whole of Tierra del Fuego. These, however, hardly affect the general aspect of the vegetation, which may now be considered as satisfactorily known.

The Falkland Islands rank next in botanical importance to Fuegia. Though lying to the northward of the main body of that country, their vegetation is so influenced by climate and by some other peculiarities common to these islands and the Patagonian plains, that they produce no tree whatever. They are situated between the parallels of 51° and 53°, and the meridians of 57½° and 61½° west, and consist of an eastern and western island, nearly equal in size, and together forming an oval, whose axis lies east and west and extends about 160 miles. The general outline is jagged, like that of Fuegia, and similarly indented by deep inlets and ramifying bays; but their level or undulating surface, never rising above 2000 feet, and the geological formation, bear no resemblance to an archipelago formed by a submerged chain of mountains. Altogether, the Botanical and other characters of the Falklands are allied to the Atlantic coast of Patagonia, opposite to the strait of Magalhaens, from whence they are only 300 miles distant.

The most evident causes for the absence of trees in the Falkland Islands are the dislocation or removal of that group from the mainland; then a comparatively plane surface, everywhere exposed to the violence of the westerly gales, and more especially to the rapid evaporation and sudden changes in temperature and in other meteorological phenomena. The southerly and westerly winds are violent, cold, and often accompanied by heavy snowstorms; the easterly and northerly arrive saturated with warmer sea vapours, which, quickly condensing over the already chilled surface of the sod, form fogs and mists that intercept the sun's rays; whilst the north-westerly winds are singularly dry and parching, from the influence of the Patagonian plains over which they blow. Such sudden alternations from heat to cold, and from damp to dry, are particularly inimical to luxuriant vegetation, and no foliage but perhaps the coriaceous growth of Australia could endure them. The characteristics both of Fuegia and Patagonia may be seen mingled in the Falklands, and except *Veronica elliptica*, which is chiefly confined to the western coasts of the western island, the plants of both these countries appear together, overspreading the whole surface of the islands. Few species are peculiar, and no genus or order predominates to any remarkable extent, unless it be the *Gramineæ*: the species themselves are well marked and do not run much into varieties. Though the want of shade is unfavourable to the fruiting of Mosses and *Hepaticæ*, there are a considerable number of species of those orders, and some are identical with those of the American mountains and of Europe.

[. . .]

Considering the distance of the Falkland Islands from the continent, their size, the extent of surface covered with vegetation, and above all, their geological formation and the nature of their climate, the number of peculiar species is very insignificant; such circumstances generally accompanying or being indicative of a concomitant change in botanical features, specific difference itself being by some attributed wholly to the operation of these causes, and the immutability of species thence called in question. The Falkland Islands appear ill adapted to the more striking vegetation of Fuegia or of Patagonia, if we may judge from the absence of trees and even of such bushes as *Berberis*, *Uscallonia*, *Fuchsia*, *Ribes*, &c., which grow in the former country and to all of which the changeable nature of the climate is injurious; while, on the other hand, the mean temperature is too low for the *Leguminosœ*, *Malvaceœ*, and other predominant Orders of Patagonia. It is more remarkable that some of the plants of each are seen, composing together the whole vegetation, yet appearing unchanged by a climate that is certainly unfavourable to the general flora of those distant regions where these very species most abound. To conclude by an example, *Sisyrinchium* and *Oxalis enneaphylla* will not associate themselves with the *Tussac* and *Empetrum* in Cape Horn, nor are *Astelia* and *Caltha appendiculata* to be found in company with *Nassauvia* and *Calceolaria Fothergillii* on the coast of Patagonia, though all these may be seen growing side by side in the Falklands in the greatest profusion.

Immediately to the south of Cape Horn are groups of islands, and possibly a larger body of land. Vegetation in the Southern Hemisphere reaches the northern shores of these inhospitable spots, where, at a distance of no less than thirty-six degrees from the actual Pole and three degrees to the northward of the Antarctic circle, the flora of the south finds its extreme limit.

[. . .]

In January 1843 I landed upon a small islet, close to the main portion of Palmer's Land, in latitude 64° 12' south, and longitude 57° west. It appeared to be the 'ultima Thule' of southern vegetation; the soil hard frozen, except on the very surface where it was thawed by a sun-heat which raised the temperature to 46°, while the sea was encumbered with pack-ice and bergs; no flowering plants were to be seen, and only eighteen belonging to the Orders Lichenes, Musci, and Alga. Beyond this latitude I believe there is no terrestrial vegetation.

The South Georgian group is situated about 1000 miles due east of Cape Horn, and exhibits a wholly different aspect from that land, being covered with perennial snows, and the harbours blocked up with everlasting glaciers; still, Captain Cook found a scanty vegetation, consisting of 'a coarse strong-bladed grass, growing in tufts, wild Burnet, and a plant like moss, which springs from the rocks'; (vide Cook's 2nd voyage). The flora of South Georgia is probably intermediate in luxuriance (if such term may be used), between the Falklands and the South Shetlands, the proximity of the Antarctic Ice being influenced by that of the large bodies of land, it approaches nearer to South Georgia than to Fuegia, and renders that climate unsuited to support even a moderate vegetation.

Sandwich Land, discovered by Captain Cook, lies further south than South Georgia, and, like Palmer's Land, is encroached upon by the perennial ice of the Atlantic Ocean. That illustrious navigator mentions two hills clear of snow, and apparently covered with a green turf, but this is all we know of their productions.

Proceeding westward from Antarctic America, the next island that requires notice, as exhibiting an Antarctic vegetation, is Tristan d'Acunha. Though only 1000 miles distant from the Cape of Good Hope, and 3000 from the Strait of

Magalhaens, the Botany of this island is far more intimately allied to that of Fuegia than Africa. Captain Carmichael's list (Linn. Trans., vol. xii. p. 483), contains twenty-eight flowering plants (I exclude *Sonchus oleraceus*); only one species of *Phylica*, and one *Pelargonium*, amounting to one-fourteenth of the whole, are Cape forms; whilst seven others, or one-fourth of the flora, are either natives of Fuegia or typical of South American Botany, and the Ferns and *Lycopodia* exhibit a still stronger affinity. There are some points in which the vegetation of Tristan d'Acunha resembles that of St. Helena and Ascension. Though these islands are separated from one another by nearly thirty degrees of latitude, they lie within eight degrees of longitude, and all are the exposed summits of ancient volcanoes, such as the highest peaks of the Andes might present, if that mighty chain were partially submerged.[3] The relation between the floras of Ascension and St. Helena[4] is evident, though to enumerate them would be out of place here; those between the latter island and Tristan d'Acunha are indicated by the genera *Phylica* and *Geranium*, and also by some of the Ferns and *Lycopodia*: as, however, it is also through those genera that the botany of Tristan d'Acunha resembles that of the Cape, it may fairly be doubted whether the apparent affinity with St. Helena is not imaginary. It is a very remarkable circumstance that while these three islands all possess some of the features of the African Flora, the predominant ones are absent; thus, whilst the St. Helena Flora is allied, and exclusively so, to that of the Cape in *Geranium*, *Melhania*, and *Phylica*, it has no representatives of entire Orders, namely *Proteaceæ*, *Rutaceæ*, *Oxalideæ*, *Crassulaceæ*, *Ericeæ*, *Restiaceæ*, and many others, far more characteristic of the African vegetation than are any of the plants inhabiting St. Helena.

The other islands whose plants will find a place in this division of the 'Antarctic Flora' are situated south of the Indian continent, widely apart from the American, and so far as geographical position is concerned, belong to Africa or India; these are, Prince Edward's and Marion Islands, the Crozets, Kerguelen's Land, and the Islands of Amsterdam and St. Paul.

Of the two first-mentioned groups the vegetation is wholly unknown; the former, Prince Edward's and Marion, are small contiguous islets in the 47th degree of latitude and 38th of east longitude; they are of rather an undulating outline and evidently volcanic formation, from a little distance they appeared covered with grass. The Crozets are a group of much larger islands, situated in the 48th degree of latitude and between the 47th and 49th meridian, east of London: they are bold rocky masses, rising to a height of 6000 feet; some, though of considerable size, are quite inaccessible, and others enveloped by eternal fogs, whence the name of Hazy Island has been given to one of the largest, of which the rocky summit alone is seen standing out in bold relief above an almost perennial fog-bank. During our passage from the Cape of Good Hope to Kerguelen's Land, Sir James Ross endeavoured to effect a landing, first upon Marion Island and afterwards upon one of the Crozets, but most unfortunately for the interests especially of Botany, our efforts were frustrated by the tempestuous weather. In one night, during which the 'Erebus' was hove to for the purpose of landing upon Marion Island, she was blown sixty miles to leeward of it; she then bore up for the Crozets, to meet a similar mishap; on this occasion, having provisions to land for a party of miserable sealers, we again beat up to Possession Island, the easternmost of the group, and after the detention of nearly a week in the most inclement season and tempestuous ocean, only arrived at the time of the brooding of another storm, which rendered it highly imprudent for any boat to leave the ship in an open roadstead. The aspect of this island was, like all the others we sighted, dreary and inhospitable to the last degree; a narrow belt of green herbage skirted its shore, above a line of black basaltic cliffs, which formed the iron-bound coast; while higher again rose crater-shaped barren hills of blue-grey or brick-red coloured rocks, utterly destitute of vegetation and alike dismal to the eye and mind. These were the first Antarctic Islands we had seen, and few of us will forget the feelings to which their desolate aspect gave rise; sensations, which for intensity afford the strongest contrast with those which an English naturalist never fails to experience during his first ramble on some tropical shore.

M. de Jussieu had the kindness to show me a small pamphlet, containing a slight account of the Crozets, drawn up from information received through the captains of sealing ships. The vegetation is described as most scanty. From the short interview which we held with a party of sealers who had been left upon one of the group, I gleaned but little information; they told me the species were few, and the famous Cabbage of Kerguelen's Land not amongst them, though another 'scurvy-grass' was abundant. The vegetation that our glasses enabled us to detect, formed, apparently, a matted carpet, extending from the shores upwards for a short distance, very similar to what we afterwards saw in Kerguelen's Land, though different from the long grass that appeared to clothe Prince Edward's Island. These two groups are situated only 500 miles south-east from the Cape of Good Hope, but being placed to the southward of the 40th degree of latitude they partake of the climate of the Antarctic Ocean. Their position between Fuegia and Kerguelen's Land and their formation being probably the same as the latter, I have little doubt their Flora, when known, will be found to prove characteristic of the extreme south of America and in no degree similar to that of Africa, with which they are even in closer proximity than is Tristan d'Acunha. Barren and inhospitable as are the shores of these islands, there are no spots on the surface of the globe whose botanical productions would be of greater interest to science, for their vegetation is wholly unknown, and is wanting to complete our otherwise pretty extensive acquaintance with the distribution of plants throughout the islands of the high southern latitudes.

Kerguelen's Land is the eastern limit to which the Fuegian Flora extends, and though placed within the 50th degree its desolate nature is proverbial. The Antarctic Expedition arrived there in May 1840, having been blown off its tempestuous coast twice, after approaching the land so nearly as to distinguish almost the nature of the vegetation which skirts the shores of the bays. The island presents a black and rugged mass of sterile mountains, rising by parallel steppes one above another in alternate slopes and precipices, terminating in frightful naked and frowning cliffs, which dip perpendicularly into the sea. The snow lying upon these slopes between the black cliffs gave a most singularly striped or banded appearance to the whole country, each band indicating a flow of volcanic matter, for the island is covered with, craters whose vents have given issue to stream upon stream of molten rock. These are worn all along the coasts into abrupt escarpments, rendering a landing impracticable, except at the heads of the sinuous bays. One bluff headland to the north end of the island is a precipice, 700 feet high, and exposes such numerous sections of horizontal deposits of red, black, and grey volcanic matter that it is difficult to count them, though overlaying one another with perfect regularity and uniformity. Sterile as Kerguelen's Land now is, it was not always so, vast beds of coal are covered by hundreds of consecutive layers of igneous and other rocks, piled to a height of one thousand feet and upwards, upon what was once a luxuriant forest. Throughout many of the lava streams are found prostrate trunks of fossil trees of no mean girth, and the incinerated remains of recent ones, which had been swallowed up simultaneously with the fossil, and these occur in strata of various ages, so that it seems impossible to reckon the period of time that must have elapsed between the origin, growth, and destruction of the successive forests now buried in one hill. A section of such a hill would display coal-beds and shale resting upon a blue basalt, at the level of the sea, covered again with whinstone, whereon are deposited successive layers of volcanic sand, baked clay-stones, porphyries, and long hues of basaltic cliffs, formed of perpendicular prisms, regularly shaped like those of Staffa or the Giant's Causeway, and along which the traveller may walk even for a mile without ascending or descending fifty feet. To calculate the time required for the original formation and following silicification of one such forest, and to multiply that by the equal number of different superincumbent strata, containing remains similar to those displayed at the north end of Kerguelen's Land, would give a startling number of years, during which periods the island must have deserved a better name than that of 'Desolation'. And if to this be added the time requisite for the deposit of the arenaceous beds containing the impressions of *Fuci*, of the clays afterwards hardened by fire, and of the prismatic cliffs, which, with the arenaceous, indicate that the land was alternately submerged and exposed as often as these successive formations occur, such a sum would bespeak an antiquity for the flora of this isolated speck on the surface of our globe far beyond our powers of calculation. If from the narrow sphere of inquiry that a few miles in extent and 1000 feet of elevation in Kerguelen's Land afford we deduce such grand results, what must be expected from the investigation of whole continents, whose culminant peaks reach nearly 30,000 feet, surrounded by an ocean perhaps as elevated above the land it rests upon, and presenting fossiliferous strata that we believe are deposited at even greater depths? On the other hand, referring to the island under consideration, as it now appears, we may regard it as the remains of some far more extended body of land. Position in longitude in the Southern Hemisphere appears to determine the amount of vegetation an island may possess. Of this we have an instance in

South Georgia, and the reason is evident; the extension of the great continents is in longitude, and the climate and other features of the islands depend upon their proximity to the land, which modifies the desolating influence of the icy ocean. The time we have granted for the formation of the various strata composing Kerguelen's Land and the forests that successively decorated them, is sufficient for the destruction of a large body of land to the northward of it, of which St. Paul's Island and Amsterdam Island may be the only remains, or for the subsidence of a chain of mountains running east and west, of which Prince Edward's Island, Marion, and the Crozets are the exposed peaks [. . .]

Notes

1 Professor E. Forbes has connected the similarity, long known to exist between the Floras of the west of Ireland and Portugal, with certain geological characteristics belonging to both these now remote, but perhaps once united countries. Thus he also connects the Alpine Flora of Scotland with that of the Scandinavian Alps, and the botany of the Channel coasts and islands with that of France (vid. 'Report of the Meetings of the British Association in Cambridge, July 1845'). Uniformity of surface is generally accompanied by a similarity of vegetation throughout an extended region. When such a surface becomes divided we are apt to conclude that the isolation of the lesser portion preceded the migration of plants from the larger; in short, that the identity of the Norfolk and Suffolk Flora with that of Holland must be due to the former having been peopled with plants by the latter, subsequently to the German Ocean having assumed its present position; and not that the two together formed an equally well clothed and extended plain, reaching, as Humboldt believes, from North Brabant to the Steppes of Asia; its western portion having been afterwards inundated by the influx of the North Sea. The uniformity of surface in the vast continent of Africa is becoming daily more evident, as the mountains of the moon recede before the intrepid explorers of the sources of the true Nile. It were natural to suppose that a barrier, such as they were conjectured to be, would exhibit changes in the vegetation, equally marked with those produced by the Cordillera, Himalayan, and other mountain chains wherever they may occur. A further proof of the suspicious nature of the reports that any very extensive and elevated land exists in Africa appears to me evident in the character of Abyssinian vegetation [. . .] Central Eastern Africa is perhaps the most interesting spot in the world for a botanist; it contains not merely Cape orders, but others typical of Madagascar, the East Indies, Arabia, both the northern and western coasts of Africa itself, and on its high mountains those even of Europe. The uniformity of the surface and Flora of Australia is equally evident.

[. . .] I have received from Baron Humboldt much highly interesting verbal information upon the distribution of organized beings in Siberia; the disappearance of some animals and plants over a vast area, and their re-appearance in another, in obedience to no known law, are very striking facts [. . .]

Many striking examples on the other hand may be instanced, of countries closely approximated in geographical position, but unlike in geological and other features, presenting widely different botanical aspects; such sudden changes in the vegetation we may observe on the east and west flanks of the Andes and on the Himalayan; in the Floras of St Helena and Ascension, and the coast of Africa; or of Tristan d'Acunha and the Cape; of New Zealand and Australia; of Juan Fernandez and the Galapagos and the coast of America; of Madagascar and South Africa; but more especially in the disparity that prevails between the Floras of the separate islands of the Galapagos and of the Sandwich group.

2 Trees allied to these seem to have characterized the ancient or fossil flora of Fuegia, for I owe to Mr Darwin's kindness impressions of the leaves of three apparently distinct species of deciduous Beech, and which are mentioned in that gentleman's journal.

3 The conjecture of these islands being the exposed culminating peaks of a submerged chain of mountains, receives confirmation from the circumstance of Sir James Ross having struck soundings in 16,002 feet in lat. 33° 21' south, and long. 9° 4' west, that is, nearly on a line between St. Helena and Tristan d'Acunha.

4 The island of St. Helena has many claims to rank as one of the most interesting botanical stations known; almost the whole of its native flowering plants and several of its genera being peculiar. Various causes have, within the memory of man, reduced this flora to a mere shadow of what it once was, for when the island was discovered, it is described as entirely clothed with forest. The greater part of this was said to be destroyed by the introduction of goats and pigs, and by the bark of the trees being stripped for tanning, so that the flora is consequently now very limited both in number of species and of individuals. During the interval that elapsed between two visits which I paid to St. Helena, one very peculiar native plant, the *Acalypha rubra*, had disappeared, and two other handsome shrubby species of *Melhania*, with particularly showy flowers, had very recently become extinct; whilst the existence of some *Wahlenbergiæ*, of a *Physalis*, and a few of the peculiar arborescent *Compositæ*, though thus far prolonged, is held upon a very precarious tenure. These plants are all well marked species, which on the destruction of the forests seem unable to accommodate themselves to their altered circumstances, perish, and are replaced by introduced species, exactly as is the case with various savage races of mankind, which do not suit themselves to the condition of the soil when altered by the European settler, but diminish in number and dwindle away even when violent measures have not been used for their extirpation. I may remark, that species in isolated islands are generally well defined; this is in part the natural consequence of another law which I have observed, that genera in islands bear a large proportion to the species, or in other words, that genera are small, seldom containing more than two or three species, and very frequently solitary representatives. It must be borne in mind that this well-marked character of the species in insular localities applies equally to mountainous as to planar islands. It might seem natural to suppose that a varied surface would have the effect of obliterating specific distinction, especially in small areas, as the Pacific Islands, the Galapagos, St. Helena, and the like, whose present contour is not the result of recent geological changes, and where time, the required element for developing such species as are the offspring of variation, has been granted.

79

ALEXANDER VON HUMBOLDT, 'THE GEOGRAPHY OF PLANTS AND ANIMALS', *COSMOS*

(London: Longman, Brown, Green and Longman and John Murray, 1849) [1845])

HAVING now passed through the entire circle of terrestrial inorganic nature, – having considered our planet in respect to its form, its internal heat, its electro-magnetic charge, its polar luminous effusions, the reaction of its interior on its variously composed crust, and finally the phænomena of its oceanic and atmospheric envelopes, – the view which we have essayed to trace in broad and general outlines might be regarded as complete, and would be so according to the limitation formerly adopted in physical descriptions of the globe. But the plan which I have proposed to myself has a more elevated aim, and I should regard the contemplation of nature as deprived of its most attractive feature, were it not also to include the sphere of organic life with its many gradations of development. The idea of life is so intimately connected with the moving, combining, forming, and decomposing forces which are incessantly in action in the globe itself, that the oldest mythical representations of many nations ascribe to these forces the production of plants and animals, and represent the epoch in which the surface of our planet was unenlivened by animated forms, as that of a primeval chaos of conflicting elements. But investigations into primary causes, or into the mysterious unresolvable problems of origin, do not enter into the domain of experience and observation; nor has the obscure commencement of the history of organisation a place in the description of the actual condition of our planet. These reservations once made, it should still be noticed in the physical description of the world, that all those substances which compose the organic forms of plants and animals are also found in the inorganic crust of the earth; and that the same powers which govern inorganic matter are seen to prevail in organic beings likewise, combining and decomposing the various substances, regulating the forms and properties of organic tissues, but acting in these cases under conditions yet unexplained, to which the vague term of 'vital phænomena' has been assigned, and which have been systematically grouped according to analogies more or less happily imagined. Hence has arisen a tendency of the mind to trace the action of physical forces to their extremest

DOI: 10.4324/9780429355653-88

limits in the development of vegetable forms, and of those organisms which are endowed with powers of voluntary motion: and here, also, the contemplation of inorganic nature becomes connected with the distribution of organic beings over the surface of the globe, *i.e.* the *geography of plants and animals*.

Without attempting to enter on the difficult question of 'spontaneous motion', or the difference between vegetable and animal life, it may be remarked that if nature had endowed us with a microscopic power of vision, and if the integuments of plants had been perfectly transparent, the vegetable kingdom would be far from presenting to us that aspect of immobility and repose which our perceptions now ascribe to it. The internal parts of the cellular structure are incessantly animated by the most various currents: ascending and descending, rotating, ramifying, and continually changing their direction, they manifest themselves by the movements of a granular mucilaginous fluid in water plants (naiades, characeæ, hydrocharideæ), and in the hairs of phenogamous land plants. Such is the peculiar molecular movement discovered by the great botanist Robert Brown (which is indeed perceptible, not only in vegetables, but also in all matter reduced to an extreme state of division); such is the gyratory current (cyclose) of globules of cambium; and, lastly, such are the articulated filamentary cells which unroll themselves in the antherides of the chara, and in the reproductive organs of liverworts and algæ, and in which Meyen, too early lost to science, believed that he recognised an analogy to the spermatozoa of the animal kingdom. If we add to these various currents and molecular agitations the phænomena of endosmose, the processes of nutrition and of growth, and internal currents of air or gases, we shall have some idea of the powers which, almost unknown to us, are incessantly in action in the apparently still life of the vegetable kingdom.

Since the time when, in an earlier work ('Ansichten der Natur', 'Tableaux de la Nature'), I attempted to describe the universal diffusion of organic life on the surface of the globe, and its distribution in height and in depth, our knowledge has been wonderfully augmented by Ehrenberg's brilliant discoveries ('über das Verhalten des kleinsten Lebens in dem Weltmeere wie in dem Eise der Polarländer'), which rest not on ingenious combinations and inferences, but on direct and exact observation. By these discoveries the sphere of animated existence – we may say the horizon of life – has expanded before our view. 'Not only is there no interruption of minute microscopic forms of animal life in the vicinity of either Pole where larger animals cannot maintain themselves, but we find among the microscopic animals of the South Polar Seas, collected in the Antarctic Expedition of Captain James Ross, a remarkable abundance of new forms, which are often of great elegance. Even in the residuum obtained from melted ice which floats in rounded fragments in lat. 78° 10' S., there have been found above fifty species of siliceous-shelled polygastrica, and even coscinodiscæ with green ovaries, which were therefore certainly living and able to resist the extreme severity of the cold. In the Gulf of the Erebus and Terror, sixty-eight siliceous-shelled polygastrica and phytolitharia, together with a single calcareous-shelled polythalamia, were brought up by the lead from depths of 1242 to 1620 English feet'.

By far the greater number of the oceanic microscopic forms hitherto observed belong to the siliceous-shelled infusoria, although the chemical analysis of sea-water has not shewn silica to be one of its essential constituents, and it could only indeed exist in water in a state of simple mixture or suspension. It is not only in particular localities, in inland waters or in the vicinity of coasts, that the ocean is thus thickly peopled with living atoms invisible to the naked eye. Samples of water taken up by Schayer in 57° S. lat., on his return from Van Diemen Island, as well as those taken between the tropics in the middle of the Atlantic, shew that the ocean water in its ordinary condition, without any appearance of discoloration, contains innumerable microscopic organisms, quite distinct from the siliceous filaments of the genus Chætoceros, floating in a fragmentary state like the oscillatoria of our fresh waters. Some polygastrica which have been found mixed with sand and excrements of penguins in the Cockburn Islands, appear to be generally distributed over the globe; other species belong to both the arctic and antarctic polar regions. Thus we see that animal life reigns in the perpetual night of the depths of the ocean, while on continents, vegetable life, stimulated by the periodical action of the solar rays, chiefly predominates. The mass of vegetation on the earth very far exceeds that of the animal creation; for what, in point of bulk, would be an assemblage of all the great cetaceæ and pachydermata living at one time, compared to the thickly-crowded colossal trunks of trees of 8 and 12 feet in diameter from the tropical forests which cover only one region of the earth, namely, that comprised between the Orinoco, the Amazons, and the Rio da Madeira? If the characteristic aspect of different portions of the earth's surface depend conjointly on all external phænomena, – if the contours of the mountains, the physiognomy of plants and animals, the azure of the sky, the form of the clouds, and the transparency of the atmosphere, – all combine in forming that general impression which is the result of the whole, yet it cannot be denied that the vegetable covering with which the earth is adorned is the principal element in the impression. Animal forms are inferior in mass, and their individual power of motion often withdraws them from our sight; vegetable forms, on the contrary, produce a greater effect by reason of their amplitude and of their constant presence. The age of trees is announced by their magnitude, and the union of age with the manifestation of constantly renewed vigour, – the ancient trunk with the fresh verdure of spring, – is a charm peculiar to the vegetable creation. In the animal kingdom (and this knowledge is also a result of Ehrenberg's discoveries) it is precisely the minutest forms which, owing to their prodigious fecundity, occupy the greatest space. The minutest infusoria (the Monadines) only attain a diameter of 1/3000th of a French line, and yet these siliceous-shelled animalcula form in humid districts subterranean strata of many fathoms in thickness.

80

WILHELM WITTICH, 'THE GULF STREAM', *CURIOSITIES OF PHYSICAL GEOGRAPHY*

(London: Charles Knight & Co., 1845)

Section VI, The Gulf Stream

THE surface of our globe is composed of land and water. By far the greater part is covered with water, as, according to a rough calculation, the seas extend over about three-fourths of it, the remaining fourth only being occupied by land. The water of the sea is salt, less so towards the poles, and more so towards the equator. This is, with some reason, considered as a wise arrangement of Providence; for if this enormous body of water were ever to fall into a state of stagnancy, it would, if consisting of sweet water, soon become putrid, the more so as it is well known that numerous and large quantities of lifeless vegetable and animal matter are found in the sea, which are in a continual state of decomposition. The noxious exhalations which in such a case would arise from this stagnant body of water would speedily infect the whole atmosphere, which would thus become the source of innumerable diseases for all living creatures. It is to be apprehended that the land would become uninhabitable for man and beast, and be converted into a desert. Though it cannot be denied that this effect is fully prevented from taking place by the saltness of the sea-water, it appears in this, as in many other cases, that nature is gifted with other means tending to produce the same effect. Even if the sea-water was sweet there could hardly be any reason for apprehending such noxious effects, when it is considered that perhaps no portion of the ocean is for any length of time in a stagnant state, and that nearly the whole of its surface is almost always in motion.

The movements to which the ocean is subject originate in different phenomena. Those which are produced by the tides are less extensive than the others. By the tides the surface of the sea is indeed alternately raised by some feet, and let down again; but it can hardly be asserted that by this remarkable operation of nature any portion of the sea is forced from its situation and compelled to change its place, except where the sea borders on land, and even there this

DOI: 10.4324/9780429355653-89

effect is circumscribed by comparatively narrow limits. The winds have a much greater power to displace the water of the surface of the sea. They propel the upper stratum in the form of waves, and wherever they continue to blow from the same quarter for any length of time with some force, the whole surface is found to be in motion to the leeward. This is especially observable in those parts of the ocean which are subject to perpetual or periodical winds. The whole surface of the oceanic regions which are swayed by the perpetual or trade winds moves without interruption to the westward, at the rate of about twelve miles every day. This movement is still greater in those parts of the sea where the monsoons and periodical winds prevail, because these winds blow in general with greater force than the trade-winds; but in these regions the motion of the sea changes, of course, with the change of the monsoons. In those parts of the ocean where the winds frequently change the quarter from which they blow, and as frequently, if not more frequently, increase or diminish in strength, the motion of the surface-stratum becomes as it were neutralized by these changes, and is rarely of such a description as to be perceptible. The motion of the surface-stratum of the sea caused by the pressure of the winds goes by the name of *drift*. It does not appear that this effect of the winds extends to a great depth under the surface. For, according to the most authentic accounts, the waves, during the greatest gales, do not exceed sixty feet in a perpendicular line in height, and as by experience it is known that the effects even of a heavy gale are not felt at a depth of more than twenty feet under the hollow of the wave, we may presume that the motion of the drift does never extend to a depth of a hundred feet under the crest of the waves.

There is, however, still another movement met with in certain parts of the ocean, which certainly involves a much larger body of water, and which apparently extends to the very bottom of the sea. It is known by the name of *current,* or *stream*. These currents are frequently compared with the larger rivers of the continent, but this comparison holds good only so far as both the currents and the rivers are volumes of water moving in a certain direction; but if the width and the depth of the former, and the immense body of moving water in them, are considered, it can hardly be said that there exists between them and the rivers any point of comparison. The direction of most of the currents is not dependent on winds, and they are found to run as frequently north and south as east and west. In running from south to north or from north to south, it happens that they bring the warm water of the intertropical seas to the higher latitudes, or the cold water from the vicinity of the poles to the regions near the equator. Therefore it is found that nearly all currents following these directions are distinguished from the surrounding seas by the temperature of their waters being either higher or lower than that of the seas contiguous to their course.

It is not more than about seventy years that these currents have attracted the attention of navigators, and that information has been collected respecting their direction, velocity, and extent, as well as respecting the temperature of their waters. Many of them are still imperfectly known, though it is obvious that an exact knowledge of them would be very desirable, as they affect, to a considerable extent, the navigation of the seas through which they run.

The best known of these currents is the *Gulf Stream*. This is a vast body of water running from the coasts of North America towards Europe, and bringing the warm water of the Gulf of Mexico to the central regions of the North Atlantic. This current has been more carefully examined, because it traverses that part of the Atlantic which is more navigated than any other part of the ocean, being besides distinguished by its velocity, its great width, and the high degree of temperature of its waters.

The Gulf Stream originates in the Gulf of Mexico, whence also its name is derived. It is properly the efflux of that sea. Its upper course describes a large segment of a circle, and lies between Florida Reef and the peninsula of the same name on one side, and the island of Cuba, the Sal Key Bank, and the Bahama Banks on the other side. With a northern course it reaches the open ocean near 28° N. lat. It continues to move in the same direction, and at no great distance from the shores of the United States, until it arrives at Cape Hatteras, north of 35° N. lat. In this its upper course the body of water in motion occupies a space varying between thirty-six and seventy-five miles, and may in this respect be compared with the wider western part of the British Channel, or with that portion of it which is west of the Bill of Portland. From Cape Hatteras the Gulf Stream runs eastward into the Atlantic, and gradually expands to two hundred miles and more in width. It continues to flow in that direction until it reaches 40° W. long., when it turns gradually south-east and south. In approaching the islands of Corvo and Flores, which are the most western of the Azores, its current becomes weak, and at the south-west of these islands it is lost, and the warm water of the stream is dispersed over the sea. This is commonly the place where the Gulf Stream terminates. But there are instances on record of the warm water of the Gulf Stream sometimes advancing as far eastward as the western coast of Portugal and Spain. The space which the Gulf Stream traverses between its efflux from the Gulf of Mexico to the islands of Corvo and Flores exceeds three thousand miles. When its water reaches the coast of Portugal and Spain it runs about a thousand miles more. These outlines show that the Gulf Stream is a phenomenon of first-rate magnitude, which deserves a more detailed description.

In its upper course, between the place where it issues from the Gulf of Mexico and that where it enters the Atlantic, the Gulf Stream may in some respects be considered as a vast river. No other current of the globe, at least like this part of the Gulf Stream, runs between banks so well defined on either side. On one side these banks are formed by the Florida Reef and the peninsula of that name, and on the other by the island of Cuba, the Sal Key Bank, and the Great and Little Bahama Banks. These banks are not *banks of sounding,* a term applied to those parts of the sea whose bottom is within the reach of the common sounding-line, which is a hundred fathoms long. The Sal Key Bank and the Bahama Banks are *proper banks*. They rise not gradually, but suddenly, and from a very deep sea. The surface, near the edges of the banks, is but a few fathoms below the common level of the sea, and towards the interior many places are found which are only a few feet under water; some are quite dry at low tide. Besides, at many places along their edges these banks are beset with rocks even with the water's edge or above its level. It is obvious, from this description, that such banks must influence the current precisely as if it were running between banks of land. Accordingly we find that the Gulf Stream, contrary to the nature of other currents, is compelled by these banks to bend in the form of a large segment of a circle. Hemmed in on either side, its waters are not only prevented from spreading, but, as these banks draw closer together as the stream proceeds farther on, the great volume of water is gradually compressed into a much narrower space, by which its velocity is so increased, that before it reaches the Atlantic it has acquired a rapidity which is much superior to that of navigable rivers, and resembles that of a mountain stream.

81

JOHN GOULD, *AN INTRODUCTION TO THE BIRDS OF AUSTRALIA*

(London: Richard and John E. Taylor, 1848)

Introduction

GEOLOGICAL investigations into the structure of the globe show that a succession of physical changes have modified its surface from the earliest period up to the present time, and that these changes have been accompanied with variations not only in the phases of animal and vegetable life, but often in the development also of organization; and as these changes cannot be supposed to have been operating uniformly over the entire surface of the globe in the same periods of time, we should naturally be prepared for finding the now existing fauna of some regions exhibiting a higher state of development than that of others; accordingly, if we contrast the fauna of the old continents of geographers with the zoology of Australia and New Zealand, we find a wide difference in the degree of organization which creation has reached in these respective regions. In New Zealand, with the exception of a *Vespertilio* and a *Mus,* which latter is said to exist there, but which has not yet been sent to this country, the most highly organized animal yet discovered, either fossil or recent, is a bird; in Australia, if compared with New Zealand, creation appears to have considerably advanced, but even here the order *Rodentia* is the highest in the scale of its indigenous animal productions; the great majority of its quadrupeds being the *Marsupiata* (Kangaroos, &c.) and the *Monotremata* (*Echidna* and *Ornithorhynchus*), which are the very lowest of the Mammalia. The ornithology of Australia is characterized by the presence of certain peculiar genera, the *Talegalla, Leipoa* and *Megapodius*; birds which do not incubate their own eggs, and which are perhaps the lowest representatives of their class, while the low organization of its botany is indicated by the remarkable absence of fruit-bearing trees, the *Cerealia,* &c.

My investigation of the natural productions of Australia induces me to believe, that at some remote period it was divided into at least two portions, since, with a few exceptions, I find the species inhabiting the same latitudes of its eastern and western divisions differing from, but representing each other. Some writers, Captain Sturt and Mr. Jukes, *e.g.* are of opinion that its subdivision was even greater,

DOI: 10.4324/9780429355653-90

and that the sandy deserts now met with in the interior were formerly the beds of the seas that flowed between the archipelago of islands of which they suppose it to have been composed. In a valuable paper by Mr. Jukes, entitled 'Notes on the Geology of the Coasts of Australia', read at the meeting of the Geological Society on the 17th of November 1847, that gentleman stated, that 'The eastern coast is occupied by a great range of high land, appearing like a continuous chain of mountains when seen from the sea, and rising in several places to 5000 feet or more above the sea-level. This chain has an axis of granite, with occasional large masses of greenstone, basalt and other igneous rocks. It is flanked on both sides by thick beds of palæozoic formations, chiefly sandstone, but also containing limestone and coal. In the northern portion of the chain Dr. Leichardt found similar formations – and especially trap and granite near the Burdekin river. In the Port Philip district there are similar igneous rocks, and on the coast tertiary formations resting on the edges of upturned palæozoic beds. In West Australia, the Darling range consists of granite below, covered by metamorphic rocks; and between it and the sea is a plain composed of tertiary beds. In the colony of North Australia there is a great sandstone plateau, rising about 1800 feet above the sea, and probably of palæozoic age; whilst on the immediate shore and round the Gulf of Carpentaria are beds supposed to belong to the tertiary period. Similar formations constitute the substratum of the central desert; in which Captain Sturt was compelled to turn, when half-way to the Gulf of Carpentaria, from the southern coast. Hence these tertiary rocks are probably continuous through the whole centre of the island, and during the tertiary period all this portion of the country was submerged, whilst the high lands on the coast rose like four groups of islands from the shallow sea'. – *Athenæum, Nov*. 24, 1847.

Whichever of these opinions be the correct one, we certainly find the natural productions of all these portions of the country composed of precisely the same types, the generality of which differ entirely from those of the islands of the Indian Archipelago on the one hand, and of New Zealand and Polynesia on the other.

With respect to the position of Australia, it will only be necessary to state that it is situated between the 10th and 45th degrees of south latitude, and the 112th and 154th degrees of longitude east from Greenwich; its extent, in round numbers, may therefore be said to be 3000 miles in length, or from west to east, and inclusive of Van Diemen's Land nearly the same in breadth, or from north to south. In its present uplifted position its form is nearly square, with a depressed centre bounded by an almost continuous range of hills and plateaux, which, varying in altitude from one to six thousand feet above the level of the sea, in some places approach the coast and present lofty and inaccessible cliffs to the ocean, while in others they trend towards the interior of the country at a distance of from twenty to eighty miles from the coast-line; but inasmuch as these elevations are all of an undulating and not of a precipitous character, no part of the country can be considered as strictly alpine. Nothing can be more different than the features of the country on the exterior and interior of this great barrier, particularly on the eastern coast, where, between the mountains and the sea, the vegetation partakes to a great extent of a tropical character; it is there, on the rich alluvial soil, formed by the debris washed down from the hills, that we find various species of *Eucalypti, Fici,* and other trees, many of which attain an immense altitude, and forests of towering palms; the surface of the ground beneath clothed with a dense and impervious underwood, composed of dwarf trees, shrubs and tree-ferns festooned with creepers and parasitic plants in the richest profusion, the continuity of which is here and there broken by rich open meadow-like districts admirably adapted for the pasturing of cattle, and to which, from the frequent occurrence of the *Angophorœ,* a tribe of trees in which the settlers see a fancied resemblance to the apple-trees of Europe, the name of Apple-tree Flats has been given. Within the ranges, on the other hand, we find immense open downs and grassy plains, studded here and there with detached belts and forests of *Eucalypti, Acaciœ,* &c., presenting a park-like appearance, to which, as we advance farther towards the interior, succeed either extensive marshes or land of a most sterile description. The face of this vast country consequently presents much variety of aspect; the infrequency of rain tends much to give a sombre brown hue to the surface of the interior, which however is relieved by the constant verdure of its trees, the peculiar lanceolate form and the pendent position of which render them almost shadowless. It is in the neighbourhood of the few rivers which intersect the country, and in the lower flats flooded by the waters, when floods occur, that we find the vegetation more luxuriant and the trees attaining a far greater size; the sides of the rivers are moreover fringed with *Casuarinœ* and other trees, which, although of large size, never arrive at the altitude of the stately *Eucalypti,* which attain, under favourable circumstances, a size and height which appear perfectly incredible. Mr. Backhouse states that one measured by him on the Lopham Road, near Emu Bay in Van Diemen's Land, which 'was rather hollow at the bottom and broken at the top, was 49 feet round at about 5 feet from the ground; another that was solid, and supposed to be 200 feet high, was 41 feet round; and a third, supposed to be 250 feet high, was 55½ feet round. As this tree spread much at the base, it would be nearly 70 feet in circumference at the surface of the ground. My companions spoke to each other when at the opposite side of this tree from myself, and their voices sounded so distant that I concluded they had inadvertently left me, to see some other object, and immediately called to them. They in answer remarked the distant sound of my voice, and inquired if I were behind the tree! When the road through this forest was forming, a man who had only about two hundred yards to go, from one company of work-people to another, lost himself: he called, and was repeatedly answered; but getting further astray, his voice became more indistinct, till it ceased to be heard, and he perished. The largest trees do not always carry up their width in proportion to their height, but many that are mere spars are 200 feet high'.

[. . .]

In a country of such vast extent as Australia, spreading over so many degrees of latitude, we might naturally expect to find much diversity in the climate, and such is really the case. Van Diemen's Land, from its isolated and more southern position, is cooler and characterized by greater humidity than Australia; its vegetation is therefore abundant, and its forests dense and difficult of access. The climate of the continent, on the other hand, between the 25th and 35th degrees of latitude, is much drier, and has a temperature which is probably higher than that of any other part of the world; the thermometer frequently rising to 110°, 120°, and even 130° in the shade; and this high temperature is not unfrequently increased by the hot winds which sweep over the country from the northward, and which indicate most strongly the parched and sterile nature of the interior. Unlike other hot countries, this great heat and dryness is unaccompanied by night dews, and the falls of rain being uncertain and irregular, droughts of many months' duration sometimes occur, during which the rivers and lagoons are dried up, the land becomes a parched waste, vegetation is burnt up, and famine spreads destruction on every side. It is easier for the imagination to conceive than the pen to depict the horrors of so dreadful a visitation. The indigenous animals and birds retire to the mountains, or to more distant regions exempt from its influence. Thousands of sheep and oxen perish, bullocks are seen dead by the road-side or in the dried-up water-holes, to which, in the hope of relief, they had dragged themselves, there to fall and die; trees are cut down for the sake of the twigs as fodder; the flocks are driven to the mountains in the hope that water may there be found, and every effort is made to avert the impending ruin; but in spite of all that can be done the loss is extreme. At length a change takes place, rain falls abundantly, and the plains, on which but lately not a blade of herbage was to be seen, and over which the stillness of desolation reigned, become green with luxuriant vegetation. *Orchideæ* and thousands of flowers of the loveliest hues are profusely spread around, as if nature rejoiced in her renovation, and the grain springing up vigorously gives promise of an abundant harvest. This change from sterility to abundance in the vegetable world is accompanied by a correspondent increase of animal life, the waters become stocked with fish, the marshy districts with frogs and other reptiles; hosts of caterpillars and other insects make their appearance, and spreading over the surface of the country commence the work of devastation, which however is speedily checked by the birds of various kinds that follow in their train. Attracted by the abundance of food, hawks of three or four species, in flocks of hundreds, depart from their usual solitary habits, become gregarious and busy at the feast, and thousands of Straw-necked Ibises (*Ibis spinicollis*), and other species of the feathered race, revel in the profusion of a welcome banquet. It must not however be imagined that this change is effected without its attendant horrors; the heavy rains often filling the river beds so suddenly, that the onward pouring flood carries with it everything that may impede its course; and woe to the unhappy settler whose house or grounds may lie within the influence of the overwhelming floods!

82

RICHARD FRANCIS BURTON, 'MALABAR', *GOA, AND THE BLUE MOUNTAINS*

(London: Richard Bentley, 1851)

Chapter XI

Malabar

THE province, now called Malabar, is part of the Kerula Rajya, the kingdom of Kerula, one of the fifty-six *deshas,* or regions, enumerated in ancient Hindoo history as forming the Bharata Khanda or Land of India. It is supposed to have been recovered from the sea by the sixth incarnation of Vishnu, who in expiation of a matricidal crime gave over to the Brahmans, particularly to those of the Moonsut tribe, the broad lands lying between Gokarna and Kanya Kumari, or Cape Comorin. The country is also known by the names of Malayalim, the 'mountain land'; Malangara and Cherun, from the Rajahs, who governed it at an early period. It is probably the kingdom of Pandion, described in the pages of the classical geographers.

By Malabar we now understand the little tract bounded on the north by Canara, to the south by the province of Cochin, having Coorg and Mysore to the east, and washed by the waves of the Indian Ocean on the west. Marco Polo (thirteenth century) speaks of it as a 'great kingdom', and Linschoten (sixteenth century) describes it as extending from Comorin to Goa. The natives assert that the old Kerula Rajya was divided into sixty-four *grama* or districts, of which only eight are included in the present province of Malabar.

The whole of this part of the coast acquired an early celebrity from the valuable exports which it dispersed over the Western World. Nelkunda, the chief port, is mentioned by Ptolemy and Pliny: and the author of the 'Periplus' places it near Barake or Ela Barake, the roadstead where vessels lay at anchor till their cargoes were brought down to the sea. Major Rennell has identified the ancient Nelkunda with the modern Nelisuram, as the latter place is situated twelve miles up the Cangerecora River – a distance corresponding with that specified in the 'Periplus'. Vincent acutely guesses Ela Barake to be the spot near Cananore, called by Marco Polo 'Eli', and by us Delhi – the 'Ruddy Mountain' of the ancients.

 DOI: 10.4324/9780429355653-91

Malabar, from remote times, has been divided into two provinces, the northern and the southern: the Toorshairoo or Cottah River forming the line of demarcation. The general breadth of the country, exclusive of the district of Wynad, is about twenty-five miles, and there is little level ground. The soil is admirably fertile; in the inland parts it is covered with clumps of bamboos, bananas, mangoes, jack-trees, and several species of palms. Substantial pagodas, and the prettiest possible little villages crown the gentle eminences that rise above the swampy rice lands, and the valleys are thickly strewed with isolated cottages and homesteads, whose thatched roofs, overgrown with creepers, peep out from the masses of luxuriant vegetation, the embankments and the neat fences of split bamboo interlaced with thorns, that conceal them. Each tenement has its own croft planted with pepper, plantains, and the betel vine, with small tufts of cocoas, bamboos, and that most graceful species of the palm, the tall and feathery areca. These hamlets are infinitely superior in appearance to aught of the kind we have ever seen in India; the houses are generally built of brick or hewn stone and mortar, and those belonging to the wealthy have been copied from the Anglo-Indian bungalow. As the traveller passes he will frequently see the natives sitting at their doors upon chairs exactly as the rustics of Tuscany would do. The quantity of rain that annually falls covers the ground with the bloom of spontaneous vegetation; cocoa-trees rise upon the very verge where land ends, and in some places the heaps of sand that emerge a few feet from the surface of the sea, look bright with a cap of emerald hue. In consequence of the great slope of the country the heaviest monsoon leaves little or no trace behind it, so that lines of communication once formed are easily preserved. Generally speaking the roads are little more than dykes running over the otherwise impassable paddy fields, and, during wet weather, those in the lower grounds are remarkably bad. Some of the highways are macadamised with pounded laterite spread in thin layers upon the sand; the material is found in great quantities about Calicut, and it makes an admirable monsoon road, as the rain affects it but little on account of its extreme hardness. The magnificent avenues of trees, which shade the principal lines, are most grateful to man and beast in a tropical climate. On all of them, however, there is one great annoyance, particularly during the monsoon, namely, the perpetual shifting to and from ferries – an operation rendered necessary by the network of lakes, rivers, and breakwaters, that intersects the country. A great public use could be made of these inconvenient streams: with very little cutting a channel of communication might be run down the coast, and thus the conveyance of goods would remain uninterrupted even during the prevalence of the most violent monsoons. Water transit, we may observe, would be a grand boon here, as carts are rare, cattle transport is almost unknown, and the transmission of merchandise by means of coolies or porters is the barbarous, slow, and expensive method at present necessarily in general use.

The practical husbandry of Malabar is essentially rude, and yet in few countries have we seen more successful cultivation. The plough is small, of simple form, and so light, that it merely scratches the ground; a pair of bullocks, or a bullock and a woman or two, are attached to the log, and whilst the labourer dawdles over his task, he chaunts monotonous ditties to Mother Earth with more pious zeal than industry. The higher lands produce the betel vine, cocoa, areca, and jack-trees, together with hill rice: the latter article is sown some time after the setting in of the heavy rains, and reaped about September or October. The lower rice-fields, lying in the valleys between the acclivities, are laid out in little plots, with raised footpaths between to facilitate passage and regulate the irrigation. They generally bear one, often two, and in some favoured spots, three crops a year; the average is scarcely more than six or seven fold, though a few will yield as much as thirty. The south-west monsoon, which lasts from June to September, brings forward the first harvest: the second is indebted to the south-east rains which set in about a month later. The Sama (Panicum Miliaceum) requires the benefit of wet weather; it is therefore sown in May, and reaped in August. The oil plant Yelloo (Sesamum Orientale) and the cooltie or horsegram cannot be put into the ground till the violence of the monsoon has abated.

The annual revenue of Malabar is about thirty lacs of rupees (300,000*l.*), land is valuable, the reason probably being that it is for the most part private, not government property.

83

JAMES LAURIE (ED.) AND JOHN HUTTON BALFOUR, 'PHYSICAL GEOGRAPHY, IN RELATION TO ORGANIZED BEINGS; OR THE GEOGRAPHICAL DISTRIBUTION OF VEGETABLES, OF ANIMALS, AND OF THE HUMAN RACE', *SYSTEM OF UNIVERSAL GEOGRAPHY*

(London: Henry G. Bohn, 1851)

§ 1. On the Geographical Distribution of Vegetables

HAVING examined the solid, liquid, and aeriform parts of the terrestrial globe, we shall now proceed to the consideration of those innumerable living beings which cover its surface. In doing so, we shall in the first place notice vegetable productions, both on account of their intimate connection with the soil, and the abundance in which they are produced. It is for the botanist to examine in detail the treasures of the vegetable kingdom, – the business of the physical geographer is only to mark its general arrangements. In the following observations, we propose to treat of botanical geography in its more enlarged and general sense, and shall afterwards, when speaking of different countries, point out the most important productions of their respective regions.

Vegetation embraces the whole extent of the globe, from one pole to the other, from the summit of the Andes, where the lichen creeps over the hardest rocks, to the very bosom of the ocean, where we meet with floating meadows of sea-weeds. Heat and cold, light and shade, fertile lands and desert plains, every place and every temperature, has its own peculiar vegetation. Palms, tree-ferns, and the

DOI: 10.4324/9780429355653-92

parasitical orchideous plants, are confined to the tropics; cruciferous and umbelliferous plants are almost exclusively found in temperate regions; while the coniferous, and many of the amentaceous tribes, flourish in more northern countries. Plants belonging to the cryptogamous class grow even upon the dark vaults of caverns, and upon the walls of the deepest mines. The temperature of the air seems to be the chief agent in limiting the range of any vegetable species. Hence the scale of atmospherical temperature serves as a scale for the progress of vegetation. In the burning climate of the torrid zone, we have only to ascend the mountains to enjoy the fruits and flowers of temperate regions. Tournefort found at the foot of Mount Ararat the ordinary plants of Armenia; a little way up, those of Italy; higher again, those which grow about Paris; afterwards the Swedish plants; and higher still, those of Lapland. Forster saw several of the plants of the Alps on the mountains of Tierra del Fuego. While the vallies of the Andes are adorned with bananas and palm-trees, the more elevated regions of that chain are covered with oaks, firs, barberries, and a number of genera common in the north of Europe.

The degree of heat and light necessary for the growth of different vegetables is very various. Some plants of the confervæ tribe, as a species of green laver, live in hot springs, at the temperature of boiling water; whilst others, such as the red snow plant *(protococcus nivalis),* vegetate amidst the perpetual snow of high mountains, or of the Arctic regions. Numerous cryptogamous plants thrive in situations where the rays of the sun never penetrate. Vigorous beech-forests are found covering the slopes of the Himalaya chain, at a height considerably greater than that of the summit of Finsterhorn, and there also shrubs grow in situations more elevated than the top of Mont Blanc; whilst, on the other hand, numerous sea-weeds send forth their fronds from the very abysses of the ocean. The difference of pressure, temperature, and light, in these instances, must be very great. The absence of moisture seems to oppose the most formidable obstacle to the growth of plants. In confirmation of this, we have only to look to those sandy deserts under the equator and towards the pole, where scarcely a drop of rain falls, and where not even a blade of grass can be seen. The mountains of the torrid zone, present often, from their base to their summit, the plants which are met with from the equator to the poles. We have been able also to cultivate, in our stoves, according to the temperature, the degree of moisture, and the nature of the soil which we employ, a vast number of plants indigenous in all climates. The geographical differences which vegetables present, depend, then, almost entirely on the different degrees of heat, light, and moisture, which they receive, as well as on the nature of the soil whence they derive nourishment, and the influence of various atmospherical phenomena which are constantly occurring. There exists, however, a great number of plants, such as chiccory, wild sorrel, and cresses, which accommodate themselves to all climates and to all localities, extending from Siberia and the frozen shores of Hudson's Bay, to those blissful islands which are scattered over the Pacific Ocean.

The dissemination of the seeds of plants over the surface of the earth is effected by four causes; water, wind, animals and man. The first of these is chiefly concerned in the propagation of aquatic plants, and those found on the sea-shore; the second, in the dissemination of cryptogamous species; and the two last, in the distribution of phænogamous plants in general. Plants may be naturalized wherever the temperature and other atmospheric phenomena are similar to those of the countries in which they are indigenous. We must not, however, allow too much latitude to the operation of these dispersing causes. The pretended migrations of plants have been greatly exaggerated. It has been stated, for example, that Europe received wheat and barley from Tartary, the walnut-tree from Persia, the olive from Syria, and the vine from the borders of the Caspian Sea; in short, historical proofs have been accumulated to show that almost all our useful plants have been brought from Asia. The observations of the ancients, however, on this subject refer only to the *cultivation* of a plant, and not to its origin. Lucullus, without doubt, was the first who brought from Cerasus, in Pontus, the cherry-trees since cultivated in Italy; but in relating this fact, Pliny tells us, that Lusitanian cherries were the most esteemed in Belgic Gaul, and that Macedonia produced a particular kind. He would not have spoken in this manner, had the cherry-trees of Macedonia and Lusitania been propagated from those of Pontus. The same author, however, seems to allow that the vine was indigenous to Gaul. Ancient traditions concur in ascribing the first cultivation of wheat to Sicily or to Attica. A kind of rye, known under the Celtic name of *arinca,* was a native of Gaul. These examples are sufficient to show that some of the farinaceous plants may be considered as indigenous in Europe.

The migrations of man, however, have had a wonderful influence on the geographical extension of plants. By him the coffee-tree has been carried from Arabia to the West Indies, and the potato from America to Europe. The accidental introduction of seeds into a bale of merchandise, has been the means of spreading many plants. In this way, some of the vegetable productions of Brazil have been transplanted into Portugal, and perhaps afterwards conveyed thence to Britain. In the dissemination of plants, there are several peculiarities which are easily accounted for. Some plants appear to live in society, and occupy exclusively large tracts of ground, from which they banish all other vegetables. Others are confined to one side of our planet. This is more especially the case with the species of the genus erica or heath, which extend from the Pole to the Cape of Good Hope, over a surface very narrow compared to its length. Some plants are propagated in the direction of the longitudes, and do not extend either to the right or the left. Thus, the two-coloured phalangium (*Phalangium bicolor*), according to Mirbel, begins to appear on the plains of Algiers, passes into Spain, crosses the Pyrenees, and ends in Brittany. The Irish menziesia (*Menziesia polifolia*), is found in Portugal, the south-western extremity of France, and in Ireland. The Pyrenean ramonda, (*Ramonda pyrenaica*), follows the valleys in the Pyrenees, which run from north to south, and there is not a single specimen of it to be found in the lateral valleys.

We see at other times singular leaps in the distribution of plants. The oak, the wild-nut, and the apple-tree, which are common in Europe, disappear towards the Uralian Mountains, and are not met with from the Tobol to Daouria. The two first of these trees, however, reappear suddenly on the banks of the Argoun and the Amour, and the last occurs anew in the Aleutian Islands.

Botanical geography is not yet in a sufficiently advanced state, to enable us to give an accurate account of the number of species, genera, and families of plants in the different regions of the globe. Willdenow, in his *Species Plantarum,* describes nearly 20,000 species of phænogamous plants, and if to these we add 7000 or 8000 cyptogamous plants which have been described, and 25,000 or 30,000 different species which have been recently discovered, we shall find that the number of plants at present known amounts to nearly 60,000.

Humboldt and Brown have given the following account of the numerical distribution of plants in different countries. In Europe, 7000; in the temperate regions of Asia, 1500; in the warmer regions of Asia and the neighbouring islands, 4500; in Africa, 3000; in North America, 4000; in South America, 13,000; in New Holland, and the islands of the South Sea, 5000. To these must be added numerous marine plants which are as yet little known, and a great variety of vegetable species which have been discovered since Humboldt and Brown's calculation was made. The view which has been given, however, sufficiently shows the small extent of our knowledge in regard to this subject; for Europe, which presents a surface much smaller than that of many of the countries mentioned, and which cannot boast of the same luxuriant vegetation as is met with in the tropics, but which has been traversed by botanists in all directions and for many years, appears, in this arithmetical distribution, to be richer than Africa and Asia, the extent of which is twice and four times as great as that of Europe. An obstacle to finding the correct number of species, is the want of correspondence among botanists as to the meaning of the terms *species* and *varieties*. Some have multiplied the number of species to an unreasonable amount, by recognising as such all plants which do not present rigorously the characters of the types described by preceding authors; others, falling into the opposite extreme, have united plants which are separated by most botanists, and have thus reduced their real number, without any philosophical reason. If we reflect that the vast continents of Asia and America have not yet been fully explored by botanists, and that the whole of the interior of Africa, of Australia, and the large islands of Oceanica, is as yet unknown alike to the geographer and the naturalist, we can easily suppose that the number of species existing on the globe may amount to nearly double those already known.

The number of known vegetable families differs in different latitudes and in different places. In going from the poles to the equator, we find the number of the *Malvaceæ* or mallow tribe, *Euphorbiaceæ* or euphorbium tribe (including the spurge, castor and croton oil plants, &c.) and *Compositæ* or compound flowers, such as dandelion, daisy, &c. increase. The *Labiatæ* or labiate plants, such as mint, thyme, &c.; *Umbelliferæ,* such as hemlock, parsley, &c.; *Amentaceæ,* such as the birch, willow, &c.; and *Cruciferæ,* such as wallflower, cresses, &c. seem to belong to temperate zones. The last disappear entirely in the torrid zone. The greater part of the European *Orchideæ* (including the various species of orchis), are found only in shady and moist woods; while the saxifrages, primroses, and gentians, prefer calcareous hilly districts.

If we examine the distribution of the great classes of the vegetable kingdom over the globe, we shall find, that cryptogamous plants, such as mosses, lichens, mushrooms, sea-weeds, &c. are to phænogamous, or common flowering plants, in the proportion of 1 to 7. These same tribes in equinoxial countries are in the proportion of 1 to 5; in New Holland or Australia, as 2 to 11; in France, as 1 to 2; while in Lapland, Greenland, Iceland, and Scotland, they are in nearly equal proportion,

Monocotyledonous plants, or plants having only one cotyledon in the seed, such as grasses, lilies, palms, &c., over the whole surface of the known globe are to dicotyledonous, or those having two cotyledons, as the majority of our common trees, &c., as 2 to 9; from the equator to the 30th degree of north latitude, as 1 to 5. In proportion as we retire from the equator, the number of dicotyledonous plants diminishes; so that it is a half less at the 60th degree of north and the 50th degree of south latitude. We have not, however, as yet collected a sufficient number of facts in regard to this subject to establish any general rules which will be applicable to all countries.

What has already been said will show the difficulties to be encountered in determining with accuracy the regions of botanical geography. The partial attempts made by Tournefort, Linnæus, Adanson, Saussure, Soulavie, Ramond and Young, and the much more important labours of Stromeyer, Treviranus, Leopold Von Buch, Wahlenberg, Hornemann, and more particularly those of Humboldt, De Candolle and Brown, have furnished facts and principles upon which M. Schow has been able to found a botanical geography of the globe. But although the valuable essay of this able naturalist has done much to promote the progress of this part of botanical science, innumerable observations are still required in order to render it complete.

[. . .]

The following are the divisions adopted by these authors.[1]

1 Maritime or saline plants. These are terrestrial plants which grow on the border of the sea or of salt lakes; as *salicornia* or saltwort, *salsola* or glasswort, and some species of *statice* or sea-pink, &c.
2 Marine plants, such as seaweeds, lavers, &c. which are either buried in the ocean or float on its surface
3 Aquatic plants, growing in fresh water, either stagnant or running; as *sagittaria* or arrowhead, *nymphœa* or white water-lily, *potamogeton* or pondweed, &c.
4 Marsh or swamp plants, living in ground which is generally submerged, but occasionally dry; as *ranunculus aquatilis* and *sceleratus,* or water and celery-leaved crowfoot; *polygonum amphibium* or amphibious persicaria, &c. The form of the plants varies according to the degree of moisture.
5 Meadow and pasture plants; as some species of *lotus,* or bird's-foot trefoil, a great number of grasses, trefoils, &c.

6 Plants found in cultivated fields. In this division are included many plants which have been introduced by man along with grain; as *centaurea cyanus,* corn blue-bottle; *sinapis arvensis,* or common wild mustard; *agrostemma,* corn-cockle; several species of *veronica,* or speedwell; *euphorbia,* or spurge, &c.

7 Rock or wall plants; as saxifrages, wall-flower, some species of *sisymbrium* or hedge-mustard, and *bromus,* or brome-grass; *linaria cymbalaria* or ivy-leaved toadflax, &c.

8 Sand plants; as *carex arenaria* or sea-carex, and *calamagrostis arenaria* or sand small-reed, which tend to fix the loose sand; *plantago arenaria,* sand plantain, &c.

9 Plants found on rubbish, or those which select the habitations of man and animals, on account of the salts and azotised substances which enter into their composition; as pellitory of the wall, nettles, and some mushrooms, &c.

10 Forest plants, including trees which live in society, as the oak, the beech, firs, &c. and the plants which grow under their shelter, as the greater part of the European orchises, some species of carex, broom-rape, &c.

11 Plants of the thickets or hedges, comprehending the small shrubs which constitute the hedge or thicket, as the hawthorn and sweet-briar, &c., and the herbaceous plants which grow at the foot of these shrubs, as tuberous moschatell, wood sorrel, violets, &c., or those which climb among their numerous branches, as bryony, black bryony, some species of everlasting pea, &c.

12 Subterranean plants, or those which live in mines and caves, almost entirely excluded from the light, as *byssus,* truffles, and some other cryptogamic plants.

13 Plants of the mountains, which M. De Candolle proposes to divide into two sections: 1. Those which grow on alpine mountains, the summits of which are covered with perpetual snow, and where during the heat of summer there is a continued and abundant flow of moisture, as numerous saxifrages, gentians, primroses, rhododendrons, &c.; 2. Those inhabiting mountains on which the snow disappears during summer, as several species of snap-dragon, among others the alpine snap dragon, umbelliferous plants, chiefly belonging to the genus *seseli* or meadow saxifrage, labiate plants, &c.

14 Parasitic plants, which derive their nourishment from other vegetables, and which, consequently may be found in all the preceding situations; as the misletoe, broom-rape, dodder, and a number of lichens, mushrooms, mosses, &c.

15 Pseudo-parasitic plants, which live upon dead vegetables, as lichens, mosses, &c., or upon the bark of living vegetables, but do not derive nourishment from them, as *epidendron,*&c.

16 Plants which vegetate in hot springs, the temperature of which ranges from 80 to 140 or 150 of Fahrenheit's thermometer; as *vitex agnus castus,* and several cryptogamous plants, as the hot-spring laver, (*ulva thermalis,*) &c.

17 Plants which are developed in artificial infusions or liquors. Among others, we mention a conserva, or sort of mould found in Madeira wine.

To cultivated plants, which have become naturalized in fields and gardens by the hands of man, we cannot assign any station or region. The greater part of them follow everywhere the human footsteps, and their native country is almost always involved in the greatest obscurity.

In general, we may say that the station of the plant above the level of the sea, varies the more that its ordinary habitation approaches to the climate of the temperate zones. Plants which grow in all latitudes also grow at all heights, and plants which only grow in a particular latitude, are found at a height above the level of the sea where the temperature corresponds to that of the latitude.

Professor Schow, in his general botanical division of the globe, characterises the regions by the most remarkable feature of their vegetation, adopting commonly used geographical terms only where he conceives that a certain division of the earth ought to constitute a distinct region, but is not sufficiently acquainted with its productions to determine and define their forms. In order that any portion of the globe may form what he proposes to call a *phyto-geographical region,* it is necessary that at least one half of the species should be indigenous in it; that a quarter of the genera should also be peculiar to it, or at least should have there a decided maximum, so that their congeners in other climates should only appear as their representatives; and finally, that individual families of plants should either be exclusively confined to the region, or have their *maxima* there. He does not, however, regard this last condition as absolutely essential, provided the two others are fulfilled. Slight degrees of difference in the vegetation characterise the divisions or provinces of a region or kingdom; he considers a quarter of the species and some genera sufficient to determine them.

The following are the twenty-two botanical regions or kingdoms into which Professor Schow divides the globe: –

1 The Region of Saxifrages and Mosses, or the Alpine Arctic Flora. This region is characterised by the abundance of mosses and lichens, the presence of the saxifrages, gentians, chickweed-tribe, sedges and willows; the total absence of tropical families; a notable decrease of the forms peculiar to the temperate zone, of the forests of firs and birches, and an absence of other forests; the small number of annual plants, and the prevalence of perennial species; and finally, a greater liveliness in their simple colours. This region is divided into two provinces: 1. The province of the *Carices* or the Arctic Flora, which comprehends all the countries within the polar circle, with some parts of America, Europe and Asia which are to the south of it, more especially Lapland, the north of Russia, Siberia, Kamschatka, New Britain, Canada, Labrador, Greenland, and the mountains of Scotland and Scandinavia; 2. The province of Primroses and Rampions, or the Alpine Flora of the south of Europe, which embraces the flora of the Pyrences, Switzerland, the Tyrol, Savoy, &c., the mountains of Greece, the Appenines, and probably the mountains of Spain.

2 The Region of the Umbelliferous and Cruciferous Plants, (to which the hemlock, parsley, walflower, cresses, &c. belong.) These tribes are here in much

greater number than in any other region; roses, crowfoots, mushrooms, amentaceous and coniferous plants, are also very numerous; the abundance of *carices,* and the fall of the leaves of almost all the trees during winter, form also the chief features of this division. It may be separated into two distinct provinces: 1. The province of the *cichoraceæ,* (including the sow-thistle, dandelion, lettuce, &c.) which embraces all the north of Europe, not comprehended in the preceding region, namely, Britain, the north of France, the Netherlands, Germany, Denmark, Poland, Hungary, and the greater part of European Russia; 2. The province of the *astragali* and *cynarocephalæ* (to which the milkvetch, burdock, thistle, &c. belong), which includes a part of Asiatic Russia, and the countries about Mount Caucasus.

3 The Region of the *Labiatæ* and *Caryophylleæ* (to which the pink, catchfly, sandworts, &c. belong,) or the Mediterranean Flora. It is distinguished by the abundance of the plants belonging to these two orders. Some tropical families are also met with, such as palms, laurels, arums, plants yielding balsam and turpentine, grasses belonging to the genus *panicum* or millet, and the true *cryperaceæ* or sedges. The forests are composed chiefly of the amentaceous and coniferous tribes, as birches, oaks, &c., the copses of *cricineæ* or heath tribe, and *terebinthaceæ,* as the mastich, &c.; we meet with a great number of evergreen trees. Vegetation never ceases entirely, but verdant meadows are more rare. M. Schow divides this region into five provinces: – 1. The province of the Cistuses, including Spain and Portugal; 2. The province of the Sage and Scabious, the south of France, Italy and Sicily; 3. The province of the Shrubby *Labiatæ,* the Levant, Greece, Asia Minor, and the southern part of the Caucasian countries; 4. The Atlantic province, the north of Africa, of which he does not yet know any distinctive character; 5. The province of the Houseleeks, the Canary Isles, and probably also the Azores, Madeira, and the north west coast of Africa. Many houseleeks, and some spurges with naked and spiny stems particularly characterise this province.

4 The Region of the *Rhamni* and *Caprifoliaceæ* (to which the buckthorn and honeysuckle belong), or the Japanese Region. This region is as yet too little known to enable us to determine accurately its characteristic features. It embraces the eastern temperate part of the old continent, namely, Japan, the north of China, and Chinese Tartary. Its vegetation appears to occupy a middle place between that of Europe, and that of North America, approaching more to the tropical than to the European.

5 The Region of *Asters* and *Solidagos,* (michelmas daisies and golden-rods.) This is marked by the great number of species belonging to these two genera, by the great variety of oaks and firs, the small number of *cruciferous* and *umbelliferous* plants, the total absence of the heath, and the presence of more numerous species of whortleberry than are to be met with in Europe. It comprehends the whole of the eastern part of North America, with the exception of what belongs to the first region. It has been divided into two provinces: – 1. That of the south, which embraces the Floridas, Alabama, Mississippi,

Louisiania, Georgia, and the Carolinas; 2. That of the north, which includes the other states of North America, such as Virginia, Pennsylvania, New York, &c.

6 The Region of *Magnolias,* comprising the most southern parts of North America. The tropical forms which show themselves more frequently than on a similar parallel of the old continent, are the chief feature in the vegetation.

7 The Region of Cactuses, Peppers, and Melastomas. These families are here predominant, both as regards the number of the species and of the individual plants. It is divided into three provinces: – 1. The province of the Ferns and Orchises, comprehending the West India Islands; 2. The province of the Palms, the lower parts of Mexico, New Granada, Guiana, and Peru; 3. Brazil also seems to form a province, and may perhaps constitute a region of itself.

8 The Region of *Cinchonœ,* or Medicinal Barks, which comprises a part of the elevated regions of South America, included in the torrid zone. The *cinchona* belongs exclusively to this region and forms its principal feature.

9 The Region of *Escallonias,* Whortleberries and Winter's Barks. It embraces the highest parts of South America. We also meet with alpine plants, as saxifrages, whitlow-grass, sandworts, sedges, and gentians. Perhaps also the mountains of Mexico belong to this region, although they may form a separate province, that of the oaks and firs.

10 The Chilian Region. The Flora of Chili differs essentially from those of New Holland, the Cape of Good Hope, and New Zealand, although an approach to them is observable in the genera, *goodenia, araucaria* (Chilian pine), *protea, gunnera,* and *ancistrum*.

11 The Region of Arborescent *Compositœ* (or arborescent plants, with flowers like the dandelion, daisy, &c.) The great number of syngenesious plants, more particularly of the family of *boopideœ,* forms the chief feature of this flora, which approaches in a remarkable manner to that of Europe, whilst it differs entirely from those of Chili, the Cape and New-Holland. This region comprehends the lower part of the basin of La Plata, and the plains which extend to the west of Buenos Ayres.

12 The Antarctic Region, formed by the countries near the Straits of Magellan. There is a considerable affinity between the vegetation here and what is seen in the north temperate zone. Polar forms, however, display themselves in the species of saxifrage, gentian, arbutus, and primrose. There is also a resemblance between the flora of this region, and those of the mountains of South America, of Chili, the Cape, and New Holland.

13 The Region of New Zealand. This flora, besides the plants peculiar to New Zealand, comprehends several others which belongs to the extremities of America, Africa, and Australia or New-Holland.

14 The Region of *Epacrides* and *Eucalypti*. It comprehends the temperate parts of New Holland, and Van Dieman's Land. Besides the two families whence it receives its name, it is characterised by the presence of a great number of *proteaceœ,* myrtles, *stylideœ, restiaceœ, diosmeœ,* acacias, &c.

15 The Region of *Mesembryanthema* or Fig Marigolds, and Stapelias. These two genera, as well as the heaths, are very abundant here. The latter family is found in greater quantity here than any where else. It embraces the southern extremity of Africa.

16 The Region of Western Africa. We are only acquainted with Guinea and Congo, the vegetation of which is a mixture of the Floras of Asia and America, though most resembling the former. This region is characterised by a considerable number of grasses and sedges, and the peculiar genus *adansonia,* the baobab, (the largest known tree in the world.)

17 The Region of Eastern Africa. In regard to the eastern coast of Africa, our knowledge is very imperfect. The region is chiefly distinguished by the genera *danais, ambora, dombeya,* and *sonacia.*

18 The Region of the *Scitamineæ,* (of the turmeric, cardamom, Indian shot, &c.) or the Indian Flora, The *scitamineæ* here are much more numerous than in America, as well as the *leguminosæ,* such as pease, broom, &c. *cucurbitaceæ,* or the cucumber tribe, and *tiliaceæ,* or the lime-tree tribe, although in a less degree. In consequence of the imperfect state of the science, we cannot subdivide this region into provinces. It comprehends India, east and west of the Ganges, the Islands of Madagascar, Bourbon and Mauritius, those between India and New Holland, and perhaps the tropical part of this last continent.

19 The Mountains of India ought to form one or two regions, the vegetation of which differs from that of the plains. These countries perhaps constitute one region with the whole of central Asia.

20 The Floras of Cochin China, Tonquin, and the north of China, notwithstanding their resemblance to that of India, present a sufficient number of peculiar indigenous plants to constitute a distinct region.

21 The Flora of Arabia and Persia, differing from that of India and the Mediterranean, forms a particular botanical region, characterised by the numerous species of *cassia* and *mimosa* (to which senna, the sensitive plant, &c. belong), which are found in it. It appears probable that Nubia and a part of Central Asia belong to it. Abyssinia, the elevated parts of which possess such a different climate, may perhaps form one of the great subdivisions, or even a totally distinct region.

22 The Islands of the South Sea, which lie within the tropics, form undoubtedly a separate region, though with but a slender degree of peculiarity. Among 214 genera, 173 are found in India, and most of the remainder are in common with America. The bread-fruit tree is among the characteristics of these islands, although it is not confined to this region.

Note

1 See Article *Geographie Botanique,* in the *Dictionnaire Classique d'Histoire Naturelle.*

84

ARTHUR HENFREY, 'ITALY', *THE VEGETATION OF EUROPE, ITS CONDITIONS AND CAUSES*

(London: Van Voorst, 1852)

LOMBARDY presents a tract of the greatest possible interest in reference to geographical botany, from the variety of conditions that prevail there; while in the kingdom of Naples, particularly the southern part, and still better in Sicily, we meet with conditions approaching more nearly to those of the equatorial countries than in any other part of Europe, with perhaps the exception of one or two points on the coast of Spain. Space being limited, we shall confine our survey of Italy to these portions of its surface.

The limits of the Lombardic region, of which we shall first treat, extend from the river Sesia, from west to east, to the Adige; on the north they run along the watershed of the Alps, forming a vast curve from Monte Rosa along the ridge between the above-named rivers, to Mont Adamo (including the canton of Tessin and a part of the Valesian Tyrol). On the south the Po forms the boundary. The country belongs to the region of that flora of Upper Italy which has been termed the zone of transition between the Alpine flora and that of the Mediterranean, and within its limits are found remarkable boundary lines between these two floras, which would be still more strongly marked had not the hand of man interfered so much and for so long a time in effacing the primitive conditions.

The combined influence of the sea surrounding Italy on three sides, the periodic winds, and the yearly and daily changes of temperature corresponding to its latitude, cause the development of the plants of the Mediterranean flora in all parts of the peninsula; but this southern flora cannot be regarded as predominant in Upper Italy. The great valley of the Po constitutes a sort of funnel, the point of which rests against the Cottian Alps, and which is enclosed on all sides by high mountains, excepting at its broad base on the east, which spreads out to the sea; from thence the hot, moist south-east winds flow freely in, till, penetrating far westward and meeting the icy currents which pour down from the vast snow-fields in the N.W., they gradually lose their mild influence, everywhere evident in the Venetian territory, but almost lost beyond the river Sesia; so that the flora there only retains its southern character in the vicinity of the Po, in the defiles of the

DOI: 10.4324/9780429355653-93

adjacent Apennines, and opposite to them. The Valley of Aosta forms a remarkable exception.

In agreement with this configuration of the country, the line of demarcation between the Alpine and Mediterranean floras follows an oblique and serpentine course from S.W. to N.E. In the neighbourhood of Turin it approaches the Po, and stretches thence S.W.-N.E. to the vicinity of Brescia. Here it coincides with the branches of the Rhætic Alps, which extend southward, converging towards it, and are immediately exposed to the influence of the maritime winds; running eastwards on the lower portion of them, it curves suddenly to the north near Selvapiana di Prandaglio, on the banks of the Chiesi, and may then be traced to a considerable height on the mountain ridges as far as Tyrol, behind Salo and Garguano, where the Mediterranean flora is developed under a true tropical temperature, only interrupted in winter by the storms of cold wind pouring down from Baldo.

The course of vegetation in Lombardy therefore deviates considerably from that of other continental tracts of central Europe. A remarkable phænomenon is also afforded by the exceptional regions around the many lakes of Upper Italy, which lie within this region.

The influence of the sea in the development of certain forms of vegetation in the countries lying around the Mediterranean depends in part upon the peculiar constituents (salts, &c.) with which the soil was in former ages and in part is still impregnated. But a much greater influence is exerted by the equable distribution of heat, caused by the vicinity of a large expanse of water, and this proposition is strongly confirmed in Lombardy around the lakes and over the tracts of rice-fields. In tracing the course of vegetation from the most western basin, the Lago d'Orta, to the most eastern, the Lago di Garda, it is readily perceived that the characters of the vegetation of the tracts surrounding them are subject to the influence of two laws, – one the general rule already announced, that the forms acquire a more southern aspect towards the east; the other a special one, compounded of three factors, namely, the relative height above the level of the sea, the exposure to the two opposite currents of air, and the position of the surrounding mountains as reflecting surfaces for the rays of heat; for these numerous lakes differ in all these respects from each other. It will suffice to mention one or two instances. Around the Lago d'Orta we find no trace of a Mediterranean flora – no olives; on Lago Maggiore at present no longer any olives, but a most luxuriant garden flora in the low Borromean islands, *Agave americana* growing and flowering in the open air. On Como the olive ascends 1600 feet above the sea, the vine nearly twice as high. On Lago di Garda, the line of vegetation of the orange (which is protected in winter from the icy N. and E. winds from the Tyrol and Baldo) rises to some 1200 feet above the sea in the closed-up valley of Bogliaco, and the line of the olive rises there to some 2000 feet; in the neighbourhood of Maderno the whole of the sloping side of a valley may be seen densely clothed with naturalized agaves. These lake districts evidently constitute isolated and exceptional regions of vegetation, betraying the vicinity of the Mediterranean flora, in the midst of the Alpine territory.

85

HENRY THOMAS BUCKLE, 'INFLUENCE EXERCISED BY PHYSICAL LAWS OVER THE ORGANIZATION OF SOCIETY AND OVER THE CHARACTER OF INDIVIDUALS', *A HISTORY OF CIVILIZATION IN ENGLAND*

(London: John W. Parker and Son, 1857)

Chapter II

Influence exercised by Physical Laws over the Organization of Society and over the Character of Individuals

If we inquire what those physical agents are by which the human race is most powerfully influenced, we shall find that they may be classed under four heads: namely, Climate, Food, Soil, and the General Aspect of Nature; by which last, I mean those appearances which, though presented chiefly to the sight, have, through the medium of that or other senses, directed the association of ideas, and hence in different countries have given rise to different habits of national thought. To one of these four classes may be referred all the external phenomena by which Man has been permanently affected. The last of these classes, or what I call the General Aspect of Nature, produces its principal results by exciting the imagination, and by suggesting those innumerable superstitions which are the great obstacles to advancing knowledge. And as, in the infancy of a people, the power of such superstitions is supreme, it has happened that the various Aspects of Nature have caused corresponding varieties in the popular character, and have imparted to the national religion peculiarities which, under certain circumstances, it is impossible to efface. The other three agents, namely, Climate, Food, and Soil, have, so far as we are aware, had no direct influence of this sort; but they have, as I am about to prove, originated the most important consequences in regard to the general organization of society, and from them there have followed many of those large

DOI: 10.4324/9780429355653-94

and conspicuous differences between nations, which are often ascribed to some fundamental difference in the various races into which mankind is divided. But while such original distinctions of race are altogether hypothetical, the discrepancies which are caused by difference of climate, food, and soil, are capable of a satisfactory explanation, and, when understood, will be found to clear up many of the difficulties which still obscure the study of history. I purpose, therefore, in the first place, to examine the laws of these three vast agents in so far as they are connected with Man in his social condition; and having traced the working of those laws with as much precision as the present state of physical knowledge will allow, I shall then examine the remaining agent, namely, the General Aspect of Nature, and shall endeavour to point out the most important divergencies to which its variations have, in different countries, naturally given rise.

Beginning, then, with climate, food, and soil, it is evident that these three physical powers are in no small degree dependent on each other: that is to say, there is a very close connexion between the climate of a country and the food which will ordinarily be grown in that country; while at the same time the food is itself influenced by the soil which produces it, as also by the elevation or depression of the land, by the state of the atmosphere, and, in a word, by all those conditions to the assemblage of which the name of Physical Geography is, in its largest sense, commonly given.

The union between these physical agents being thus intimate, it seems advisable to consider them not under their own separate heads, but rather under the separate heads of the effects produced by their united action. In this way we shall rise at once to a more comprehensive view of the whole question; we shall avoid the confusion that would be caused by artificially separating phenomena which are in themselves inseparable; and we shall be able to see more clearly the extent of that remarkable influence which, in an early stage of society, the powers of Nature exercise over the fortunes of Man.

Of all the results which are produced among a people by their climate, food, and soil, the accumulation of wealth is the earliest, and in many respects the most important. For although the progress of knowledge eventually accelerates the increase of wealth, it is nevertheless certain that, in the first formation of society, the wealth must accumulate before the knowledge can begin. As long as every man is engaged in collecting the materials necessary for his own subsistence, there will be neither leisure nor taste for higher pursuits; no science can possibly be created, and the utmost that can be effected will be an attempt to economize labour by the contrivance of such rude and imperfect instruments as even the most barbarous people are able to invent.

In a state of society like this, the accumulation of wealth is the first great step that can be taken, because without wealth there can be no leisure, and without leisure there can be no knowledge. If what a people consume is always exactly equal to what they possess, there will be no residue, and therefore, no capital being accumulated, there will be no means by which the unemployed classes may be maintained. But if the produce is greater than the consumption, an overplus arises, which, according to well-known principles, increases itself, and eventually becomes a fund out of which, immediately or remotely, every one is supported who does not create the wealth upon which he lives. And now it is that the existence of an intellectual class first becomes possible, because for the first time there exists a previous accumulation, by means of which men can use what they did not produce, and are thus enabled to devote themselves to subjects for which at an earlier period the pressure of their daily wants would have left them no time.

Thus it is that of all the great social improvements the accumulation of wealth must be the first, because without it there can be neither taste nor leisure for that acquisition of knowledge on which, as I shall hereafter prove, the progress of civilization depends. Now, it is evident that among an entirely ignorant people, the rapidity with which wealth is created will be solely regulated by the physical peculiarities of their country. At a later period, and when the wealth has been capitalized, other causes come into play; but until this occurs, the progress can only depend on two circumstances: first on the energy and regularity with which labour is conducted, and secondly on the returns made to that labour by the bounty of nature. And these two causes are themselves the result of physical antecedents. The returns made to labour are governed by the fertility of the soil, which is itself regulated partly by the admixture of its chemical components, partly by the extent to which, from rivers or from other natural causes, the soil is irrigated, and partly by the heat and humidity of the atmosphere. On the other hand, the energy and regularity with which labour is conducted, will be entirely dependent on the influence of climate. This will display itself in two different ways. The first, which is a very obvious consideration, is, that if the heat is intense, men will be indisposed, and in some degree unfitted, for that active industry which in a milder climate they might willingly have exerted. The other consideration, which has been less noticed, but is equally important, is, that climate influences labour not only by enervating the labourer or by invigorating him, but also by the effect it produces on the regularity of his habits. Thus we find that no people living in a very northern latitude have ever possessed that steady and unflinching industry for which the inhabitants of temperate regions are remarkable. The reason of this becomes clear, when we remember that in the more northern countries the severity of the weather, and, at some seasons, the deficiency of light, render it impossible for the people to continue their usual out-of-door employments. The result is, that the working-classes, being compelled to cease from their ordinary pursuits, are rendered more prone to desultory habits; the chain of their industry is as it were broken, and they lose that impetus which long-continued and uninterrupted practice never fails to give. Hence there arises a national character more fitful and capricious than that possessed by a people whose climate permits the regular exercise of their ordinary industry. Indeed, so powerful is this principle, that we may perceive its operation even under the most opposite circumstances. It would be difficult to conceive a greater difference in government, laws, religion, and manners, than that which distinguishes Sweden and Norway on the one hand, from Spain and Portugal on the other. But these four countries have one great point in common. In all of them, continued agricultural industry is impracticable. In the two southern countries, labour is interrupted by the heat, by the dryness of the weather, and by the consequent state of the soil. In the two northern countries, the same effect is produced by the severity of the winter and the shortness of the days. The consequence is, that these four nations, though so different in other respects, are all remarkable for a certain instability and fickleness of character; presenting a striking contrast to the more regular and settled habits which are established in countries whose climate subjects the working-classes to fewer interruptions, and imposes on them the necessity of a more constant and unremitting employment.

These are the great physical causes by which the creation of wealth is governed. There are, no doubt, other circumstances which operate with considerable force, and which, in a more advanced state of society, possess an equal, and sometimes a superior, influence. But this is at a later period; and looking at the history of wealth in its earliest stage, it will be found to depend entirely on soil and climate: the soil regulating the returns made to any given amount of labour; the climate regulating the energy and constancy of the labour itself. It requires but a hasty glance at past events, to prove the immense power of these two great physical conditions. For there is no instance in history of any country being civilized by its own efforts, unless it has possessed one of these conditions in a very favourable form. In Asia, civilization has always been confined to that vast tract where a rich and alluvial soil has secured to man that wealth without some share of which no intellectual progress can begin. This great region extends, with a few interruptions, from the east of Southern China to the western coasts of Asia Minor, of Phœnicia, and of Palestine. To the north of this immense belt, there is a long line of barren country which has invariably been peopled by rude and wandering tribes, who are kept in poverty by the ungenial nature of the soil, and who, as long as they remained on it, have never emerged from their uncivilized state. How entirely this depends on physical causes, is evident from the fact that these same Mongolian and Tartarian hordes have, at different periods, founded great monarchies in China, in India, and in Persia, and have, on all such occasions, attained a civilization nowise inferior to that possessed by the most flourishing of the ancient kingdoms. For in the fertile plains of Southern Asia, nature has supplied all the materials of wealth; and there it was that these barbarous tribes acquired for the first time some degree of refinement, produced a national literature, and organized a national polity; none of which things they, in their native land, had been able to effect. In the same way, the Arabs in their own country have, owing to the extreme aridity of their soil, always been a rude and uncultivated people; for in their case, as in all others, great ignorance is the fruit of great poverty. But in the seventh century they conquered Persia; in the eighth century they conquered the best part of Spain; in the ninth century they conquered the Punjaub, and eventually nearly the whole of India. Scarcely were they established in their fresh settlements, when their character seemed to undergo a great change. They, who in their original land were little else than roving savages, were now for the first time able to accumulate wealth, and, therefore, for the first time did they make some progress in the arts of civilization. In Arabia they had been a mere race of wandering shepherds; in their new abodes they became the founders of mighty empires, – they built cities, endowed schools, collected libraries; and the traces of their power are still to be seen at Cordova, at Bagdad, and at Delhi.

Part 9

EVOLUTIONARY THOUGHT BEFORE *ORIGIN OF SPECIES*

Evolutionary Thought Before *Origin of Species*

VOLUMES I–II are structured around the dates 1780 to 1858 because 1859 marks the publication of Charles Darwin's *Origin of Species*, the long nineteenth century's most important contribution to the environmental sciences, a text that generated an intense, often hostile, and frequently painful period of debate. *Origin* marks a symbolic and real upheaval in mid-Victorian culture, but the pressures it placed on religious cosmogonies and human self-estimation were only an intensification of strains that science had been placing on scripturally informed conceptualisations of the world for more than half a century. It is also true that while *Origin* represented a formidable iteration of evolutionary theory, it did not emerge in a vacuum. As this section will trace, evolutionary thought has a respectable pre-*Origin* pedigree: a series of contributions to its gradual development are its subject. The following extracts provide an overview of the steps and mis-steps that characterised various attempts to reach explanations for the variability of existing species and the confusing fact that the geological record disclosed not only that species became extinct but that others had seemingly emerged. The hostility and derision that met most of these proto-evolutionary ideas motivated Darwin to delay publication of a work outlining his theories. While his theory began to clearly emerge by 1838, he spent another twenty years accumulating supporting evidence in the hope that this would somewhat protect the eventual 'big book' from the unfavourable reception that met its predecessors. Indeed, Darwin only chose to finally announce his discovery after learning that Alfred Russel Wallace had independently reached the same conclusions. The following section fittingly ends with the joint paper they submitted to the Linnean Society in 1858 that publicly announced the 'development hypothesis' and 'natural selection', but the prehistory of evolutionary thought is worthy of closer scrutiny.

Those wishing to fully trace this history would benefit from reading Darwin's 'Historical Sketch of the Progress of Opinion on the Origin of Species, previously to the Publication of This Work' (see Further Reading), a paper added to the 1861 edition of *Origin* to answer criticisms that he had inadequately acknowledged previous contributions. Darwin's comprehensive overview also provides opportunities to read some works not included in the present selection. Darwin traces the emergence of questions about the status of species to French eighteenth-century naturalists, citing Buffon, Lamarck, and St-Hilaire, while Gavin de Beer has argued (see Further Reading) that 'the subject of mutability of species' had earlier been considered by Charles-Louis Montesquieu, Pierre Louis Maupertuis, and Denis Diderot. Others trace such ideas back further, to *Of the Origin and Progress of Language* (1773–92) by the Edinburgh eccentric James Burnett (Lord Monboddo).

Certainly, the seemingly endless gradations between species led many eighteenth-century naturalists to question the long-held notion that species were immutable and to argue that the concept of species was artificial. The emergence of recognisably modern evolutionary ideas, however, is generally traced to the work of Darwin's grandfather, Erasmus Darwin. *Zoonomia*, the first of two of his extracts in this section, is also often regarded as the first formal statement of an evolutionary theory. Darwin's attempts to understand reproduction of the 'embryon' (embryo) lead him to consider the idea that species could evolve into one another, and although he offers no causal mechanism for species change, his ideas anticipate natural selection in observing that contests between males of a species mean that 'the strongest and most active animal should propagate the species which should thence be improved'. Famously libidinous, he also argues that 'the three great objects of desire, which have changed the forms of many animals by their exertions to gratify them, are those of lust, hunger, and security', thus auguring a key element of subsequent evolutionary theory, the influence of environment in prompting competition and adaptation. Darwin is led to his ingenious speculations by observing the existence of distinct life cycles in organisms such as frogs and butterflies, by changes wrought by humans to domesticated plants and animals, and by noticing that there were distinct homological similarities between the physiological structures of different vertebrates: all areas that would be scrutinised by subsequent evolutionists. Finally, his work is advanced in its acceptance of a lengthy geological timescale, his estimate of the existence of 'millions of ages before the commencement of the history of mankind' pre-dating Charles Lyell (see Part 3 of this volume) by three decades. Given the parallels between the work of grandfather and grandson, it is unsurprising that *Zoonomia* stimulated Charles Darwin's researches after his return from the *Beagle* expedition, but it should also be noted that after *Origin*, Charles distanced himself from his grandparent because the latter was associated with Lamarckism.

Erasmus Darwin's evolutionary ideas are also evident in his poetry, which forms the next extract. While there are intimations of such ideas in his earlier poems, *The Loves of the Plants* (1789) and *The Botanic Garden* (1791), the posthumously published *The Temple of Nature* (1804) provides their fullest iteration. The forty lines excerpted from Canto 1 represent the ideas of *Zoonomia* dramatically in a grand iambic pentameter, transporting readers across lands and beneath the waves to witness the 'successive bloom' of generations which 'new powers acquire, and larger limbs assume', and arguing that higher plant and animal organisms all 'arose from rudiments of form and sense,/An embryon point, or microscopic ens!' The impact of environment in changing the forms of individuals is evident in the underwater visions of creatures in which 'each muscle quickens, and each sense improves'.

The notion that the changes that occur within an individual's lifetime can be passed on to its progeny is also evident in – and indeed encapsulates the central theory of – Jean-Baptiste Lamarck's *Zoological Philosophy* (1809), the first substantial evolutionary work by an established scientist. Lamarck was contributing to that significant debate of the French academy between those (such as Cuvier) who argued for the immutability of species that maintained the Great Chain of Being (see Part 2 of this volume) and those (like St-Hilaire) who believed species were mutable and that human systems of classification were artificial (the position of Buffon within this debate, it should be noted, was more ambiguous). Toby Appel's work (Further Reading) explores this debate in detail. Lamarck firmly sides with St-Hilaire, forwarding a mechanism of species change in order to explain

> why the increasing complexity of the organisation of animals from the most imperfect to the most perfect exhibits only an irregular gradation, in the course of which there occur numerous anomalies or deviations with a variety in which no order is apparent.

Centrally important is his argument that organisms adapt to their environment and that the frequent use of particular organs gradually strengthens them, while the disuse of organs eventually leads to their disappearance. Crucially, Lamarckism goes further by arguing that lifetime mutations are passed on to the following generation.

This theory of the inheritance of *acquired* characteristics was partially accepted as a factor by Charles Darwin, but his theory of natural selection provides a more robust and coherent model that ultimately superseded Lamarck's. Darwinism argues that those animals born with adaptations liable to be useful in the 'struggle for existence' are more likely to breed and therefore more likely to pass on their favourable adaptations and that in time, the accumulation of a particular favoured adaptation within a species would predominate and, in some circumstances, lead to the formation of distinct new species. Natural selection, thus understood, does not claim that acquired characteristics are passed on or place emphasis on them. While Darwin wrote warmly of Lamarck's work, he was alert to the vehemence with which it was received and that Lamarck's failure to provide a clear body of supporting evidence did long-term damage to the evolutionary cause. Lamarck's ideas, however, are not mere historical curios: to Soviet science, they offered evolutionary ideas stripped of the alleged Darwinian focus on competition, while recent epigenetic biologists have returned to and partially vindicated aspects of Lamarckism.

The next extract is at once a significant contribution to evolutionary theory and an example of historical irony. As Darwin acknowledged in his 'Historical Sketch', Patrick Matthew in 1831 'gives precisely the same view on the origin of species as that . . . propounded by Mr Wallace and myself'. Diplomatically, Darwin added that Matthew had presented this theory 'very briefly in scattered passages in an Appendix to a work on a different subject, so that it remained unnoticed until Mr Matthew himself drew attention to it' in 1860. Matthew's *On Naval Timber and Arboriculture* is obscure, rambling, diffuse, and intemperate. It was not widely read, and it is unlikely that many readers persisted to Matthew's appendices, where his précis of evolutionary theory is buried.

Geology provided much of the supporting evidence on which Darwin drew in making his claims, and in particular, he found the expanding timescales of geological history amenable to theories about natural selection. As well as these major observations, much specific evidence arose out of the 'new science', including claims such as Roderick Murchison's (see Part 3 of this volume) that 'although the older fossiliferous strata often contain *vast quantities* of organic remains, *the number of species is much smaller* than in the more recent deposits'. Observations like Murchison's provided evidence of extinctions, the appearance of new species, and ever-greater biological complexity over time. In speaking of 'the amount of zoological mutation throughout a large portion of these ancient fossiliferous strata', Murchison argues that 'the inference we obtain is, that most of the groups which graduate into each other by lithological and geological characters, exhibit also a true zoological transition', but he does not describe this in ways that imply a process of evolutionary change. Nonetheless, twenty years before *Origin of Species*, Murchison tantalisingly notes that

> while some of the Lower Silurian animals lived on to those days when the Upper Silurian beds were deposited, scarcely a single species which existed when the Silurian æra commenced, can be detected in the strata which mark its close.

The fifth extract is from Charles Lyell's seminal *Principles of Geology* and is valuable here for a number of reasons. Lyell's promotion of Uniformitarian geology provided an expanded account of earth history – the space, essentially, in which evolution became theoretically possible. His work was therefore considerably important to Darwin, a fellow member of the Geological Society who took a copy of Lyell's recently published *Principles* on his *Beagle* expeditions. While Lyell enabled the co-announcement of evolutionary theory by Darwin and Wallace, he had religiously-based anxieties about the theory, an irony given frequent clerical attacks on his work. His rootedness in earlier Natural Theological traditions is evident in Lyell's responses to Lamarckism in the extract provided. It is a marker of Lyell's broad knowledge that a major geological treatise devotes considerable space to questions of zoology and botany, but the space given to their refutation – spread across two chapters of the second volume – also signals the dangers Lyell perceived in Lamarck. Lyell, it should be noted, changed his view on evolutionary matters, if not Lamarckism: the 1867 11th edition of *Principles* returned in Chapter 35 to the subject to offer his qualified support for Darwinism (see Volume III of this anthology).

The extract, from Chapter 2 of the original 1832 edition, is typical of the extended argument pursued, which often turns to Lamarck's own evidence in order to refute it. Accepting Lamarck's observations about wide-ranging variations within and between species, Lyell emphatically defends the immutability of species. To give way to Lamarckism and mutability, he says in this unexcerpted quotation from the chapter, opens a dangerous path to the observer:

> Henceforth his speculations know no definite bounds; he gives the rein to conjecture, and fancies that the outward form, internal structure,

> instinctive faculties, nay, that reason itself, may have been gradually developed from some of the simplest states of existence,– that all animals, that man himself, and the irrational beings, may have had one common origin; that all may be parts of one continuous and progressive scheme of development from the most imperfect to the more complex; in fine, he renounces his belief in the high genealogy of his species, and looks forward, as if in compensation, to the future perfectibility of man in his physical, intellectual, and moral attributes.

With unintended prescience, Lyell anticipates many of the main features of Darwinism and its consequences: the idea of descent from common ancestors, the troubling lack of uniqueness of *Homo sapiens*, dreams of evolutionary perfectibility (i.e., eugenics), and debates about common ancestors.

Lyell is scathing about Lamarckism, describing it as 'tread[ing] in the footsteps of the naturalists of the middle ages, who believed the doctrine of spontaneous generation to be applicable to all those parts of the animal and vegetable kingdoms' and arguing that it is founded on a false confusion due to the proliferation of new botanical and zoological specimens: more science, he argues, will make the relationships between the seemingly chaotic multiplicity of species clear and clarify Natural Theology's 'graduated scale of being'. Admitting some degree of variation, and that this can be passed on to progeny, he argues that God sets limits to ensure that divinely wrought distinctions between species ultimately remain. Lyell concedes that considerable variation has been achieved in human breeding of domesticated animals and plants, but even amongst dogs, he claims, there has been no evidence of Lamarck's belief in 'the growth of new organs and the gradual obliteration of others'. Like Lamarck's contemporary French critics, Lyell draws on evidence of Egyptian mummified animals to argue that they closely resemble their modern ancestors. Even the apparently strong evidence of the considerable variation of modern cultivated brassicas descended from a common ancestor (sea kale) is unconvincing to Lyell, who argues that the cultivated varieties quickly revert if not kept in nutrient-rich conditions and prevented from cross-pollinating: while variation is possible, its long-term maintenance is reliant on human effort that is not reproducible in the wild, he argues.

The following two extracts are closely linked and should be considered together. Having turned in previous sections of this volume to Darwin's *Journal of Researches into the Geology and Natural History of the Various Countries Visited by H.M.S. Beagle* (1839), we turn now to his account in that work of his time in the Galápagos archipelago because of their significance to the formation of his ground-breaking theory. Alongside this, and existing in a state of dialogue with it, is an extract from a sister publication to *Journal of Researches* – Volume 3 of *Zoology of the Voyage of H.M.S. Beagle* (1841), a collaboration between Darwin and John Gould.

It took some time for Darwin to realise the evolutionary implications of his observations about differences between the animals of the Galápagos and mainland

South America and between different species or sub-species on different islands, but in the *Journal*, there are tantalising glimpses of his later postulation that these were the result of natural selection occurring within the isolated habitats of the different islands. In speculating whether rats resident on the islands were identical to the common rat, for example, he adds 'I cannot help suspecting that it is only the same altered by the peculiar conditions of its new country'.

In his summary of the island's fauna in the *Journal*, Darwin refers to the authority of Gould, a pivotal figure in Darwin's journey from *Beagle* to *Origin of Species*. After returning from the voyage in 1837, Darwin presented his animal specimens to the Zoological Society of London, and Gould was entrusted with the birds. His findings were subsequently published in *Zoology*, where Darwin's Advertisement notes that 'when I presented my collection of Birds to the Zoological Society, Mr Gould kindly undertook to furnish me with descriptions of the new species and names of those already known'. The results of this collaboration were swift: as early as 10 January 1837, Gould suggested to a meeting of the Zoological Society that the Galapagos birds Darwin had described as blackbirds, finches, and grossbills were a unique series of twelve species of ground finches. Moreover, the species that Darwin had identified as a 'Galapagos wren' was, Gould informed him, yet another finch species. Darwin's first inkling of the evolutionary implications of his hastily collected finches arose because of his collaboration with Gould. In *Zoology*, Darwin refers back to the *Journal*, noting that it had 'given my reasons for believing that in some cases the separate islands possess their own representatives of the different species' but acknowledging that 'I did not suspect this fact until it was too late to distinguish the specimens from the different islands of the group'. It had not occurred to Darwin to label his finches according to their island locations, but others had done so. In the light of Gould's insights, Darwin acquired labelled specimens collected by Captain FitzRoy and others.

Gould's confirmation that the species were indeed unique to particular islands made a vital step towards 'the development hypothesis'. *Journal of Researches* was published in 1839, meaning that Darwin was able to reflect Gould's helpful findings in the summary of birds that is extracted here, but there is still a sense of the provisional nature of these engagements. The finches are now recognised as 'the most singular of any in the archipelago', and Darwin is now aware of the need for further study: 'They all agree in many points', he notes, but

> it is very remarkable that a nearly perfect gradation of structure in this one group can be traced in the form of the beak, from one exceeding in dimensions that of the largest gros-beak, to another differing but little from that of a warbler.

The rest of the first extract includes Darwin's lengthy remarks on the giant tortoises of the Galapagos. Again, there is no understanding at this stage of the evolutionary implications of the distinct island populations that he observed, but this would soon change. These passages certainly illustrate Darwin's observant, sympathetic gaze and his customary attempt to provide a holistic overview of an animal's environmental behaviour and relationships. Given the still-precarious status of the various island populations, Darwin's descriptions of vast numbers of tortoises is painfully elegiac. He describes their role as a source of meat for the islanders, and this, alongside loss of habitats, led tortoise numbers to decline from estimates of over 250,000 in the sixteenth century to a historic low in the 1970s of around 3,000. While this low was accompanied by the extinction of three species, it has been somewhat remedied by more recent conservation work.

The accompanying extract from *Zoology* includes two examples of the bird descriptions. Collaboratively written, Gould provides taxonomic descriptions and ornithological authority, while Darwin is responsible for fascinating accounts of the habits, range, and behaviours of the birds. This makes the *Zoology* a curious, rewarding combination of travel literature, ornithology, and biogeography, a traveller's tale condensed and synthesised for the delectation of a Victorian public eager for insights into 'exotic' worlds but also a major contribution to the foundations of the most important scientific breakthrough of the century.

The eighth extract is from perhaps the most controversial work of evolutionary theory in the period, *Vestiges of the Natural History of Creation* (1844), published anonymously but eventually acknowledged as the work of Scottish editor and journalist Robert Chambers. If Darwin was nervously aware of the hostile reception of Lamarck, the vituperation and derision that met Chambers' book was a key reason for ongoing delays to the publication of *Origin*. The next extract will turn to such criticisms.

Chambers's misfortune was to produce a broadly sound theory without providing a coherent body of supporting evidence. *Vestiges* draws together a series of suggestive phenomena: that organic life through the course of the earth's history demonstrates increasing complexity; that there are homological similarities between vertebrates generally and mammals in particular; that embryos pass through stages of development that successively recapitulate the life forms of more 'primitive' species (homology and embryology are key issues in Part 4 of this volume); and, echoing Lamarck, that there are many examples of 'rudimentary organs' (organs which have become disused or atrophied over time). Unfortunately, what Chambers creates from this accumulation of sound evidence is a theory that is along the right lines but lacking a mechanism and specific details: '*the simplest and most primitive type*', he argues, '*gave birth to the type next above it, that this again produced the next higher, and so on to the very highest*'. Chambers's theory appears weirdly rooted in the Great Chain of Being, as he imagines 'that the place vacated by one species was immediately taken by the next in succession', and he is at pains to stress that he sees the process of species change as a divinely ordained mechanism. Lacking anything like Darwin's detailed explanation of the processes of natural selection, Chambers's theory appears insufficiently supported and was roundly condemned as such. A best-seller, and initially well received, it was quickly denounced, first by clergymen and then professional scientists (including Sir David Brewster, Adam Sedgwick, and Sir John Herschel). Darwin, however, saw much to praise in the work, identifying in it further areas for investigation but also levelling criticisms of the details in the introduction to *Origin*. Part of the problem was the sheer range of evidence, some of which – including a lengthy disquisition on the mathematical theories of Charles Babbage – appears amateurish and eccentric. Even the publication of *Explanations: A Sequel* (1845) to answer specific criticisms could not rescue *Vestiges*' wider reputation, but it ran through twelve editions up to 1884 (when Chambers publicly acknowledged authorship) and was fiercely debated.

Thomas Monck Mason's *Creation by the Immediate Agency of God, as Opposed to Creation by Natural Law; Being a Refutation of the Work Entitled Vestiges of the Natural History of Creation* (1845) has been selected as an example of critical responses to Chambers, and specifically of religious responses to the work. Mason's 'Introductory Remarks' highlight the stakes at play in the 1840s and 1850s in venturing into scientific theories that were (rightly) seen as a challenge to scripture. If *Vestiges* were accepted, Mason contends, 'the Bible is a fiction' and 'the mighty structure of the Christian faith, under the shadow of which all our institutions have grown up' is 'destined to vanish'. The vehemence of reactions to geological and evolutionary theories during the period can be traced to the fundamental schism opened up between science and scripture and its results traced in the gradual decay of Natural Theology, and there is a strong resemblance between Mason's tone and that of Henry Cole in geological debates a decade earlier (see Part 2 of this volume). The shrillness and alarm of Cole and Mason are testament to the rearguard resistance offered to a very real assault on orthodoxy. At stake, in a broader environmental context, are different visions of environment and the place of humanity within it: while a religious cosmogony firmly retains the place of *Homo sapiens* at the apex of divine creation, evolutionary theory was widely seen to diminish the human role, positing instead a more complex network of organic and inorganic relationships that would ultimately give rise both to the notion of the biosphere and fuller understanding of the damaging ecological impacts of our species – a subject to be taken up more fully from Volume II onwards.

Although Mason devotes lengthy chapters to point-by-point refutations of *Vestiges*, his methodology largely consists of suggesting that no firm conclusions can be drawn from its limited evidence base – a point echoed by many scientific critics. One example is given here, involving Mason's rejection of Chambers' homological arguments: while not denying 'that all organic creation should appear to be formed upon . . . a definite number of models', he argues that this 'can afford no grounds for any conclusion whatever'. Illustrating the Natural Theological tendency to set up inarguable religious foundations for arguments, he '[takes] it for granted that it was the design of the Almighty that His works . . . should observe a certain uniformity of character' and that the similarities between species that Chambers sees as grounds for an evolutionary reading are simply 'a necessary consequence of the fulfilment of the design'.

The final extract brings us forward thirteen years to the pre-*Origin* announcement by Charles Darwin and Alfred Russel Wallace of 'On the Tendency of Species to Form Varieties; and on the Perpetuation of Varieties and Species by Natural Means of Selection' (1858), the joint paper read at the Linnean Society referred to at the start of this headnote. In contrast to earlier manifestations of evolutionary theory, there is at last a persuasive (if not yet conclusive) mechanism to explain species change. Readers wishing to trace in more detail the gradual development of Darwin's ideas between *Beagle* and *Origin* might start with the Darwin correspondence edited by his son, Francis (Further Reading).

The joint paper, read by the Society's secretary, comprises an introductory letter by Lyell and Joseph Dalton Hooker outlining the contents and contexts and ensuring that Darwin's prior discovery of the theory is acknowledged. Darwin's letter to Asa Gray, referred to in this letter, is not extracted here, but I have provided selections from the Darwin and Wallace papers. Darwin's brief essay, from materials initially written between 1839 and 1844, is a précis of the main points of *Origin*, in particular its emphasis on the effects of environment and competition for scarce resources in ensuring 'that any minute variation in structure, habits, or instincts' that better adapted an individual to new environmental conditions would lead to 'a better chance of surviving' for the individual 'and those of its offspring which inherited the variation'. As well as this statement of natural selection, Darwin foreshadows the theory of sexual selection that would be the subject of *The Descent of Man* (1871) (see Volume III of this anthology).

Wallace's paper is remarkably similar in its approach and claims, and it is a credit to Darwin that he acknowledged the younger naturalist's right to be credited alongside him. Wallace, a long-term believer in the transmutation of species, had read Lamarck and Chambers favourably and used much of his time as a travelling naturalist – in South America and the Far East – seeking a credible mechanism. An 1855 paper in *Annals and Magazine of Natural History* supported transmutation, but only in the Malay Archipelago in 1858 did a theory of natural selection occur to him (his autobiography, *My Life*, provides invaluable insights into its emergence – see Further Reading). As the extract shows, he was finally able to move beyond Lamarck's theory of the inheritance of acquired characteristics. A staunch supporter of Darwin's priority as founder of evolutionary science, Wallace's significance lies in the fact that his own independent researches lent some weight to Darwin during the period of intense bombardment that greeted the publication of *Origin* in 1859. As Volume III will show, however, Wallace's future career saw him remain in the field of biogeographical natural science that he had previously pursued rather than continuing evolutionary studies.

Even by the tenth anniversary of that work, evolutionary theory had not been fully established, and its critics remained. The various implications and interpretations of Darwinism will be scrutinised in the next three volumes. The complexity of Darwin's idea make it susceptible to wildly different readings. Many, including Social Darwinists, emphasised what they saw as Darwin's Malthusian focus on competition. The 'struggle for existence' was taken up in order to justify laissez-faire economics; eugenic race theories; Hegelian dialectics; revolutionary politics; and a world view that saw nature, in Tennyson's famous phrase, as 'red in tooth and claw'. In such readings, humanity often retains its challenged position at the apex of organic life because of its supposedly superior development. In this sense, Darwinism could maintain a beleaguered anthropocentric vision. At the same time, however, the combined work of the various writers featured in this section gradually and very slowly augured in a new vision of relationships between species and environments in ways that re-imagined the role of *Homo sapiens* and human culture after mid-century. Most importantly for us, evolutionary theory provided the conceptual framework that encouraged ecologists, geographers, biogeographers, and others to analyse the complex webs of organic and inorganic networks that make up life on planet earth. While anthropocentrism continues to reign over our benighted and stubborn species, nineteenth-century evolutionary theories opened up other arguments and perspectives.

Further reading

Appel, Toby A. *The Cuvier-Geoffroy Debate: French Biology in the Decades Before Darwin*. Oxford: Oxford University Press, 1987.

Barber, Lynn, *The Heyday of Natural History 1820–1870* (London: Cape, 1980).

Beer, Gavin de, *Charles Darwin: a Scientific Biography* (Pennsylvania State University: Doubleday, 1965).

Bowler, Peter J., *Evolution: The History of an Idea* (Oakland: University of California Press, 2009).

Colp, Ralph, 'The Relationship of Charles Darwin to the Ideas of his Grandfather, Dr Erasmus Darwin', *Biography* 9:1 (Winter 1986), 1–24.

Costa, James T., *Wallace, Darwin, and the Origin of Species* (Cambridge, MA: Harvard University Press, 2014).

Darwin, Charles, 'An Historical Sketch of the Progress of Opinion on the Origin of Species, Previously to the Publication of This Work', *Origin of Species*, 3rd ed. (London: John Murray, 1861).

Darwin, Francis (ed.), *The Life and Letters of Charles Darwin, Including an Autobiographical Sketch*, 3 vols, vol. 2, 7th revised ed. (London: John Murray, 1888).

Fara, Patricia, *Erasmus Darwin: Sex, Science, and Serendipity* (Oxford: Oxford University Press, 2012).

Fichman, Martin, *An Elusive Victorian: the Evolution of Alfred Russel Wallace* (Chicago: University of Chicago Press, 2004).

Gissis, Snait B. and Jablonka, Eva (eds.), *Transformations of Lamarckism: From Subtle Fluids to Molecular Biology* (Cambridge: MIT Press, 2011).

Griffiths, Devin, *The Age of Analogy: Science and Literature Between the Darwins* (Baltimore: Johns Hopkins University Press, 2016).

Keynes, Richard, *Charles Darwin's Beagle Diary* (Cambridge: Cambridge University Press, 2001).

Secord, James A., *Victorian Sensation: The Extraordinary Publication, Reception, and Secret Authorship of Vestiges of the Natural History of Creation* (Chicago: University of Chicago Press).

Stott, Rebecca, *Darwin and the Barnacle: The Story of One Tiny Creature and History's Most Spectacular Scientific Breakthrough* (London: Faber and Faber, 2003).

———, *Darwin's Ghosts: In Search of the First Evolutionists* (London: Bloomsbury, 2013).

Wallace, Alfred Russel, 'On the Law Which has Regulated the Introduction of New Species', *Annals and Magazine of Natural History* (September 1855), 184–96.

———, *My Life* (London: Chapman & Hall, 1905).

Ward, Peter, *Lamarck's Revenge: How Epigenetics Is Revolutionizing Our Understanding of Evolution's Past and Present* (New York: Bloomsbury, 2018).

Wyhe, John van, 'Where Do Darwin's Finches Come From?', *The Evolutionary Review* 3: 1 (2012), 185–95.

86

ERASMUS DARWIN, *ZOONOMIA, OR, THE LAWS OF ORGANIC LIFE*, 2ND CORRECTED ED., VOL. 1

(London: J. Johnson, 1796)

Sect. XXXIX, Of Generation

[. . .]

I. The ingenious Dr. Hartley in his work on man, and some other philosophers, have been of opinion, that our immortal part acquires during this life certain habits of action or of sentiment, which become for ever indissoluble, continuing after death in a future state of existence; and add, that if these habits are of the malevolent kind, they must render the possessor miserable even in heaven. I would apply this ingenious idea to the generation or production of the embryon, or new animal, which partakes so much of the form and propensities of the parent.

Owing to the imperfection of language the offspring is termed a new animal, but is in truth a branch or elongation of the parent; since a part of the embryon-animal is, or was, a part of the parent; and therefore in strict language it cannot be said to be entirely new at the time of its production; and therefore it may retain some of the habits of the parent-system.

At the earliest period of its existence the embryon, as secreted from the blood of the male, would seem to consist of a living filament with certain capabilities of irritation, sensation, volition, and association; and also with some acquired habits or propensities peculiar to the parent: the former of these are in common with other animals; the latter seem to distinguish or produce the kind of animal, whether man or quadruped, with the similarity of feature or form to the parent. It is difficult to be conceived, that a living entity can be separated or produced from the blood by the action of a gland; and which shall afterwards become an animal similar to that in whose vessels it is formed; even though we should suppose with some modern theorists, that the blood is alive; yet every other hypothesis concerning generation rests on principles still more difficult to our comprehension.

[. . .]

DOI: 10.4324/9780429355653-96

The natural history of butterflies, and moths, and beetles, and gnats, is full of curiosity; some of them pass many months, and others even years, in their caterpillar or grub state; they then rest many weeks without food, suspended in the air, buried in the earth, or submersed in water; and change themselves during this time into an animal apparently of a different nature; the stomachs of some of them, which before digested vegetable leaves or roots, now only digest honey; they have acquired wings for the purpose of seeking this new food, and a long proboscis to collect it from flowers, and I suppose a sense of smell to detect the secret places in flowers, where it is formed. The moths, which fly by night, have a much longer proboscis rolled up under their chins like a watch spring; which they extend to collect the honey from flowers in their sleeping state; when they are closed, and the nectaries in consequence more difficult to be plundered. The beetle kind are furnished with an external covering of a hard material to their wings, that they may occasionally again make holes in the earth, in which they passed the former state of their existence.

But what most of all distinguishes these new animals is, that they are now furnished with the powers of reproduction; and that they now differ from each other in sex, which does not appear in their caterpillar or grub state. In some of them the change from a caterpillar into a butterfly or moth seems to be accomplished for the sole purpose of their propagation; since they immediately die after this is finished, and take no food in the interim, as the silk-worm in this climate; though it is possible, it might take honey as food, if it was presented to it. [. . .]

From hence I conclude, that with the acquisition of new parts, new sensations, and new desires, as well as new powers, are produced; and this by accretion to the old ones, and not by distention of them. And finally, that the most essential parts of the system, as the brain for the purpose of distributing the power of life, and the placenta for the purpose of oxygenating the blood, and the additional absorbent vessels for the purpose of acquiring aliment, are first formed by the irritations above mentioned, and by the pleasurable sensations attending those irritations, and by the exertions in consequence of painful sensations, similar to those of hunger and suffocation. After these an apparatus of limbs for future uses, or for the purpose of moving the body in its present natant state, and of lungs for future respiration, and of testes for future reproduction, are formed by the irritations and sensations, and consequent exertions of the parts previously existing, and to which the new parts are to be attached.

[. . .]

6. From this account of reproduction it appears, that all animals have a similar origin, viz. from a single living filament; and that the difference of their forms and qualities has arisen only from the different irritabilities and sensibilities, or voluntarities, or associabilities, of this original living filament; and perhaps in some degree from the different forms of the particles of the fluids, by which it has been at first stimulated into activity. And that from hence, as Linnæus has conjectured in respect to the vegetable world, it is not impossible, but the great variety of species of animals, which now tenant the earth, may have had their origin from

the mixture of a few natural orders. And that those animal and vegetable mules, which could continue their species, have done so, and constitute the numerous families of animals and vegetables which now exist; and that those mules, which were produced with imperfect organs of generation, perished without reproduction, according to the observation of Aristotle; and are the animals, which we now call mules. [. . .]

8. When we revolve in our minds, first, the great changes, which we see naturally produced in animals after their nativity, as in the production of the butterfly with painted wings from the crawling caterpillar; or of the respiring frog from the subnatant tadpole; from the feminine boy to the bearded man, and from the infant girl to the lactescent woman; both which changes may be prevented by certain mutilations of the glands necessary to reproduction.

Secondly, when we think over the great changes introduced into various animals by artificial or accidental cultivation, as in horses, which we have exercised for the different purposes of strength or swiftness, in carrying burthens or in running races; or in dogs, which have been cultivated for strength and courage, as the bull-dog; or for acuteness of his sense or smell, as the hound and spaniel; or for the swiftness of his foot, as the greyhound; or for his swimming in the water, or for drawing snow-sledges, as the rough-haired dogs of the north; or lastly, as a play-dog for children, as the lap-dog; with the changes of the forms of the cattle, which have been domesticated from the greatest antiquity, as camels, and sheep; which have undergone so total a transformation, that we are now ignorant from what species of wild animals they had their origin. Add to these the great changes of shape and colour, which we daily see produced in smaller animals from our domestication of them, as rabbits, or pigeons; or from the difference of climates and even of seasons; thus the sheep of warm climates are covered with hair instead of wool; and the hares and partridges of the latitudes, which are long buried in snow, become white during the winter months; add to these the various changes produced in the forms of mankind, by their early modes of exertion; or by the diseases occasioned by their habits of life; both of which became hereditary, and that through many generations. Those who labour at the anvil, the oar, or the loom, as well as those who carry sedan-chairs, or who have been educated to dance upon the rope, are distinguishable by the shape of their limbs; and the diseases occasioned by intoxication deform the countenance with leprous eruptions, or the body with tumid viscera, or the joints with knots and distortions.

Thirdly, when we enumerate the great changes produced in the species of animals before their nativity; these are such as resemble the form or colour of their parents, which have been altered by the cultivation or accidents above related, and are thus continued to their posterity. Or they are changes produced by the mixture of species as in mules; or changes produced probably by the exuberance of nourishment supplied to the fetus, as in monstrous births with additional limbs; many of these enormities of shape are propagated, and continued as a variety at least, if not as a new species of animal. I have seen a breed of cats with an additional claw on every foot; of poultry also with an additional claw, and with wings to their feet; and of others without rumps. Mr. Buffon mentions a breed of dogs without tails, which are common at Rome and at Naples, which he supposes to have been produced by a custom long established of cutting their tails close off. There are many kinds of pigeons, admired for their peculiarities, which are monsters thus produced and propagated. [. . .]

When we consider all these changes of animal form, and innumerable others, which may be collected from the books of natural history; we cannot but be convinced, that the fetus or embryon is formed by apposition of new parts, and not by the distention of a primordial nest of germs, included one within another, like the cups of a conjurer.

Fourthly, when we revolve in our minds the great similarity of structure, which obtains in all the warm-blooded animals, as well quadrupeds, birds, and amphibious animals, as in mankind; from the mouse and bat to the elephant and whale; one is led to conclude, that they have alike been produced from a similar living filament. In some this filament in its advance to maturity has acquired hands and fingers, with a fine sense of touch, as in mankind. In others it has acquired claws or talons, as in tygers and eagles. In others, toes with an intervening web, or membrane, as in seals and geese. In others it has acquired cloven hoofs, as in cows and swine; and whole hoofs in others, as in the horse. While in the bird kind this original living filament has put forth wings instead of arms or legs, and feathers instead of hair. In some it has protruded horns on the forehead instead of teeth in the fore part of the upper jaw; in others tushes instead of horns; and in others beaks instead of either. And all this exactly as is daily seen in the transmutations of the tadpole, which acquires legs and lungs, when he wants them; and loses his tail, when it is no longer of service to him.

Fifthly, from their first rudiment, or primordium, to the termination of their lives, all animals undergo perpetual transformations; which are in part produced by their own exertions in consequence of their desires and aversions, of their pleasures and their pains, or of irritations, or of associations; and many of these acquired forms or propensities are transmitted to their posterity. [. . .]

As air and water are supplied to animals in sufficient profusion, the three great objects of desire, which have changed the forms of many animals by their exertions to gratify them, are those of lust, hunger, and security. A great want of one part of the animal world has consisted in the desire of the exclusive possession of the females; and these have acquired weapons to combat each other for this purpose, as the very thick, shield-like, horny skin on the shoulder of the boar is a defence only against animals of his own species, who strike obliquely upwards, nor are his tushes for other purposes, except to defend himself, as he is not naturally a carnivorous animal. So the horns of the stag are sharp to offend his adversary, but are branched for the purpose of parrying or receiving the thrusts of horns similar to his own, and have therefore been formed for the purpose of combating other stags for the exclusive possession of the females; who are observed, like the ladies in the times of chivalry, to attend the car of the victor.

The birds, which do not carry food to their young, and do not therefore marry, are armed with spurs for the purpose of fighting for the exclusive possession of the females, as cocks and quails. It is certain that these weapons are not provided for their defence against other adversaries, because the females of these species are without this armour. The final cause of this contest amongst the males seems to be, that the strongest and most active animal should propagate the species, which should thence become improved.

[. . .]

From thus meditating on the great similarity of the structure of the warm-blooded animals, and at the same time of the great changes they undergo both before and after their nativity; and by considering in how minute a portion of time many of the changes of animals above described have been produced; would it be too bold to imagine, that in the great length of time, since the earth began to exist, perhaps millions of ages before the commencement of the history of mankind, would it be too bold to imagine, that all warm-blooded animals have arisen from one living filament, which THE GREAT FIRST CAUSE endued with animality, with the power of acquiring new parts, attended with new propensities, directed by irritations, sensations, volitions, and associations; and thus possessing the faculty of continuing to improve by its own inherent activity, and of delivering down those improvements by generation to its posterity, world without end!

87

ERASMUS DARWIN, 'PRODUCTION OF LIFE', *THE TEMPLE OF NATURE; OR THE ORIGIN OF SOCIETY*

(London: T. Bensley, 1803)

V. 'Organic Life beneath the shoreless waves
Was born and nurs'd in Ocean's pearly caves;
First forms minute, unseen by spheric glass,
Move on the mud, or pierce the watery mass;
These, as successive generations bloom,
New powers acquire, and larger limbs assume;
Whence countless groups of vegetation spring,
And breathing realms of fin, and feet, and wing.

'Thus the tall Oak, the giant of the wood,
Which bears Britannia's thunders on the flood;
The Whale, unmeasured monster of the main,
The lordly Lion, monarch of the plain,
The Eagle soaring in the realms of air,
Whose eye undazzled drinks the solar glare;
Imperious man, who rules the bestial crowd,
Of language, reason, and reflection proud,
With brow erect, who scorns this earthy sod,
And styles himself the image of his God;
Arose from rudiments of form and sense,
An embryon point, or microscopic ens!

'Now in vast shoals beneath the brineless tide,
On earth's firm crust testaceous tribes reside;
Age after age expands the peopled plain,
The tenants perish, but their cells remain;
Whence coral walls and sparry hills ascend
From pole to pole, and round the line extend.

DOI: 10.4324/9780429355653-97

'Next when imprison'd fires in central caves
Burst the firm earth, and drank the headlong waves;
And, as new airs with dread explosion swell,
Form'd lava-isles, and continents of shell;
Pil'd rocks on rocks, on mountains mountains raised,
And high in heaven the first volcanoes blazed;
In countless swarms an insect-myriad moves
From sea-fan gardens, and from coral groves;
Leaves the cold caverns of the deep, and creeps
On shelving shores, or climbs on rocky steeps.
As in dry air the sea-born stranger roves,
Each muscle quickens, and each sense improves;
Cold gills aquatic form respiring lungs,
And sounds aerial flow from slimy tongues.

88

JEAN-BAPTISTE LAMARCK, *ZOOLOGICAL PHILOSOPHY, OR EXPOSITION WITH REGARD TO THE NATURAL HISTORY OF ANIMALS: THE DIVERSITY OF THEIR ORGANISATION AND THE FACULTIES WHICH THEY DERIVE FROM IT*, TRANS. HUGH ELLIOT

(London: Macmillan & Co., 1914 [first published in French 1809])

Chapter 7. Of the influence of the environment on the activities and habits of animals, and the influence of the activities and habits of these living bodies in modifying their organisation and structure

WE are not here concerned with an argument, but with the examination of a positive fact – a fact which is of more general application than is supposed, and which has not received the attention that it deserves, no doubt because it is usually very difficult to recognise. This fact consists in the influence that is exerted by the environment on the various living bodies exposed to it.

It is indeed long since the influence of the various states of our organisation on our character, inclinations, activities and even ideas has been recognised; but I do not think that anyone has yet drawn attention to the influence of our activities and habits even on our organisation. Now since these activities and habits depend entirely on the environment in which we are habitually placed, I shall endeavour to show how great is the influence exerted by that environment on the general shape, state of the parts and even organisation of living bodies. [. . .]

The influence of the environment as a matter of fact is in all times and places operative on living bodies; but what makes this influence difficult to perceive is

DOI: 10.4324/9780429355653-98

that its effects only become perceptible or recognisable (especially in animals) after a long period of time. Before setting forth to examine the proofs of this fact, which deserves our attention and is so important for zoological philosophy, let us sum up the thread of the discussions that we have already begun.

In the preceding chapter we saw that it is now an unquestionable fact that on passing along the animal scale in the opposite direction from that of nature, we discover the existence, in the groups composing this scale, of a continuous but irregular degradation in the organisation of animals, an increasing simplification in their organisation, and, lastly, a corresponding diminution in the number of their faculties.

This well-ascertained fact may throw the strongest light over the actual order followed by nature in the production of all the animals that she has brought into existence, but it does not show us why the increasing complexity of the organisation of animals from the most imperfect to the most perfect exhibits only an irregular gradation, in the course of which there occur numerous anomalies or deviations with a variety in which no order is apparent.

Now on seeking the reason of this strange irregularity in the increasing complexity of animal organisation, if we consider the influence that is exerted by the infinitely varied environments of all parts of the world on the general shape, structure and even organisation of these animals, all will then be clearly explained.

[. . .]

The environment affects the shape and organisation of animals, that is to say that when the environment becomes very different, it produces in course of time corresponding modifications in the shape and organisation of animals.

[. . .] Great alterations in the environment of animals lead to great alterations in their needs, and these alterations in their needs necessarily lead to others in their activities. Now if the new needs become permanent, the animals then adopt new habits which last as long as the needs that evoked them. This is easy to demonstrate, and indeed requires no amplification.

It is then obvious that a great and permanent alteration in the environment of any race of animals induces new habits in these animals.

Now, if a new environment, which has become permanent for some race of animals, induces new habits in these animals, that is to say, leads them to new activities which become habitual, the result will be the use of some one part in preference to some other part, and in some cases the total disuse of some part no longer necessary.

[. . .]

Among individuals of the same species, some of which are continually well fed and in an environment favourable to their development, while others are in an opposite environment, there arises a difference in the state of the individuals which gradually becomes very remarkable. How many examples I might cite both in animals and plants which bear out the truth of this principle! Now if the environment remains constant, so that the condition of the ill-fed, suffering or sickly individuals becomes permanent, their internal organisation is ultimately modified, and these acquired modifications are preserved by reproduction among the individuals in question, and finally give rise to a race quite distinct from that in which the individuals have been continuously in an environment favourable to their development.

[. . .]

Every botanist knows that plants which are transported from their native places to gardens for purposes of cultivation, gradually undergo changes which ultimately make them unrecognisable. Many plants, by nature hairy, become glabrous or nearly so; a number of those which used to lie and creep on the ground, become erect; others lose their thorns or excrescences; others again whose stem was perennial and woody in their native hot climates, become herbaceous in our own climates and some of them become annuals; lastly, the size of their parts itself undergoes very considerable changes. [. . .]

How many different races of our domestic fowls and pigeons have we obtained by rearing them in various environments and different countries; birds which we should now vainly seek in nature? Those which have changed the least, doubtless because their domestication is of shorter standing and because they do not live in a foreign climate, none the less display great differences in some of their parts, as a result of the habits which we have made them contract. Thus our domestic ducks and geese are of the same type as wild ducks and geese; but ours have lost the power of rising into high regions of the air and flying across large tracts of country; moreover, a real change has come about in the state of their parts, as compared with those of the animals of the race from which they come.

[. . .]

Where in natural conditions do we find that multitude of races of dogs which now actually exist, owing to the domestication to which we have reduced them? Where do we find those bull-dogs, greyhounds, water-spaniels, spaniels, lap-dogs, etc., etc.; races which show wider differences than those which we call specific when they occur among animals of one genus living in natural freedom?

No doubt a single, original race, closely resembling the wolf, if indeed it was not actually the wolf, was at some period reduced by man to domestication. That race, of which all the individuals were then alike, was gradually scattered with man into different countries and climates; and after they had been subjected for some time to the influences of their environment and of the various habits which had been forced upon them in each country, they underwent remarkable alterations and formed various special races. [. . .] The crossing of these races by reproduction then gave rise in turn to all those that we now know.

[. . .]

It is known that localities differ as to their character and quality, by reason of their position, construction and climate: as is readily perceived on passing through various localities distinguished by special qualities; this is one cause of variation for animals and plants living in these various places. But what is not known so well and indeed what is not generally believed, is that every locality itself changes in time as to exposure, climate, character and quality, although with such extreme slowness, according to our notions, that we ascribe to it complete stability.

[. . .]

It is obvious then that as regards the character and situation of the substances which occupy the various parts of the earth's surface, there exists a variety of environmental factors which induces a corresponding variety in the shapes and structure of animals, independent of that special variety which necessarily results from the progress of the complexity of organisation in each animal.

In every locality where animals can live, the conditions constituting any one order of things remain the same for long periods: indeed they alter so slowly that man cannot directly observe it. It is only by an inspection of ancient monuments that he becomes convinced that in each of these localities the order of things which he now finds has not always been existent; he may thence infer that it will go on changing.

Races of animals living in any of these localities must then retain their habits equally long: hence the apparent constancy of the races that we call species, – a constancy which has raised in us the belief that these races are as old as nature.

[. . .]

Now the true principle to be noted in all this is as follows:

1 Every fairly considerable and permanent alteration in the environment of any race of animals works a real alteration in the needs of that race.
2 Every change in the needs of animals necessitates new activities on their part for the satisfaction of those needs, and hence new habits.
3 Every new need, necessitating new activities for its satisfaction, requires the animal, either to make more frequent use of some of its parts which it previously used less, and thus greatly to develop and enlarge them; or else to make use of entirely new parts, to which the needs have imperceptibly given birth by efforts of its inner feeling [. . .]

Thus to obtain a knowledge of the true causes of that great diversity of shapes and habits found in the various known animals, we must reflect that the infinitely diversified but slowly changing environment in which the animals of each race have successively been placed, has involved each of them in new needs and corresponding alterations in their habits. This is a truth which, once recognised, cannot be disputed. Now we shall easily discern how the new needs may have been satisfied, and the new habits acquired, if we pay attention to the two following laws of nature, which are always verified by observation.

First Law

In every animal which has not passed the limit of its development, more frequent and continuous use of any organ gradually strengthens, develops and enlarges that organ, and gives it a power proportional to the length of time it has been so used;

while the permanent disuse of any organ imperceptibly weakens and deteriorates it, and progressively diminishes its functional capacity, until it finally disappears.

Second Law

ALL the acquisitions or losses wrought by nature on individuals, through the influence of the environment in which their race has long been placed, and hence through the influence of the predominant use or permanent disuse of any organ; all these are preserved by reproduction to the new individuals which arise, provided that the acquired modifications are common to both sexes, or at least to the individuals which produce the young.

Here we have two permanent truths, which can only be doubted by those who have never observed or followed the operations of nature.

89

PATRICK MATTHEW, 'APPENDIX: NOTE B', *ON NAVAL TIMBER AND ARBORICULTURE*

(London: Longman, Rees, Orme, Brown, and Green, 1831)

Note B

THERE is a law universal in nature, tending to render every reproductive being the best possibly suited to its condition that its kind, or that organized matter, is susceptible of, which appears intended to model the physical and mental or instinctive powers, to their highest perfection, and to continue them so. This law sustains the lion in his strength, the hare in her swiftness, and the fox in his wiles. As Nature, in all her modifications of life, has a power of increase far beyond what is needed to supply the place of what falls by Time's decay, those individuals who possess not the requisite strength, swiftness, hardihood, or cunning, fall prematurely without reproducing – either a prey to their natural devourers, or sinking under disease, generally induced by want of nourishment, their place being occupied by the more perfect of their own kind, who are pressing on the means of subsistence. The law of entail, necessary to hereditary nobility, is an outrage on this law of nature which she will not pass unavenged – a law which has the most debasing influence upon the energies of a people, and will sooner or later lead to general subversion, more especially when the executive of a country remains for a considerable time efficient, and no effort is needed on the part of the nobility to protect their own, or no war to draw forth or preserve their powers by exertion.

DOI: 10.4324/9780429355653-99

90

CHARLES LYELL, *PRINCIPLES OF GEOLOGY, BEING AN ATTEMPT TO EXPLAIN THE FORMER CHANGES OF THE EARTH'S SURFACE, BY REFERENCE TO CAUSES NOW IN OPERATION*, 3 VOLS, VOL. 2

(London: John Murray, 1830–3), Vol. 1 (1830), Vol. 2 (1832), Vol. 3 (1833)

Chapter 2

LAMARCK has somewhat misstated the idea commonly entertained of a species, for it is not true that naturalists in general assume that the organization of an animal or plant remains absolutely constant, and that it can never vary in any of its parts. All must be aware that circumstances influence the habits, and that the habits may alter the state of the parts and organs. But the difference of opinion relates to the extent to which these modifications of the habits and organs of a particular species may be carried.

Now let us first inquire what positive facts can be adduced in the history of known species, to establish a great and permanent amount of change in the form, structure, or instinct of individuals descending from some common stock. The best authenticated examples of the extent to which species can be made to vary, may be looked for in the history of domesticated animals and cultivated plants. It usually happens that those species, both of the animal and vegetable kingdom, which have the greatest pliability of organization, those which are most capable of accommodating themselves to a great variety of new circumstances, are most serviceable to man. These only can be carried by him into different climates, and can have their properties or instincts variously diversified by differences of nourishment and habits. If the resources of a species be so limited, and its habits and faculties be of such a confined and local character, that it can only flourish in a few particular spots, it can rarely be of great utility.

 DOI: 10.4324/9780429355653-100

We may consider, therefore, that in perfecting the arts of domesticating animals and cultivating plants, mankind have first selected those species which have the most flexible frames and constitutions, and have then been engaged for ages in conducting a series of experiments, with much patience and at great cost, to ascertain what may be the greatest possible deviation from a common type which can be elicited in these extreme cases.

The modifications produced in the different races of dogs, exhibit the influence of man in the most striking point of view. These animals have been transported into every climate, and placed in every variety of circumstances; they have been made, as a modern naturalist observes, the servant, the companion, the guardian, and the intimate friend of man, and the power of a superior genius has had a wonderful influence, not only on their forms, but on their manners and intelligence. Different races have undergone remarkable changes in the quantity and colour of their clothing: the dogs of Guinea are almost naked, while those of the Arctic circle are covered with a warm coat both of hair and wool, which enables them to bear the most intense cold without inconvenience. There are differences also of another kind no less remarkable, as in size, the length of their muzzles, and the convexity of their foreheads.

But if we look for some of those essential changes which would be required to lend even the semblance of a foundation for the theory of Lamarck, respecting the growth of new organs and the gradual obliteration of others, we find nothing of the kind. For in all these varieties of the dog, says Cuvier, the relation of the bones with each other remain essentially the same; the form of the teeth never changes in any perceptible degree, except that in some individuals, one additional false grinder occasionally appears, sometimes on the one side, and sometimes on the other. The greatest departure from a common type, and it constitutes the maximum of variation as yet known in the animal kingdom, is exemplified in those races of dogs which have a supernumerary toe on the hind foot with the corresponding tarsal bones, a variety analogous to one presented by six-fingered families of the human race.

Lamarck has thrown out as a conjecture, that the wolf may have been the original of the dog, but he has adduced no data to bear out such an hypothesis. 'The wolf', observes Dr Prichard, 'and the dog differ, not only with respect to their habits and instincts, which in the brute creation are very uniform within the limits of one species; but some differences have also been pointed out in their internal organization, particularly in the structure of a part of the intestinal canal'.

It is well known that the horse, the ox, the boar and other domestic animals, which have been introduced into South America, and have run wild in many parts, have entirely lost all marks of domesticity, and have reverted to the original characters of their species. But the dog has also become wild in Cuba, Haiti, and in all the Caribbean islands. In the course of the seventeenth century, they hunted in packs from twelve to fifty, or more in number, and fearlessly attacked herds of wild-boars and other animals. It is natural, therefore, to enquire to what form they reverted? Now they are said by many travellers to have resembled very nearly the shepherd's dog; but it is certain that they were never turned into wolves. They were extremely savage, and their ravages appear to have been as much dreaded as those of wolves, but when any of their whelps were caught, and brought from the woods to the towns, they grew up in the most perfect submission to man.

As the advocates of the theory of transmutation trust much to the slow and insensible changes which time may work, they are accustomed to lament the absence of accurate descriptions, and figures of particular animals and plants, handed down from the earliest periods of history, such as might have afforded data for comparing the condition of species, at two periods considerably remote. But fortunately, we are in some measure independent of such evidence, for by a singular accident, the priests of Egypt have bequeathed to us, in their cemeteries, that information, which the museums and works of the Greek philosophers have failed to transmit.

For the careful investigation of these documents, we are greatly indebted to the skill and diligence of those naturalists who accompanied the French armies during their brief occupation of Egypt: that conquest of four years, from which we may date the improvement of the modern Egyptians in the arts and sciences, and the rapid progress which has been made of late in our knowledge of the arts and sciences of their remote predecessors. Instead of wasting their whole time as so many preceding travellers had done, in exclusively collecting human mummies, M. Geoffroy and his associates examined diligently, and sent home great numbers of embalmed bodies of consecrated animals, such as the bull, the dog, the cat, the ape, the ichneumon, the crocodile, and the ibis.

[. . .]

Among the Egyptian mummies thus procured were not only those of numerous wild quadrupeds, birds, and reptiles, but, what was perhaps of still greater importance in deciding the great question under discussion, there were the mummies of domestic animals, among which those above mentioned, the bull, the dog, and the cat, were frequent. Now such was the conformity of the whole of these species to those now living, that there was no more difference, says Cuvier, between them than between the human mummies and the embalmed bodies of men of the present day. Yet some of these animals have since that period been transported by man to almost every variety of climate, and forced to accommodate their habits to new circumstances, as far as their nature would permit. The cat, for example, has been carried over the whole earth, and, within the last three centuries, has been naturalized in every part of the new world, from the cold regions of Canada to the tropical plains of Guiana; yet it has scarcely undergone any perceptible mutation, and is still the same animal which was held sacred by the Egyptians.

Of the ox, undoubtedly there are many very distinct races; but the bull Apis, which was led in solemn processions by the Egyptian priests, did not differ from some of those now living. The black cattle that have run wild in America, where there were many peculiarities in the climate not to be found, perhaps, in any part of the old world, and where scarcely a single plant on which they fed was of precisely the same species, instead of altering their form and habits, have actually reverted to the exact likeness of the aboriginal wild cattle of Europe.

In answer to the arguments drawn from the Egyptian mummies, Lamarck said that they were identical with their living descendants in the same country, because the climate and physical geography of the banks of the Nile have remained unaltered for the last thirty centuries. But why, we ask, have other individuals of these species retained the same characters in so many different quarters of the globe, where the climate and many other conditions are so varied?

The evidence derived from the Egyptian monuments was not confined to the animal kingdom; the fruits, seeds, and other portions of twenty different plants, were faithfully preserved in the same manner; and among these the common wheat was procured by Delille, from closed vessels in the sepulchres of the kings, the grains of which retained not only their form, but even their colour, so effectual has proved the process of embalming with bitumen in a dry and equable climate. No difference could be detected between this wheat and that which now grows in the East and elsewhere, and similar identifications were made in regard to all the other plants.

And here we may observe, that there is an obvious answer to Lamarck's objection, that the botanist cannot point out a country where the common wheat grows wild, unless in places where it may have been derived from neighbouring cultivation. All naturalists are well aware that the geographical distribution of a great number of species is extremely limited, and that it was to be expected that every useful plant should first be cultivated successfully in the country where it was indigenous, and that, probably, every station which it partially occupied, when growing wild, would be selected by the agriculturist as best suited to it when artificially increased. [. . .]

If we are to infer that some one of the wild grasses has been transformed into the common wheat, and that some animal of the genus *Canis,* still unreclaimed, has been metamorphosed into the dog, merely because we cannot find the domestic dog, or the cultivated wheat, in a state of nature, we may be next called upon to make similar admissions in regard to the camel; for it seems very doubtful whether any race of this species of quadruped is now wild.

But if agriculture, it will be said, does not supply examples of extraordinary changes of form and organization, the horticulturist can, at least, appeal to facts which may confound the preceding train of reasoning. The crab has been transformed into the apple; the sloe into the plum: flowers have changed their colour and become double; and these new characters can be perpetuated by seed,– a bitter plant with wavy sea-green leaves has been taken from the sea-side where it grew like wild charlock, has been transplanted into the garden, lost its saltness, and has been metamorphosed into two distinct vegetables as unlike each other as is each to the parent plant – the red cabbage and the cauliflower, These, and a multitude of analogous facts, are undoubtedly among the wonders of nature, and attest more strongly, perhaps, the extent to which species may be modified, than any examples derived from the animal kingdom. But in these cases we find, that we soon reach certain limits, beyond which we are unable to cause the individuals, descending from the same stock, to vary; while, on the other hand, it is easy to show that these extraordinary varieties could seldom arise, and could never be perpetuated in a wild state for many generations, under any imaginable combination of accidents. They may be regarded as extreme cases brought about by human interference, and not as phenomena which indicate a capability of indefinite modification in the natural world.

[. . .]

The different races of cabbages afford, as we have admitted, an astonishing example of deviation from a common type; but we can scarcely conceive them to have originated, much less to have lasted for several generations, without the intervention of man. It is only by strong manures that these varieties have been obtained, and in poorer soils they instantly degenerate. If, therefore, we suppose in a state of nature the seed of the wild Brassica oleracea to have been wafted from the sea-side to some spot enriched by the dung of animals, and to have there become a cauliflower, it would soon diffuse its seed to some comparatively steril soils around, and the offspring would relapse to the likeness of the parent stock, like some individuals which may now be seen growing on the cornice of old London bridge.

But if we go so far as to imagine the soil, in the spot first occupied, to be constantly manured by herds of wild animals, so as to continue as rich as that of a garden, still the variety could not be maintained, because we know that each of these races is prone to fecundate others, and gardeners are compelled to exert the utmost diligence to prevent cross-breeds. The intermixture of the pollen of varieties growing in the poorer soil around, would soon destroy the peculiar characters of the race which occupied the highly-manured tract; for, if these accidents so continually happen in spite of us, among the culinary varieties, it is easy to see how soon this cause might obliterate every marked singularity in a wild state.

Besides, it is well-known that although the pampered races which we rear in our gardens for use or ornament, may often be perpetuated by seed, yet they rarely produce seed in such abundance, or so prolific in quality, as wild individuals; so that, if the care of man were withdrawn, the most fertile variety would always, in the end, prevail over the more steril.

91

CHARLES DARWIN, *JOURNAL OF RESEARCHES INTO THE GEOLOGY AND NATURAL HISTORY OF THE VARIOUS COUNTRIES VISITED BY H.M.S. BEAGLE UNDER THE COMMAND OF CAPTAIN FITZROY, R.N. FROM 1832 TO 1836*

(London: Henry Colburn, 1839)

Chapter 29, September to October 1835, Galapagos Archipelago

October 3rd

Of mammalia a large kind of mouse forms a well-marked species. From its large thin ears and other characters, it approaches in form a section of the genus, which is confined to the sterile regions of South America. There is also a rat which Mr Waterhouse believes is probably distinct from the English kind; but I cannot help suspecting that it is only the same altered by the peculiar conditions of its new country.

In my collections from these islands, Mr Gould considers that there are twenty-six different species of land birds. With the exception of one, all probably are undescribed kinds, which inhabit this archipelago, and no other part of the world. Among the waders and waterfowl it is more difficult, without detailed comparison, to say what are new. But a water-rail which lives near the summits of the mountains, is undescribed, as perhaps is a Totanus and a heron. The only kind of gull which is found among these islands, is also new; when the wandering habits of this genus are considered, this is a very remarkable circumstance. The species most closely allied to it, comes from the Strait of Magellan. Of the other aquatic birds, the species appear the same with well-known American birds.

DOI: 10.4324/9780429355653-101

The general character of the plumage of these birds is extremely plain, and like the Flora possesses little beauty. Although the species are thus peculiar to the archipelago, yet nearly all in their general structure, habits, colour of feathers, and even tone of voice, are strictly American. The following brief list will give an idea of their kinds. 1st. A buzzard, having many of the characters of Polyborus or Caracara; and in its habits not to be distinguished from that peculiar South American genus; 2d. Two owls; 3d. Three species of tyrant-flycatchers – a form strictly American. One of these appears identical with a common kind (*Musicapa coronata*? Lath.), which has a very wide range, from La Plata throughout Brazil to Mexico; 4th. A sylvicola, an American form, and especially common in the northern division of the continent; 5th. Three species of mocking-birds, a genus common to both Americas; 6th. A finch, with a stiff tail and a long claw to its hinder toe, closely allied to a North American genus; 7th. A swallow belonging to the American division of that genus; 8th. A dove, like, but distinct from, the Chilian species; 9th. A group of finches, of which Mr Gould considers there are thirteen species; and these he has distributed into four new sub-genera. These birds are the most singular of any in the archipelago. They all agree in many points; namely, in a peculiar structure of their bill, short tails, general form, and in their plumage. The females are gray or brown, but the old cocks jet-black. All the species, excepting two, feed in flocks on the ground, and have very similar habits. It is very remarkable that a nearly perfect gradation of structure in this one group can be traced in the form of the beak, from one exceeding in dimensions that of the largest grosbeak, to another differing but little from that of a warbler. Of the aquatic birds I have already remarked that some are peculiar to these islands, and some common to North and South America.

We will now turn to the order of reptiles, which forms, perhaps, the most striking feature in the zoology of these islands. The species are not numerous, but the number of individuals of each kind, is extraordinarily great. There is one kind both of the turtle and tortoise; of lizards four; and of snakes about the same number.

I will first describe the habits of the tortoise (*Testudo Indicus*) which has been frequently alluded to. These animals are found, I believe, in all the islands of the Archipelago; certainly in the greater number. They frequent in preference the high damp parts, but likewise inhabit the lower and arid districts [. . .] Some individuals grow to an immense size: Mr Lawson, an Englishman, who had at the time of our visit charge of the colony, told us that he had seen several so large, that it required six or eight men to lift them from the ground; and that some had afforded as much as two hundred pounds of meat. The old males are the largest, the females rarely growing to so great a size. The male can readily be distinguished from the female by the greater length of its tail. The tortoises which live on those islands where there is no water, or in the lower and arid parts of the others, chiefly feed on the succulent cactus. Those which frequent the higher and damp regions, eat the leaves of various trees, a kind of berry (called guayavita) which is acid and austere, and likewise a pale green filamentous lichen, that hangs in tresses from the boughs of the trees.

The tortoise is very fond of water, drinking large quantities, and wallowing in the mud. The larger islands alone possess springs, and these are always situated towards the central parts, and at a considerable elevation. The tortoises, therefore, which frequent the lower districts, when thirsty, are obliged to travel from a long distance. Hence broad and well-beaten paths radiate off in every direction from the wells even down to the sea-coast; and the Spaniards by following them up, first discovered the watering-places. When I landed at Chatham Island, I could not imagine what animal travelled so methodically along the well-chosen tracks. Near the springs it was a curious spectacle to behold many of these great monsters; one set eagerly travelling onwards with outstretched necks, and another set returning, after having drunk their fill. When the tortoise arrives at the spring, quite regardless of any spectator, it buries its head in the water above its eyes, and greedily swallows great mouthfuls, at the rate of about ten in a minute. The inhabitants say each animal stays three or four days in the neighbourhood of the water, and then returns to the lower country; but they differed in their accounts respecting the frequency of these visits. The animal probably regulates them according to the nature of the food which it has consumed. It is, however, certain, that tortoises can subsist even on those islands where there is no other water, than what falls during a few rainy days in the year.

I believe it is well ascertained, that the bladder of the frog acts as a reservoir for the moisture necessary for its existence: such seems to be the case with the tortoise. For some time after a visit to the springs, the urinary bladder of these animals is distended with fluid, which is said gradually to decrease in volume, and to become less pure. The inhabitants, when walking in the lower district, and overcome with thirst, often take advantage of this circumstance, by killing a tortoise, and if the bladder is full, drinking its contents. In one I saw killed, the fluid was quite limpid, and had only a *very slightly* bitter taste. The inhabitants, however, always drink first the water in the pericardium, which is described as being best.

The tortoises, when moving towards any definite point, travel by night and day, and arrive at their journey's end much sooner than would be expected. The inhabitants, from observations on marked individuals, consider that they can move a distance of about eight miles in two or three days. One large tortoise, which I watched, I found walked at the rate of sixty yards in ten minutes, that is 360 in the hour, or four miles a day,– allowing also a little time for it to eat on the road.

During the breeding season, when the male and female are together, the male utters a hoarse roar or bellowing, which it is said, can be heard at the distance of more than a hundred yards. The female never uses her voice, and the male only at such times; so that when the people hear this noise, they know the two are together. They were at this time (October) laying their eggs. The female, where the soil is sandy, deposits them together, and covers them up with sand; but where the ground is rocky she drops them indiscriminately in any hollow. Mr Bynoe found seven placed in a line in a fissure. The egg is white and spherical; one which I measured was seven inches and three-eighths in circumference. The young animals, as soon as they are hatched, fall a prey in great numbers to the buzzard, with the habits of the Caracara. The old ones seem generally to die from accidents, as from falling down precipices. At least several of the inhabitants told me, they had never found one dead without some such apparent cause.

The inhabitants believe that these animals are absolutely deaf; certainly they do not overhear a person walking close behind them. I was always amused, when overtaking one of these great monsters as it was quietly pacing along, to see how suddenly, the instant I passed, it would draw in its head and legs, and uttering a deep hiss fall to the ground with a heavy sound, as if struck dead. I frequently got on their backs, and then, upon giving a few raps on the hinder part of the shell, they would rise and walk away;– but I found it very difficult to keep my balance.

The flesh of this animal is largely employed, both fresh and salted; and a beautifully clear oil is prepared from the fat. When a tortoise is caught, the man makes a slit in the skin near its tail, so as to see inside its body, whether the fat under the dorsal plate is thick. If it is not, the animal is liberated; and it is said to recover soon from this strange operation. In order to secure the tortoises, it is not sufficient to turn them like turtle, for they are often able to regain their upright position.

It was confidently asserted, that the tortoises coming from different islands in the archipelago were slightly different in form; and that in certain islands they attained a larger average size than in others. Mr Lawson maintained that he could at once tell from which island any one was brought. Unfortunately, the specimens which came home in the Beagle were too small to institute any certain comparison. This tortoise, which goes by the name of *Testudo Indicus*, is at present found in many parts of the world. It is the opinion of Mr Bell, and some others who have studied reptiles, that it is not improbable that they all originally came from this archipelago. When it is known how long these islands have been frequented by the bucaniers, and that they constantly took away numbers of these animals alive, it seems very probable that they should have distributed them in different parts of the world.

92

JOHN GOULD AND CHARLES DARWIN, 'FAM. COCCOTHRAUSTINÆ', *ZOOLOGY OF THE VOYAGE OF H.M.S. BEAGLE*, VOL. 3: BIRDS

(London: Smith, Elder and Co., 1841)

FAM. – COCCOTHRAUSTINÆ.

GENUS, GEOSPIZA, *Gould.*

Corporis figura brevissima et robusta.

Rostrum magnum, robustum, validum, altitudine longitudinem præstante; culmine arcuato et capitis verticem superante, apice sine denticulo, lateribus tumidis.

Naribus basalibus et semitectis plumis frontalibus.

Mandibulâ superiori tomiis medium versus sinum exhibentibus, ad mandibulæ inferioris processum recipiendum. Mandibula inferior ad basin lata, hoc infra oculos tendente. Alæ mediocres remige primo paulo breviore secundo, hoc longissimo.

Cauda brevissima et æqualis.

Tarsi magni et validi, digito postico, cum ungue robusto et digito intermedio breviore; digitis externis inter se æqualibus at digito postico brevioribus. Color in maribus niger, in fœm. fuscus.

THIS singular genus appears to be confined to the islands of the Galapagos Archipelago. It is very numerous, both in individuals and in species, so that it forms the most striking feature in their ornithology. The characters of the species of Geospiza, as well as of the following allied subgenera, run closely into each other in a most remarkable manner.

In my Journal of Researches, p. 475, I have given my reasons for believing that in some cases the separate islands possess their own representatives of the different species, and this almost necessarily would cause a fine gradation in their characters. Unfortunately I did not suspect this fact until it was too late to distinguish the specimens from the different islands of the group; but from the collection

DOI: 10.4324/9780429355653-102

made for Captain FitzRoy, I have been able in some small measure to rectify this omission.

In each species of these genera a perfect gradation in colouring might, I think, be formed from one jet black to another pale brown. My observations showed that the former were invariably the males; but Mr. Bynoe, the surgeon of the Beagle, who opened many specimens, assured me that he found two quite black specimens of one of the smaller species of Geospiza, which certainly were females: this, however, undoubtedly is an exception to the general fact; and is analogous to those cases, which Mr. Blyth has recorded of female linnets and some other birds, in a state of high constitutional vigour, assuming the brighter plumage of the male. The jet black birds, in cases where there could be no doubt in regard to the species, were in singularly few proportional numbers to the brown ones: I can only account for this by the supposition that the intense black colour is attained only by three-year-old birds. I may here mention, that the time of year (beginning of October) in which my collection was made, probably corresponds, as far as the purposes of incubation are concerned, with our autumn. The several species of Geospiza are undistinguishable from each other in habits; they often form, together with the species of the following subgenera, and likewise with doves, large irregular flocks. They frequent the rocky and extremely arid parts of the land sparingly covered with almost naked bushes, near the coasts; for here they find, by scratching in the cindery soil with their powerful beaks and claws, the seeds of grasses and other plants, which rapidly spring up during the short rainy season, and as rapidly disappear. They often eat small portions of the succulent leaves of the *Opuntia Galapageia,* probably for the sake of the moisture contained in them: in this dry climate the birds suffer much from the want of water, and these finches, as well as others, daily crowd round the small and scanty wells, which are found on some of the islands. I seldom, however, saw these birds in the upper and damp region, which supports a thriving vegetation; excepting on the cleared and cultivated fields near the houses in Charles Island, where, as I was informed by the colonists, they do much injury by digging up roots and seeds from a depth of even six inches.

93

[ROBERT CHAMBERS], 'HYPOTHESIS OF THE DEVELOPMENT OF THE VEGETABLE AND ANIMAL KINGDOMS', *VESTIGES OF THE NATURAL HISTORY OF CREATION*

(London: John Churchill, 1844)

While the external forms of all these various animals are so different, it is very remarkable that the whole are, after all, variations of a fundamental plan, which can be traced as a basis throughout the whole, the variations being merely modifications of that plan to suit the particular conditions in which each particular animal has been designed to live. Starting from the primeval germ, which, as we have seen, is the representative of a particular order of full-grown animals, we find all others to be merely advances from that type, with the extension of endowments and modification of forms which are required in each particular case; each form, also, retaining a strong affinity to that which precedes it, and tending to impress its own features on that which succeeds. This unity of structure, as it is called, becomes the more remarkable, when we observe that the organs, while preserving a resemblance, are often put to different uses. For example: the ribs become, in the serpent, organs of locomotion, and the snout is extended, in the elephant, into a prehensile instrument.

It is equally remarkable that analogous purposes are served in different animals by organs essentially different. Thus, the mammalia breathe by lungs; the fishes, by gills. These are not modifications of one organ, but distinct organs. In mammifers, the gills exist and act at an early stage of the fœtal state, but afterwards go back and appear no more; while the lungs are developed. In fishes, again, the gills only are fully developed; while the lung structure either makes no advance at all, or only appears in the rudimentary form of an air-bladder. So, also, the baleen of the whale and the teeth of the land mammalia are different organs. The whale, in embryo, shews the rudiments of teeth; but these, not being wanted, are not developed, and the baleen is brought forward instead. The land animals, we may also

DOI: 10.4324/9780429355653-103

be sure, have the rudiments of baleen in their organization. In many instances, a particular structure is found advanced to a certain point in a particular set of animals, (for instance, feet in the serpent tribe,) although it is not there required in any degree; but the peculiarity, being carried a little farther forward, is perhaps useful in the next set of animals in the scale. Such are called rudimentary organs.

[. . .]

These facts clearly shew how all the various organic forms of our world are bound up in one – how a fundamental unity pervades and embraces them all, collecting them, from the humblest lichen up to the highest mammifer, in one system, the whole creation of which must have depended upon one law or decree of the Almighty, though it did not all come forth at one time. After what we have seen, the idea of a separate exertion for each must appear totally inadmissible. The single fact of abortive or rudimentary organs condemns it; for these, on such a supposition, could be regarded in no other light than as blemishes or blunders – the thing of all others most irreconcilable with that idea of Almighty Perfection which a general view of nature so irresistibly conveys. On the other hand, when the organic creation is admitted to have been effected by a general law, we see nothing in these abortive parts but harmless peculiarities of development, and interesting evidences of the manner in which the Divine Author has been pleased to work.

[. . .]

The idea, then, which I form of the progress of organic life upon the globe – and the hypothesis is applicable to all similar theatres of vital being – is, *that the simplest and most primitive type, under a law to which that of like-production is subordinate, gave birth to the type next above it, that this again produced the next higher, and so on to the very highest,* the stages of advance being in all cases very small – namely, from one species only to another; so that the phenomenon has always been of a simple and modest character. Whether the whole of any species was at once translated forward, or only a few parents were employed to give birth to the new type, must remain undetermined; but, supposing that the former was the case, we must presume that the moves along the line or lines were simultaneous, so that the place vacated by one species was immediately taken by the next in succession, and so on back to the first, for the supply of which the formation of a new germinal vesicle out of inorganic matter was alone necessary. Thus, the production of new forms, as shewn in the pages of the geological record, has never been anything more than a new stage of progress in gestation, an event as simply natural, and attended as little by any circumstances of a wonderful or startling kind, as the silent advance of an ordinary mother from one week to another of her pregnancy. Yet, be it remembered, the whole phenomena are, in another point of view, wonders of the highest kind, for in each of them we have to trace the effect of an Almighty Will which had arranged the whole in such harmony with external physical circumstances, that both were developed in parallel steps – and probably this development upon our planet is but a sample of what has taken place, through the same cause, in all the other countless theatres of being which are suspended in space.

94

THOMAS MONCK MASON, *CREATION BY THE IMMEDIATE AGENCY OF GOD, AS OPPOSED TO CREATION BY NATURAL LAW; BEING A REFUTATION OF THE WORK ENTITLED VESTIGES OF THE NATURAL HISTORY OF CREATION*

(London: John W. Parker, 1845)

WITH these facts before us it would be mere affectation to hesitate any longer as to the light in which we are to regard the present work in its relation to the Christian Scriptures. Without meaning to infer which of the two is most entitled to our acceptance, certain it is that one or the other is false – one or the other *alone* is true. Of the importance of the interest at stake there can be no doubt. It is, in fact, to those who believe, nothing less than a question of life and death. If the theory of the *Vestiges of the Natural History of Creation* be true, then are all the hopes of the Christian illusory and vain; the Bible is a fiction; the light we have been following from infancy is a mere *ignis fatuus*; the mighty structure of the Christian faith, under the shadow of which all our institutions have grown up, and with reference to which all the moral, civil, and political arrangements of the civilized world have been devised, is but the 'baseless fabric of a vision', destined to vanish before the light of approaching day, and 'leave not a wreck behind'.

Nor are the consequences confined to the Christian believer only. The whole world has a paramount interest in the result; for as certainly as there is no Bible, so surely is there no religion or morality of any kind. For what religious obligations can bind man to a God who only makes him as a matter of course, and when made takes no concern in him? Or what moral relations can exist between individuals independent of any common tie but that of interest (p. 383), who are mere 'electro-chemical' machines (p. 204), created with no view but that of their own personal gratification and enjoyment? (p. 379).

DOI: 10.4324/9780429355653-104

That all organic creation should appear to be formed upon one, or more properly, a definite number of models, supposing it an established truth, is, of itself simply, a fact that can afford no grounds for any conclusion whatever in respect of the mode in which that creation may have been effected. Taking it for granted that it was the design of the Almighty that His works of a certain class should observe a certain uniformity of character, (and that it was so is proved in the result,) the mere fulfilment of that design can never be construed to favour one mode of operation more than another, supposing both to be equally available for the purpose.

The fact that all the objects in each class should be capable of being arranged so as to present the phenomena of a certain number of regular successions of forms, is a necessary consequence of the fulfilment of the design, and therefore liable to the same exception. A number of *different* things constituted upon *one* model *must* be susceptible of such a mode of arrangement, however they may be supposed to have been produced; and the individuals, or the species which they may be taken to represent, must, as a matter of course, exhibit some of the features belonging to each of the adjacent terms of the series, as in the cases referred to, of the struthionidæ and ornithorhyncus, connecting the classes of aves and mammalia.

That the resemblance between the objects standing next one another in the succession should be very strong, is merely a consequence, and a necessary one, of the *number* and *equable distribution* of the species, and has no reference to any other kind of connexion between them.

95

CHARLES DARWIN AND ALFRED RUSSEL WALLACE, 'ON THE TENDENCY OF SPECIES TO FORM VARIETIES; AND ON THE PERPETUATION OF VARIETIES AND SPECIES BY NATURAL MEANS OF SELECTION', *JOURNAL OF THE PROCEEDINGS OF THE LINNEAN SOCIETY OF LONDON, ZOOLOGY* 3:9

(1858)

[Read July 1st, 1858.]

London, June 30th, 1858.

My Dear Sir,– The accompanying papers, which we have the honour of communicating to the Linnean Society, and which all relate to the same subject, viz. the Laws which affect the Production of Varieties, Races, and Species, contain the results of the investigations of two indefatigable naturalists, Mr. Charles Darwin and Mr. Alfred Wallace.

These gentlemen having, independently and unknown to one another, conceived the same very ingenious theory to account for the appearance and perpetuation of varieties and of specific forms on our planet, may both fairly claim the merit of being original thinkers in this important line of inquiry; but neither of them having published his views, though Mr. Darwin has for many years past been repeatedly urged by us to do so, and both authors having now unreservedly placed their papers in our hands, we think it would best promote the

DOI: 10.4324/9780429355653-105

interests of science that a selection from them should be laid before the Linnean Society.

Taken in the order of their dates, they consist of:–

1 Extracts from a MS. work on Species, by Mr. Darwin, which was sketched in 1839, and copied in 1844, when the copy was read by Dr. Hooker, and its contents afterwards communicated to Sir Charles Lyell. The first Part is devoted to 'The Variation of Organic Beings under Domestication and in their Natural State'; and the second chapter of that Part, from which we propose to read to the Society the extracts referred to, is headed, 'On the Variation of Organic Beings in a state of Nature; on the Natural Means of Selection; on the Comparison of Domestic Races and true Species'.
2 An abstract of a private letter addressed to Professor Asa Gray, of Boston, U.S., in October 1857, by Mr. Darwin, in which he repeats his views, and which shows that these remained unaltered from 1839 to 1857.
3 An Essay by Mr. Wallace, entitled 'On the Tendency of Varieties to depart indefinitely from the Original Type'. This was written at Ternate in February 1858, for the perusal of his friend and correspondent Mr. Darwin, and sent to him with the expressed wish that it should be forwarded to Sir Charles Lyell, if Mr. Darwin thought it sufficiently novel and interesting. So highly did Mr. Darwin appreciate the value of the views therein set forth, that he proposed, in a letter to Sir Charles Lyell, to obtain Mr. Wallace's consent to allow the Essay to be published as soon as possible. Of this step we highly approved, provided Mr. Darwin did not withhold from the public, as he was strongly inclined to do (in favour of Mr. Wallace), the memoir which he had himself written on the same subject, and which, as before stated, one of us had perused in 1844, and the contents of which we had both of us been privy to for many years. On representing this to Mr. Darwin, he gave us permission to make what use we thought proper of his memoir, &c.; and in adopting our present course, of presenting it to the Linnean Society, we have explained to him that we are not solely considering the relative claims to priority of himself and his friend, but the interests of science generally; for we feel it to be desirable that views founded on a wide deduction from facts, and matured by years of reflection, should constitute at once a goal from which others may start, and that, while the scientific world is waiting for the appearance of Mr. Darwin's complete work, some of the leading results of his labours, as well as those of his able correspondent, should together be laid before the public.

We have the honour to be yours very obediently,

CHARLES LYELL.
Jos. D. HOOKER.

[. . .]

I. Extract from an unpublished Work on Species, by C. DARWIN, Esq., consisting of a portion of a Chapter entitled, 'On the Variation of Organic Beings in a state of Nature; on the Natural Means of Selection; on the Comparison of Domestic Races and true Species'.

De Candolle, in an eloquent passage, has declared that all nature is at war, one organism with another, or with external nature.

Seeing the contented face of nature, this may at first well be doubted; but reflection will inevitably prove it to be true. The war, however, is not constant, but recurrent in a slight degree at short periods, and more severely at occasional more distant periods; and hence its effects are easily overlooked. It is the doctrine of Malthus applied in most cases with tenfold force. As in every climate there are seasons, for each of its inhabitants, of greater and less abundance, so all annually breed; and the moral restraint which in some small degree checks the increase of mankind is entirely lost. Even slow-breeding mankind has doubled in twenty-five years; and if he could increase his food with greater ease, he would double in less time. But for animals without artificial means, the amount of food for each species must, on an average, be constant, whereas the increase of all organisms tends to be geometrical, and in a vast majority of cases at an enormous ratio. [. . .]

Many practical illustrations of this rapid tendency to increase are on record, among which, during peculiar seasons, are the extraordinary numbers of certain animals; for instance, during the years 1826 to 1828, in La Plata, when from drought some millions of cattle perished, the whole country actually swarmed with mice. Now I think it cannot be doubted that during the breeding-season all the mice (with the exception of a few males or females in excess) ordinarily pair, and therefore that this astounding increase during three years must be attributed to a greater number than usual surviving the first year, and then breeding, and so on till the third year, when their numbers were brought down to their usual limits on the return of wet weather. Where man has introduced plants and animals into a new and favourable country, there are many accounts in how surprisingly few years the whole country has become stocked with them. This increase would necessarily stop as soon as the country was fully stocked; and yet we have every reason to believe, from what is known of wild animals, that all would pair in the spring. In the majority of cases it is most difficult to imagine where the checks fall – though – generally, no doubt, on the seeds, eggs, and young; but when we remember how impossible, even in mankind (so much better known than any other animal), it is to infer from repeated casual observations what the average duration of life is, or to discover the different percentage of deaths to births in different countries, we ought to feel no surprise at our being unable to discover where the check falls in any animal or plant. It should always be remembered, that in most cases the checks are recurrent yearly in a small, regular degree, and in an extreme degree during unusually cold, hot, dry, or wet years, according to the constitution of the being in question. Lighten any check in the least degree, and the geometrical powers of increase in every organism will almost instantly increase the average number of the favoured species. Nature may be compared to a surface on which rest ten thousand sharp wedges touching each other and driven inwards by incessant blows. [. . .] Reflect on the enormous multiplying power inherent and annually in action in all animals; reflect on the countless seeds scattered by a hundred ingenious contrivances, year after year, over the whole face of the land; and yet we have every reason to suppose that the average percentage of each of the inhabitants of a country usually remains constant. Finally, let it be borne in mind that this average number of individuals (the external conditions remaining the same) in each country is kept up by recurrent struggles against other species or against external nature (as on the borders of the Arctic regions, where the cold checks life), and that ordinarily each individual of every species holds its place, either by its own struggle and capacity of acquiring nourishment in some period of its life, from the egg upwards; or by the struggle of its parents (in short-lived organisms, when the main check occurs at longer intervals) with other individuals of the same or different species.

But let the external conditions of a country alter. If in a small degree, the relative proportions of the inhabitants will in most cases simply be slightly changed; but let the number of inhabitants be small, as on an island, and free access to it from other countries be circumscribed, and let the change of conditions continue progressing (forming new stations), in such a case the original inhabitants must cease to be as perfectly adapted to the changed conditions as they were originally. It has been shown in a former part of this work, that such changes of external conditions would, from their acting on the reproductive system, probably cause the organization of those beings which were most affected to become, as under domestication, plastic. Now, can it be doubted, from the struggle each individual has to obtain subsistence, that any minute variation in structure, habits, or instincts, adapting that individual better to the new conditions, would tell upon its vigour and health? In the struggle it would have a better chance of surviving; and those of its offspring which inherited the variation, be it ever so slight, would also have a better chance. Yearly more are bred than can survive; the smallest grain in the balance, in the long run, must tell on which death shall fall, and which shall survive. Let this work of selection on the one hand, and death on the other, go on for a thousand generations, who will pretend to affirm that it would produce no effect? [. . .]

Besides this natural means of selection, by which those individuals are preserved, whether in their egg, or larval, or mature state, which are best adapted to the place they fill in nature, there is a second agency at work in most unisexual animals, tending to produce the same effect, namely, the struggle of the males for the females. These struggles are generally decided by the law of battle, but in the case of birds, apparently, by the charms of their song, by their beauty or their power of courtship, as in the dancing rock-thrush of Guiana. The most vigorous and healthy males, implying perfect adaptation, must generally gain the victory in their contests. This kind of selection, however, is less rigorous than the other; it does not require the death of the less successful, but gives to them fewer descendants. The struggle falls, moreover, at a time of year when food is generally abundant, and perhaps the effect chiefly produced would be the modification of the secondary sexual characters, which are not related to the power of obtaining food, or to defence from enemies, but to fighting with or rivalling other males. The result of this struggle amongst the males may be compared in some respects to that produced by those agriculturists who pay less attention to the careful selection of all their young animals, and more to the occasional use of a choice mate.

[. . .]

III. On the Tendency of Varieties to depart indefinitely from the Original Type. By ALFRED RUSSEL WALLACE.

[. . .]

The life of wild animals is a struggle for existence. The full exertion of all their faculties and all their energies is required to preserve their own existence and provide for that of their infant offspring. The possibility of procuring food during the least favourable seasons, and of escaping the attacks of their most dangerous enemies, are the primary conditions which determine the existence both of individuals and of entire species. These conditions will also determine the population of a species; and by a careful consideration of all the circumstances we may be enabled to comprehend, and in some degree to explain, what at first sight appears so inexplicable – the excessive abundance of some species, while others closely allied to them are very rare.

The general proportion that must obtain between certain groups of animals is readily seen. Large animals cannot be so abundant as small ones; the carnivora must be less numerous than the herbivora; eagles and lions can never be so plentiful as pigeons and antelopes; the wild asses of the Tartarian deserts cannot equal in numbers the horses of the more luxuriant prairies and pampas of America. The greater or less fecundity of an animal is often considered to be one of the chief causes of its abundance or scarcity; but a consideration of the facts will show us that it really has little or nothing to do with the matter. Even the least prolific of animals would increase rapidly if unchecked, whereas it is evident that the animal population of the globe must be stationary, or perhaps, through the influence of man, decreasing. Fluctuations there may be; but permanent increase, except in restricted localities, is almost impossible. For example, our own observation must convince us that birds do not go on increasing every year in a geometrical ratio, as they would do, were there not some powerful check to their natural increase. Very few birds produce less than two young ones each year, while many have six, eight, or ten; four will certainly be below the average; and if we suppose that each pair produce young only four times in their life, that will also be below the average, supposing them not to die either by violence or want of food. Yet at this rate how tremendous would be the increase in a few years from a single pair! A simple calculation will show that in fifteen years each pair of birds would have increased to nearly ten millions! whereas we have no reason to believe that the number of the birds of any country increases at all in fifteen or in one hundred and fifty years. With such powers of increase the population must have reached its limits, and have become stationary, in a very few years after the origin of each species. It is evident, therefore, that each year an immense number of birds must perish – as many in fact as are born; and as on the lowest calculation the progeny are each year twice as numerous as their parents, it follows that, whatever be the average number of individuals existing in any given country, twice that number must perish annually, – a striking result, but one which seems at least highly probable, and is perhaps under rather than over the truth. It would therefore appear that, as far as the continuance of the species and the keeping up the average number of individuals are concerned, large broods are superfluous.

[. . .]

It appears evident, therefore, that so long as a country remains physically unchanged, the numbers of its animal population cannot materially increase. If one species does so, some others requiring the same kind of food must diminish in proportion. The numbers that die annually must be immense; and as the individual existence of each animal depends upon itself, those that die must be the weakest – the very young, the aged, and the diseased, – while those that prolong their existence can only be the most perfect in health and vigour – those who are best able to obtain food regularly, and avoid their numerous enemies. It is, as we commenced by remarking, 'a struggle for existence', in which the weakest and least perfectly organized must always succumb.

Now it is clear that what takes place among the individuals of a species must also occur among the several allied species of a group, – viz. that those which are best adapted to obtain a regular supply of food, and to defend themselves against the attacks of their enemies and the vicissitudes of the seasons, must necessarily obtain and preserve a superiority in population; while those species which from some defect of power or organization are the least capable of counteracting the vicissitudes of food, supply, &c., must diminish in numbers, and, in extreme cases, become altogether extinct. Between these extremes the species will present various degrees of capacity for ensuring the means of preserving life; and it is thus we account for the abundance or rarity of species. [. . .]

Most or perhaps all the variations from the typical form of a species must have some definite effect, however slight, on the habits or capacities of the individuals. Even a change of colour might, by rendering them more or less distinguishable, affect their safety; a greater or less development of hair might modify their habits. More important changes, such as an increase in the power or dimensions of the limbs or any of the external organs, would more or less affect their mode of procuring food or the range of country which they inhabit. It is also evident that most changes would affect, either favourably or adversely, the powers of prolonging existence. An antelope with shorter or weaker legs must necessarily suffer more from the attacks of the feline carnivora; the passenger pigeon with less powerful wings would sooner or later be affected in its powers of procuring a regular supply of food; and in both cases the result must necessarily be a diminution of the population of the modified species. If, on the other hand, any species should produce a variety having slightly increased powers of preserving existence, that variety must inevitably in time acquire a superiority in numbers. These results must follow as surely as old age, intemperance, or scarcity of food produce an increased mortality. In both cases there may be many individual exceptions; but on the average the rule will invariably be found to hold good. All varieties will therefore fall into two classes – those which under the same conditions would never reach the population of the parent species, and those which would in time obtain and keep a numerical superiority. Now, let some alteration of physical conditions occur in the district – a long period of drought, a destruction of vegetation by locusts, the irruption of some new carnivorous animal seeking 'pastures new' – any change in fact tending to render existence more difficult to the species in question, and tasking its utmost powers to avoid complete extermination; it is evident that, of all the individuals composing the species, those forming the least numerous and most feebly organized variety would suffer first, and, were the pressure severe, must soon become extinct. The same causes continuing in action, the parent species would next suffer, would gradually diminish in numbers, and with a recurrence of similar unfavourable conditions might also become extinct. The superior variety would then alone remain, and on a return to favourable circumstances would rapidly increase in numbers and occupy the place of the extinct species and variety.

The variety would now have replaced the species, of which it would be a more perfectly developed and more highly organized form. It would be in all respects better adapted to secure its safety, and to prolong its individual existence and that of the race. Such a variety could not return to the original form; for that form is an inferior one, and could never compete with it for existence. Granted, therefore, a 'tendency' to reproduce the original type of the species, still the variety must ever remain preponderant in numbers, and under adverse physical conditions again alone survive. But this new, improved, and populous race might itself, in course of time, give rise to new varieties, exhibiting several diverging modifications of form, any of which, tending to increase the facilities for preserving existence, must, by the same general law, in their turn become predominant. Here, then, we have progression and continued divergence deduced from the general laws which regulate the existence of animals in a state of nature, and from the undisputed fact that varieties do frequently occur. [. . .] All we argue for is, that certain varieties have a tendency to maintain their existence longer than the original species, and this tendency must make itself felt; for though the doctrine of chances or averages can never be trusted to on a limited scale, yet, if applied to high numbers, the results come nearer to what theory demands, and, as we approach to an infinity of examples, become strictly accurate. Now the scale on which nature works is so vast – the numbers of individuals and periods of time with which she deals approach so near to infinity, that any cause, however slight, and however liable to be veiled and counteracted by accidental circumstances, must in the end produce its full legitimate results.

[. . .]

The hypothesis of Lamarck – that progressive changes in species have been produced by the attempts of animals to increase the development of their own organs, and thus modify their structure and habits – has been repeatedly and easily refuted by all writers on the subject of varieties and species, and it seems to have been considered that when this was done the whole question has been finally settled; but the view here developed renders such an hypothesis quite unnecessary, by showing that similar results must be produced by the action of principles constantly at work in nature. The powerful retractile talons of the falcon- and the cat-tribes have not been produced or increased by the volition of those animals; but among the different varieties which occurred in the earlier and less highly organized forms of these groups, those always survived longest which had the greatest facilities for seizing their prey. Neither did the giraffe acquire its long neck by desiring to reach the foliage of the more lofty shrubs, and constantly stretching its neck for the purpose, but because any varieties which occurred among its antitypes with a longer neck than usual at once secured a fresh range of pasture over the same ground as their shorter-necked companions, and on the first scarcity of food were thereby enabled to outlive them. [. . .]

We believe we have now shown that there is a tendency in nature to the continued progression of certain classes of varieties further and further from the original type – a progression to which there appears no reason to assign any definite limits – and that the same principle which produces this result in a state of nature will also explain why domestic varieties have a tendency to revert to the original type. This progression, by minute steps, in various directions, but always checked and balanced by the necessary conditions, subject to which alone existence can be preserved, may, it is believed, be followed out so as to agree with all the phenomena presented by organized beings, their extinction and succession in past ages, and all the extraordinary modifications of form, instinct, and habits which they exhibit.

Ternate, February, 1858.

Part 10

AGRICULTURAL SCIENCE AND LAND MANAGEMENT

Agricultural Science and Land Management

THIS section is included because of the obvious and wide-ranging relationship between environments and human management of land. In a current context, in which agriculture is clearly identified as a significant driver of climate change, it is instructive to look back to a period in which an ongoing process of intensification of agriculture was accelerating – in terms of habitat loss, land change, plant and animal breeding, and the gradual emergence of an agricultural science that drew on research developments, particularly in chemistry and biology. As we noted in the Part 1 headnote, agriculture has, from its institution in the Neolithic period, been a key driver in human exploitation of land and of environmental disasters arising from practices such as deforestation. It is mistaken, therefore, to imagine that agriculturally induced ecocrises emerged only in the nineteenth or twentieth centuries. At the same time, it is worth examining the specifics of this period and to gather together some examples of writings on agriculture and land management from the 1780s to the 1850s. The focus will be largely British but with a nod to Britain's colonial agricultural ventures, which will feature more prominently in subsequent volumes. The relative shortness of this section is partly because of further coverage in Volume II of this anthology but also a reflection of the relatively slow emergence of a recognisable agricultural science in the period: many of the extracts are amateur rather than professional, works because most of the contributions to the 'advancement' of agriculture were informal and rooted in the work of practitioners.

In terms of tracking environmental contexts in Britain, as the most 'advanced' economy of the period, a number of factors need to be mapped, and the following overview is drawn from relevant chapters in I.G. Simmons's *Environmental History of Great Britain* and Oliver Rackham's *History of the British Countryside* (see Further Reading). Estimates suggest that the U.K.'s human population roughly doubled from 5 million in 1700 to 9.2 million in 1801. By the 1850s, it exceeded 25 million. Accelerating population growth clearly had implications in terms of requiring the cultivation of additional land at home and abroad and in an impetus to increase productivity. In Britain, the Enclosures Acts, of which 2,800 were passed between 1760 and 1850, were transforming landscapes: increasing productivity was accompanied by the consolidation of land in fewer hands, intensification of farming methods, increases in hedgerows and boundary walls, and diminution of traditional commons rights. Different areas of the country increasingly specialised in arable or pastoral, while new strains of cereals and new breeds of animals were introduced. Agricultural 'improvers' turned their attention to converting 'waste lands' (mountains, moors, heaths, marshes, bogs) to production, with mixed results for agriculture and a general decline in native flora and fauna. The countryside, it should be noted, was not purely agricultural: commercial plantations, mines, quarries, and sites of industrial production were also increasing features of eighteenth- and nineteenth-century landscapes, with accompanying problems of pollution, poverty, and hunger accompanying such changes: the social, political, and cultural aspects of agricultural lands will form a focus in Volume II of this anthology. Traditional landed estates in the period, often supported by colonial investments in slave economies overseas, continued to dominate rural societies, and at the same time, there was a proliferation of landscape gardening, and a growing appetite for landscape travel accompanied interests in Romanticism, art, and nature. Relevant volumes of Joan Thirsk's monumental study of British agricultural history (Further Reading) are worth consulting in this regard.

It is also worth briefly considering the direct scientific influences on agriculture in this period, in particular those relating to chemistry. A key figure in the establishment of chemistry as a scientific discipline in the eighteenth century, Antoine-Laurent de Lavoisier also understood its agricultural applications and established and served as the head of a Royal Commission on Agriculture, reporting in 1788 on his own agricultural experiments. Justus von Leibig's *Organic Chemistry in Its Application to Agriculture and Physiology* (1840) argued that chemistry could provide agriculture with increased yields and reduced outlays, while his promotion of Karl Sprengel's 'law of the minimum' was influential in pointing out that plant growth is determined by the scarcest available resource. Leibig's work led to a wider focus on the role of nitrogen, potassium, and phosphorus, and attempts to find sources for use in agriculture. In this period, exploited nutrient sources included sodium nitrates imported into Europe from Chile's Atacama Desert; enormous deposits of sea bird guano (rich in phosphorus and potassium), which became significant from the 1830s; potash from timber burnt during deforestation; crushed bonemeal from meat manufacture; and processed fossilised coprolites, rich in phosphorus. The development of these various organic and inorganic fertilisers saw strong increases in vegetable yields and helped Britain to emerge from the 'Hungry Forties' into a period of agricultural prosperity that lasted for two decades.

That agriculture might be environmentally problematic was becoming understood during this period, but it was far more common to see it as positive, progressive, and an aspect of national identity. This view is evident in the (unexcerpted) Preface of Nicholas Turner's *An Essay on Draining and Improving Peat Bogs; In Which Their Nature and Properties Are Fully Considered* (1784):

> The most discerning part of mankind have in all ages considered agriculture of the greatest importance, and the wisest men of every civilized nation have studied the Improvement of this most necessary art, on which the strength, wealth, and prosperity of their country depended.

In the decades that followed Turner's publication, Britain's arable and pastoral lands became as much a symbol of patriotic pride as its oak woodlands (concomitant with the Royal Navy's 'wooden walls': see Part 1). For the ways in which Britain's landscapes accrued patriotic iconography, see Daniels and Cosgrove (Further Reading).

The first three extracts cover the subject of bog 'reclamation'. Bog habitats, and the species on which they rely, are now amongst the most marginalised, threatened, and fragile in Britain and Europe. That they have almost entirely disappeared from Britain is due to the processes and attitudes exemplified by Turner's *Essay*. Turner wrote at a time when the form and formation of peat bogs were imperfectly understood, and his identification of the role of accumulated water in combination with sphagnum moss marks a new phase of investigation. While acknowledging the value of peat as a fuel, his remarks largely centre on how to convert bogs to tillage. He recommends draining, ditching, and the application of stones to the surface to press water from the bogs. The work of Turner and others provided a roadmap for a wider process of transformation concomitant with the monoculturalisation and impoverishment of modern farming.

A Method of Raising Hops in Red Bogs (1800) and *An Essay on Peat or Turf, and on Turf and Wood Ashes, as a Manure* (1800) explore different ways to exploit bogs. These texts were published by the Dublin Society, a centre of agricultural improvement ideas around the turn of the century. The latter draws on Turner's essay, Robert Jameson's *System of Mineralogy* (1804), and the *Transactions of the Agricultural Society of Amsterdam*, indicating that the nascent agricultural sciences participated in wider European networks. The first extract is in the form of correspondence on the subject from an unnamed member of the society. It begins with the statement that 'it must be matter of concern to all, to see great tracts of land lie entirely useless in a country' like Ireland. The immediate concern was the 'extensive bogs to be met with every where', which provide only 'poor coarse pasture'. No cognizance of the environmental benefits of these unique habitats is evident in this work.

The practicalities of raising hops on reclaimed bogs form the bulk of the text, with the author recommending methods of ditching and draining that were already well established during the extensive reclamation of fens in Bedfordshire, East Anglia, and Lincolnshire from the seventeenth to the nineteenth centuries. These processes, it should be noted, involve the oxidisation of peat and the release of methane, a major driver of climate change. At the time, such examples suggested that previously 'unproductive' or barren lands could become amongst the most fertile: today, East Anglia represents almost half of the 'best' (Grade 1) agricultural land in England. The more rugged and inaccessible landscapes of Ireland, alongside its status as a colonial territory, rendered its 'improvement' more difficult. *Method* involved 'ploughing, harrowing, fallowing, and digging' on small plots of land, alongside manuring and hoeing. The author insists that the resulting produce repays the expense.

The opening parts of *Essay on Peat or Turf* reference Turner's speculations on the origins and composition of peat and largely agree with his conclusions while going further in exploring the chemical processes of de-oxidisation. Latter parts delve into the relative benefits of peat, turf, and wood ashes as 'manures' (fertilisers) for different soils – a reminder that such issues were pressing in the period before artificial fertilisers, particularly in those districts of the British Isles where animal manure supplies were limited. The brief extract included here, however, outlines the ways in which bog 'improvement' had become a serious object of study and experimentation and advocates a method of applying 'calcareous earth' (crushed limestone) to wetlands, thus reducing acidity and water content and accelerating decomposition of vegetable matter.

Management of wetlands for crops is a key issue in a very different context in the next extract: 'The Hindoo Method of Cultivating the Sugar Cane' from the 1803 *Annual Register*. The article is taken verbatim from an article published that year by Scots botanist William Roxburgh in William Tennant's *Indian Recreations*. Roxburgh begins with a patronising overview of 'native' agriculture as 'exceedingly slow' in development and 'marked by a strong disinclination to depart from the beaten path established from time immemorial' or to draw on the superior 'example of Europeans' unless 'a certain prospect of gain' is allied to 'little additional trouble': the implication of native laziness here justifies vigorous and 'progressive' interventions in Indian environments and rural communities.

As the article continues, however, it becomes clear that the author admires the sugar cane cultivation methods of the Rajamundry farmers in northern India. The colonial and economic background is clearly evident: while Britain received produce from 'some of the best of the West India islands', there was an urgent need to find 'new sources from whence that wholesome commodity may be procured, at the cheapest rate'. The exploitation of 'Indian husbandry', then, was 'of national importance'. The patriotic duty of ensuring food supplies to the home country was a particular issue at a time in which Britain's reliance on imported cereals was growing: the article appeared only three years before Napoleon Bonaparte instigated the 'Continental Blockade' of British imports. More generally, the article indicates the ways in which the British economy was enmeshed in global networks and dependent on many global environments but also its growing appetite for once-luxury products like sugar.

Roxburgh's advice is directed at the East India Company, the powerful trading body given its first licence to trade in the Indian Ocean region in 1600 and later used to govern large parts of India before the imposition of direct British rule in 1858. According to Anthony Farrington (2002), at the time Roxburgh was writing, the East India Company accounted for approximately half of world trade. As Bosme (2013, 44–8) has shown, however, Indian sugar had struggled to compete with 'slave sugar' for decades, partly because of discriminatory duties that protected West Indian produce, and partly because of famines and social unrest. To increase sugar production, Roxburgh recommended that the Company focus on areas where it was already cultivated and bring European expertise to bear to 'increase the culture, and improve, if necessary, the quality'. Attempts to introduce West Indian plantation methods had already proved expensive and ineffective, while a growing Abolitionist movement gave impetus to the British government to consider future alternatives to 'slave sugar'. In these contexts, we can analyse Roxburgh's account of sugar cultivation in Rajamundry and Pethapur, which, he insists, involve lower costs (because of the simplicity of native agriculture) than 'slave sugar'. What follows is a remarkable social history of Indian agricultural methods as 'simple', 'efficacious', sustainable, and responsive to changeable conditions. Unlike advocates of the introduction of intensive West Indian grinding and milling methods, Roxburgh suggests only minor improvements to the existing Indian field mills. Roxburgh's sympathies for the 'ryut' (ryot, peasant cultivator) become increasingly clear, particularly in his closing criticisms of the high rents demanded of them by 'zemindars' (zamindars, landowners). Thus, we gain insights into rural environments that were overseen by a combination of Mughal and British administrative structures. While Roxburgh's musings are avowedly imperial, they also offer a relatively nuanced response to India and its inhabitants.

The next extract returns to British agriculture and to one of the most significant works of the period, Arthur Young's 45-volume *Annals of Agriculture and Other Useful Arts*. Young's position as Secretary of the Board of Agriculture from 1793 made him an influential figure, a sponsor of county agricultural societies and other organisations that exchanged ideas for 'improvement'. As a key vehicle of this, *Annals* compiled annual contributions from a range of authors, many of them farmers, with myriad reflections on current practices and new innovations. Nothing, apart from his status as a farmer or landowner, is known of Edward Powys, the author of 'On Feeding Cattle With Green Food' (1808), but his advocacy of agricultural improvement as patriotic duty is clear in opening remarks that 'the principal object respecting agriculture, in the present state of this country, is to procure the greatest possible supply of the necessaries of life within the kingdom'. He proposes others adopt his practice of increasing grain and turnip production and to use grass and other feeds to keep cattle indoors for parts of the year to increase manure production. The extract includes descriptions of his more intensified practice, but his closing remarks about the value to 'the cottager' (renters of very small areas of land) of grassland for a cow to graze are socially progressive and in line with a wider interest in the period in the plight of cottagers and labourers. Jeremy Burchardt's invaluable study of the early allotment movements of the eighteenth and nineteenth century (Further Reading) is worth reading in this regard.

The same desire to turn a forensic gaze on the state of the land and its productivity motivates our next extract, from Young's *General Report On Enclosures* (1808), a voluminous work treating many aspects of a complex, volatile subject with care and tact. Ultimately in favour of this controversial process that had for more than a century been transforming the British countryside, changing the Commons and open-field practices of the middle ages into modern, privatised, and capitalised production, Young is nonetheless sympathetic to its critics, most notably those who were unable to maintain the smallholdings allotted to them by the Enclosures Act and declined into an expanding class of poor agricultural labourers. The heated historical debates on this vexed issue are worthy of study, and readers may turn to the Further Reading for various perspectives.

The extract examines the enclosure of 'waste lands', which are defined as land lacking value. Land, from this perspective, can only have monetary rather than environmental value. Young examines the effects of enclosures on farmers, landlords, the poor, and the public (his coverage of the church is not included). Celebrating the gains to farmers and landlords, Young accepts the difficulties presented to the poor by enclosures and suggests some measures for amelioration. His closing thoughts on public benefits are clear, however, and form a strong thread of his support for enclosures. The wider population, he argues, can only gain:

> To convert tracts really or nearly waste, into profitable farms; to change ling for turnips; gorse for barley; and overstocked and rotting grass lands to wheat [. . .] is a real acquisition of territory, pregnant with every advantage attending the husbandry of the kingdom; food, population, wealth and strength.

Against a backdrop of rising populations of hungry mouths and the upheavals associated with the French Revolutionary era and a series of bad harvests, Young's attitude is understandable, but one must also note the almost colonial tone in which he speaks of the accumulation of virgin territories ripe for conquest. Land in the nineteenth century would be exploited with great intensity, and the spirit of 'improvement' motivating many writers in this section would in time be alloyed to the power of Victorian technology and directed towards defying Malthusian anxieties (see Part 8 of this volume) by unleashing the soil's productivity.

The next two extracts are examples of another important phenomenon in the development of more empirical studies of British agriculture, the *General View* county survey series that began in the 1790s and saw second editions in the 1810s. Initiated by Arthur Young's Board of Agriculture, they chimed with a wider appetite to gather knowledge of the U.K. economy and are at one with the spirit of rational investigation that spurred the institution of the census in 1801 and other forms of social surveillance in the decades that followed. Endeavours to gather and disseminate information were meant to spur farming practitioners and government policy makers. A 1794 Cheshire *General View* by Thomas Wedge was superseded by Henry Holland's edition, while a three-volume edition (1811–17) of John Farey's Derbyshire *General View* followed a 1794 edition. In a pattern replicated in most of these surveys, the *General Views* cover the geography, parish divisions, climate, soils, and minerals of the county before offering a detailed survey of its agriculture under various headings, then turning to 'Improvements', 'Live Stock', 'Rural Economy', 'Political Economy', and other subjects specific to particular counties. Three sample selections are given here from *Cheshire* – the first a survey of county topography, the second a favourable overview of estates management, and the third some brief suggestions for further improvement of a county characterised as prosperous but facing challenges.

The longer title of Farey's *Derbyshire*, placing particular emphasis on the county's minerals, reflects his expertise in geology, much of it gained from working under William Smith (see Part 3 of this volume). Much of Farey's first volume is devoted to a detailed and brilliant geological survey and is characterised by a wider eagerness to gather and present data. He provides extensive lists of the county's hills, valleys, parishes, and settlements, for example, as well as offering detailed meteorological data.

Farey is particularly keen to show the applications of Smith's recent understanding of geological stratification to agriculturalists and landowners: the 'important consequences' of geology to 'Mining, Agriculture, and indeed to numerous others of the useful Arts' are economic gain, he argues in a section not extracted here. To understand

> the general excavation of Valleys . . . by which the Great Creator has in so admirable and perfect a manner fitted it, for the various uses of Man and the numerous other organized Beings with which he has since stored its surface and waters

is to engage in a profitable branch of Natural Theology. One of the particular advantages of using Smith's stratigraphical approach was to improve one's chances of discovering useful mineral deposits on one's lands. The extract includes Farey's informed and justified speculations about the existence of coal measures in parts of the county. While surface mining of various kinds had been a feature of the Derbyshire countryside for centuries, the vast expansion of deep mining in the county was a few decades away. For landowners formerly reliant on their own farmlands, rents, and investments, the discovery of coal at a time of rapidly expanding industrialisation would create fortunes. Farey's work indicates the degree to which agriculture was only one element of the rural economy, as well as foreshadowing the enormous social and environmental impacts of mines, quarries, and industrial production in rural communities. Chalkin and Wordie's study of the role of landowners in the British economy (Further Reading) is a useful introduction to related issues.

The subtitle of Joseph Hayward's *On the Science of Agriculture* (1825), which is not fully rendered in the heading of the extract due to its inordinate length, gives a strong indication of the spirit in which the work was written. Written in response to the 'Agricultural Chemistry of Mr Kirwan and Sir Humphry Davy' and the 'Code of Agriculture of Sir John Sinclair, Sir Joseph Banks, and Other Authors', it aims to show

> That there is not only a discrepance in the opinions of those Authors, on many of the most important operations of Agriculture; but that this arises from their inferences and conclusions being erroneous; and their principles unfounded, or inapplicable; and particularly on the subject of breeding, and the nature, preparation, and application of manures.

Hayward's vituperative confidence is further evident in his 'Advertisement and Dedication':

> I am not a professor of chemistry, nor an extensive practical agriculturalist; nor the member of any learned society: and as it is the fashion of the times to attach great importance to such authority, some may consider me arrogant, presumptuous, and invidious, in attempting to intrude on the public my commentaries on the works of such established characters.

Hayward's remarks indicate the degree to which science was becoming both professional and popular in the period – the celebrity of Davy, Banks, and Michael Faraday being matched in the period only by Lord Byron and Walter Scott. Hayward's irritation speaks of exclusion from elite scientific circles: his application to join the Horticultural Society, of which Davy was a leading member, was politely refused, and the society did not acknowledge or review Hayward's *The Science of Horticulture* (1818). That the grounds of Hayward's attacks on Davy and others may have been more personal than scientific is clear. Rather than exploring Hayward's various grievances, however, the extract from his introduction offers a fascinating definition of agriculture, an account of its purposes and benefits, and an endorsement of its growing relationship with chemistry. Perhaps most fascinating about Hayward's summary is the emphasis placed on the co-dependence of plants and animals and in the similarities of their reproductive functions. The focus remains firmly on 'supplying the wants of mankind' rather than straying into proto-ecological territory, however, and repeated references to the 'value' of plants and animals are always directed towards 'the peculiar object desired' by humans in order to direct agriculturalists 'to the adapting their means to the end in view'. Hayward's work in this sense somewhat resembles that of Thomas Ewbank's vision of the world as a manufacturing workshop (see Parts 2, 5, and 6 of this volume). 'Animals and vegetables administer to the wants of mankind in various ways', he argues, almost – and disturbingly – implying voluntary self-sacrificial agency on their part, as they provide 'their flesh or the immediate substance of their bodies', 'their offspring and seed', or 'their exterior covering'. Nature, too, is 'ever kind and liberal' in ministering to humans and other creatures but also a powerful force that humans must not abuse. Hayward's thoughts are far from environmentalist and indeed lapse into familiar platitudes: the place of *Homo sapiens* within 'an all-bountiful Providence' is assured, and it is unclear what Hayward means when he advocates 'deference' to nature over 'opposition' to or interference in nature's laws. All the same, his warnings that 'attempts to force or oppose her' will 'produce disorder, and often destruction' are more prescient than he could have imagined.

Hayward's work reminds us that agriculture is a significant and charged intersection between the human and non-human. It is also amongst the most destructive because of the inbuilt urge to exercise environmental sovereignty that it discloses and enacts. The next extract takes us forward a decade, and it is worth pausing to consider some of the changes to British agricultural period in the period from 1780 to 1835 and to examine some key trends of the broader period. Human population pressures are of course an important issue. From a roughly 11 million population in 1801, Britain had doubled by 1851. Napoleonic upheavals and a period of unpredictable harvests were followed by a return, after 1815, to unrestricted imports of grain and other foodstuffs, which in turn led to the imposition of tariffs (the Corn Laws) to protect British farmers unable to compete with foreign competition. Despite the confident and progressive tone of many of the pre-1830s works we have selected, the 'Hungry Forties' would bring a decline in agriculture, food shortages, famine in Ireland and elsewhere, and a rise in social conflict and political activity in the countryside.

The penultimate extract, *Hints on Agriculture* (1835) by A Practical Farmer (the untraced 'J.M.' of the title page) is a reminder that despite various advances in agricultural science, day-to-day activities were undertaken by ordinary farmers who were not always willing or able to follow the latest developments in the (ahem) field. A self-proclaimed 'practical' agriculturalist, J.M. published his volume of 'hints' on good practice via a Hull publisher, and his work offers an unfussy balance between traditional and progressive practices. The volume is also an unintentional social history, a detailed account of how a farm operated at this period and the challenges it faced. A few examples of J.M.'s hints are included, turning on planting, ploughing, furrowing, drainage, and crop cultivation.

The final extract of this section and volume takes us up towards the end of our period. John Lauris Blake's *Lessons in Modern Farming: Or, Agriculture for Schools* (1852) also offers our first American contribution to this section, a curious and fascinating work that indicates a desire for agricultural education to become more widespread, scientific, and organised. The introduction explains Blake's desire to avoid what he describes as the parlous recent history of American farming, in which 'farms have become so worn out and unproductive, as not to yield an equivalent for the labor of cultivation'. Destitution and abandonment of individual farms were the initial results, he claims, but other more serious perils attended them:

> When whole districts of country were affected in the same way, when it was seen that thousands of farms were thus becoming worthless – that the calamity, if it might be so called, was every year making fearful progress westward, threatening in the end the insufficiency of the earth to support its inhabitants – verily, a gloomy spectacle was presented to the imagination.

Blake references the wastefully destructive practices of many pioneer farmers who headed west to establish claims but quickly exhausted the fertility of virgin lands such as the prairies by over-exploiting them, before moving further west only to repeat the same poor practices. Louis Hacker's studies (Further Reading) of this period, written just before the Dust Bowl crisis that they foreshadow, suggest that a subsequent wave of more experienced farmers largely repaired the damage done by the earliest settlers by more sensitively cultivating the land. Having painted an apocalyptic vision of soil exhaustion, Blake's introduction then praises such farmers, and the application of science, in maintaining 'the self-sustaining energy of the soil to all the purposes for which it was designed'. Like Hayward, Blake refers to 'those physical laws which govern material nature', but his approach is far more scientific and also suggests more understanding of the catastrophes attendant on bad practices: in this respect, Blake's work, though wildly different in many ways, is close in spirit to George Perkins Marsh's ground-breaking environmentalist study, *Man and Nature*, which would be published in 1864 (see Volume III of this anthology). In a rousing tone, however, Blake voices the confidence in the progression of agriculture in alliance with science that we have witnessed in many of the extracts in this section, but instead of directing his work at farmers or scientists, his 'magnificent scheme' is to educate 'the sons of farmers' in order to provide 'not simply to the generations of now living men; but to the long succession of unborn generations, to the end of time, luxuriating in that abundance which the earth may thus be made to yield'.

As a 'humble agent' in a glorious project, Blake attempts 'to introduce into the common schools of the country a taste at least for scientific agriculture and rural literature'. To do so effectively, he argues, one must not throw large quantities of indigestible science at either teachers or pupils but rather mix these with other materials, including examples of 'rural literature' (poetry and prose). The science itself is presented in informal, informative catechisms, examples of which are included to illustrate his methods. Not only will all of this obviate the absurdity of farmers' sons receiving 'an educational smattering on almost every thing, save that on which they are to depend for a reputable subsistence', but also reflect 'that the business of agriculture in all its various details, requires an amount of general intelligence far beyond what is required for conducting the business of most mechanical trades'. Rooting American national identity in the soil and in farming communities, Blake promotes a pervasive progressivism and optimism. The inclusion of this work as the final extract of the volume is also an indicator of the greater focus on U.S. writings on environment that will follow, particularly in Volumes III–IV, reflecting the growth of American economic power and of its increasing status in the natural sciences.

Further reading

Agar, Nigel E., *Behind the Plough: Agrarian Society in Nineteenth-Century Hertfordshire* (Hatfield: University of Hertfordshire Press, 2005).

Bosma, Ulbe, 'East Indian Sugar Versus Slave Sugar', *The Sugar Plantation in India and Indonesia: Industrial Production, 1770–2010* (Cambridge: Cambridge University Press, 2013), pp. 47–87

Burchardt, Jeremy, *The Allotment Movement in England, 1793–1873* (Woodbridge: Boydell, 2011).

Chalkin, C.W. and J.R. Wordie (eds.). *Town and Countryside: The English Landowner in the National Economy, 1660–1860* (London: Unwin Hyman, 1989).

Daniels, Stephen and Denis Cosgrove (eds.), *The Iconography of Landscape: Essays on the Symbolic Representation, Design and Use of Past Environments* (Cambridge: Cambridge University Press, 1988).

Farrington, Anthony, *Trading Places: The East India Company and Asia 1600–1834* (London: British Library, 2002).

Hacker, Louis Morton 'Western Land Hunger and the War of 1812: A Conjecture', *Mississippi Valley Historical Review* 10:4 (1924) 365–95.

Hammond, J.L. and Hammond, Barbara, *The Village Labourer 1760–1832: A Study of the Government of England Before the Reform Bill* (London: Longman, Green, & Co., 1911).

Holderness, B.A. and Michael Turner (eds.). *Land, Labour and Agriculture: Essays for Gordon Mingay* (London and Rio Grande, Ohio: Hambledon Press, 1991).

McDonald, Donald, *Agricultural Writers from Sir Walter of Henley to Arthur Young, 1200–1800* (Los Angeles: B. Franklin, 1968).

Mingay, G.E., *Enclosure and the Small Farmer in the Age of the Industrial Revolution* (London: Macmillan, 1968).

Rackham, Oliver, *The History of the British Countryside* (London: Dent, 1987).

Rosenman, Ellen, 'On Enclosure Acts and the Commons', *BRANCH: Britain, Representation and Nineteenth-Century History*, ed. Dino Franco Felluga, *Extension of Romanticism and Victorianism on the Net* (May 2014).

Simmons, I.G., *An Environmental History of Great Britain from 10,000 Years Ago to the Present* (Edinburgh: Edinburgh University Press, 2006).

Thirsk, Joan (ed.), *Agrarian History of England and Wales*, 8 vols (Cambridge: Cambridge University Press, 1978–2011).

Turner, M. *Enclosures in Britain, 1750–1830* (London: Macmillan, 1984).

Young, Arthur, *Annals of Agriculture and Other Useful Arts*, 45 vols (London: Richard Phillips, 1785–1809).

96

NICHOLAS TURNER, *AN ESSAY ON DRAINING AND IMPROVING PEAT BOGS*

(Chichester: Dennett Jacques, 1784)

[. . .]

My opinion is, that peat bogs originate from these waters, and that the vegetable part of peat is a species of marsh moss, whose character I have nearly described before, which grows spontaneously in waters thus impregnated; for all vegetables have no doubt a regular structure, having vessels for receiving and perspiring different fluids; and the nutriment they imbibe is changed into new juices peculiar to themselves, by being secreted and elaborated in the body of the plant by a power of which we know only the effect; thus each kind of soil or water abounds in plants natural to it, and which degenerate or cannot exist in others.

[. . .]

Peat bogs are in many places of a great value for the fuel they afford, which is esteemed wholesome, and adapted to all the culinary uses of the middle station of life; and when charred in order to dissipate its fuliginous and sulphureous parts, it is used in the nicer chemical operations. Another very profitable use to which it may be applied, is lime-burning, as it gives a very strong, tho' not a brisk heat, which makes the operation slower; but that is not material in a perpetual kiln, or indeed in a common one that is under cover and confined at the top to keep in the heat, and that has got a quick draft. The lime will be fit for the husbandman, tho' not quite so saleable for building, as it will contain much bass of red dust from the foulness of the fuel.

[. . .]

The cheapest method of converting a bog is, if there is any river or stream that can be brought above its level, when drained, to make proper dams, heads, and cuts, to water it, which will scour away all its impurities, add to the surface, and make it good meadow; even the most limpid waters will have this effect; and indeed any except such as are chalybeate, which are always more or less enemies to vegetation. The more rapid the water the sooner it will make the surface sound; and if in addition to this a good coat of lime is laid on, it will, by acting as a cement, make it still sounder, and by rendering the oily viscid particles of the bog miscible with water, and proper for vegetation, the improvement will be the quicker.

DOI: 10.4324/9780429355653-107

[. . .]

To any distance within half a mile from the edge of the bog, if earth, sand, clay, gravel, or even stones, are carried on, and three loads and a half laid on a square perch, each load containing one square yard, it will cover the surface over just four inches thick. The improvement will be astonishing from the pressure of five hundred and sixty tons on an acre. Trifling as this weight may seem on so much space, yet it will have great effect; for as a bog may be compared to a cheese when first put into a press, so if it is served in this manner, it will be in quite as different a state as a cheese when taken out. There is no doubt but weight alone will cure a peat bog, and that frequent open trenches are the most proper, on the first draining, to receive the water from its lateral pores; and that three tons and a half weight on a square rod acts just as a press does on a cheese, and effectually cures this soil, so remarkable for its retentive quality. It will render it so firm as to admit of tillage, and also under draining. A course of crops may be taken which will be sufficient to repay all expences, great as they may appear to be; but should this not be the case, from the failure of crops or any other cause, yet the increased, I may say the new created value of the soil at the price that land bears, will leave such a profit as no expenditure of the same sum can ever produce in any other application.

97

A METHOD OF RAISING HOPS IN RED BOGS, PUBLISHED BY ORDER OF THE RT. HON. THE DUBLIN SOCIETY

(Graisberry & Campbell, 1800)

'It must be matter of concern to all, to see great tracts of land lie entirely useless in a country, which has the utmost reason to husband all advantages with care; such are the many and extensive bogs to be met with every where; which, except a poor coarse pasture on the better kinds, afford no other profit to the owner, than what can be made by burning the soil of them in turf. I hope therefore it will be an attempt agreeable to gentlemen of your public spirit, to introduce a culture of them, which, at a small expence, will turn to great account, and to make those unprofitable lands, without much labour in reclaiming them, bear a good and valuable crop. The crop I mean is hops; and the bogs, in which I have reared them with most success, the worst and most useless of all others – the red bogs. The profit has for many years fully answered my expence, and what has turned to my advantage, will do so with every body else in the same method of improvement.

'I own, that notwithstanding these precautions, this improvement is expensive; but raising hops in any ground is so, and, I am sure, greater in the most favourable upland situation, than in bog. A very little arithmetic will shew, that ditching and inclosing, which in bog is no expence, the turf made at the same time being equal to the charge; that ploughing, harrowing, fallowing, and digging, which in my method are intirely saved, with the additional articles of dunging, hoeing, and paring the allies in uplands, are more than an equivalent for all the labour and expence attendant on bog-hops; and from fifteen years experience, I can venture to affirm, that the produce from the latter is as great in quantity, and, in quality, as good. Many reasons might be given why it should be so; some of them I beg leave to lay before your readers: they may be necessary to remove the prejudices, which generally attend new projects, and to make this improvement as common in this kingdom as, I am sure, it will be beneficial whenever it becomes so.

DOI: 10.4324/9780429355653-108

98

AN ESSAY ON PEAT OR TURF, AND ON TURF AND WOOD ASHES, AS A MANURE

(Dublin: W. Sleater, 1800)

Improvement of Moss Land. – The method of improving peat moss, by means of calcareous earth, is certainly one of the most useful discoveries of the present day, and will, no doubt, form an important æra in the annals of agriculture. Not long since peat lands were only considered as useful, on account of the fuel which they afforded: the case is now widely altered; the moss grounds are eagerly sought after, and in the west of Scotland, this mode of improvement is carried on with great spirit, and is repaying the judicious manager very amply.[1] To give a detailed account of the methods which are followed (and that is the only one that can properly be given) would require a volume, and after all, would be little more than a repetition of what is already known. I shall therefore content myself with endeavouring to apply the preceding experiments, to explain the mode of action of the calcareous earth. We have already observed that peat contains the suberique acid, or one nearly allied to it, which appears to be formed in greater quantity the longer the peat is exposed to the action of the air; thus assisting in retarding the decomposition of the peat, of course preventing its being useful in vegetation. Marle, shells, and limestone, are useful, in a triple capacity: in the first place, by removing the acid, the vegetable matter is allowed to decompose more rapidly; secondly, the combination of this acid with the lime, forms a compound, which may assist in vegetation; and lastly, this acid having a stronger attraction for lime than the carbonic, will disengage it in considerable quantity, when it will assist vegetation. I shall conclude, with remarking, that there is a considerable prejudice in favour of quicklime; if the explanation now given have any plausibility, it is plain that carbonat will answer as well, if not better, in the improvement of moss lands.

Note

1 The Highland Society of Scotland, with their accustomed liberality and desire for improvement, have offered a piece of plate of twenty guineas, for the most approved

 DOI: 10.4324/9780429355653-109

essay, on the following subject. 'An account of the best and easiest method of converting the various kinds of peat or peat moss, into a manure, either by reducing them to ashes, or by making them undergo the putrid fermentation. Also a medal, value five guineas, for the best Essay on the most proper method of cutting peats, so as to prevent the abuse of mosses; and also the particular seasons of the year when the mosses ought to be flooded, to encourage their growth and renewal'.

99

[WILLIAM ROXBURGH], 'THE HINDOO METHOD OF CULTIVATING THE SUGAR CANE. FROM TENNANT'S "INDIAN REVELATIONS"', *THE ANNUAL REGISTER; OR A VIEW OF THE HISTORY, POLITICS, AND CULTURE FOR THE YEAR 1803*

(1803)

Among the natives of India the transitions from one stage of improvement to another are exceedingly slow, as scarce to deserve the name, except it be the few who have benefited from the examples of Europeans. They naturally possess a strong disinclination to depart from the beaten path established from time immemorial; however, when they see a certain prospect of gain, with little additional trouble, they have frequently been known to adopt our practices. We ourselves ought now generally to keep in view, and to instil into their minds this maxim, that every new proposition, merely on account of its novelty, must not be rejected, otherwise our knowledge would no longer be progressive and every kind of improvement must cease.

At a period, like the present, when the importation of East India has become so much an object of importance to Britain, in consequence of the present state of some of the best of the West India islands, every inquiry that may tend to open new sources from whence that wholesome commodity may be procured, at the cheapest rate, is of national importance.

I believe there are few districts in the company's extensive domains where there will not be found large tracts of lands fit for the culture of the sugar cane: yet I know, from experience, the introduction of a new branch of agriculture, among the natives, to be attended with infinite trouble; therefore, where we find a province or district, in which the working of sugar has been in practice from time immemorial, there we may expect, without much exertion, to be able to increase the culture, and improve, if necessary, the quality.

DOI: 10.4324/9780429355653-110

[. . .]

This branch of agriculture, in [Rajamundry] is chiefly carried on in the Peddapore, and Pettapore, along the banks of the Elyseram river, which, though small, has a constant flow of water in it the whole year round, sufficiently large, not only to water the sugar plantations during the dry seasons, but also, a great variety of other productions; such as paddy, ginger, turmeric, yams, chillies [. . .]

In these two zemindaries, from 350 to 700 *Vissums*; or from 700 to 1400 acres of land [. . .] is annually employed for rearing the sugar cane [. . .] They could, and would with pleasure, if they were certain of a market, grow and manufacture more than ten times the usual quantity. It is very profitable; and there is an abundance of very proper land; all they want is a certain market for their sugar.

[. . .]

The method of cultivating the cane and manufacturing the sugar by the natives, hereabouts, is, like all their other works, exceedingly simple. The whole apparatus, a few pair of bullocks excepted, does not amount to more than fifteen or twenty pagodas; as many thousand pounds is generally, I believe, necessary to set out the West India planter.

The soil that suits the cane best, in this climate, is a rich vegetable earth, which on exposure to the air, crumbles down into a very fine mould: it is also necessary for it to be of such a level as allows it to be watered from the river by simply damning it up, which almost the whole land adjoining to the rover admits of, and yet so high, as to be easily drained during heavy rains.

Such a soil, and in such a situation, having been well meliorated, by various crops of leguminous plants, of fallowing for two or three years, is slightly manured, or has had cattle pent upon it [. . .] During the months of April and May, it is repeatedly stirred with the common Hindoo plough, which soon brings this rich loose soil into very excellent order. About the end of May or beginning of June, the rains usually set in, by frequent heavy showers. Now is the time to plant the cane [. . .]

The method is most simple: labourers with baskets, of the cuttings, with one or two joints each, arrange themselves along one side of the field; they walk side by side in as straight a line as their eye or judgment enables them, dropping the sets at the distance of about eighteen inches in the rows, and four feet asunder from row to row; other labourers follow, and, with the foot, press the set about two inches in the soft mud-like soil: this, with a sweep or two with the sole of the foot, they most easily and readily cover: nothing more is done if the weather is moderately showery, till the young shoots are some two or three inches high; the earth is then loosened a few inches around them, with a small weeding iron [. . .] Should the season prove dry, the field is occasionally watered from the river, continuing to weed, and to keep the earth loose about the stools.

In August, two or three months from the time of planting, small trenches are cut through the field, at short distances, and so contrived to drain off the water, should the season prove too wet for the canes, which is frequently the case, and would render their juices weak and unprofitable; the farmer, therefore, never fails to have his field plentifully and judiciously intersected with drains, while the cane is small, and before the time of the violent rains. Should the season prove too dry, these drains serve to conduct the water from the river, through the field, and also to carry off what does not soak into the earth in a few hours [. . .] They reckon these drains indispensably necessary; and, on their being well-contrived, depends, in a great measure, their future hopes of profit. Immediately after the field is trenched, the canes are all propped; this is an operation I do not remember to have seen mentioned by any writer on the subject, and is, perhaps, peculiar to these parts [. . .] A small strong bamboo, eight or ten feet long, is then stuck into the earth in the middle of each stool, and the canes are tied to it; this secures them in an erect position, and gives the air free access around every part [. . .] Tying the leaves so carefully round the cane, they say, prevents them from cracking and splitting with the sun, helps to render the juice richer, and prevents their branching out [. . .]

In January and February the canes are ready to cut, which is about nine months from the time of planting; of course I do need not describe it. Their height when standing in the field, will now be from eight to ten feet, foliage included; and the naked cane from an inch to an inch and a quarter in diameter.

A mill or two, or even more, according to the size of the field, is erected, when wanted, in the open air; generally under the shade of large mango trees, of which there are great abundance hereabouts. The mill is small, exceedingly simple, and at the same time efficacious. The juice, as fast as expressed, is received into common earthen pots, strained, and put into boilers, which are, in general of an oval form, composed of ill-made thick plates, of country iron, rivetted. These boilers hold from eighty to one hundred gallons [. . .]

The liquor is never here removed into a second boiler, but is in the same boiled down to a proper consistence, which they guess by the eye, and by the touch; the fire is then withdrawn, and, in the same vessel, allowed to cool a little. When it becomes pretty thick, they stir it about with stirring sticks, till it begins to take the form of sugar; it is then taken out and put upon mats made of the leaves of the palmira tree (*Borassus flabelliformis*), when the stirring is continued till it is cold: it is then put up in pots, baskets, &c. till a merchant appears to buy it.

The Hindoo name of this sugar is Pansadurry; the colour is fairer than most of the raw sugars made our West India islands; but it is of a clammy unctuous nature [. . .]

In this country, the canes, while growing, are subject to fewer accidents than in the West Indies [. . .]

The lands occupied with the sugar cane in the zemindaries of Peddapore and Pettapore, exclusive of those islands formed by the mouths of the Godavery, amount to five hundred and fifty *Vissums*, or eleven hundred acres, and their annual produce is forty-four hundred weight per acre [. . .] It is acknowledged by all that this quantity might be increased to any extent, with advantage to the zemindar, the farmer, and the government [. . .]

All that seems necessary in these immense tracts, is to open a market to the ryut, and secure to him a strict agreement to his lease with the zemindar.

Transgressions in this point are the great bar to Indian husbandry; for, in a good season, the zemindar raises his demands, and makes the farmers of all denominations pay, probably, a fourth more than the rent agreed upon. Custom has rendered this iniquity common, and the farmer has no idea of obtaining redress of an evil, which to him appears as irremediable as the ravages of the elements.

100

EDWARD POWYS, 'ON FEEDING CATTLE WITH GREEN FOOD', ARTHUR YOUNG (ED.), *ANNALS OF AGRICULTURE AND OTHER USEFUL ARTS*, 45 VOLS (1785–1808), VOL. 45

(London: Richard Phillips, 1808)

On Feeding Cattle with Green Food

Mr Edward Powys.
Communicated by the Earl of Galloway.

I CONCEIVE the principal object respecting agriculture, in the present state of this country, is to procure the greatest possible supply of the necessaries of life within the kingdom itself; and one principal means of doing this, is to raise the greatest produce from a given quantity of land.

To effect this, every encouragement should be given by land-owners to the cultivation of grain and turnips; because I look upon the produce of an acre of grain to be, to the produce of an acre of grass, in the proportion of at least fifteen to two, in furnishing the necessaries of life. I suppose the grain made into bread, and the grass digested by a feeding beast, and changed into an increase of weight.

One great means of increasing the growth of grain and turnips, I think, would be to encourage the farmer to make as much manure as possible. This would be effected by allowing him to sell all his wheat and rye straw, with the restriction of laying out the whole price in manure; and by gentlemen who have land in their hands, trying the experiment of keeping their cattle and horses in the house upon green food great part of the summer.

For these last six years I have sold all the wheat straw I did not want for thatching and the beds of certain kinds of horses, and can assure you, that the same farm has produced for some years back one-third more grain, and keeps double the live-stock it did six years ago.

 DOI: 10.4324/9780429355653-111

As a proof that what I say, of keeping cattle in the house in summer upon green food, is not matter of theory only, but of practice, I shall mention my own experience.

For these last five years I have kept eight or ten waggon-horses in the stable upon clover, cut and carried for them once a day; the small waste that they made was thrown into a low cratch (or receptacle, with staves on each side) for my pigs, which have generally been from 25 to 40. My horses and pigs, thus fed, have eaten, between the beginning of May and corn-harvest, from two and a half to three and a half acres, according to the goodness of the clover. My horses have been, by this means, in much better condition than if turned into a field; there has been a saving of at least eight or ten acres of clover for other stock; a great deal of the richest manure has been made (much more, and richer, than in the same time in winter), and the additional daily expence has been, one man less than half his time, in cutting, raking, and carrying with a horse and cart, one load each day.

[. . .]

The first year I tried the experiment, the manure made was estimated by a good farmer at 20*l*.; but I wish to make allowance for the value of the straw, and the manure that would have been made by the horses standing in the stable the usual hours in summer.

I must endeavour to remove an objection that may perhaps be made to this experiment, by observing, that I cannot think land injured any more by the green food being cut by the scythe, than by cattle or horses; and as to the dung that is dropped in summer, it breeds flies, and does more harm than good.

I have ever thought land exhausted infinitely more by its produce being suffered to ripen and seed, than by its being cut in a green state. The advantage I had derived from this experiment, induced me, last summer, to try whether cattle might not be treated in the same way.

I began with putting into stalls 19; I afterwards increased my stock fed in this manner to 50, consisting of horses, feeding-cattle, milking-cows, and colts, besides a large quantity of pigs.

The horses, as usual, answered well.

The feeding-cattle came on much faster than I ever saw them in summer.

The milking-cows fed very much, and milked very well.

The colts did well, and lived chiefly upon the refuse of the cattle.

The pigs, as usual, eat the refuse of the horses.

The quantity of land run over with the scythe for this purpose was,

Fourteen acres of trefoil, very moderate, on account of the clover root having died in winter.

Two acres of vetches, very moderate, on account of the severe winter.

Five acres of very good grass.

The cattle were turned out late at night for about six or seven hours.

The trefoil caused some trouble, on account of the cattle sometimes swelling, but brought them on very well, though they throve best upon the winter vetch or tares, and upon the grass.

The daily expense was one old man of more than 70, to feed and clean them; another young man, to cut, rake, and carry the food with a single-horse cart.

If this stock had been turned out, I should suppose they would have run over at least 60 acres, if the crop had been good, and much more, if the indifferent trefoil is considered.

[. . .]

I have before observed, that I never saw cattle in summer come on so fast. I speak this, not only from my own observation, but from that also of several farmers and butchers, who came, through curiosity or business, frequently to visit them. The most feeding green food is winter vetches; and the most advantageous mode of cultivating them, I think, is to plough up a clean stubble (that is intended for turnips), manure it, and sow it with vetches soon after corn harvest. When the vetches are all cut in May and June, or rather in the latter month, the field may be ploughed and sown with turnips for a winter crop.

[. . .]

I repeat it once more, that the interests of the public, of the landlord, and tenant (for I know of no distinction, when many years are taken into consideration), are united in the greatest produce of the necessaries of life; and that if arable land is kept clean and full of manure, it receives no injury from producing the greatest quantity of grain. – The increased produce of land benefits the public in too obvious a manner to enlarge upon. It benefits the landlord, by his being able, at the expiration of certain fair intervals, to raise the rent of his farm; and the tenant or occupier, by getting more profit from a given quantity of land, and with nearly a given capital.

[. . .]

Let me notice the great benefit and comfort that the common workman, in any line, derives from sufficient grass land being attached to his dwelling, to keep a cow in summer and winter. The landlord will also receive benefit, as well as self-satisfaction, from being the cause of the plenty that the produce of a cow makes in a large and poor family.

I can, from experience, assert, that the cottager can afford to give his landlord one-third, if not one-half more, for that small quantity of land, than a farmer. The value of the cow is generally more than one year's rent, and the addition of a small cow-house is a trifling expence.

I cannot help recommending this the more strongly, because I know well, from experience, the astonishing comfort and advantage that a poor family receives from the produce of its cow, and that it is also for the interest, as well as inward satisfaction, of the landlord.

101

ARTHUR YOUNG, *GENERAL REPORT ON ENCLOSURES, DRAWN UP BY ORDER OF THE BOARD OF AGRICULTURE*

(London: B. McMillan, 1808)

§ III. – The Benefits that Result from Enclosing Waste Lands

WHERE the value while in common is so low, that many good judges have questioned whether all the commons in the kingdom are worth a groat to the public, the enclosure, division, and cultivation of them in any degree, must be a great advantage indeed. The Board possesses ample proofs of this fact, from which a very cursory selection will leave not the slightest doubt. As the subject, however, should be clearly understood in all its bearings, and the degree of the merit of these measures ascertained, even where most decisive, it will be necessary to examine this result in relation to,

1 The Farmer.
2 The Landlord.
3 The Poor.
[. . .]
5 The Public.

All these are interested in the measure.

1. – The Farmer

THERE have been cases in which one or two great farmers, whose lands were conveniently situated adjoining a large, dry, and valuable common, who might possibly make a greater profit by sweeping off the food, and starving all other stock, by flocks of folding wethers, than such few individuals could receive from the share afterwards allotted to their farms; but as such benefit was always gained by an equal degree of loss to every one beside, it in no case merited the smallest

DOI: 10.4324/9780429355653-112

attention, however clamorous such individuals might be against the enclosures. These cases also are rare, and demand no more than admitting the possibility of their existence. In all others, the division is absolutely beneficial, and even singularly so; for the farmers who in consequence of an enclosure act, receive the addition of allotments, cultivate them without adding to their capitals, perhaps not to their teams, and scarcely to their trouble or attention; the benefit is pure, and unmixed with any drawback, giving them a larger profit than that proportion of advantage which they derived from their original farms. The benefit to this class, therefore, cannot admit of any doubt.

[. . .]

2. – The Landlord

THE case of the landlord is dispatched in a moment; for in no case whatever can he be supposed to have derived one hundredth part of the benefit from commons, which their enclosure gives him, by subjecting his portion to a rent per acre, proportionably to the benefit derived by the farmer. If he received any thing from the common by means of plantations, it must have been by enclosure under certain acts, and therefore is excluded from this consideration.

[. . .]

3. – The Poor

THE benefit in this case is by no means unmixed, and therefore demands a more careful examination: general assertions are of very little value, and especially when they come from persons directly interested in the measure. Many arguments have been framed, as well as assertions advanced, to prove that enclosing commons has been *universally* beneficial to the poor; but as this is directly in the teeth of their own feelings and positive assertions, as well as of those of many other most respectable eye-witnesses; – as the amount of the evil, if it exists at all, can never prove a reason against enclosing, but at most, call merely for a more tender attention to their interest, these papers would be very incomplete, if the question was not fairly examined. The national benefits of enclosing are much too great and decided, to want the smallest concealment in any point.

In the year 1800 a journey was made of above 1600 miles, in which the effect of enclosing, on the spot where the enclosures had taken place, was examined, without trusting to the reports of the poor only, but of the clergy, farmers, and even commissioners who had been employed; and it appeared, that in many cases the poor had unquestionably been injured. In some cases, many cows had been kept without a legal right, and nothing given for the *practice*. In other cases, where allotments were assigned, the cottagers could not pay the expense of the measure, and were forced to sell their allotments. In others, they kept cows by right of hiring their cottages, or common rights, and the land going of course to the proprietor, was added to the farms, and the poor sold their cows: this is a very common case.

The causes were in this manner various, but the result the same. The Board will observe that these injuries, *in fact,* though not in *legal right,* are not mentioned to shew that such enclosures should not have taken place; nor to assert that an increase of regular employment by the cultivation of the farmers might not more than make amends for them, which is another question, but to state the facts really as they are.

[. . .]

5. – The Public

THE objections which have hitherto been offered against the enclosure of commons, have been so weak and ill-founded, as not to merit any particular attention, when urged as motives to prevent the attempt. At all events, the public must be benefited; to convert tracts really or nearly waste, into profitable farms; to change ling for turnips; gorse for barley; and overstocked and rotting grass lands to wheat, and every other useful production, is a real acquisition of territory, pregnant with every advantage attending the husbandry of the kingdom; food, population, wealth and strength.

102

HENRY HOLLAND, *GENERAL VIEW OF THE AGRICULTURE OF CHESHIRE; WITH OBSERVATIONS DRAWN UP FOR THE CONSIDERATION OF THE BOARD OF AGRICULTURE*

(London: Richard Phillips, 1808)

Sect. IV, Soil and Surface

THE general appearance of Cheshire is that of an extended plain, thickly covered with wood; so that from some points of view the whole country resembles one vast and continued forest. The most elevated part of the county is on the eastern border, where a chain of barren hills, connected with the mountainous ridge that divides Cheshire from Derbyshire, extends from Lawton, on the borders of Staffordshire, to the north-eastern extremity of the county, a distance of about 30 miles. From Macclesfield, in a north-westerly direction, the surface is irregular and hilly; but this continues only as far as Alderley, five or six miles from Macclesfield. Here we find a singular hill called Alderley Edge, rising gradually from the S.S.E., and falling down abruptly towards the north. On the western side of the county, a broken and irregular range of hills presents itself, originating near Malpas, running northwards across Delamere forest, and terminating, by a bold promontory, not far from Frodsham. The length of this ridge, which however meets with several interruptions, is more than twenty miles. The most singular feature in it, is the insulated rock of Beeston, situated about two miles to the south of Tarporley; which forms a most striking object from the whole of the surrounding country, and even from the neighbourhood of Liverpool. This rock, which on one side rises almost perpendicularly to the height of 366 feet, is composed of sandstone. Its summit is crowned with the ruins of Beeston castle, a fortress erected A.D. 1220, and formerly esteemed impregnable.

[. . .]

DOI: 10.4324/9780429355653-113

Chap. II.

STATE OF PROPERTY.

Sect. I, Estates and their Management

THERE are few counties of equal extent with Cheshire, in which the number of wealthy land-owners seems so considerable. Whether the revenue derived from the soil is in itself greater, or that men of fortune reside more on their estates in this county than in others, may be a question; but from various accounts which I have received, it appears that not less than fifty noblemen and gentlemen are now resident in Cheshire, in possession of property within it of from 3 to 10,000£. a year; and that there are at least as many others with properties of from 1 to 3,000£. a year. At the same time the number of smaller land-owners is not apparently less than in other counties. The description of this latter class has however been very much altered of late years. From the advantages which have been derived from trade; and from the effects of the increase of taxes, which have prevented a man living with the same degree of comfort on the same portion of land he could formerly; many of the old owners have been induced to sell their estates; and new proprietors have spread themselves over the county, very different in their habits and prejudices. It may be doubtful whether the change on the whole has been disadvantageous. Land, when transferred, is generally improved by its new possessor. With a new, and often a more enlightened view of its advantages and resources, he brings with him the means and the disposition to try experiments, and give to his new acquisition its greatest value. He feels the want of comforts and conveniences, which custom had rendered familiar to a former occupier; he builds, drains, and plants; and by his spirit and example stimulates all around him to increased exertions.

There does not appear to be any peculiarity in the management of the estates in Cheshire. Some gentlemen of fortune, who are fond of agricultural pursuits (and I am happy to state that the number of these seems to be increasing) not only farm to a considerable extent themselves, but take an interest and pleasure in the management of their estates, and in every object conducive to their improved cultivation. Other gentlemen employ as their stewards either attornies, or respectable yeomen who are conversant in the value and management of land.

[. . .]

Conclusion

Means of Improvement

FROM what has been said in the foregoing pages, it will have appeared that the prosperity of the county of Chester has been, of late, rapidly progressive. Trade flourishes, and agriculture is improving. With an increase of population, industry has increased, and discovered new means of obtaining wealth: waste lands are annually inclosed; and new and substantial buildings are erected in various parts of the county. Much, however, yet remains to be done; and I have endeavoured to point out some means by which the cultivation of the land may be carried to a greater degree of perfection. Amongt these the most important have been the introduction of green crops, alternately with corn, into the course of tillage, and the irrigation of meadows; practices which have already been adopted to a certain extent, and which will, in all probability, become much more frequent. Their consequences are indeed so eminently beneficial, and especially to the dairy farmer, that it is surprising attention to individual interests has not sooner led to their more general introduction.

Another subject to which I have endeavoured to call the attention of the gentlemen and land-owners, for the improvement of the country, has been the advantage to be derived from the increase of plantations on the wastes, and in all situations in which the growth of trees might be encouraged, without injury to agriculture. The decrease of timber is a most serious evil, which hereafter will be severely felt, unless new supplies are provided for posterity by the spirited exertions of the present generation.

103

JOHN FAREY, *GENERAL VIEW OF THE AGRICULTURE AND MINERALS OF DERBYSHIRE; WITH OBSERVATIONS ON THE MEANS OF THEIR IMPROVEMENT, DRAWN UP FOR THE CONSIDERATION OF THE BOARD OF AGRICULTURE AND INTERNAL IMPROVEMENT*

(London: B. McMillan, 1811)

THE extended surface of Red Marl which I have mentioned above, as occupying almost all the southern part of Derbyshire, &c. and the Gravel coverings, prevent the tracing of the yellow Limestone strata in connection, further south than Radford and Strelly, near Nottingham: they seem however to make their appearance again, by denudation, at the N E and N border of the Ashby-de-la-Zouch Coal district (Green), [through] a line of detached Limestone Hills, at Grace Dieu (close to, and apparently almost underlaying, the Slate of Charnwood Forest in the Red Marl), Osgathorp Village, Barrow Hill, Clouds Hill, Breedon Hill, and Stanton Park, in Leicestershire, and at Calke and Ticknall, in Derbyshire, where they contain Entrochi. And from numerous circumstances, I am inclined to believe, that the flat blue beds, such as appear at Grace Dieu, Osgathorp, Stanton, Calke, and Ticknall, have a regular continuance from Grace Dieu to Decoy Wood, ½*m*. SSW of Bretby Church, as upper measures to all the Coal-measures S of them. The rearing or almost vertical yellow Limestone strata, composing Breedon Hill, Clouds Hill, and Barrow Hill, I am inclined not to consider as the effects of Faults or dislocations of the strata, but as contortions or original humps, to which these strata were liable, in particular places, at the time of their formation; since exactly similar rearing measures are to be seen at Wild Park, in Derbyshire, in a long chain of Hills near Abberley, in Worcestershire, and in others stretching from Wolverhampton to Dudley in Staffordshire, where, at Wrens'-nest Hill in particular, it

DOI: 10.4324/9780429355653-114

seems impossible to account for the almost vertical position of these strata by Faults, since they lap round a long pyramidical hill, without breaks at the angles!

At Wild Park, N E of Brailsford, and at Birchwood Park, E of Roston, two other patches of the yellow Limestone strata appear, and they seem to me, very probable indications of a Coal-field concealed by Gravel, extending for some miles to the S E, S, and S W of Ashburne, to which the Limestone strata of which these are contortions or humps (as above explained), are upper measures, though all are concealed by the Gravel, except at these two places. The Coal-measures, proved in the trials at Sprinx in Ednaston some years ago, and at Darley Moor, near Yeveley, although no Coal-seams of importance were reached, add greatly to the probability of my supposition, as to the existence of a Coal-field on the south of Ashburne, which being a populous district, far from Coal-pits, Navigations, or extensive Woods, for the supply of fuel, it would seem, that a more extensive and scientific search ought to be made, and the expense, perhaps, defrayed from a fund, supplied by a general subscription of the Inhabitants and Land-owners, who all have an interest in the discovery of Coal in their neighbourhood; at the same time, that the concealment which the Gravel here occasions, renders it a doubtful speculation for any individual Land-owner, to undertake the necessary search in his Estate; because, should he succeed, the probability is, that the adjoining proprietors collectively, or perhaps some of them individually, would profit more than himself, by the discoveries he might make; and if he failed, most others would but laugh at him, as is too commonly the case. In the Woodwardian Collection of Fossils at Cambridge, are Ironstone-balls from Brailsford (Catalogue, vol. II. p. 86), most probably from Sprinx, in this Parish, which if so, is a further confirmation of my opinions respecting this district.

104

JOSEPH HAYWARD, *ON THE SCIENCE OF AGRICULTURE, COMPRISING A COMMENTARY ON AND COMPARATIVE INVESTIGATION OF THE AGRICULTURAL CHEMISTRY OF MR KIRWAN AND SIR HUMPHRY DAVY*

(London: Longman, Hurst, Rees, Orme, Brown, and Green, 1825)

Introduction

AGRICULTURE is defined by Mr Kirwan to be 'the art of making the earth produce the largest crops of useful vegetables at the smallest expense'; but this conveys only a contracted and partial idea of that which must be comprehended in the science of husbandry. Vegetables, and animals which feed on vegetables, constitute that produce of the earth which is essential to the existence, and requisite to the comforts, of mankind. The art of husbandry is, no doubt, simple, if it be considered as limited to manual operations only; but the science of husbandry or agriculture is more properly, a knowledge of the laws of nature which determine the existence of both animals and vegetables, and particularly of those, which influence and govern them in their sexual intercourse and propagation, and also in their feeding, lodging, &c.

Sir Humphry Davy very justly observes, 'It is scarcely possible to enter upon any investigation in agriculture without finding it connected, more or less, with doctrines or elucidations derived from chemistry'. And a chemical examination shows that the earth is but little concerned in vegetation, other than as a medium or vehicle, bed or couch, in and on which, the most important operations of nature are conducted and performed.

We are commonly led to consider vegetables as the chief produce of the earth; but vegetables and animals are so completely dependent upon each other, that

DOI: 10.4324/9780429355653-115

before we can affect in any degree the produce of the one, we must comprehend the influence they have each on the other. From a great similarity to themselves in nature, mankind were very early enabled to form a clear comprehension of the general function of animals, in regard to their food, and as they are influenced by climate; also, of the difference in the sexes, and the natural result of their intercourse: but it was not understood until a much later period, that the general functions of vegetables are in every respect similar to those of animals, and that the operations of nature regarding both, are regulated by much the same laws; this, however, is now clearly demonstrated.

As some portions of the earth, in the production of animals and vegetables, conduce more to the supplying the wants of mankind than others, and not only one species more so than another, but some varieties of the same species are more valuable than others; therefore it must be obvious, that as well as enquiring into the general nature and various qualities of the earth, we must also enquire into and ascertain, the qualities of those varieties of animals and vegetables which are the most conducive to our wants, and what causes produce those varieties, or diminish or increase their peculiar qualities.

General View of the Subject

[. . .]

In the production of variety in animals and vegetables, no doubt climate has the preponderating influence, and next to this food and lodging; but in the general course of nature, these three grand principles operate in unison, and when all concur in one effect, the greatest distinctions are produced.

Food being the most effective and essential cause of variety in the value of animals, this has commanded more attention than the operation of climate and lodging in respect to animals: and climate and lodging being the more immediate, effective, and obvious cause of the variety and value of vegetables, these have commanded more attention than food in respect to vegetables. But to avail ourselves of the full advantage allowed by nature, food, climate, and lodging, both as they affect the one and the other, must be clearly understood and equally attended to.

[. . .]

Animals and vegetables administer to the wants of mankind; some by their flesh, or the immediate substance of their bodies; others by their offspring and seed, or the food provided by nature for their offspring; others, again, by their exterior covering, and others by assisting man in his labour, and contributing to his pleasures: therefore, in breeding and feeding both animals and vegetables, due regard must be paid to the peculiar object desired. And as nature ever determines the end to the means, the attention of the agriculturalists must be directed to the adapting their means to the end in view.

Nature is ever kind and liberal in providing for the necessities of her creatures; and being inclined to make an exuberant return in her productions, for extra aid, she thus gives mankind an opportunity to avail themselves of such propensities: by removing obstructions and favouring and protecting the general operations of nature, and supplying the deficiency of any needful support, they may increase their means of subsistence and enjoyment.

But although mankind are thus blessed by an all-bountiful Providence, their power is prescribed, and they are not permitted to act in opposition to the laws of nature with impunity; whenever, therefore, they presume to interfere with the operations of nature, with a view to produce any beneficial or certain effect, they must pay all due deference to her laws; all attempts to produce sudden and abrupt changes, and wide extremes, must be avoided: by assisting nature certain objects may be obtained, but attempts to force or oppose her, generally produce disorder, and often destruction.

105

'A PRACTICAL FARMER' (J.M.), *HINTS ON AGRICULTURE*

(Hull: James Purdon, 1835)

To make a good fallow on an exhausted soil, already much infected with annual weeds, the following method is recommended. A very light ploughing in the *early* part of the autumn, with a harrowing in the spring, will cause the seeds to vegetate; and your purpose will be promoted by a second light ploughing and thorough harrowing. In May, plough to a proper depth, with a sufficient number of ploughings, say not less than seven, and occasional harrowings; and a good fallow will be obtained, if the season be at all favourable. If the impoverished soil has been exhausted by successive wheat crops, plough it up at the latter end of the year; and let it remain over for Barley. Sow the Barley-stubble with winter Tares, on which feed Sheep the ensuing Summer; sow Wheat the following Michaelmas, and Seeds in the Spring. This process, in conjunction with sufficient lime and manure, will restore the land to a vigorous and improved condition. The Seeds should be, by all means, well harrowed in, and rolled down, otherwise they fail in Wheat; but, if properly worked, will be found perhaps preferable to Barley for the Seeds. I have known six or even eight harrowings in a place wherein both the Wheat and the Seeds have succeeded well. In most cases it is proper to harrow *once* before the Seeds are sown.

The customary management of land, is first to lay it dry by surface or other drains. If springy, it should be under-drained. This method, which the author employed forty years ago, is now frequently adopted. If the land be clay and level, when fallowed, cross-plough two or three times in flats or ridges from fifteen to forty yards wide, as may be requisite to its proper drainage. Let the furrows always fall in the same place to grip on; and, in every future cross-ploughing, follow the same process. One or more deep furrows should also be made by ploughing out three or four times lengthwise on the field for a drain, as the water cannot well be conducted from the middle of a field of more than 150 yards across. Proper attention to the above in the course of the summer fallow will greatly facilitate the draining; and to plough the other way in lands of not more than four or five-bout ridges; six, or seven and a half feet, then they may harrow, drill, horse-hoe, scruffle, &c.; the horses to walk in the furrows, if the land be rather wet, would certainly be of important benefit. In large high-ridged lands, make a land

DOI: 10.4324/9780429355653-116

of four-bouts between each; then, when the lands are taken up, you will have two furrows, and when ploughed down, one furrow for the water to drain off.

On all lands let there be at least from two to five acres of Swede Turnips in every hundred; and, by all means, allow a double dressing of manure. The best time to sow is from the 12th to the 20th of May; and rather than have strong land, with all the moisture worked out and remaining cloddy, open the ridges in March; manure, and close them to remain to the time of sowing; harrow them down a little, and drill, &c.; or, plough and harrow well after harvest; manure and plough before the winter, and drill on the level in the month of May.

Most lands should be ploughed, if possible, as soon as the previous crop is off; which, with a good harrowing, is not unfrequently of equal value to a whole summer's fallow; and by taking up the land after seed-time, or half-ploughing it by throwing one furrow on the other, and going across the lands; and when large, high-ridged lands, plough two bouts in each furrow, to drain off the water, and let the land be exposed to the frost. This process affords excellent preparation for Swede Turnips, Potatoes, Peas, Beans, Mustard, &c. on almost any spring crop.

106

JOHN LAURIS BLAKE, *LESSONS IN MODERN FARMING: OR, AGRICULTURE FOR SCHOOLS, CONTAINING SCIENTIFIC EXERCISES FOR RECITATION; AND ELEGANT EXTRACTS FROM RURAL LITERATURE FOR ACADEMIC OR FAMILY READING*

(New York: Newman and Ivison, 1852 [1851])

Introduction

THE fact has been long apparent, that there has been a constant decline in most agricultural products; that in numerous instances, farms have become so worn out and unproductive, as not to yield an equivalent for the labor of cultivation. The consequence was a natural one, the tenants being reduced to great destitution, the farms were abandoned, and new lands taken up for culture. A few such cases, land in this country being abundant, did not necessarily create general alarm; but when whole districts of country were affected in the same way, when it was seen that thousands of farms were thus becoming worthless – that the calamity, if it might be so called, was every year making fearful progress westward, threatening in the end the insufficiency of the earth to support its inhabitants – verily, a gloomy spectacle was presented to the imagination.

To the philosopher and the political economist, this was indeed matter of deep alarm, and requiring profound investigation. Their investigations were attended with satisfactory results. Science, as in other departments of knowledge, came to our relief, demonstrating the self-sustaining energy of the soil to all the purposes for which it was designed, provided we keep it under the dominion of those physical laws which govern material nature. The principle suggested was a simple one, easy of comprehension; the remedial process was natural, and attended with no serious obstacles. It was simply to restore to the earth such fertilizing agents as are taken from it in each successive crop. Indeed, with a small expense and with

DOI: 10.4324/9780429355653-117

materials found in unlimited profusion, this original capacity may be greatly augmented. To do this is the aim of scientific men who are turning their attention to agriculture. Hence, the praiseworthy efforts that have latterly been made to dispel the gloom that had been gathering around the husbandman – to scatter light in his path – to alleviate his labors – to augment his profits – and to make his occupation appear to him, what it is designed to be, and what it may be, among the most honorable and useful among men.

Among the instrumentalities for achieving these laudable objects, has been the more extensive production of books on agriculture, the circulation of periodical literature, and the delivery of scientific lectures on the subject; and, especially, the establishment of agricultural schools, where the sons of farmers and all others may receive an agricultural education analagous to the education provided for young men designed for the various other professions in useful labor. This is the great enterprise of scientific agriculturists of the present day. The scheme is a magnificent one. It embraces within its telescopic range the welfare of the whole human family. It looks not simply to the generations of now living men; but to the long succession of unborn generations, to the end of time, luxuriating in that abundance which the earth may thus be made to yield. This scheme, to be complete and efficacious, will embrace of necessity such of those simple elements of scientific agriculture as may be imparted in the common schools of the country, so that all our youth and, ultimately, our whole population – every laboring man that tills the ground, shall be as familiar with them, as at present with the elements of general education, reading, penmanship, arithmetic, geography, and grammar, or history.

Let this be done – well done – thoroughly done – learned professors on agriculture, in all our colleges, with model farms at control, where all the young men of the country, educated for the learned professions, shall also be made competent to become teachers in agricultural chemistry; and every district school provided with good teachers and good text books on the subject, and we shall be likely to hear no more of exhausted soils and diminished crops – no more of social and mental degradation in connection with agricultural labor – no more of necessary poverty as the result of rural occupation. If such a scheme of enlightened philanthropy does not originate in minds of the highest grade of moral excellence; if it does not array itself with the most gifted embellishments of the human intellect; and if it does not now commend itself to the sympathies of the good, and the wise, and the powerful of the age; it is difficult to imagine of what scheme this can be affirmed.

The present volume is designed to be an humble agent in elevating and enlightening the agricultural community; to be one of the instruments used in the foregoing scheme for rendering the soil more productive, and those who cultivate it better remunerated. In this volume an effort is made to introduce into the common schools of the country a taste at least for scientific agriculture and rural literature. It seems absurd that children in the country should receive an educational smattering on almost every thing, save that on which they are to depend for a reputable subsistence. If a boy is to be taught how to make shoes, or pantaloons, or hats – to use the jack-plane and hand-saw, or the mason's hammer and trowel – to blow the bellows of a blacksmith, or to build wagons, or to repair watches and clocks, he must spend an apprenticeship of five or six years; but if he is to be a farmer, he is treated as though by instinct he may perform every act required of him in the occupation, as the new dropped calf, or colt, or pig, instinctively turns to the teats of its dam.

The absurdity of this policy is manifest, when we consider that the business of agriculture, in all its various details, requires an amount of general intelligence far beyond what is required for conducting the business of most mechanical trades. The farmer who understands the composition of soils, and the theory of manuring them, the physiology of vegetables and animals, the general nature of tillage, and the principles of cropping and raising stock, must not only possess general intelligence, but must be a man of some science, or he is obliged to perform his labors without understanding the nature of them, as parrots talk, or persons sometimes will sing by rote, without any knowledge of the principles of music. The farmer, too, should be a man of sound judgment, and of habits of philosophical observation, thus being enabled to take advantage of his own experience, and of every agricultural and meteorological incident presented to view. Without this mental endowment he may perhaps raise potatoes, and pork, and Indian corn, sufficient to keep his family alive; but he no more deserves the name of a farmer, than the mason's hod-carrier deserves the name of architect or master-builder.

[. . .]

An exigency like this may be rendered more clear. It is known that most of our district school-teachers have had no classical instruction in the chemistry of agriculture. This is probably true of ninety-nine out of every hundred. Now, if a text-book were put into their hands from which to instruct their pupils, abounding in all the scientific array of terms common in books used by college professors, the masters would feel compelled to decline the task. The sight of the book would lead them to do so. They would feel their incompetency to act in any other way. It would prevent their doing, or attempting to do, what they might be induced to do from a text-book differently prepared. In this case the text-book should be as much adjusted to the taste and the ability of the master as to the scholars; indeed, it should be one from which he might, and from which he would be disposed to be the instructor of himself as well as of his pupils. With such a work, though he had previously known nothing of the subject, he would easily keep in advance of them. An Agricultural Text-Book, therefore, for common schools, must, with teachers, as well as scholars, be attractive and not repulsive, in its very appearance; and, so far as the topics proposed for study are concerned, it should contain those which are the most simple, and which to the common mind are easily rendered most interesting. And even these topics may be made additionally winning from being combined freely with ingredients for which they had no scientific affinity – with literary sweetmeats.

It is possible that persons may be found who will take half an hour, and turn over the leaves of this volume from beginning to end, and because they do not find page after page filled with decimal fractions giving the relative proportions of vegetable or animal constituency, as imposing as Bowditch's Navigator; or, because they do not find column after column of chemical terms, drawn up in formidable order, like the columns of an army about entering battle with their guns and bayonets glittering in the sun, will imagine that this work is not sufficiently scientific. Now, it has verily been our deliberate intention to place in ambush throughout the work, in small squads, whatever of science we have cooked up, so as not to frighten the unlearned out of their wits, with a martial exhibition of our forces. We think we are not mistaken in our philosophy upon this point.

[. . .]

What are some of the most obvious examples of the science, as thus defined?

THE detection of alloys in counterfeit coin; of poison mixed with other substances, either before being used, or when found in the stomach; of the component parts of any medicine, and hence its effect on the animal constitution or any particular disease; of the materials used in dye-stuffs, and hence the measure of their durability; of the elementary constituents forming any particular description of food, and hence its adaptation and power for sustaining animal life and growth. Indeed, there is seemingly no end to the benefits derived from chemical knowledge. By it we learn that grease or oil will prevent friction in machinery, or rust on metals; that the contact of the atmosphere on certain liquids will produce vinegar – on butter will cause it to become rancid, and on eggs will hasten putrefaction; and that yeast mixed with kneaded flour will occasion fermentation. Hence it may readily be seen how chemistry may be made beneficial to agriculture.

What are some of the cases that first occur to you of the beneficial effects of chemistry to agriculture?

TO remedy defects in soil. If the soil is too light to preserve fertilizing agents, mix with it clay or other substances to increase its power of retaining them. If it is too cold, mix with it sand to render it warm. If clayey and impervious, mix with it substances that will cause it to be mellow and light. If sour, mix with it what will neutralize this property in it. Or, if wanting in some particular agent needful to the growth of a particular crop, cast upon it that description of manure which contains this agent. So, if the soil is found to be naturally too moist, chemistry will admonish the farmer to resort to draining; or, if naturally too dry, to resort to irrigation.

But most of this is probably known by most farmers, who never studied chemistry; how then do the cases named apply to the subject?

TRUE, some of these things may have been learned from one's own experience in agriculture; some from observing the practice of others, or from the exercise of mere common sense. Nevertheless, they involve chemical principles; and if portions of a science are thus learned, as it were, from necessity, it shows the importance of pursuing the study more systematically in other matters that will be found equally valuable, although less obvious, to the uneducated farmer. It is by no means the fact that chemistry is learned from books only at school; but the fact that portions of it are learned under such disadvantageous circumstances, and by a slow process, shows the advantage, to farmers especially, of making it a part of elementary or common school education to their sons.

How can the necessity for a rotation of crops be explained on principles of agricultural chemistry?

IT is easily done. Here the knowledge imparted by the science is of immense value. It has ever been known that, in raising a succession of the same crop for a long course of years, there would be gradually and regularly a diminution in the amount of the product. The fact was apparent, because it was constantly observed by the most intelligent of uneducated farmers; but the reason of it was a perfect mystery. Now chemistry explains the mystery. It tells them that one crop requires mainly in its growth one particular fertilizing agent; that another crop in the same way requires mainly another fertilizing agent; and so of a third, and a fourth. Thus, for instance, potatoes being planted one year; then Indian corn; then oats; then wheat; and then grass – or any analogous rotation – there will be no lack or diminution of production: when the same, or a similar alternation of culture, may be followed to any indefinite extent, and with similar results. All this is made plain by chemistry, as the most familiar process with which one can be acquainted.

How is it that chemistry is able to comprehend these facts, unless by a succession of experiments, like those of the common farmer?

THE chemist, in the first place, ascertains of what particular substance each particular vegetable to be raised is composed, or what material from the soil enters into its growth. He then ascertains of what elementary substances any particular portion of soil is composed. This being done, if the soil is deficient or destitute of any one elementary substance required in the growth of a particular vegetable, that vegetable cannot be produced on it. Thus four different vegetables being examined and found to require each mainly for itself a particular elementary substance; and that the soil contains these four substances, but only enough of each for one crop, it follows that either one of them planted more than one year in succession would fail of receiving nourishment, but if each one were planted one year only and in succession or rotation, there would be nourishment enough in the soil for all four of them, and there would be a good crop of each.

PEOPLE INDEX

The following are either extract authors or people mentioned in extracts or headnotes. The extracts authored by figures are included in square brackets.

Abildgaard, Nicolai Abraham (1743–1809): Danish artist, sculptor, and architect of 'the Golden Age of Danish Painting', who also provided illustrations for works of natural history.

Abu Hanifa Dinawari (d. 895): Persian Islamic Golden Age polymathic scholar and author, whose natural history work included books on plants and geography which are partially lost.

Adanson, Michel (1727–1806): French botanical geographer and taxonomist, author of *Familles Naturelles des Plantes* (1763).

Agassiz, Louis (1807–73): Swiss-American biologist, comparative anatomist, glaciologist, and palaeontologist, author of *Poissons Fossiles* (1833–43), and pioneer of studies of extinct marine species. Professor of Biology and Geology at Harvard University, his work is criticised for its scientific racism. A Natural Theologian, he was opposed to evolutionary theory. His works include *Contributions to the Natural History of the United States of America* (1857–62), *Geological Sketches* (1866), and *A Journey in Brazil* (1868). [41]

Al-Asma'i (740–c. 833): Arabic grammarian, botanist, zoologist, and anatomist of the Basra school, author of *Asma'iyyat* and *Book of the Wild Animals*.

Aldrovandi, Ulisse (1522–1605): Italian botanist and ornithologist, keeper of Bologna's pioneering botanical garden.

Alexander the Great of Macedonia (356–323 BCE): illustrious military leader of the Hellenic era, famous for his conquests in Asia and relatively early death.

Anaxagoras (c. 500–480 BCE): pre-Socratic Greek philosopher known for his interest in scientific subjects.

Anaximander (c. 610–545 BCE): pre-Socratic Greek philosopher with a particular interest in the sciences, creator of one of the earliest world maps, author of *On Nature* and *The Sphere*.

Anderson, Rev Dr John (1796–1864): Cleric and amateur geologist, author of *The Course of Creation* (1850) and *Dura Den: A Monograph of the Yellow Sandstone* (1859).

Ángelis, Pedro de (1784–1859): Neapolitan historian, journalist, and diplomatic representative in Argentina of the Kingdom of the Two Sicilies, and founder of the newspaper *El Lucero*.

Anning, Mary (1799–1847): English palaeontologist and geologist responsible for many remarkable fossil discoveries. Because of her gender, her contributions were only belatedly acknowledged and she faced exclusion from scientific societies. Announcements of her early findings did not credit her, but **William Buckland** helped ensure that her significance was understood. Her Lyme Regis fossil shop became a pilgrimage site for many leading scientists, but Anning faced financial hardships later in life. [29]

Ansell, John (d. 1847): English botanist and collector, member of the Horticultural Society of London, and assistant botanist on the 1841 expedition described in *Flora Niger* by **W. J. Hooker**.

Argand, François Pierre Amédée (1750–1803): Swiss physicist and chemist, inventor of the Argand lamp, an improvement on traditional oil lamps that was also useful in microscopy.

Aristotle (384–322 BCE): leading and enduringly influential Greek philosopher whose polymathic studies comprised logic, ethics, aesthetics, literature, and science.

Bacon, Francis (Lord Bacon) (1561–1626): English Renaissance scholar, lawyer, and politician. A ruthless, corrupt Lord Keeper, Bacon was banished from court shortly before his death. *A Treatise on the Advancement of Learning* (1605) and *Novum Organum* (1620), promoting new methods for acquiring and organising knowledge, were a major influence on Enlightenment science.

Backhouse, James (1794–1869): English botanist and Quaker missionary, whose work in Australia is drawn upon by **John Gould**.

Balbi, Adriano (1782–1848): Venetian physicist, geographer, and customs officer, Professor of Geography at San Michele College (Murano), and author of *Prospetto Politico-geografico dello Stato Attuale del Globo* (1808) and *Abrégé de Geographie* (1832).

Balfour, John Hutton (1808–84): Scottish botanist, Professor of Botany at the Universities of Glasgow and Edinburgh, and author of *A Manual of Botany* (1851). [83]

Banks, Sir Joseph (1743–1820): influential English naturalist and patron of the sciences, Banks achieved fame as a result of his involvement in Captain James Cook's first voyage on *HMS Endeavour* (1768–71) to Brazil, Tahiti, New Zealand, and Australia. His earlier voyage as a botanist was in 1766, on *HMS Niger*, travelling to Labrador and Newfoundland. After returning to England after *Endeavour*, Banks became Britain's principal patron of the natural sciences, accruing extensive collections ultimately bequeathed to the Natural History Museum. He was a founding member of the Linnaean Society, the Royal Institution, and the Royal Horticultural Society, and the longest-serving President of the Royal Society. [64]

Baker, Charles (1820–94): microscope maker, and after 1851 head of the firm of C Baker, of High Holborn which had a reputation for quality microscopes and accessories.

Barrington, Daines (1727–1800): English antiquary, lawyer, and naturalist, one of the correspondents of **Gilbert White**'s Selborne letters. A Vice President of the Royal Society with wide-ranging natural history interests, Barrington shared White's fascination with birds. White drew upon Barrington's attempts to standardise tasks in natural history, including meteorological recording, creating calendars of floral activity, and analysing birdsong.

Barrow, Sir John (1764–1848): English geographer, traveller, linguist, and Second Secretary to the Admiralty (1804–45). His works include *Travels in China* (1804), *A Voyage to Cochinchina in the Years 1792 and 1793* (1806), *Travels into The Interior of Southern Africa* (1806), and *A Description of Pitcairn's Island and its Inhabitants* (1845).

Batsch, August Johann Georg Karl (1761–1802): German naturalist, author of *Provisional Guide to the Knowledge, Development, and History of the Animals and Minerals* (1788–9).

Becher, Johann Joachim (1635–82): German physician, alchemist, and chemist, who forwarded the influential but ultimately discredited phlogiston theory of combustion.

Bell, Thomas (1792–1880): English surgeon and zoologist, author of *A History of British Quadrupeds* (1836), *A History of British Reptiles* (1838), and vol. 5 of **Charles Darwin**'s *Zoology of the Voyage of the Beagle* (1842–3).

Benett, Etheldred Anna Maria (1776–1845): English geologist, palaeontologist, stratigrapher, and antiquary, commonly regarded as the first female geologist. Author of *A Catalogue of the Organic Remains of the County of Wilts* (1831), her brilliance and financial independence overcame some of the barriers facing aspiring women scientists, but she was barred from membership of the Geological Society despite the fact that her extensive geological collection was widely used by its members. Tsar Nicholas I granted her an honorary doctorate in recognition of her achievements. [26]

Benett, George (1804–93): Anglo-Australian physician and botanist, author of *Gatherings of a Naturalist in Australasia* (1860).

Bentham, Jeremy (1748–1832): English utilitarian philosopher, social reformer, and animal rights campaigner, author of *Principles of Morals and Legislation* (1780).

Berger, Alexander Malachias (1737–1805): Swedish disciple and collaborator of **Linnaeus**, author of *Calendarium Florae* (1756).

Bewick, Thomas (1753–1828): English wood-engraver and natural historian, author of *A History of Quadrupeds* (1790) and the wildly popular *A History of British Birds* (1791–1804).

Bichat, Marie François Xavier (1771–1802): French comparative anatomist and pathologist, known as the father of histology. Alongside **Georges Cuvier** he created the foundations of modern comparative anatomy. Although a Vitalist, Bichat was committed to materialist investigations of anatomy, and his researches in the microscopic anatomy of biological tissues was all the more remarkable for working without microscopes. [33]

Bicheno, James Ebenezer (1785–1851): British colonial official, author, and amateur naturalist.

Blackwood, Francis Price (1809–54): Royal Navy Officer and explorer, involved in expeditions to Australia, Papua New Guinea, and nearby regions.

Blake, Rev. John Lauris (1831–99): American cleric, lawyer, Republican congressman, author of *Lessons in Modern Farming* (1852) and *Farmer's Every Day Book* (1854). [106]

Blainville, Henri Marie Ducrotay de (1777–1850): French anatomist, taxonomist, and zoologist who coined the term palaeontology in 1822.

Blumenbach, Johann Friedrich (1752–1840): German physician, naturalist, and race theorist, regarded as a principal founder of zoology and anthropology.

Bonaparte, Napoleon (1769–1821): French emperor and military leader in the post-Revolutionary period, conqueror of much of Europe, ultimately defeated and exiled.

Bonpland, Aimé (1773–1858): French botanist, co-author (with **Alexander von Humboldt**) of *Personal Narrative of Travels to the Equinoctial Regions of the New Continent* (1814–29). He was closely associated with many key figures of French Enlightenment science, but from 1816 he spent much of his life in South America. [66]

Both, Jan (c.1610–52): Dutch landscape painter and etcher of the Italianate school.

Bougainville, Louis Antoine de (1729–1811): French admiral and explorer who circumnavigated the globe, leading a scientific exploration in the 1760s of the Falkland Islands, Papua New Guinea, Australia, the Pacific, and South America.

Boullay, Pierre François Guillaume (1777–1869): French botanist, chemist, and pharmacist.

Boyle, Robert (1627–91): influential Anglo-Irish chemist, physicist, and religious writer often regarded as a founder of modern chemistry, author of *The Sceptical Chymist* (1661) and proposer of Boyle's Law (on gas pressure).

Bremer, Sir Gordon (1786–1850): Royal Navy officer who served in numerous military campaigns, known for his early exploration of the coast of Australia.

Brewster, Sir David (1781–1868): British scientist and inventor who experimented in physical optics, and the polarization of light, and particularly remembered for the discovery of 'Brewster's angle'. A photographic pioneer, he invented the first portable 3D-viewing device, the binocular camera, and the kaleidoscope. He was the author of *A Treatise on New Philosophical Instruments, for Various Purposes in the Arts and Sciences* (1813).

Broderip, William (1789–1859): Lawyer and naturalist and lawyer, co-author (with Nicholas A. Vigors), of *Guide to the Gardens of the Zoological Society* (1829).

Brongniart, Alexandre (1770–1847): French chemist, geologist, mineralogist, palaeontologist and zoologist, collaborator (with **Georges Cuvier**) on *Essai sur la Géographie Minéralogique des Environs de Paris* (1811), and the first to attempt to classify Tertiary formations.

Brown, Rev. James Mellor (1796–1867): British 'scriptural geologist', author of *Reflections on Geology* (1838).

Brown, Robert (1773–1858): Scottish botanist, palaeobotanist, and pioneer of microscopy who observed cell nuclei, cytoplasm, and pollination, and first distinguished gymnosperms and angiosperms. Expanding knowledge of Australian botany during his 1801–5 expeditions with **Matthew Flinders**, his *Prodromus Florae Novae Hollandiae* (1810) was the earliest scientific account of Australian flora. He also authored *Observations, Systematical and Geographical, on the Herbarium Collected by Professor Christian Smith, in the Vicinity of the Congo* (1818).

Bru de Ramon, Juan Bautista (1740–99): Spanish artist and dissector who worked on reconstructing the Megatherium from fossilised remains: mounted in Madrid's Royal Cabinet of Natural History, in 1789, it was probably the first extinct mammal skeleton to be displayed.

Bruckner, John (1726–1804): Dutch Walloon cleric, Natural Theologian, and naturalist, author of *A Philosophical Survey of the Animal Creation, an Essay* (1768). [8]

Brunel, Isambard Kingdom (1806–59): British engineer, celebrated for his railroads and bridges, and for the transatlantic steamships *S.S. Great Western* and *S.S. Great Britain*.

Buch, Christian Leopold von (1774–1853): Prolific Prussian geologist and palaeontologist who worked across many subjects and defined the Jurassic period.

Buckland, Rev. William (1784–1856): Anglican theologian and geologist, author of the best-selling *Reliquiæ Diluvianæ, or, Observations on the Organic Remains attesting the Action of a Universal Deluge* (1820) and a volume in the *Bridgewater Treatises* (1836). A leading figure of the early Geological Society, Buckland promoted 'gap creationism' and wrote the first full account of a fossilised dinosaur, the *Megalosaurus*. A Fellow of Corpus Christi College, Oxford, where he was Reader of Mineralogy, he later became Dean of Westminster. His son, Francis Buckland was a surgeon and popular natural historian. [15, 27]

Buckle, Henry Thomas (1821–62): English historian, author of the unfinished *History of Civilization* (1857). While Buckle is sometimes described as 'the father of scientific history' his work has strongly subjective Eurocentric and colonial aspects. [85]

Buffon, Georges Louis Leclerc de (1707–88): French naturalist, catastrophist geologist, and encyclopaedist, author of *Histoire Naturelle* (1749–1804) and *Theorie de la Terre* (1749). From 1739, head of the Parisian Jardin du Roi, a powerful position within the French academy, his pre-eminence within French science was unrivalled during his lifetime, and he influenced later biologists, geologists, and biogeographers. [6]

Bugg, George (1769–1851), Anglican deacon and enemy of Tractarianism and modern geology, author of *Scriptural Geology* (1826).

Bulkley, Edward (d. 1714): East India Company surgeon, naturalist, and specimen collector.

Bunge, Alexander Von (1803–90): Russian-German botanist who undertook expeditions into Siberia and present-day Mongolia in the 1820s and 1830s, author of *Plantarum Mongolica-Chinensium* (1835).

Burnet, Thomas (c. 1635–1715): English theologian and cosmogonist, author of *Telluris Theoria Sacra* (1681).

Burnett, James (Lord Monboddo) (1714–99): Scottish linguist, and proto-evolutionary natural philosopher, author of *The Origin and Progress of Language* (1773–92).

Burns, Robert (1759–96): Scottish poet strongly associated with Scotland's national identity. Born to a humble agricultural background, his works include 'Auld Lang Syne', To a Mouse', 'Tam o'Shanter' and 'Ae Fond Kiss'.

Burton, Richard Francis (1821–90): British explorer, author, and military officer who served in India and the Crimean War before being engaged by the Royal Geographical Society to investigate East Africa. He was the author of numerous books, as well as the English translator of *Kama Sutra*. He spoke 29 languages. [82]

Buxton, Sir Thomas Fowell (1786–1845): English baronet, MP, abolitionist, and social reformer, founder of the African Civilization Society and sponsor of the African voyage described in *Flora Niger* (edited by **W.J. Hooker**).

Bynoe, Mr (1803–65): English Royal Navy ship's surgeon on *HMS Beagle*, working alongside **Charles Darwin** in the collection of specimens, and whose notes were used in *Origin of Species* and John Stokes's *Discoveries in Australia* (1846).

Caesar, Julius (100–44 BCE): Roman general and statesman, whose military achievements in the Gallic Wars considerably extended Roman territory. His role in the Civil War led to the end of the Roman republic and augured in the Imperial era. He was assassinated by Senatorial opponents.

Caley, George (1770–1829): English botanist and explorer, active in Australia for much of his career, and closely associated with **Sir Joseph Banks**, Kew Gardens, and **Matthew Flinders**.

Cambessèdes, Jacques (1799–1863): French botanist best known for his *Excursions dans les îles Baléares* (1826)

Candolle, Augustin Pyramus de (1778–1841): French-Swiss botanist, an early and influential proponent of the 'natural system' of classification. He investigated daily leaf movements of plants and the competitive nature of vegetation, coining a term ('nature's war') that influenced **Charles Darwin**. His *Organographie Végétale ou Description Raisonnee Des Organes Des Plantes* (1827) was crucial to **Johann Wolfgang von Goethe**'s botany.

Cesalpino, Andrea (1524–1603): Italian physician, botanist, and philosopher who explored what would become the field of palaeontology. His accurate theories on fossil remains were, like those of **Robert Hooke**, far ahead of popular consensus. Director of the Botanical Garden at Pisa, he pioneered classification of flora by fruits and seeds, instead of medicinal properties or astrology.

Chalmers, Dr Thomas (1780–1847): Scottish minister and theologian who led both the Church of Scotland and the Free Church of Scotland and who was closely associated with St Andrew's University and the Royal Society of Edinburgh. He was perhaps the earliest proponent of 'gap creationism' and contributed an 1833 volume to the *Bridgewater Treatises*.

Chambers, Robert (1802–71): Scottish publisher, journalist, editor of *Chambers's Edinburgh Journal*, geologist, and evolutionary theorist whose controversial *Vestiges of the Natural History of Creation* (1844) was published anonymously. [93]

Clark, William (1770–1838): American explorer, military officer, and governor who, along with **Meriwether Lewis** led the Lewis and Clark expedition (1803–6) to the Pacific coast.

Clayton, Rev. Edward (d. 1895): Missionary, Honorary Canon of Chester, and Anglican minister, friend and correspondent of **John Ruskin**.

Clift, William (1775–1849): British museum curator and scientific illustrator, whose palaeontological assistance was acknowledged by **Georges Cuvier**, **Gideon Mantell**, and **Richard Owen**.

Cole, Rev Henry (1792–1858): Anglican curate and translator of John Calvin. His critique of contemporary geology, and in particular **Adam Sedgwick**, appears in *Popular Geology Subversive of Divine Revelation!* (1834). [14]

Collins, David (1756–1810): British military officer and administrator of Britain's early Australian colonies in New South Wales, Victoria, and Van Diemen's Land.

Commerson, Philibert (1727–73): French naturalist who took part in **Louis Antoine de Bougainville**'s voyage of circumnavigation (1766–9).

Conybeare, William Daniel (1787–1857): English cleric, geologist, palaeontologist and early member of the Geological Society, who published the first scientific description of the plesiosaur (without acknowledging the role of **Mary Anning** as its discoverer). His best-known work is *Outlines of the Geology of England and Wales* (1822). [24]

Cook, Captain James (1728–79): British navigator, explorer, and cartographer who led three voyages in the 1760s and 1770s to the Pacific Ocean and Australasia. His voyages on *HMS Endeavour* supported scientific discovery and included **Sir Joseph Banks** and **Daniel Solander** as botanists. He was killed on Hawaii in 1779.

Copernicus, Nicolaus (1473–1543): Polish astronomer and mathematician best known for his controversial argument in *On the Revolutions of the Celestial Spheres* (1543) that the planets revolved around the sun. Copernicus did much to foster subsequent Enlightenment science and the work of **Galileo.**

Culpeper, Nicholas (1616–54): English astrological and medicinal botanist, author of *The English Physitian* (1652), (later titled *The Complete Herbal*), which has remained in print to this day.

Cunnington, William (1754–1810): English self-educated merchant, antiquary and archaeologist who promoted more sensitive methods of archaeological investigation and worked extensively in Wiltshire

Curtis, John (1791–1862): English entomologist, engraver, and illustrator, author of *British Entomology* (1824) and *Farm Insects: Being the Natural History and Economy of the Insects Injurious to the Field Crops of Great Britain and Ireland* (1860). [58]

Curtis, William (1746–99): English botanist, entomologist, and apothecary, employed during the 1770s as a demonstrator of plants at the Chelsea Physic Garden. He established botanic gardens in Lambeth and Brompton, authored *Flora Londinensis* (1777–98), and established *The Botanical Magazine* in 1787. [42]

Cuvier, Baron Georges (1769–1832): French comparative anatomist, naturalist, and catastrophist geologist, Professor at Paris's National Museum of Natural History, and contemporary of **Georges Buffon**, **Jean-Baptiste Lamarck** (whose evolutionary theories he opposed), and **Étienne Geoffroy Saint-Hilaire**. Cuvier was a major figure in comparative anatomy, and in promoting understanding animals as functioning organisms within environments. [21, 34, 35, 36]

D'Alton, Joseph Wilhelm Eduard (1772–1840): German engraver and naturalist, best known for his zoological illustrations.

Darwin, Charles (1809–82): English biologist, geologist, and key promotor of evolutionary theory (sharing credit with **Alfred Russel Wallace**). His role as naturalist on *HMS Beagle* (1831–6), captured in *Journal of Researches* (1839) and *Zoology of the Voyage of HMS Beagle* (1838–43), gained him fame. His experiences of the Galapagos subsequently informed his understanding of natural selection, published as *On the Origin of Species* (1859). His other major works include *The Various Contrivances by Which Orchids are Fertilised by Insects* (1862), *The Variation of Animals and Plants Under Domestication* (1868), and *The Descent of Man* (1871), which forwarded his theory of sexual selection. [51, 61, 69, 78, 91, 92, 95]

Darwin, Erasmus (1731–1802): English physician, natural philosopher, inventor, poet, and abolitionlist, grandfather of **Charles Darwin**. As a translator of **Linnaeus,** he authored *A System of Vegetables* (1783–5), *The Families of Plants* (1787) and coined many now-common English plant names. Other works include *Zoonomia* (1794–6) and *The Temple of Nature* (1804), both replete with proto-evolutionary ideas. [86, 87]

Davies, Rev. Hugh (1739–1821): Anglesey botanist, clergyman, and author of *Welsh Botanology* (1813), a work providing a county guide to the flora of Anglesey with the

earliest cross-references to Welsh and scientific plant names. Davies collaborated with **Thomas Pennant** and **James Sowerby**.

Davy, Sir Humphry (1778–1829): English chemist and inventor of the Davy Lamp and other innovations, remembered for using electricity to isolate a range of novel elements.

De la Beche, Henry Thomas (1796–1855): English geologist and palæontologist, President of the Palaeontological Society, the first director of the Geological Survey of Great Britain, and author of handbooks on geological methods, including *A Geological Manual* (1831), *Researches in Theoretical Geology* (1834), and *How to Observe Geology* (1835). De La Beche played a prominent role in the Great Devonian Controversy, as the principle opponent of **Roderick Murchison** and **Adam Sedgwick**.

Delile, Alire Raffeneau (1778–1850): French botanist and Egyptologist, Director of the Cairo Botanical Garden, and Professor of natural history at the University of Montepellier. He named more than 400 species, specializing in non-flowering plants.

Demosthenes (384–322 BCE): Greek orator, rhetorician, lawyer, and statesman.

Derham, William (1657–1735): English Renaissance clergyman, naturalist, and Natural Theologian, author of *Physico-Theology* (1713). A Royal Society Fellow he studied natural history, physics, astronomy, and meteorology, publishing in *Philosophical Transactions* for over thirty years, and produced the earliest measurement of the speed of sound.

Descartes, René (1596–1650): French philosopher, mathematician, and scientist, inventor of analytic geometry and founder of the influential dualist model of philosophy. His works include *Discourse on the Method of Rightly Conducting One's Reason* (1637) and *Principles of Philosophy* (1644).

Diderot, Denis (1713–84): French Enlightenment philosopher and art critic, editor and contributor to *Encyclopédie* (1751–72), and author of *Letter on the Blind* (1749).

Diodurus the Sicilian (1st C BCE): ancient Greek historian, author of the forty volume *Bibliotheca Historica* (60–30 BCE).

Dioscorides, Pedanius (40–90 CE): Greek physician, herbalist, and botanist, author of *De Materia Medica*.

Don, David (1799–1841): Scottish botanist and taxonomer, Professor of Botany, King's College, London, who wrote the first British descriptions of the Coast Redwood.

D'Orbigny, Alcide Charles Victor Marie Dessalines (1802–57): French naturalist who made significant contributions to archaeology, anthropology, geology, palaeontology, and zoology.

Dughet, Gaspard (1615–75): influential French-Italian 'Old Master', (also known as Gaspard or Gaspar Poussin) who worked primarily on landscapes of the countryside around Rome, where he was born.

Duméril, André Marie Constant (1774–1860): French zoologist, palaeontologist, and anatomist who collaborated with **Georges Cuvier**. He published *Zoologie Analytique* (1801), *L'Erpétologie Generale des Reptiles* (1834–54), and (with his son Auguste) *Catalogue Méthodique de la Collection des Reptiles* (1853).

Durer, Albrecht (1471–1528): German Renaissance artist and printmaker who also wrote theoretical treatises on mathematics and perspective.

d'Urville, Jules Sébastien César Dumont (1790–1842): French naval officer, cartographer, explorer, and naturalist.

Ehrenberg, Christian Gottfried (1795–1876): German naturalist, comparative anatomist, geologist, and zoologist. While Professor of Medicine at Berlin he accompanied **Alexander von Humboldt** on his Russian journeys. Subsequently pursuing diverse scientific enquiries, he made substantial contributions to microscopy.

Evelyn, John (1620–1706): English author, gardener, and diarist, author of *Fumifugium* (1661) and *Sylva* (1664), and one of the earliest critics of the environmental impacts of deforestation and coal industries. [1, 2]

Ewbank, Thomas (1792–1870): English writer of humble origins who emigrated to the U.S. in 1819 to pursue a career in manufacturing. Retiring in 1836, he wrote on mechanics and travel writing (*Life in Brazil*, 1846). After being appointed U.S. Commissioner of Patents (1849–52), he wrote *The World a Workshop, or the Physical Relation of Man to the Earth* (1855), a curious mixture of materialist science and Natural Theology, and *Thoughts on Matter and Force* (1858). [18, 55, 63]

Ewing, Rev Thomas James (1813–82): Anglican clergyman and ornithologist who emigrated to Tasmania in the 1830s.

Faber, G.S. (1773–1854): English Evangelical and theologist, who opposed uniformitarianist geology. He authored *The Origins of Pagan Idolatry* (1816), *A Treatise of the Three Dispensations* (1823), and *The Difficulties of Infidelity* (1824).

Faraday, Michael (1791–1867): brilliant and influential English chemist and physicist, discoverer of electromagnetic and electrochemical principles, and of benzene.

Farey, John (1766–1826): English geologist and land agent whose career as surveyor and geologist was due to the guidance of **William Smith**, whose breakthroughs he promoted. Farey authored *General View of the Agriculture and Minerals of Derbyshire* (1811–17), scientific papers, and numerous entries for encyclopaedias. [103]

Fitzroy, Robert (1805–65): English Vice-Admiral, voyager, colonial governor, and meteorologist, captain of *H.M.S. Beagle* during its second expedition, made famous by **Charles Darwin**'s contributions as naturalist. He authored *Narrative of the Surveying Voyages of H.M.S. Adventure and Beagle* (1839).

Fleming, Rev. Prof. John (1785–1857): Scots minister, Natural Theologian, and naturalist with a particular interest in the mollusca. He was closely associated with the University of Aberdeen and the Free Church's New College, Edinburgh.

Flinders, Matthew (1774–1814): English navigator, explorer, and cartographer, who led the first inshore circumnavigation of Australia alongside **Robert Brown**. His journal, *A Voyage to Terra Australis*, published shortly after his death, was enormously popular.

Forbes, Edward. (1815–54): Manx naturalist and pioneer biogeographer, Professor of botany at King's College, London, and curator of the Geological Society of London's museum. His brother was the mineralogist, David Forbes.

Forman, Walter (?): Royal Navy captain and author of *Treatises on Several Very Important Subjects in Natural Philosophy* (1832)' in which he argued for the singularity of the Noachian deluge in geological history.

Fouilloux, Jacques du (1519–80): French author of the hunting guide, *La Vénerie* (1560).

Forster, Georg (1754–94): Prussian naturalist, travel writer, ethnologist, and revolutionary who took part in the second expedition of **James Cook**, as described in *A Voyage Round the World in His Britannic Majesty's Sloop Resolution* (1777), and influenced **Alexander von Humboldt**.

Franklin, Lady Jane (1791–1875): Second wife of the English explorer **Sir John Franklin**, renowned for her philanthropic work and travels during her husband's tenure as Lieutenant-Governor of Van Diemen's Land.

Franklin, Sir John (1786–1847): Royal Navy officer and colonial governor best known for his command of the ill-fated arctic expedition of 1845. By 1847, Franklin and his entire crew succumbed to starvation, scurvy, and hypothermia. The wrecks of his ships, *HMS Erebus* and *HMS Terror* were rediscovered in 2014.

Gaffarel, Jacques (1601–81): French astrologer and natural philosopher, best known for his work on Persian occultism.

Galilei, Galileo (1564–1642): Italian astronomer, engineer, and physicist whose championing of the heliocentric theories of **Nikolaus Copernicus** in *Dialogue Concerning the Two Chief World Systems* (1632) led to his conviction for heresy by the Inquisition.

Galvani, Luigi (1737–98): Italian physicist, physician, and biologist who discovered animal electricity and pioneered the study of bioelectricity and bioelectromagnetics. His findings were disputed by **Alessandro Volta**.

Gardner, George (1810–49): Scottish biologist and botanist, who travelled extensively in Brazil, author of *Pocket Herbarium of British Mosses* (1836) and *Travels in the Interior of Brazil* (1846). [70]

Garriga, Jose (?): Spanish naturalist and palaeontologist, author of *Descripción de un Quadrupedo Muy Corpulento y Raro, que se Conserva en el Real Gabinete de Historia Natural de Madrid* (1796).

Gawler, George (1795–1869): English military officer and administrator, Governor of South Australia (1838–41).

Gerard, John (1545–1612): English botanist whose illustrated *Herball, or Generall Historie of Plantes* (1597) was the most influential British work of botany and herbalism for a century. Superintendent at the Hertfordshire gardens of the Queen's Lord High Treasurer, William Cecil, for twenty years, his own herbal garden in Holborn, London was the source of the many remedies proscribed in his role as a leading herbalist.

Gilbert, John (1812–45): English naturalist, taxidermist and explorer, closely associated with his employer, the ornithologist **John Gould**. Gilbert collected and prepared many specimens for Gould for works such as *Birds of Australia* (1840–8) and is credited with being the first European to collect many Australian species.

Gipps, Sir George (1790–1847): English administrator, Governor-General of New South Wales (1838–46) and First Governor of New Zealand.

Gleichen, Wilhelm Friedrich (1717–83): German biologist who pioneered new methods in microscopy, author of *Treatise on Seed and Infusoria* (1778).

Glenelg, Lord: *see Grant, Charles.*

Goethe, Johann Wolfgang von (1749–1832): leading German Romantic poet, playwright, novelist, philosopher, and commentator, who also produced important works of natural history, particularly in the field of botany, including *Metamorphosis of Plants* (1790). [44]

Goring, Dr C. R. (1792–1840): English physician and amateur microscopist closely associated with **Andrew Pritchard**, for whom Goring produced illustrations for works including *Microscopic illustrations of a Few New, Popular and Diverting Living Objects with their Natural History* (1830).

Gosse, Philip Henry (1810–88): English naturalist and scientific populariser closely associated with the aquarium and rockpool crazes of the 1840s and 1850s, and as an opponent of evolutionary theory. His works include *Canadian Naturalist* (1840), *The Birds of Jamaica* (1847), *A Naturalist's Sojourn in Jamaica* (1851), *The Aquarium: an Unveiling of the Wonders of the Deep Sea* (1854), *Tenby* (1856), and *Omphalos* (1857). [31]

Gould, Elizabeth (1804–41): British artist and illustrator, wife of naturalist and author **John Gould**. She produced countless illustrations for ornithological works, including plates in **Charles Darwin**'s *The Zoology of the Voyage of H.M.S. Beagle* (1838–43), and John Gould's *Birds of Australia* (1840–8). Scholarship tends to argue that her contribution to Gould's work was under-attributed.

Gould, John (1804–81): English ornithologist whose work was scientifically significant and wildly popular. He was pivotal in enabling **Charles Darwin** to realise the implications of the bird specimens collected in the Galapagos Islands in evolutionary terms. His works include *Birds of Europe* (1832–7), the third volume of Darwin's *Zoology of the Voyage of the Beagle* (1838–41), *Birds of Australia* (1840–8), *Birds of Asia* (1850–83), and *Birds of Great Britain* (1875). [62, 81, 92]

Grant, Charles (Baron Glenelg) (1778–1866): Scottish politician and colonial administrator, whose offices included Chief Secretary for Ireland, President of the Board of Trade, and Secretary of State for War and the Colonies.

Greenough, George Bellas (1778–1855): English gentleman-geologist, and a founder of the early Geological Society. Greenough produced a rival geological map to that of **William Smith**, and during the Great Devonian Controversy was amongst the opponents of **Roderick Murchison** and **Adam Sedgwick**.

Grey, Sir George (1812–98): British explorer, colonial administrator, and writer on Maori culture whose offices included Governor of South Australia, Governor of New Zealand, Governor of Cape Colony, and Premier of New Zealand.

Hadley, George (1685-1768): English lawyer, meteorologist, and Fellow of the Royal Society who studied the atmospheric conditions governing trade winds.

Hall, Sir James (1761–1832): Scottish Baronet, MP, geologist, geophysicist, and President of the Royal Society of Edinburgh. A supporter of **James Hutton**, he was also influenced by **Antoine-Laurent de Lavoisier**.

Haller, Albrecht von (1708–77): Swiss anatomist, naturalist, and encyclopaedist, often called 'the father of modern physiology'.

Halley, Edmond (1656–1742): English Astronomer Royal, mathematician, physicist, and meteorologist, after whom Halley's Comet is named. His meteorological work focused on trade winds, monsoons, and atmospheric phenomena.

Harding, James Duffield (1798–1863): influential British watercolourist and landscape painter.

Hassall, Arthur Hill (1817–94): British chemist, physician, and microscopist famous for his contributions to public health and food safety, author of *A Microscopical Examination of the Water Supplied to the Inhabitants of London and the Suburban Districts* (1850).

Hayward, Joseph (?): British author of *The Science of Horticulture* (1818) and *On the Science of Agriculture* (1825). [104]

Hemans, Felicia (1793–1835): Anglo-Welsh poet, author of the heroic patriotic ballad, *Casabianca* (1836) and *England and Spain, or, Valour and Patriotism* (1808).

Henfrey, Arthur (1819–59): British botanist and surgeon, author of *The Rudiments of Botany* (1849) and *The Vegetation of Europe, Its Conditions and Causes* (1852). [84]

Henslow, John Stevens (1796–1861): English priest, geologist, and botanist who was a formative influence on the career of the young **Charles Darwin**, who studied under Henslow at the University of Cambridge. Henslow authored *A Catalogue of British Plants* (1829) and was an early proponent of widening access to museums.

Herbert, William (1778–1847): British clergyman, botanist, poet, and MP.

Herschel, Sir John Frederick William (1792–1871): English baronet, mathematician, astronomer, chemist, botanist, inventor, and experimental photographer. Author of *Preliminary Discourse on the Study of Natural Philosophy* (1830).

Hipparchus (190–120 BCE): Greek astronomer, mathematician, and geographer particularly associated with the study of equinoxes.

Hitchcock, Edward (1793–1864): Massachusetts pastor, geologist, and Amherst College Professor and President, author of *Geology of the Connecticut Valley* (1823), *Catalogue of the Plants within Twenty Miles of Amherst* (1829), *Reports on the Geology of Massachusetts* (1833), *and The Religion of Geology and Its Connected Sciences* (1851). As Massachusetts and Vermont State Geologist, he produced some of the first U.S. geological surveys. His geology sought to reconcile scripture and science, and opposed Darwinism. He was married to **Orra Hitchcock**. [17]

Hitchcock, Orra White (1796–1863): American naturalist and illustrator, who contributed to the lectures and works of her husband, **Edward Hitchcock**, and was also recognised for her own scientific and artistic work. White produced many paintings and more than a thousand illustrations: the Hitchcocks created an illustrated *Herbarium Parvum, Pictum* (1817–21), comprising floral specimens and watercolours, and Orra produced *Fungi Selecti Picti* (1821).

Hoare, Sir Richard Colt, 2nd Baronet (1758–1838): English landowner, antiquarian, historian, and archaeologist, famed for his contributions to these fields within Wiltshire.

Hobbema, Meindert (1638–1709): Dutch Golden Age landscape artist whose work was little known in his lifetime.

Holland, Sir Henry (1788–1873): English physician, travel writer, and agricultural writer who contributed to germ theory, author of *General View of the Agriculture of Cheshire* (1808). [102]

Home, Sir Everard (1756–1832): British surgeon, anatomist, and palaeontologist who was first to describe the ichthyosaur discovered by **Mary Anning** (who he did not credit). His anatomical work was mired in accusations of plagiarism. [22]

Homer (?): Ancient Greek playwright. Homer is the name attributed to the author of the epic poems, *The Iliad* and *The Odyssey*.

Hooke, Robert (1635–1703): polymathic English Renaissance natural philosopher, physicist, astronomer, botanist, zoologist, and cartographer. Hooke made his fortune by producing architectural surveys after the Great Fire of London (1665) and devoted his time thereafter to science. He is credited as the first to identify micro-organisms via microscopy, for being one of the earliest figures to recognise that fossils were preserved animal remains, and for his experimental contributions to the works of **Robert Boyle** and **Sir Isaac Newton**. [4]

Hooker, Joseph Dalton (1817–1911): significant British botanist, biogeographer, and explorer. He succeeded his father, **Sir W. J. Hooker**, as Director of Kew Botanical Gardens, and was Charles Darwin's closest friend. He travelled to the Antarctic, Himalayas, Sri Lanka, Palestine, Morocco, and the U.S.A.. His works include *Flora Antarctica* (1844), *Himalayan Journals* (1854), *British Flora* (1858), and *Genera Plantarum* (1862–83) [53, 64, 78]

Hooker, Sir William Jackson (1785–1865): English botanist, botanical illustrator, and scientific traveller, father of **Joseph Dalton Hooker**. In the 1820s, Regius Professor of Botany at the University of Glasgow and Director of its Botanic Gardens, he became Director of the newly state-owned Royal Gardens at Kew in 1831 (a position held until his death), where he oversaw considerable expansion of the gardens and professionalisation of the herbarium. His works include *Journal of a Tour in Iceland* (1813), *Flora Scotica* (1821), *The British Flora* (1833), *Flora Niger* (1849), and *The British Ferns* (1861). [46, 54, 71]

Horace (Quintus Horatius Flaccus) (65–8 BCE): Roman lyric poet whose career coincided with Rome's transition from Republic to Empire.

Horneman, Friedrich (1772–1801): German explorer, author of *The Journal of Frederick Horneman's Travels, from Cairo to Mourzouk* (1802).

Horsfield, Thomas (1775–1859): American naturalist and explorer, author of *Zoological Researches in Java and the Neighbouring Islands* (1824).

Hübner, Jacob (1761–1826): Bavarian naturalist and a founder of entomology, author of *Samnlung Europäischer Schmetterlinge* (1796–1805).

Hudson, William (1730–93): British botanist, apothecary, and Fellow of the Royal Society, author of *Flora Anglica* (1761).

Humboldt, Alexander von (1769–1859): Prussian geographer, cartographer, naturalist, explorer, meteorologist, and geologist, one of the most famous Europeans of his generation, and a founder of biogeography. His works describing South American explorations were scientifically ground-breaking and highly popular, and he is often credited as the first to identify anthropogenic climate change. His works include *Views of Nature* (1807), *Political Essay on the Kingdom of New Spain* (1810) *Personal Narrative of Travels to the Equinoctial Regions of the New Continent* (1814–29, with **Aimé Bonpland**), *Aspects of Nature* (1849), and *Kosmos* (1845–62). [66, 75, 79]

Hunnis, William (d. 1597): English poet and writer, author of *Certayne Psalms* (1549), *A Hive full of Hunnye* (1578), *Seven Sobbes of a Sorrowful Soule for Sinne* (1583), and *Hunnies Recreations* (1588).

Hutt, John (1795–1880): English administrator, Governor of Western Australia (1839–46).

Hutton, James (1726–97): Scottish physician, chemical manufacturer, geologist, agriculturalist, and naturalist widely regarded as the founder of British geology, author of *Theory of the Earth* (1788), the groundbreaking work of uniformitarianism, and pioneer of the concept of 'deep time'. His theories were at odds with the Neptunist geology of **Abraham Gottlob Werner**. As a joint founder of the Royal Society of Edinburgh in

1783, he mixed with stars of the Scottish Enlightenment, including Adam Smith, David Hume, and **John Playfair**. [19, 20]

Ibn al-Baytar (1197–1248): Andalusian physician, pharmacist, and botanist, author of *Compendium on Simple Medicaments and Foods*.

Ibn Ishaq, Muhammad (d. 767): Arab historian, author of an early biography of the prophet Muhammad (*Sīrat Rasūl Allāh*).

Ibn Juljul, Abu Dawud Sulayman ibn Hassan (944–94): Andalusian Arab scholar, physician, and pharmacist, whose botanical studies intersected with his devotion to medicine.

Isocrates (436–338 BCE): ancient Greek rhetorician, author of *Antidosis*, *Panatheniacus*, and *Panegyricus*.

Jameson, Robert (1774–1854): Scottish naturalist and mineralogist, student of **Abraham Gottlob Werner**, Regius Professor at the University of Edinburgh, and founder of the Wernerian Natural History Society. Initially Britain's foremost proponent of Neptunist geology, before switching to the Uniformitarian and Plutonist theories of **James Hutton**, his works include *System of Mineralogy* (1804) and *Manual of Mineralogy* (1821).

Jukes, Joseph (1811–69): English geologist and explorer, author of *Excursions in and about Newfoundland* (1842) and *A Sketch of the Physical Structure of Australia* (1850).

Jussieu, Antoine Laurent de (1748–1836): French botanist, author of *Genera plantarum* (1789) and the earliest to publish a work based on the 'natural system' of taxonomy.

Kane, Lady Katherine Sophia (1811–86): Irish botanist whose *Irish Botany* (1833) was the first major work published on the nation's wild plants and rapidly became a standard work for Irish university botanical classes. She was the first woman elected to the Botanical Society of Edinburgh, and oversaw a substantial plant collection in Dublin. [48]

K'Eogh, John (1681–1754): Irish naturalist and chaplain, author of *Botanologia Universalis Hibernica* (1735).

Kepler, Johannes (1571–1630): German astronomer, astrologer, and mathematician, whose laws of planetary motion and contributions to Newtonian theories of gravity are his principal achievement. His works include *Astronomia Nova* (1609) and *Pitome Astronomiae Copernicanae* (c. 1618).

King, William (1809–86): Anglo-Irish geologist credited with the discovery of *Homo neanderthalis*.

Kirwan, Richard (1733–1812): Irish geologist, chemist, and meteorologist, author of *Elements of Mineralogy* (1784), *The Manures Most Advantageously Applicable to the Various Sorts of Soils* (1796), and *Geological Essays* (1799).

Lamarck, Jean-Baptiste (1744–1829): French soldier and naturalist who coined the term biology, and, in *Philosophie Zoologique* (1809), proposed early evolutionary views that were widely dismissed, and that differed from those of **Charles Darwin** in emphasising inheritance of acquired characteristics. A member of the French Academy of Sciences, he was closely involved with the Jardin des Plantes and became Chair of Botany and Professor of Zoology. His other works include *Flore Françoise* (1778) and *Système des Animaux Sans Vertèbres* (1801). [88]

Lankester, Edwin (1814–74): English surgeon and scientist closely associated with John Snow's attempts to identify the waterborne nature of cholera in the 1850s.

Latham John (1740–1837): English physician and naturalist, author of *A General Synopsis of Birds* (1781–1801) and *General History of Birds* (1821–8). Described as the 'grandfather of Australian ornithology', Latham examined specimens of Australian birds, and is credited with many of their common names, including emus, cockatoos, and lyrebirds.

Laurenti, Josephus Nicolaus (1735–1805): Austrian-Italian zoologist who formed the class Reptilia (reptiles) in *Specimen Medicum, Exhibens Synops in Reptilium Emendatam cum Experimentis circa Venena* (1768).

Laurie, James (?): Author-editor of *System of Universal Geography* (1851). [83]

Laurillard, Charles Léopold (1783–1853): French palaeontologist and zoologist who acted as personal secretary to **Georges Cuvier** and continued his work on comparative anatomy. Laurillard's notes on the Gigantic Sloth in vol. 8 the 1836 edition of Cuvier's *Recherches sur Les Ossemens Fossiles* were used by **Richard Owen**.

Lavoisier, Antoine-Laurent de (1743–94): French nobleman who revolutionised chemistry in the eighteenth century. He recognised and named oxygen and hydrogen, and discredited the phlogiston theories of **Joachim Becher**. Lavoisier helped create the metric system and supported the role of sciences as forces of progress in post-revolutionary society.

Lawson, Nicholas (1790–1851): Norwegian administrator who, while vice-governor of the Ecuadorian Galapagos Islands (1832–7), contributed to **Charles Darwin**'s understanding of the islands' zoology.

l'Écluse, Charles de (Carolius Clusius) (1526–1609): Dutch doctor, pioneering botanist, and horticulturalist, who oversaw the establishment of the botanical gardens at Leiden, author of *Rariorum Alioquot Stirpium per Hispanias* (1576) and *Rariorum Plantarum Historia* (1601).

Leeuwenhoeck, Antonie van (1632–1723): Self-educated Dutch entrepreneur and naturalist, often referred to as 'the Father of Microbiology' because of his pioneering work in microscopy. He coined the term 'animalcules' to describe micro-organisms, and was the first to see and describe infusoria (protists), bacteria, spermatozoa, blood flow in capillaries, and the banded pattern of muscular fibres.

Leichhardt, Friedrich Wilhelm Ludwig (1813–c. 1848): German naturalist and explorer who disappeared during his third expedition into the interior of Australia. His *Journal of an Overland Expedition in Australia, from Moreton Bay to Port Essington* was published a year before his disappearance.

Lenz, Aemilius Hristianovich (1804–65): Russian physicist, geographer, and Fellow of the St Petersburg Academy of Science remembered for his expeditions to Caspian Sea and Dagestan. He particularly specialised in electromagnetism.

Lewin, John William (1770–1819): English-Australian artist who illustrated early works of Australian natural history, including his own *Birds of New Holland* (1808).

Lewis, Meriwether (1774–1809): American explorer, military officer, and politician, leader of the Lewis and Clark expedition (1803–6) to reach the Pacific coast.

Liebig, Baron Justus von (1803–73): University of Giessen Professor, and a founder of organic and biological chemistry. His emphasis on the importance of nitrogen, and 'law of the minimum' made him key to the development of agricultural science. Liebig developed beef extract manufacturing processes, and permitted the establishment of the Liebig Extract of Meat Company which became well known for Oxo brand beef stock cubes.

Liebniz, Gottfried Wilhelm (1646–1716): prolific and brilliant polymathic German philosopher, mathematician, and scientist, closely associated with Enlightenment rationalism. His works include *Discours de Metaphysique* (1686) and *Théodicée* (1710).

Lindley, John (1799–1865): English botanist and promotor of the French 'natural system' of classification in the U.K. Supported by **Sir Joseph Banks** and **Sir W.J. Hooker**, he became Chair of Botany, University College, London. His works include *Monographia Rosarum* (1820), *A Synopsis of British Flora* (1830), *An Introduction to the Natural System of Botany* (1830), and contributions to the *Botanical Register* and *The Gardeners' Chronicles*. [47]

Linnaeus (Carl von Linne) (1707–88): Swedish naturalist, botanist, and taxonomer, and founder of the influential 'sexual system' of classification who is often credited as an early proto-ecological writer. [5]

Lorrain, Claude (1600–82): French Baroque 'Old Master' artist, still extremely popular in the nineteenth century, but also one of the main targets of criticisms by **John Ruskin**.

Lucullus, Lucius Licinius (118–56 BCE): Roman Republic politician and military leader.

Lund, Peter Wilhelm (1801–80): Danish archaeologist, palaeontologist, and zoologist, whose life and career were largely centred on Brazil. A prominent early discover of many prehistoric megafauna, including the sabre-toothed cat, Lund, a Christian, supported **Georges Cuvier**'s catastrophism, and his 1843 discovery of prehistoric human skulls alongside long-extinct species may have led him to retire from science.

Luxford, George (1807–54): Printer, publisher, journalist, and botanist, member of the Botanical Societies of London and Edinburgh, and editor of *The Phytologist*, author of *A Flora of the Neighbourhood of Reigate, Surrey* (1838). [49]

Lyall, David (1817–95): Scottish botanist, assistant surgeon on *HMS Terror* during **Sir James Clark Ross**'s voyage, and lifelong friend of its botanist, **Joseph Dalton Hooker**.

Lyell, Charles (1797–1875): Scottish naturalist, and the most significant geologist of his generation, author of *Principles of Geology* (1830–3), and promoter of uniformitarianist geology. His understanding of 'deep time' was pivotal in the emergence of the evolutionary theories of **Charles Darwin**, and he also offered explanations of climate change and deepened understanding of the Tertiary period and seismology. [25, 90]

MacCulloch, John (1773–1835): Scottish geologist who made the first geological maps of Scotland. His works, include *A Geological Classification of Rocks with Descriptive Synopses of the Species and Varieties, comprising the Elements of Practical Geology* (1821) and *A System of Geology* (1831).

Machonochie, Alexander (1787–1860): Scottish naval officer, geographer, and penal reformer, who after service in the Napoleonic Wars turned to geography, becoming first professor of that subject at University College, London. His interest in penal reform, first gained while in Tasmania alongside **Sir John Franklin**, were given expression during his enlightened management of Norfolk Island penal colony in the 1840s.

Maillet, Benoît de (1656–1738): remarkable and widely-travelled French naturalist and diplomat who produced a very early evolutionary hypothesis and who anticipated Uniformitarian theories of 'deep time'. He was the author of *Telliamed* (1748).

Malte-Brun, Conrad (1775–1826): Danish-French geographer, co-author (with Edme Mentelle) of *Géographie Mathématique, Physique et Politique de Toutes les Parties du Monde* (1803–12). Malte-Brun coined the terms Indo-China and Oceania.

Malthus, Rev. Thomas Robert (1766–1834): English cleric and demographer, author of the controversial and influential *Essay on the Principle of Population* (1798) which inspired the creation of the national census in 1801 and is the fountainhead of intense debate on human demographics in the centuries that followed. [10, 74]

Mantell, Gideon (1790–1852): Geologist, palaeontologist, and obstetrician who pioneered the early scientific study of saurians (dinosaurs), after discovering teeth and skeletons of the *Iguanodon*. In the early 1820s, inspired by his contemporary **Mary Anning**, Mantell found many fossil remains in his native Sussex. An authority on prehistoric reptiles, he named the genus *Hylaeosaurus*, and discovered four of the five earliest-known dinosaur genera. He was the author of *The Fossils of the South Downs* (1822), *Illustrations of the Geology of Sussex, containing figures and descriptions of the fossils of Tilgate Forest* (1827), *The Geology of the South-East of England* (1833), and *Thoughts on a Pebble* (1836). [30]

Martens, Conrad (1801–78): Anglo-Australian landscape painter on board *HMS Beagle* alongside **Charles Darwin** during 1833–4. Arriving in Australia in 1835, he painted there until his death in 1878. His work, on the *Beagle* and during his prolific career in Australia, is a fascinating social document of British colonial exploration and governance.

Martius, Karl Friedrich Philipp von (1794–1868): German botanist, explorer, and Keeper of the Munich Botanical Gardens, who travelled extensively in Brazil.

Mason, Thomas Monck (1803–89): Irish flautist, author, and balloonist, author of *Account of the Late Aeronautical Expedition from London to Wellburg* (1836) and *Creation by the Immediate Agency of God, as Opposed to Creation by Natural Law* (1845). [94]

Matthew, Patrick (1790–1874): Scottish merchant, landowner, silviculturalist, and horticulturalist, author of *On Naval Timber and Arboriculture* (1831), which included one of the earliest statements of natural selection prior to **Charles Darwin**. [89]

Maupertuis, Pierre Louis (1698–1759): French mathematician, philosopher, and natural historian, author of *The Figure of the Earth* (1738).

McArthur, John (1791–1862): English military officer, administrator and amateur artist, Commandant at Port Essington (Victoria settlement) (1839–49).

Melville, Alexander (1819–1901): Irish comparative anatomist and Professor of Natural History at Queen's College, Galway who supported **Gideon Mantell** in palaeontological disputes with **Richard Owen**. He co-authored *The Dodo and Its Kindred* (1848) with Hugh Edwin Strickland.

Mendel, Gregor (1822–84): Moravian biologist, mathematician, and meteorologist, generally regarded the founder of genetics, whose work on pea plants provided vital supporting evidence for evolutionary theory.

Menzies, Archibald (1754–1842): Scottish Royal Navy surgeon, naturalist, and explorer, who, during the Vancouver Expedition (1791–5), visited five continents and circumnavigated the globe.

Miller, Rev. Hugh (1802–56): Self-educated Scottish stonemason, geologist, palaeontologist, folklorist, and Evangelical, author of *The Old Red Sandstone* (1841), *The Foot-prints of the Creator* (1849), and *Testimony of the Rocks* (1857). [32]

Milton, John (1608–74): leading English epic poet, thinker, and Commonwealth civil servant, author of *Comus* (1634), *Areopagitica* (1644), *Poems* (1645), *Eikonoklastes* (1649), *Paradise Lost* (1667), *Paradise Regained* (1671), and *Samson Agonistes* (1671).

Montesquieu, Charles-Louis (1689–1755): French judge, author, historian, and philosopher, author of *The Spirit of the Laws* (1748).

Müller, Otto Friedrich (1730–84): Danish naturalist and scientific illustrator who oversaw the fourth and fifth volumes of *Flora Danica*, and authored *Fauna Insectorum Fridrichsdaliana* (1764) and *Animalcula infusoria* (1786).

Murchison, Sir Roderick Impey (1792–1871): English geologist, author of *The Silurian System* (1839). From 1834, he and **Adam Sedgwick** were ranged successfully against **Henry de la Beche** and **George Greenhough** in geological periodisation debates known as the Great Devonian Controversy. Although Murchison argued for a longer timespan of earth history than literal Biblical exegesis allowed, he was a Christian scientist, and opposed **Charles Darwin**'s evolutionary theory. His works include *Outline of the Geology of the Neighbourhood of Cheltenham* (1834) and *Geology of Russia* (1845). [28]

Newton, Sir Isaac (1642–1727), English mathematician, physicist, and astronomer, and one of the key figures of Enlightenment science. His works include *Mathematical Principles of Natural Philosophy* (1687), *Opticks* (1704), and *Universal Arithmetic* (1707).

Nolan, Frederick (1784–1864): Irish Theologian and fierce critic of Tractarianism and modern geology, author of *Analogy of Revelation and Science Established* (1833).

Ochsenheimer, Ferdinand (1767–1822): German lepidopterist and actor, author of *Die Schmetterling von Europa* (1806–25), as well as a number of dramatic works.

Oeder, Georg Christian (1728–91): German-Danish botanist, economist, and doctor, Professor of Botany at Copenhagen's Botanic Garden and proposer and first editor of *Flora Danica* (from 1761).

Oken, Lorenz (1779–1851): German botanist and zoologist, a leading exponent of Naturphilosophie and homology.

Oppel, Nicolaus Michael (1782–1820): Bavarian naturalist, student and assistant to **André Marie Constant Duméril**, at the Muséum of Natural History, Paris, where he catalogued and classified reptile specimens. Oppel published *The Orders, Families, and Types of Reptiles* (1811).

Ørsted Hans Christian (1777–1851): Danish chemist and physicist who first established the link between magnetism and electricity by demonstrating that electric currents create magnetic fields (Ørsted's law).

Owen, Sir Richard (1804–92): leading British comparative anatomist, famous for his reconstruction of the extinct Moa from a single thigh bone, and for a wealth of palaeontological investigations. Consulted in attempts to reconstruct Megalosaurs and other early dinosaurs from the available finds of the period, he was frequently at odds with **Gideon Mantell**. Owen was the first to coin the term *Dinosauria* ('terrible lizard'). Orthodox and conservative, he remained fiercely critic of Darwin. He was pivotal in establishing the Natural History Museum, London, and shaped museum policies for decades. [37, 38, 39, 40]

Paley, Rev. William (1743–1805): English cleric and highly influential Natural Theologian, author of *The Principles of Moral and Political Philosophy* (1785) and *Natural Theology* (1802). [11]

Pallas, Peter Simon (1741–1811): Prussian naturalist, after whom palladium and several species of animals are named, and who was the first to name and scientifically describe the Harvest Mouse first identified by **Gilbert White**. Working in Russia under the patronage of Catherine the Great, his works include *Miscellanea Zoologica* (1766) and *Travels Through the Southern Provinces of the Russian Empire* (1771–6).

Pander, Christian Heinrich (1794–1865): Russian-German biologist, embryologist, palaeontologist, and explorer.

Parāśara (?): Indian Maharshi, astrologer, and botanist, writer of ancient Hindu texts, including verses in the *Rgveda* and works of botany (the *Vṛkṣāyurveda*).

Parish, Sir Woodbine (1796–1882): British diplomat who served in Argentina from 1823 to 1832. He was a correspondent of **Charles Darwin**.

Park, Mungo (1771–1806): Scottish West African explorer whose *Travels in the Interior Districts of Africa* (1799) inspired subsequent generations of explorers and European colonial ambitions in Africa.

Parkinson, James (1755–1824): English radical, surgeon, geologist, and palaeontologist after whom Parkinson's Disease is named. His geology pursued non-literal readings of *Genesis*, and his palaeontology led to *Organic Remains of a Former World* (1804–11).

Parkinson, John (1567–1650): English botanist and herbalist, apothecary to James I of England, founder of the Worshipful Society of Apothecaries, and author of *The Botanical Theatre* (1640).

Penn, Granvill (1761–1844): English scholar, Biblical critic, scriptural geologist, and supporter of veterinarianism who published *A Comparative Estimate of the Mineral and Mosaical Geologies* (1822).

Pennant, Thomas (1726–98): acclaimed Welsh naturalist, antiquarian, and travel writer, author of *British Zoology* (1766–77), *History of Quadrupeds* (1781), and *Western Hindoostan* (1798). He was a correspondent of **Linnaeus**, **Gilbert White**, **Sir Joseph Banks**, **Georges Buffon**, and many other Enlightenment naturalists.

Petiver, Jacob (1665–1718): London apothecary, specimen collector, and Fellow of the Royal Society.

Phillip, Admiral Arthur (1738–1814): English Royal navy officer who established the Botany Bay penal colony and became first Governor of New South Wales.

Phillips, William (1775–1828): British geologist, founding member of the Geological Society, and co-author, with **William Daniel Conybeare**, of *Outlines of the Geology of England and Wales* (1822).

Pillans, James (1778–1864): Scottish educator, Professor of Humanity and Laws at University of Edinburgh, author of *Outlines of Geography* (1847) and *Elements of Physical and Classical Geography* (1854), and inventor of coloured chalks. [86]

Pindar (c. 518–438 BCE): ancient Greek lyric poet and critic, author of many 'victory odes'.

Playfair, John (1748–1819): Scottish mathematician, scientist, and Church of Scotland minister. An influential professor of natural philosophy at the University of Edinburgh, his *Illustrations of the Huttonian Theory of the Earth* (1802) provided a summary of **James Hutton**'s theories that helped disseminate uniformitarian ideas and facilitated their development by **Charles Lyell** in the 1830s.

Pliny the Elder (Gaius Plinius Secundus) (23/24–79 CE): Roman author, military commander, naturalist, and philosopher, whose pioneering encyclopedia, *Naturalis Historia* (77 CE), remained a valued resource for centuries.

Plutarch (46–119 CE): Greek-Roman Platonist philosopher, historian, biographer, and priest mainly remembered for *Parallel Lives* (2nd c. CE).

Polybius (c. 200–118 BCE): ancient Greek historian, author of *The Histories*.

Poussin, Gaspar: *see Dughet, Gaspard*

Powys, Edward (?): farmer or landowner, author of 'On Feeding Cattle With Green Food' (1808). [100]

Prichard, James Cowles (1786–1848): British physician and ethnologist whose *Physical History of Mankind* has a marginal place in the development of evolutionary theory.

Pritchard, Andrew (1804–82): British microscopist and naturalist, involved in the development of microscopes and the scientific study of micro-organisms, author of *The Natural History of Animalcules* (1834). [59]

Ptolemy, Claudius (c. 100–c. 170 CE): Greek mathematician, geographer, astronomer, and astrologer, author of *Almagest* and *Geographia* (c. 150 CE).

Pusey, Edward Bouverie (1800–82): English theologian and leading figure in the Oxford Movement, Regius Professor of Hebrew at the University of Oxford.

Pye-Smith, Dr John (1774–1851): Sheffield Congregational theologian and tutor, author of *On the relation between the Holy Scriptures and some parts of geological science* (1840).

Ramond de Carbonnières, Louis (1755–1827): French, botanist, geologist, mountaineer, and politician, author of *Observations Made in the Pyrenees* (1789).

Ray, John (1627–1705): English Renaissance scholar, naturalist, and Natural Theologian. A Cambridgeshire blacksmith's son who studied at the University of Cambridge, Ray lectured at Trinity in Greek, Mathematics, and Humanities. He pursued natural history independently, producing a Flora of Cambridgeshire in 1660, and writing *Observations in the Low Countries* (1673) and *The Wisdom of God Manifested in the Work of Creation* (1691). [3]

Rennell, Major James (1742–1830): English geographer, cartographer, oceanographer, and historian, and a founder of the Royal Geographical Society. Closely associated with British India, his works include *Bengal Atlas* (1779) and *Comparative Geography of Western Asia* (1831).

Richter, Henry Constantine (1821–1902): English zoological illustrator and groundbreaking lithographer, closely associated with the works of **John Gould**.

Ritter, Carl (1779–1859): German geographer, first chair of Geography at the University of Berlin, and, with **Alexander von Humboldt**, the founder of modern geography. His 21-volume *Erdkunde im Verhältnis zur Natur und zur Geschichte des Menschen oder allgemeine, vergleichende Geographie, als sichere Grundlage des Studiums und Unterricts in physicalischen und historischen Wissenschaften* (1817–52) remains one of the most impressive achievements of the discipline.

Roget, Peter Mark (1779–1869): British physician, naturalist, and lexicographer, best known for creating *Thesaurus of English Words and Phrases* (1852). A Natural Theologian and Fellow of the Royal Society, he contributed a volume to the *Bridgewater Treatises*. [12]

Rosa, Salvator (1615–73): Italian Baroque 'Old Master' artist and poet, frequently critiqued by **John Ruskin**.

Ross, Sir James Clark (1800–62): British Royal Navy officer involved in Arctic expeditions early in his career, and leader of Antarctic expeditions (1839–43), during which he

named the Ross Sea, Victoria Land, Mount Erebus, and Mount Terror. In 1848 he led an unsuccessful attempt to discover the lost arctic explorer, **Sir John Franklin**.

Roxburgh, William (1751–1815): Scottish surgeon, botanist, and meteorologist, who worked closely with **Sir Joseph Banks**. Pioneer of British botany in India, he authored *Hortus Bengalensis* (1814), the catalogue of the Calcutta Botanical Gardens which he oversaw. [99]

Ruskin, John (1819–1900): prolific, brilliant, and influential British polymath, author of numerous works of art and architectural criticism, science and natural history, political economy, and literary criticism. His wide-ranging engagements with nature and his status as an early environmentalist and proto-ecologist, make him a significant figure in natural history. Initially a keen reader of a range of sciences, Ruskin ultimately rejected evolutionary theory and vehemently opposed the theories of **Charles Darwin**. [16, 52]

Russell, Anna Worsley (1807–76): leading British female botanist of her generation, who contributed extensively to H.C. Watson's *New Botanist's Guide* (1835), authored *Catalogue of Plants Found in the Neighbourhood of Newbury* (1839), and published papers on mycology. A member of the Botanical Society of London, she produced several hundred botanical illustrations subsequently bequeathed to the British Museum. [50]

Rutty, John (1697–1775): Irish naturalist and doctor, author of *History of Dublin* (1772).

Sainte-Hilaire, Étienne Geoffroy (1772–1844): French Enlightenment naturalist and supporter of **Jean-Baptiste Lamarck**, author of *Cours de L'Histoire des Mammifères* (1829).

Saussure, Nicolas Théodore (1767–1845): Swiss chemist, phytochemist, and botanist who pioneered the study of photosynthesis, author of *Chemical Research on Plant Growth* (1804).

Saussure, Horace Bénédict de (1740–99): Swiss geologist, meteorologist, and Alpine mountaineer, a pioneering figure in Alpinism and meteorology, and author of *Voyages dans les Alpes* (1796–1808). An early mountaineer, he was the third person to climb Mt Blanc, in 1787.

Scheuchzer, Johann Jakob (1672–1733): Swiss scholar and naturalist, author of *Beschreibung der Naturgeschichte des Schweitzerlandes* (1706–8).

Schouw, Joakim Frederik (1789–1852): Danish botanist and politician, Professor of botany at the University of Copenhagen, and specialist in phytogeography.

Scopoli, Giovanni Antonio (1723–88): Austrian naturalist and physician, author of *Flora Carniolica* (1760) and *Entomologia Carniolica* (1763).

Sedgwick, Rev. Adam (1785–1873): British Geologist and clergyman, a significant figure in the establishment of modern, scientific geology, closely associated with the establishment of the Cambrian and Devonian periods. Although critical of scriptural geology, he opposed the evolutionary theories of **Charles Darwin**. He was the author of *On the Studies of the University* (1834). [13]

Shakespeare, William (1564–1616): English playwright, poet, and actor, widely regarded as the greatest writer in human history. His works include nearly 40 plays and 154 sonnets.

Shaw, Dr George (1751–1813): English botanist and zoologist, and Keeper of the Natural History department of the British Museum. Co-founder of the Linnean Society in 1788 and Fellow of the Royal Society, his works include *Zoology of New Holland* (1794) and *General Zoology, or Systematic Natural History* (1809–26).

Shelley, Percy Bysshe (1792–1822): Leading English romantic poet, essayist, and radical. His works include *Mont Blanc* (1816), 'Ozymandias' (1818), 'The Mask of Anarchy' (1819), *Prometheus Unbound* (1820) and *A Defence of Poetry* (1821/1840). His advocacy of vegetarianism is expressed in *A Vindication of Natural Diet* (1813).

Silliman, Professor Benjamin (1779–1864): American chemist, geologist, Yale University educator, and abolitionist, famous for being the first to distil petroleum. A proponent

of materialist science and Christianity, Silliman promoted 'old-earth creationism' to reconcile science and scripture.

Sinclair, Sir John, 1st Baronet (1754–1835): British politician and author who coined the term 'statistics', and whose *Statistical Account of Scotland* (1799) was part of his contribution to scientific agriculture.

Sloane, Hans (1660–1753): Anglo-Irish naturalist, collector, and physician, Fellow of the Royal Society and the Royal College of Physicians. Sloane's contributions to natural history were not due to particular investigations or expertise, but to his patronage of science. His vast collection, bequeathed to the nation, formed the basis of what would become the British Library, British Museum, and Natural History Museum.

Smellie, William (1740–95): Scottish printer, antiquary, and naturalist, first editor of *Encyclopaedia Britannica* (1768–71), translator of **Georges Buffon**, and author of *The Philosophy of Natural History* (1791), a work sometimes described as a precursor to Darwinism for anticipating 'the struggle for existence'. Smellie was co-founder of The Royal Society of Edinburgh and the Society of Antiquaries of Scotland. [9, 57]

Smith, James Edward (1759–1828): English botanist whose specialist work on bryophytes were included in *English Botany* by **James Sowerby**, a work to which Smith contributed more widely. A founder of the Linnean Society and Fellow of the Royal Society, his works include *Flora Britannica* (1800–4), *The Introduction to Physiological and Systematic Botany* (1807), and articles in Abraham Rees's *Cyclopaedia* (1819). [45]

Smith, James (1782–1867): Scottish antiquarian, archaeologist, geologist, Bible critic, and merchant. Smith was Fellow of the Royal Society of Edinburgh, President of the Geological Society of Glasgow and the Archaeological Society of Glasgow, and a member of the Wernerian Society and the Highland Society. He is the author of *Researches in New Pliocene and Post-Tertiary Geology* (1862).

Smith, William (1769–1839): influential English geologist, cartographer, and engineer, who produced the world's first geological map (1815), a 6 x 8 ft masterpiece delineating the outcropping strata of the United Kingdom. The original is in Burlington House, London. Smith's works include *Geological Atlas of England and Wales* (1815) and *Strata Identified by Organized Fossils* (1816). Although his contributions were belatedly recognised, his work revolutionised geology by showing the link between strata, fossils, and periodisation. [23]

Solander, Daniel Carlsson (1733–82), Swedish botanist, leading disciple of Linnaeus, and fellow botanist (with **Sir Joseph Banks**) during Captain **James Cook**'s voyages on *HMS Endeavour* (1768–71).

Soulavie, Jean-Louis (1752–1813): French cleric, geologist, and pioneer of biogeography, author of *Histoire Naturelle de la France Meridionale* (1780–4).

Sowerby, James (1757–1822): English naturalist, publisher, illustrator, and mineralogist, who worked closely with **William Curtis** and **James Edward Smith**. His *English Botany* (1790–1835) is amongst the most important and popular works of British botany. [45]

Sparrman, Anders (1748–1820): Swedish naturalist and South African explorer, follower of **Linnaeus**, and member of the Royal Swedish Academy of Sciences. His works include *A Voyage to the Cape of Good Hope* (1789) and *Ornithology of Sweden* (1806).

Sprengel, Karl (1787–1859): Hanoverian botanist and agriculturalist, associated with the 'theory of minimum', later promulgated by **Justus von Liebig**.

Sprengel, Kurt Polycarp Joachim (1766–1833): German physician and botanist, who produced medical histories, as well as pioneering work in microscopic studies of plant tissues.

Stahl, Georg Ernst (1659–1734): German chemist, philosopher, and physician, an influential figure behind the theory of Vitalism, which argued that that life (unlike inanimate matter) was created and sustained by a vital spark, or 'élan vital'. Stahl was

also a key promoter of **Johann Joachim Becher**'s phlogiston theory, ultimately refuted by **Antoine Lavoisier**.

Stanley, Captain Owen (1811–50): Royal Navy officer, surveyor, and amateur naturalist who commanded *HMS Britomart* during its voyage to Australia and New Zealand (1838–43).

Stephenson, George (1781–1848): engineer, whose pioneering contributions to the early railway industry have become legendary.

Stillingfleet, Benjamin (1702–71): English naturalist, author of *A Calendar of Flora* (1755).

Stokes, Admiral John Lort (1811–85): Royal Navy officer who served on *HMS Beagle* for eighteen years, including the journeys of **Captain Fitzroy** and **Charles Darwin**, author of *Discoveries in Australia* (1846).

Stokes, Pringle (1793–1828): British naval officer best known for his command of *HMS Beagle* on its first voyage of exploration in the south Atlantic (1825–8), during which he committed suicide in the Straits of Magellan.

Strange, Frederick (c. 1810–54): British plant and animal collector associated with the early era of Australian colonisation, and killed on Percy Island during a conflict with aborigines. His extensive ornithological data was subsequently disseminated by **John Gould**, and his botanical work by **W. J. Hooker**.

Strong, Captain John (?): English naval officer who, as captain of *HMS Welfare*, led a 1689–91 expedition to South America and explored and named the Falkland Islands.

Sturmius, John Christopher (1635–1703): German mathematician, author of *Mathesis Juvenilis*.

Sturt, Charles Napier (1795–1869): British officer and explorer, author of *Narrative of an Expedition into Central Australia* (1849).

Swammerdam, Jan (1637–80): Dutch microscopist, entomologist, and biologist who pioneered understanding of insect life cycles and first observed red blood cells. His *Bybel der Natuure* was published posthumously in 1737.

Tamerlaine: *see Timur*.

Tasman, Abel Janszoon (1603–59): Dutch East India Company merchant and explorer, the first westerner to reach New Zealand, Fiji, and Tasmania.

Theophrastus (371–287 BCE): Greek philosopher and botanist, often regarded as the father of botany, author of *Enquiry into Plants* and *On the Causes of Plants*.

Thomson, Thomas (1817–78): British East India Company surgeon and botanist, friend of **Joseph Dalton Hooker**, whom he assisted in *Flora Indica* (1855). He was the author of *Western Himalaya and Tibet; A Narrative of a Journey Through the Mountains of Northern India, During the Years 1847–8* (1852). [72].

Threlkeld, Caleb (1676–1728): Irish dissenting minister, author of the first Irish flora, *Synopsis Stirpium Hibernicarum* (1727).

Thunberg, Carl Peter (1743–1828): Swedish naturalist, Linnaean geographical botanist, and Professor of Medicine and Botany at Uppsala University, author of *Travels in Europe, Africa, and Asia Made Between the Years 1770 and 1779* (1795) and *Flora Japonica* (1784). [65]

Timur or Tamerlaine (1336–1405): Turco-Mongol nomadic conqueror and founder of the Timurid Empire.

Titian or Tiziano Vecelli (or Vecellio) (c. 1488/90–1576): influential Venetian Renaissance artist.

Tournefort, Joseph Pitton de (1656–1708): French botanist and explorer of Turkey, Armenia, and Georgia, author of *Eléments de Botanique* (1694).

Trotter, Henry Dundas (1802–59): Scottish Royal Navy officer, commander of the 'Niger Expedition' launched in 1841 and described in *Flora Niger* by **Sir W. J. Hooker**.

Tupaia (c. 1725–70): Tahitian navigator and arioi from Ra'iatea (the Society Islands). His talents and Pacific knowledge were invaluable to Captain **James Cook**, during

the Australian and Pacific sections of the *HMS Endeavour* voyage of 1768–71. Tupaia travelled with Cook to New Zealand, acting as the expedition's interpreter to the Polynesian Māori, and Australia. He died from a shipborne illness while in Batavia.

Turner, J.M.W. (1775–1851): English Romantic landscape artist, and **John Ruskin**'s artistic hero.

Turner, Nicholas (?): author of *An Essay on Draining and Improving Peat Bogs* (1784). [96]

Ussher, Bishop George (1581–1656): Church of Ireland Primate, author of *Annals of the World* (1654), who used Biblical exegesis to determine the date of the earth's creation as 4004 BCE.

Vattel, Emmerich de (1714–67): Swiss jurist and writer, author of *The Law of Nations* (1758), and counsellor to Frederick Augustus II of Saxony. [7]

Vicq d'Azyr, Félix (1748–94): French anatomist, physician, and pioneer of comparative anatomy who proposed the biological theory of homology, demonstrating structural similarities in the anatomy of birds, fishes, mammals, and reptiles.

Vieillot, Louis Jean Pierre (1748–1830): pioneering French ornithologist and collector, whose expeditions to the Americas led him to name several species of birds. Vieillot promoted the study of living birds and of plumage changes. *Analyse d'une Nouvelle Ornithologie Élémentaire. d'Éterville*, (1816) offered a new system of bird classification.

Vigors, Nicholas Aylward (1785–1840): Irish zoologist and politician, co-founder of the Zoological Society of London and Fellow of the Linnean Society and Royal Society.

Vitruvius (Marco Vitruvius Pollo) (c. 80–c. 15 BCE): Roman architectural writer and engineer, author of the influential *De Architectura* (c. 30 BCE).

Vogel, Dr J. R. T. (Theodore) (1812–41): German naturalist, Chief Botanist of the 'Niger Expedition' launched in 1841 and described in *Flora Niger* by **Sir W. J. Hooker**.

Volta, Alessandro (1745–1827): Italian chemist and physicist, associated with the voltaic pile, the first electrical battery to provide continuous electric current to a circuit. Its invention allowed a subsequent series of rapid discoveries, including electrolysis of water into oxygen and hydrogen, and the isolation of various chemical elements.

Wahlenberg, Göran (1780–1851): Swedish botanical geographer who succeeded **Carl Peter Thunberg** as Professor of Medicine and Botany at Uppsala University, author of *Flora Lapponica* (1812).

Wallace, Alfred Russel (1823–1913): British biogeographer, biologist, explorer, social activist, and spiritualist, credited (with **Charles Darwin**) as co-discoverer of evolutionary theory by natural selection, and author of many works, including *The Malay Archipelago* (1869), *The Geographical Distribution of Animals* (1876), and *Man's Place in the Universe* (1903).

Washington, John (1800–63): Royal Navy officer, hydrographer, and a founding member of the Royal Geographical Society.

Waterhouse, George Robert (1810–88): English naturalist, Curator of the museum of Zoological Society of London, and author of *A Natural History of the Mammalia* (1846–8). He contributed to **Charles Darwin**'s *The Zoology of the Voyage of HMS Beagle* (1838–43) and worked with **Louis Agassiz**.

Waterton, Charles (1782–1865): English naturalist, ornithologist, explorer, and conservationist, author of *Wanderings in South America* (1825) and *Essays in Natural History* (1838). He turned his Walton Hall estate into the world's first nature reserve and fought campaigns against industrial pollution. [60, 67, 68]

Wedge, Thomas (1760–1854): English agriculturalist, author of *A General View of the Agriculture of the County Palatine of Chester* (1794).

Werner, Abraham Gottlob (1749–1817): the 'father of German Geology', Werner argued for the successive, chronological arrangement of the earth's strata, and was a key proponent of the ultimately-discredited theory of Neptunism which argued that the earth's rocks were formed by the crystallisation of minerals in deep oceans.

Whewell, Rev. Dr William (1794–1866): prolific English theologian, astronomer, economist, and natural philosopher, widely credited with coining the term 'scientist', author of *The Philosophy of the Inductive Sciences* (1841) and many other works, and a contributor to the *Bridgewater Treatises*.

Whiston, William (1667–1752): English theologian and historian, author of *New Theory of the Earth* (1708), a Creationist account of earth history.

White, Rev. Gilbert (1720–93): English clergyman and naturalist, author of *A Natural History of Selborne* (1788–9), amongst the most significant of all works of English natural history. He is often regarded as a pioneer of ecology and environmentalism. [43, 56, 73]

White John (c. 1756–1832): Irish surgeon, naturalist, and botanical collector, Surgeon-General of New South Wales, whose *A Journal of a Voyage to New South Wales* (1790) provided many of the first descriptions of Australian species.

Wickham, John Clements (1798–1864): Scottish Royal Navy officer and explorer best known as first officer, alongside **Charles Darwin** on *HMS Beagle* during its 1831–6 voyage.

Willughby, Francis (1635–72): English ornithologist and ichthyologist, whose works were posthumously published by his friend, **John Ray**.

Wittich, Wilhelm (1742–1848): German educationalist and Professor of German at University College, London, author of *German for Beginners* (1838), *Curiosities of Physical Geography* (1845) and *A Visit to the Western Coast of Norway* (1848). [80]

Woodward, John (1665–1728): early English naturalist, geologist, and antiquarian, author of *Essay towards the Natural History of the Earth* (1695).

Wren, Sir Christopher (1632–1723): English Enlightenment architect, anatomist, mathematician, and astronomer who rebuilt numerous London churches after the Great Fire (1666), including St Paul's. Founder and president of the Royal Society, his scientific work was highly influential.

Young, Arthur (1741–1820): British agricultural writer and improver, secretary of the Board of Agriculture, and travel writer, author of *A Course of Experimental Agriculture* (1770), *Tours in Ireland* (1780), *Annals of Agriculture* (1784–1808), and *Travels in France* (1792) [101]

INDEX

Note: The apparatus for each volume also includes indexes prepared by the Routledge editorial team and not by the editor of these volumes.